Metal Based Thin Films for Electronics

Klaus Wetzig and
Claus M. Schneider (Eds.)

Metal Based Thin Films
for Electronics

Second, Revised and Enlarged Edition

Klaus Wetzig and Claus M. Schneider (Editors)

WILEY-
VCH

WILEY-VCH Verlag GmbH & Co. KGaA

Editors:

Klaus Wetzig
IFW Dresden, Germany
e-mail: k.wetzig@ifw-dresden.de

Claus M. Schneider
Forschungszentrum Jülich, Germany
e-mail: c.m.schneider@fz-juelich.de

All books published by Wiley-VCH are carefully produced.
Nevertheless, authors, editors, and publisher do not warrant
the information contained in these books, including this
book, to be free of errors. Readers are advised to keep in
mind that statements, data, illustrations, procedural details
or other items may inadvertently be inaccurate.

Library of Congress Card No.: applied for

British Library Cataloguing-in-Publication Data
A catalogue record for this book is available from the
British Library

**Bibliographic information published by
Die Deutsche Bibliothek**
Die Deutsche Bibliothek lists this publication in the
Deutsche Nationalbibliografie; detailed bibliographic
data is available in the Internet at <http://dnb.ddb.de>

© 2006 WILEY-VCH Verlag GmbH & Co. KGaA,
Weinheim

Printed in the Federal Republic of Germany
Printed on acid-free paper

Composition: B. Präßler-Wüstling
Printing: betz-druck GmbH, Darmstadt
Binding: J. Schäffer GmbH, Grünstadt

ISBN-13: 978-3-527-40650-0
ISBN-10: 3-527-40650-6

Contents

Metal Based Thin Films for Electronics, Second Edition. Klaus Wetzig and Claus M. Schneider (Eds.)
Copyright © 2006 WILEY-VCH Verlag GmbH & Co. KGaA, Weinheim
ISBN: 3-527-40650-6

Preface

This up-to-date handbook covers the main issues of the preparation, characterization, and properties of complex metal-based layer systems. The authors – an outstanding group of researchers – discuss advanced methods for structural, chemical, and electronic state characterization with reference to the properties of thin functional layers, such as metallization and barrier layers for microelectronics, magnetoresistive layers for GMR, TMR, and spin injection, layer structures for photovoltaics, sensor and resistance layers. As such the book addresses materials specialists in industry, especially in microelectronics, as well as scientists. It can also be recommended for advanced studies in materials science, analytics, surface and solid state science.

The first edition of this book was published in 2003 and was well received by both readers and reviewers, thus becoming exhausted in less than two years. This success motivated the editors and the publisher to release a revised and corrected second edition. We also took the opportunity to account for the rapid progress taking place in the field of functional thin layers for microelectronics. We incorporated recent results and findings to bring the individual chapters to the current state-of-the-art. This includes, for example, investigations of the interface between barrier layer and dielectrics in metallization systems, spin-torque-induced magnetic switching in spintronics, the development of new metallization technologies for migration-resistant SAW devices, and advancements in the nanoanalytics of thin functional layers.

Furthermore, we included several new sections, highlighting new and timely research topics. In the field of acoustoelectronics this refers to Section 2.2.7 "Interdigital Transducers with Piezoelectric Layers" and Section 2.2.8 "Metal Strips with Dielectric Coatings". Section 2.6 "Metallic Layers for Photovoltaics" now provides insight into this rapidly developing research field.

Also, this second edition would not have been possible without the fruitful and very constructive cooperation of the authors of the individual chapters and we would like to express our sincere gratitude. A particular acknowledgement goes to Mrs. B. Präßler-Wüstling for her tireless support in writing and assembling the chapters of this second edition. Finally, we are grateful for the help and support provided by Mrs. C. Wanka, Dr. A. Grossmann and Dr. P. Capper from Wiley-VCH during this project.

Dresden and Jülich, January 2006

Klaus Wetzig
Claus M. Schneider

Metal Based Thin Films for Electronics, Second Edition. Klaus Wetzig and Claus M. Schneider (Eds.)
Copyright © 2006 WILEY-VCH Verlag GmbH & Co. KGaA, Weinheim
ISBN: 3-527-40650-6

List of Contributors

Prof. Dr. rer. nat. habil. Winfried Blau
Europäische Forschungsgesellschaft
Dünne Schichten
Gostritzer Strasse 61–63
01217 Dresden
Germany
(Chapter 2.6)

Dr. rer. nat. Stefan Braun
FhI Werkstoff- und Strahltechnik (IWS)
Abt. EUV- und Röntgenoptiken
Winterbergstrasse 28
01277 Dresden
Germany
(Chapters 2.5, 4.6, 5.4)

Dr. rer. nat. habil. Winfried Brückner
IFW Dresden
Institut für Festkörperforschung
Abt. Dünnschichtsysteme und Nanostrukturen
Helmholtzstrasse 20
01069 Dresden
Germany
(Chapter 3.5)

Dr. rer. nat. Dieter Elefant
IFW Dresden
Institut für Festkörperforschung
Abt. Dünnschichtsysteme und Nanostrukturen
Helmholtzstrasse 20
01069 Dresden
Germany
(Chapter 4.5)

Dr. rer. nat. Michael Hecker
AMD Saxony LLC & Co. KG Dresden
Materials Analysis Department
Wilschdorfer Landstrasse 101
01109 Dresden
Germany
(Chapter 3.3)

Dr. rer. nat. habil. Hermann Mai
FhI Werkstoff- und Strahltechnik (IWS)
Abt. EUV- und Röntgenoptiken
Winterbergstrasse 28
01277 Dresden
Germany
(Chapters 2.5, 4.6, 5.4)

Dr.-Ing. Siegfried Menzel
IFW Dresden
Institut für Festkörperanalytik und
Strukturforschung
Abt. Oberflächen- und Mikrobereichsanalytik
Helmholtzstrasse 20
01069 Dresden
Germany
(Chapter 4.3)

Prof. Dr. rer. nat. habil. Claus M. Schneider
FZ Jülich
Institut für Festkörperforschung
52425 Jülich
Germany
(Chapters 1, 2.4, 4.4, 6)

Dr. rer. nat. Joachim Schumann
IFW Dresden
Institut für Festkörperforschung
Abt. Dünnschichtsysteme und Nanostrukturen
Helmholtzstrasse 20
01069 Dresden
Germany
(Chapters 2.3, 4.7, 5.5)

Prof. Dr. rer. nat. Ralph Spolenak
ETH Zurich
Laboratory for Nanometallurgy
Department of Materials
Wolfgang-Pauli-Strasse 10
8093 Zürich
Switzerland
(Chapters 2.1, 4.1)

Metal Based Thin Films for Electronics, Second Edition. Klaus Wetzig and Claus M. Schneider (Eds.)
Copyright © 2006 WILEY-VCH Verlag GmbH & Co. KGaA, Weinheim
ISBN: 3-527-40650-6

Dr. rer. nat. Jürgen Thomas
IFW Dresden
Institut für Festkörperanalytik und
Strukturforschung
Abt. Oberflächen- und Mikrobereichsanalytik
Helmholtzstrasse 20
01069 Dresden
Germany
(Chapters 3.2, 4.4, 4.7)

Dr. rer. nat. Christoph Treutler
Robert-Bosch GmbH
Zentralbereich Forschung und
Vorausentwicklung
Postfach 106050
70059 Stuttgart
Germany
(Chapter 5.3)

Dr. rer. nat. Hartmut Vinzelberg
IFW Dresden
Institut für Festkörperforschung
Abt. Dünnschichtsysteme und Nanostrukturen
Helmholtzstrasse 20
01069 Dresden
Germany
(Chapter 4.5)

Dr. rer. nat. habil. Manfred Weihnacht
IFW Dresden
Institut für Festkörperforschung
Abt. Oberflächen und Grenzschichten
Helmholtzstrasse 20
01069 Dresden
Germany
(Chapters 2.2, 5.2)

Dr. rer. nat. Horst Wendrock
IFW Dresden
Institut für Festkörperanalytik und
Strukturforschung
Abt. Oberflächen- und Mikrobereichsanalytik
Helmholtzstrasse 20
01069 Dresden
Germany
(Chapter 4.1)

Dr. rer. nat. Christian Wenzel
TU Dresden
Institut für Halbleiter- und Mikrosystemtechnik
Mommsenstrasse 13
01069 Dresden
Germany
(Chapter 3.1)

Prof. Dr. rer. nat. habil.
Dr. h. c. Klaus Wetzig
IFW Dresden
Institut für Festkörperanalytik und
Strukturforschung
Helmholtzstrasse 20
01069 Dresden
Germany
(Chapters 1, 3.2, 3.4, 4.1, 4.3, 6)

Dr. rer. nat. habil. Ehrenfried Zschech
AMD Saxony LLC & Co. KG Dresden Materials
Analysis Department
Wilschdorfer Landstrasse 101
01109 Dresden
Germany
(Chapters 2.1, 4.2, 5.1)

1 Introduction

1.1 Prologue

Electronic devices have found widespread use in our everyday lives. The applications cover many areas such as consumer electronics, information technology, engineering, automotive application, transportation, medical diagnostics and treatments, etc. The construction of these devices and their building blocks involves elaborate fabrication processes which are based on a thorough understanding of materials science and solid state physics. The device functionality may involve conventional microelectronic, acoustoelectronic, optoelectronic, or future spinelectronic elements, or a combination of these (Fig. 1.1). The functionality is achieved by a carefully engineered and complex combination of metallic, semiconducting, and insulating layers. These layers are often micro- and nanostructured by sophisticated lithography techniques in order to achieve the desired properties. Sometimes, as in the case of microprocessors, the structuring involves several levels. The individual feature sizes created by the structuring processes may be as small as 100 nm and are expected to become even smaller in the future in leading edge applications.

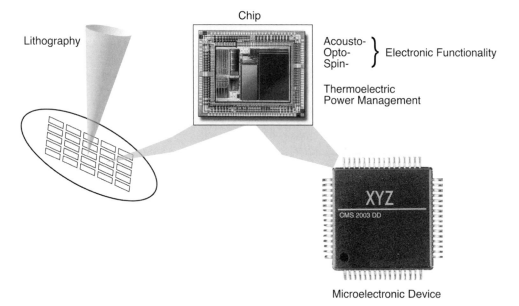

Figure 1.1: The application regimes of metal-based thin films in the microelectronics area

Metal Based Thin Films for Electronics, Second Edition. Klaus Wetzig and Claus M. Schneider (Eds.)
Copyright © 2006 WILEY-VCH Verlag GmbH & Co. KGaA, Weinheim
ISBN: 3-527-40650-6

The fabrication of these electronic devices requires a very good control of the materials properties. This concerns not only the physical material parameters, but also the film structure and morphology. The latter are largely determined by the details of the deposition process and postgrowth processing procedures. In addition, the interfaces between different materials and material classes are also becoming of crucial importance. In this situation, a wide variety of analysis tools must be used to ensure a reliable process control and – if necessary – a precise failure analysis. These tools include not only different real space (microscopy) and reciprocal space (diffraction) techniques, but also spectroscopic techniques, electrical transport measurements, stress and strain analyses, migration investigations, etc.

Novel device technologies are often closely linked to the use of new materials or material classes. One recent example is the replacement of the conventional Al interconnects in microprocessors by Cu ones. This step not only involves new fabrication procedures, such as the "damascene" technique, but also requires new barrier layers to avoid the mixing of Cu and Si. Another example is the emerging technology of magneto- or spinelectronics. In its present state it employs complex magnetic units composed of metal or metal/insulator layer stacks. In addition to the electrical properties, the layers must also provide a distinct magnetic functionality. Since all of the classical ferromagnets Fe, Co, Ni and many antiferromagnets used in magnetoelectronics are metals, this adds another and very exciting facet to the application of metal-based films in electronics.

From the above considerations follows quite clearly that metal-based thin films play a central role in the different steps of the fabrication and for the specific functionality of electronic devices. The most evident use concerns conducting lines and interconnects. Less obvious is their employment as barrier layers against interdiffusion and segregation. Also very important are metallization layers, for example, in acoustoelectronic devices. In chemically complex systems, the physical properties can be conveniently changed by the chemical composition. This is particularly true for the conductivity and is exploited in silicides for thermoelectric applications. Metal-based films are also very important for X-ray optical techniques used to fabricate (X-ray lithography) and analyze (X-ray diffraction and spectroscopy) electronic device structures.

Since metal-based films have such a widespread use in the different areas of microelectronics, knowledge of the respective properties and phenomena is distributed over various fields of physics and materials science. As a consequence, one usually has to consult many different sources in order to get the desired piece of information or a broader overview of a specific issue. Considering the importance of metal-based films in the field of electronics it is thus justified to describe and discuss these systems, the associated effects and phenomena, and their applications in one place.

1.2   Organization, Aim and Content of This Book

The main purpose of this book is two-fold. On the one hand, it is meant to serve as a compendium for metal-based thin film systems and their usage in electronics technology. As such, it addresses both the scientist and the research engineer. On the other hand, the book also includes a more tutorial part which is intended to bridge the gap between fundamental phenomena and their technological applications. It may therefore also serve as a textbook for advanced students in solid state physics, materials science, and electronics engineering.

The book is organized into several chapters covering the range from principal aspects and phenomena over contemporary challenges in materials science to actual device concepts and applications. We thereby mainly concentrate on the relevant fields of interconnects, acousto-electronics, thermoelectrics, magnetoelectronics, and X-ray optics.

In *Chapter 2* we review the various fundamental aspects of metal and metal-based films with respect to the individual fields and applications addressed in this book. This chapter is mainly intended to convey background information for the advanced student in a more tutorial form. It forms a basis for the discussion of the future challenges and the device-related topics in the subsequent chapters. The first section is devoted to a key aspect in microelectronics, namely the means to transfer and distribute information and power in a microelectronic device, for example, in a microprocessor. This is achieved by means of metallic interconnects which are usually arranged in very complicated and delicate three-dimensional networks. The contribution discusses both Al and Cu-based technologies for interconnects and highlights the specific implications and problems associated with each technology. A somewhat less familiar, though not less eminent area of microelectronics is acoustoelectronics. Acoustoelectronic devices are based on the exploitation of phenomena involving the generation, transport, and filtering of surface acoustic waves. Their functionality is largely determined by the interaction between a piezoelectric substrate and a metallic film serving as an electrode. Surface acoustic wave devices play a strategic role in telecommunication and other high frequency applications. A rather novel facet of microelectronics is called magnetoelectronics or "spintronics" which is the topic of the third section of Chapter 2. Spintronics is still an emerging technology which is based on the transport of spins and charges, rather than just charges. It thus combines magnetic functionalities and materials with established microelectronics concepts. Current spintronics applications concern read heads in hard disk drives, magnetocouplers, or nonvolatile magnetic random access memories (MRAM). In the long run, reprogrammable magnetic logic circuits or active magnetoelectronic devices, such as a spin transistor, may be expected. The section reviews the fundamental aspects of spin-dependent transport and magnetic coupling phenomena in thin films and layer stacks. It also discusses the basic thin film arrangements and their specific properties with respect to a particular device functionality. Thermoelectricity exploits the conversion of thermal energy into electrical energy and vice versa for power generators, cooling devices, and temperature or radiation sensors. The particular relevance of thermoelectrical systems for microelectronics arises – among other reasons – from the increasing need for efficient thermal and power management of chip devices. The implementation of Peltier elements in the chip architecture can provide on-chip cooling facilities. The recovery of excess heat and its conversion into electrical energy may help to reduce the overall power consumption and represents a step towards future self-sufficient systems. The different material systems and thermoelectric concepts which are currently under consideration are treated in the fourth section. Particular emphasis is put on the role of the various materials properties with respect to the thermoelectrical efficiency parameters and figure of merit. The final section of the chapter deals with the role of metallic layers and multilayer systems for X-ray optics. The connection of X-ray optics to microelectronics comes from the progress in optical lithography techniques which aim at feature sizes well below 100 nm. Because of the smaller wavelengths, the novel lithography approaches can no longer be based on transmission optics, but have to use reflective optics instead. Metal thin film systems are therefore needed to realize the appropriate optical elements (mirrors, gratings, etc.). The section discusses the fundamental aspects of X-ray optics with respect to thin film systems based on reflection and diffraction.

Chapter 3 is devoted to the deposition techniques used to prepare thin film systems and to the main analytical approaches employed to study their behavior. The analysis involves microscopy, spectroscopy, or diffraction techniques and gives access to different properties, such as the film morphology, chemical composition, crystallographic and electronic structure. Deposition techniques for thin metallic films exist in a wide variety and are described in Section 3.1. Today vacuum based physical and also chemical deposition techniques play the dominant role in the preparation of thin metallic films, but non-vacuum based deposition methods such as electroplating or the modified CVD technique ALD (atomic layer deposition) are also of growing interest and will therefore be discussed in this book. Both transmission electron microscopy (TEM) and electron diffraction are strong techniques for studying micro- and nanostructures in metal based thin films. Furthermore, with enhancement of an analytical TEM by spectroscopic attachments for such as energy dispersive X-ray spectroscopy (EDXS) and electron energy-loss spectroscopy (EELS) it is also possible to receive chemical information (element distribution and chemical binding) in the nanometer range of thin films. A powerful method for the immediate study of electrical thin film properties is in situ scanning electron microscopy (SEM). In situ SEM methods allow the investigation of potential contrast, ferroelectric domains, electron beam induced current (EBIC) and processes of electro- or acoustomigration respectively. X-ray scattering techniques are discussed as a widely-used tool for structural information on thin films. Both the possibilities and limitations of wide angle diffraction, reflectometry, soft X-rays and magnetic scattering are discussed. Spectroscopic techniques allow the element distribution and depth profile analysis of thin films. They can be carried out by electrons, X-rays or ions and are frequently used in connection with imaging techniques such as scanning or transmission electron microscopy. In contrast to bulk materials, thin films on substrates are usually under mechanical stress. Thus, stress measurement methods play an important role in the characterization of thin films for electronics. Different techniques such as the substrate curvature and the $\sin^2 \Psi$ method are discussed under application aspects.

As one of the core parts of this book *Chapter 4* addresses current challenges in the investigation and application of metal-based thin films. These include the aspects of thermal stability, acousto- and electromigration, barrier and nucleation layers, functional electric and magnetic layers and multilayers for X-ray optical purposes. These topics represent the forefront of the current research in materials science and solid state physics. Because of the continuing downscaling in the architecture of integrated circuits electromigration is a life- limiting process in metallization layers. The damage analysis is discussed both for Al and Cu interconnects. The introduction of copper as the conducting material for interconnects requires effective diffusion barriers since copper readily diffuses into silicon oxide and silicon. The optimization of barriers and new barrier/ seed concepts are therefore the focus of attention. Migration effects are also observed in surface acoustic waves (SAW) structures, as a result of diminishing structure dimensions (< 1 μm) and increasing electrical input power values (> 1 W) which cause very high power density levels and therefore high stress loading of metallization. Thus, new metallization concepts have to be discussed in detail. Spintronic applications of functional magnetic layers, such as for sensors and MRAMs, may be realized by thin film systems which may be grouped into multilayers, spin valves and tunnel junctions. These systems excel in a precisely defined functionality which is strongly influenced by temperature. Therefore the thermal stability plays a dominant role in both the manufacture and operation of functional magnetic layers, as will be demonstrated for magnetoresistive layer stacks. A further group of thin film based components with growing importance are multilayers for X-ray optical purposes, e.g. as reflectors for X-rays. Finally, the last part of Chapter 4 is focused on functional electric layers with well–defined electronic and electrical transport properties. Such thin film materials are used as resistance layers, thermoelectric sensors and generator devices. The optimization of the electrical and thermoelectric film parameters will be discussed in depth.

The application of metal-based thin film systems in electronic and microelectronics-related devices is the focus of *Chapter 5*. The diversity of the devices treated in this chapter highlights the widespread application areas of metal film systems. The first section deals with interconnect technologies for memory and logic products. Of particular interest are the combination of Cu interconnects and low-*k* dielectrics. The subsequent section on surface acoustic wave devices gives examples of high frequency filters, resonators and delay lines. The device concepts range from relatively simple transversal bandpass filters to programmable phase shift keying filters. The magnetic and magnetoelectronic sensor devices are mainly related to automotive applications and thus emphasize one of the growing future markets for microelectronics products. There, magnetic sensors are employed to measure positions, angles, rotational speeds, or torques with the aim of improving fuel economy, vehicle and passenger safety, and driving comfort in the present and new generations of automobiles. Reducing the energy consumption and improving the energy efficiency is also the driving force in the development of thermoelectric sensors and transducers. The devices discussed range from thermal converters for high precision AC measurements to low power thermoelectric generators and microcoolers for applications in microelectronics. The chapter closes with a section describing several examples of X-ray optical elements based on metal thin film systems. The applications cover not only X-ray telescopes and microscopes, but also recent developments in the area of extreme ultraviolet lithography (EUVL) instrumentation. The latter will be the fundamental tool for a future downscaling in microelectronics.

The final *Chapter 6* of the book gives an overview of the developments to be expected in the field of metal-based thin films and their implementation into microelectronic circuits and devices. The future will not only see the use of new materials and device concepts, but also the fusion of distinct areas to achieve improved or novel functionalities. This concerns, for example, the possible implementation of optical interconnects which may be seen as a combination of standard microelectronics and optoelectronics. Another example is the incorporation of nonvolatile functions on the basis of magnetic components, i.e., the synthesis of conventional microelectronics and spinelectronics. The major driving forces behind these activities are not only the expected revenue, but also the opportunity for new discoveries and developments which may completely change our current picture of microelectronics in the future.

Acknowledgement

Particular thanks go to Mrs. B. Präßler-Wüstling, Mrs. V. Haase, Mrs. C. Singer and Mrs. K. Schmiedel for their skillful processing of the various manuscripts during the preparation of the book. We also would like to thank Mrs. V. Palmer and Mrs. C. Wanka from Wiley-VCH for their support and help during this project.

2 Thin Film Systems: Basic Aspects

2.1 Interconnects for Microelectronics

2.1.1 Introduction

Interconnects are the means of transportation of information within a microelectronic circuit. When one takes apart an old radio, one finds that all the active components are connected by single wires, each of which has been soldered to transistors, resistances, capacities etc. Nowadays this apparent chaos of wires is replaced by printed circuit boards where a polymeric substrate is covered with copper and channels are etched to create wires. A three-dimensional wire structure has been converted into a planar structure and only now and then has an additional wire to be added to form a bridge. The aspect ratio of the copper wires is small i.e. they are much wider than high. Low cost wet etch techniques can be employed.

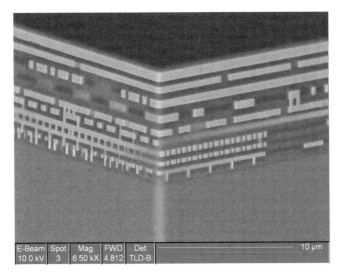

Figure 2.1: SEM cross section of AMD's microprocessors with nine levels of copper interconnects [2.1]

When one takes apart a modern microelectronic chip one will find a completely different scenario. Unlike the printed circuit board there are several up to ten different metal layers in an integrated circuit. The active components are all situated in the substrate, which is usually made of silicon. The aspect ratios are higher, in extreme cases even significantly greater than one (i.e. the lines are higher than wide). In addition to the conductor lines made out of copper or aluminum, one introduces several other layers, such as dielectrics, etch stops, anti-

Metal Based Thin Films for Electronics, Second Edition. Klaus Wetzig and Claus M. Schneider (Eds.)
Copyright © 2006 WILEY-VCH Verlag GmbH & Co. KGaA, Weinheim
ISBN: 3-527-40650-6

reflective coatings, diffusion barriers and vertical connections (so-called plugs). The barriers will be discussed at the end of this section. All of these layers have special functions and properties. Figure 2.1 shows a cross-section of a state of the art microprocessor revealing the complicated layers of wiring that will be described in detail later in this section.

In order to create a working circuit the engineer has to overcome two basic challenges: manufacturability and reliability. The first, even if sometimes very complicated, is usually straightforward. It is apparent whether a circuit works or not. However, sometimes the search for the reason for failure can be tedious. In the case of reliability, on the other hand, it is difficult to predict whether a circuit will work ten years from now. This task is usually addressed by a combination of the following concepts. Lifetime tests have to be accelerated by increasing temperature and the current densities that the wires have to carry. Obviously these conditions are not realistic and thus have to be extrapolated to service conditions. Empirical models that are based on variations in temperature and current density are relatively easily to formulate, however, they bear the risk of being too aggressive or too conservative when the extrapolation is made. It is now the task of materials science to promote understanding of the mechanisms responsible for failure, investigate their temperature and current density dependence and formulate mechanistic models that can be employed for lifetime prediction.

The next sections will focus on the fabrication of interconnect lines, their dimensionality and function, the materials science applicable to them and finally will elucidate future perspectives. The last section will focus on function and materials choice of diffusion barriers.

2.1.2 Metallization Layers

Function of On-Chip Interconnects and Materials Selection

The function of an interconnect system is to distribute clock and other signals and to provide power/ground, by connecting the various circuit/systems functions of a chip. The fundamental development requirement for the on-chip interconnects is to meet the high-speed needs of chips to transmit clocks and signals despite further down-scaling of feature sizes. In particular, the so-called RC (resistance × capacitance) time delay has to be minimized using a smart interconnect design and new technologies and materials. This task includes the development and the implementation of conducting material with low resistivity for interconnects and of dielectric material with low permittivity as the isolating material between them. Additionally, numerous other, mostly ultrathin films of the back-end-of-line layer stack have to be considered since they all contribute to the overall performance and reliability of the interconnect system: barriers, capping layers, etch stop layers, hard masks, etc.

The following requirements for the conductor material have been derived from the performance and reliability requirements of the on-chip interconnect system:

– low resistivity (high electrical conductivity)
– high thermal conductivity
– high melting point (and thus low diffusivities)
– materials compatibility to the isolating dielectric material and to barrier and capping layers
– technology compatibility to the back-end-of-line process.

The first level of metallization is the contacting plug that provides the connection to the metal-oxide-semiconductor field effect transistors (MOSFETs). Until now, tungsten has been widely used for the contacting plug to the devices and for the so-called local interconnects in microprocessors and dynamic random access memories (DRAMs). One challenge will be high aspect ratio (A/R) stacked capacitor DRAM contacts with A/R up to 100.

For the further interconnect system, metallic films with lower resistivity, e.g. aluminum and copper, are usually used. For more than 30 years, the back-end-of-line manufacturing in the semiconductor industry was dominated by the "metal PVD and metal wet-etch" wiring technology. That means, aluminum and aluminum alloys were deposited using physical vapor deposition (PVD) followed by a wet subtractive etching. Al(Cu) alloys have been used since the late 1960s to alleviate electromigration concerns associated with the Al(Si) metallurgy. Thin layers of refractory metals like Ti above and/or below the Al(Cu) interconnects and the formation of Al_3Ti films reduced the contact resistance and improved the electromigration stability.

Both performance and reliability of high-performance microprocessors (HP MPU) is increasingly determined by design, technology and materials for interconnects and dielectric interlayers which result in lower RC products. The need for new conductor and dielectric materials that would be necessary to meet the projected overall technology requirements has been described in the Technology Roadmap for Semiconductors since 1994 [2.2., 2.3]. As the dimension of the interconnect lines continues to shrink, aluminum-based interconnects and CVD oxide/nitride interlayer dielectrics are being replaced by inlaid copper with reduced electrical resistance and by low-k dielectrics. Besides the higher conductivity, inlaid copper lines also offer the advantages of improved electromigration performance and reduced cost of manufacturing [2.4–2.6].

Table 2.1: Materials properties for aluminum and copper

Physical property	Aluminum	Copper
Specific electrical resistivity ($\mu\Omega$ cm)	2.72	1.71
Thermal conductivity (W m^{-1} K^{-1})	238	327.7
Melting point (K)	933.5	1358

Table 2.1 compares materials properties for aluminum and copper. The resistivity of copper is about 40 % lower than that of aluminum, which is generally mentioned as the major advantage in introducing this material, since it improves the product performance of microprocessors and memories due to the direct impact on the RC product. The high melting point of copper is advantageous for the interconnect reliability, since all diffusion-controlled atomic transport processes are slower, and consequently, the current-carrying capacity is higher.

The introduction of a planar multilevel metallization architecture with inlaid copper interconnects was first reported by IBM in September 1997 [2.4]. Copper was deposited using an electrochemical deposition process (ECD) into trenches and vertical contacts (vias). These inlaid structures had been etched into silicon oxide layers deposited using chemical vapor deposition (CVD) from tetraethylorthosilicate (TEOS). Since that time, copper has mainly been used for microprocessor applications.

Although chips with inlaid copper interconnects in silicon dioxide (dielectric constant $k = 4.0$) were introduced in 1998, we have witnessed the start of the change to new insulator materials with lower dielectric constant only recently. Fluorine-doped silicon dioxide (FSG, $k = 3.7$) combined with the copper dual inlaid process has been in production since the 180 nm technology node. Next potential materials for the 90 nm technology node and below are lower-k materials ($k = 2.6$–3.0) like organic spin-on-polymers (SOP) and plasma-enhanced CVD (PECVD) inorganic/organic hybrid materials [2.7, 2.8]. Potential commercial materials for this k range could be Applied Material's Black Diamond organosilicate material, Coral organosilicate glass (OSG) film from Novellus, Flowfill CVD from Trikon Technologies and SiLK spin-on low-k material from Dow Chemicals. Eventually, the nitride etch-stop layers and even the dielectric antireflection coating (ARC) layers/hard masks (with $k \sim 7$–9) will probably be replaced by SiC-based films (with $k < 5$) offered by Novellus and Applied Materials.

Ultra-low-k materials (ULK, $k < 2.5$) are in development. Integration efforts focus on solving the problems of these reduced density materials with their compromised thermal and mechanical properties. The ideal ULK material will have a closed pore structure and uniformly distributed pores with a maximum pore size, tied to, and decreasing with, the technology node. A tight pore size distribution is also desirable. Porous ULK materials need even more planarization development efforts than nonporous materials.

The introduction of these new conductor and isolator materials, along with the reduced thickness and higher conformity requirements for barriers and nucleation layers, is a difficult integration challenge. The most challenging integration modules are dielectric etching, integrated cleaning, chemical-mechanical polishing (CMP) and packaging. A primary integration challenge with the low-k materials is the adhesion failure of barrier or capping materials with the dielectric during planarization. Process integration and device related aspects are described in Section 5.1.

Table 2.2: Selected interconnect technology requirements from the 2003 ITRS [2.2]

Technology node	90 nm	45 nm	22 nm
Number of metal levels	9	10	11
Local wiring pitch (nm)	214	108	54
Local wiring aspect ratio, Cu	1.7	1.8	2.0
Conductor effective resistivity, Cu intermediate wiring ($\mu\Omega$ cm)	2.2	2.2	2.2
Interlevel insulator effective dielectric constant (k)	3.1–3.6	2.3–2.6	<2.0

Table 2.2 shows some 2003 ITRS requirements for on-chip interconnect systems [2.2]. Important parameters are the minimum trench width and the aspect ratio (A/R) for interconnects. According to the roadmap, the minimum trench width for interconnects will decrease from today's > 100 nm to 27 nm in 2016.

Fabrication

In principle, there are two routes that are currently applied: (a) the subtractive method (aluminum), and (b) the inlaid method (copper).

Up to 1998, all microelectronic devices were fabricated with aluminum metallization. Most of the chips, except for high end applications like microprocessors or fast RAM, are still built that way. Traditionally, the fabrication of interconnects and dielectrics belongs to the so-called back-end-of-line process, because they are physically located at the 'back-end' of a fabrication line. All transistor fabrication processes belong to the front end.

Aluminum lines are fabricated by depositing a homogenous thin film by magnetron sputtering. Usually the deposition of Al is preceded by a thin layer of titanium or titanium nitride as a diffusion barrier and followed by the same material as an antireflection coating for lithography purposes. Subsequently, a layer of photoresist is deposited and cured. The photoresist is then exposed to UV light and a pattern is transferred from a mask to the photoresist. After dissolution of the exposed areas the photoresist has the same pattern as the aluminum will eventually have. In the next step the photoresist serves as an etch mask to remove Al by a reactive ion etch process (RIE). Usually chlorine is used as an etch gas, which prevents under-etch by forming a passivating layer on the sidewalls. The photoresist is then removed by an oxygen plasma. In the next step a dielectric, usually a glass, is deposited and planarized by a polishing step if several layers of metallization are anticipated to follow. In order to allow for a 3D structure, the dielectric is patterned and filled with so-called tungsten plugs by a CVD process. This process (Fig. 2.2) is repeated iteratively in order to generate a multilayer structure.

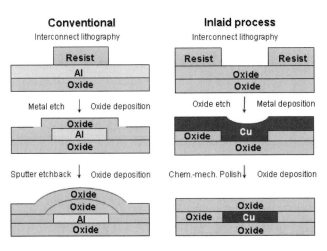

Figure 2.2: Comparison of Al (conventional) and Cu (inlaid) process flow

As mentioned before, with the demand for higher and higher clock frequencies the time delay associated with the interconnect structure (RC time delay) eventually became time limiting. This, among other factors, was the driving force in replacing the combination aluminum-glass by the combination copper-polymer. The change in material, however, proved to be a significant step for the industry as it incorporated significant changes in processing in a previously only marginally researched field. Unlike for aluminum, there is no adequate RIE process for mass-production of copper. Therefore, the entire process had to be inverted (Fig. 2.2).

After the front end processes a layer of dielectric is first deposited. This layer is then patterned to form trenches in which the copper is to be deposited except for the first layer, which is in direct contact with the silicon substrate. In this case tungsten plugs are deposited by a CVD process. Different deposition techniques are required for several reasons: (a) the sputter process cannot fill high aspect ratio trenches; (b) as copper is a highly undesirable contaminant for silicon, the copper line has to be encapsulated from all sides with diffusion barriers; (c) an additional silicon nitride etch stop is needed for subsequent layers.

The solution to these issues is a sputter deposition of a diffusion barrier (Ta, TaN) and a thin copper seed layer. In the next step the wafer is immersed into a copper electroplating bath and the trenches are overfilled so that a relatively homogenous copper surface is established. This process is called electrochemical deposition (ECD). Now, however, there is a short circuit between all interconnects. This dilemma is solved by borrowing an ancient artisan technique from Damascus. The excess copper is removed by a chemical mechanical polishing step (CMP) that confines the metal to the corresponding trenches ("Damascene" technique). The encapsulation is then completed by a thin layer of silicon nitride which serves as an etch stop for the subsequent layers as well as a diffusion barrier. Again the process is repeated iteratively to form a 3D structure. A general trend that should be noted is the increase in the line widths from the layer closest to the silicon to the top (Figure 2.1). The layer thickness generally stays the same. So far the metal in these interconnects has been treated as a homogenous material. However, in reality the continuous line is formed by a polycrystalline copper material. The next section will address the consequences of thin film crystallinity.

2.1.3   Materials Science of Metallic Interconnects

Microstructural Effects

When a thin film is deposited onto an amorphous substrate, the microstructure that evolves depends on a variety of deposition parameters. The key thin film parameters that one tries to influence/optimize are grain size, texture, defect density and roughness. Again the cases of aluminum and copper have to be differentiated.

The key parameters in aluminum sputter deposition are deposition rate, substrate temperature, argon pressure and residual gas pressure. For high throughput and high purity high deposition rates are desirable. An elevated substrate temperature enhances grain growth but leads to higher roughness. Usually, aluminum films are also thermally annealed after the deposition process to trigger grain growth and heal defects. For aluminum the equilibrium grain size is given by the film thickness according to the *Mullins* criterion [2.9], where grain boundary grooving limits further grain growth.

The preferred crystallographic orientation of grains (texture) that one tries to establish for reliability reasons (see Section 4.1), is the (111) out of plane orientation. The ⟨111⟩ axes of all grains are closely aligned to the substrate normal. Details of measurement and quantification of textures can be found in Section 3.3. For aluminum films this texture is easily achieved as the (111) surfaces have the lowest surface energy and thus the energy of the system is minimized.

In the case of electroplated copper, on the other hand, several new effects can be observed. Electroplated copper tends to self anneal at room temperature [2.10, 2.11]. This effect is rather surprising as the homologous temperature of copper (temperature normalized by the melting temperature, usually a measure of diffusion) is significantly lower than for aluminum. The grains grow abnormally (Fig. 2.3) that means that a minority of grains 'eats up' all the grains around them. In normal grain growth all grains would grow at a similar rate. Usually, the self-annealing effect is attributed to the bath chemistry of the electroplating bath that appears as contaminants in the copper film.

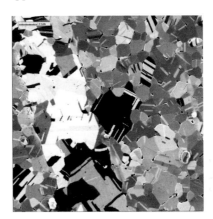

Figure 2.3: FIB (focused ion beam) images of electroplated and sputtered 1 µm Cu films at 0° tilt (scale bar: 3 µm) [2.12]

The scenario is further complicated as copper interconnects exist in the form of lines and contacts with geometry- and process-dependent microstructure and properties. The microstructure of copper interconnects is essential for both product performance and reliability. In particular, degradation mechanisms in copper interconnects can be influenced by their microstructure parameters such as grain size, texture and stress. From the materials point of view, alternative barrier and capping layer materials pose additional challenges (see Section 4.2).

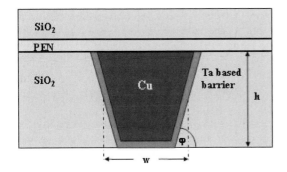

Figure 2.4: Copper line cross section (schematically); w, h = width and height of the copper line; φ = sidewall angle [2.13]

Test structures with arrays of parallel single inlaid copper lines with varied line width allow for the quantification of the geometry dependence of the copper microstructure. In [2.13, 2.14], trenches with a height of 0.45 µm and a width between 0.35 and 1.0 µm were etched into silicon oxide. Their sidewalls were not fully vertical, i.e., the sidewall angle φ was about 85°, with the smaller width at the trench bottom. The trenches were filled using

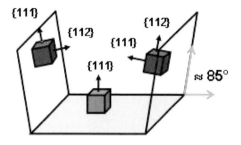

ECD after deposition of Ta barrier and Cu seed layers. After CMP, a thin silicon nitride film was deposited onto the samples (see Fig. 2.4).

Due to the geometry of the trenches additional texture components can be observed (Section 3.3). One can find ⟨111⟩ texture components that are perpendicular to the sidewalls of the trenches. With decreasing line width, the relative contribution of the texture component from sidewall oriented copper grains to the total texture increases. The preferred crystallographic orientation of both bottom oriented and sidewall oriented grains in narrow copper lines is shown schematically in Fig. 2.5 [2.13]. Again surface energy minimization is the driving force. As copper has a relatively low stacking fault energy compared to aluminum (especially when not very pure), plastic deformation is accommodated by mechanical twinning. This is a process in which all atoms in a crystal region are moved in unison to form a twin boundary that corresponds to a stacking fault. By this process the macroscopic shape of the crystal is changed and plastic strain is accommodated. One can envision this process by taking a single crystal, cutting it in half along a (111) plane, rotating it by 180° and gluing it together again. In terms of texture a (115) component is added to the already existing ⟨111⟩ component [2.12–2.14]. Another attribute of copper is its high elastic anisotropy. This also has consequences [2.15] regarding the evolution of texture in copper metallization. We have explained that ⟨111⟩ texture in aluminum is driven by surface energy minimization. If a thin copper film, however, is under mechanical stress an additional term comes into play. Thus we have a competition between surface energy minimization and strain energy minimization, an energy term that scales with the volume. Thus it is apparent that for thicker films the strain energy term starts to dominate and we observe a texture change from ⟨111⟩ to ⟨100⟩ out of plane.

In the last paragraph we have explained a texture component by the existence of mechanical stresses in thin films. But what is the origin of these stresses? One can imagine that in a sputter process when energetic argon ions are impinging on the film in formation compressive deposition stresses arise. However, as usually the metallization is annealed after deposition, deposition stresses are of secondary importance. The most important stresses are the so-called thermal stresses induced by the mismatch in the thermal expansion coefficients (CTE) of the film and substrate, usually of the order of tens of ppm K^{-1}. When a thin film is now annealed at high temperatures, where diffusive processes are active and stresses can be relaxed, and subsequently cooled, tensile thermal stresses are observed as the metal has a much higher CTE than the semiconducting substrate. As these stresses are very important for interconnect reliability great care is taken in their measurement (also see Section 3.5).

At room temperature these films are usually at their tensile yield stress [2.16]. As the yield stress of thin films scales inversely with their thickness and grain size [2.17–2.19], thinner films or narrower lines experience higher and higher stresses. These stresses can be responsible for failure by stress voiding due to thin film creep processes and institute a reliability issue. Recently, a fairly broad distribution of local stresses has been found by X-ray microdiffraction techniques [2.20] (see Section 3.3).

Another case where local stresses are important is electromigration which will be briefly introduced later. There, stresses may even have a beneficial effect as their gradients counteract the electromigration driving force (for details see Section 4.1).

The detailed mechanical stresses in copper lines, however, are affected by the inlaid process and arise from the differences in the linear thermal expansion coefficients, not only between the metallic interconnect and the substrate, but also the dielectric that rigidly confines it, as well as the line aspect ratio. But both grain size and texture interact with stress in inlaid copper lines too, and the stress level can be changed by the same process parameters which influence other microstructure features [2.13, 2.14]. The stress level in inlaid copper lines was measured by *Besser et al.* [2.14, 2.21], *Du et al.* [2.22] and *Prinz et al.* [2.23, 2.24]. They found that the stress in copper lines increases with increasing post-plating anneal temperature. *Vinci et al.* and others [2.25–2.27] observed stress-induced voiding in passivated copper lines, which institutes an additional failure mode.

Since the atomic transport processes which can cause degradation of copper interconnects depend on interconnect geometry and copper microstructure, the measurement of microstructure parameters like grain size, texture and stress on a routine basis is a reasonable monitor of the process stability. This approach of copper microstructure monitoring was reported by *Zschech et al.* [2.28]. Statistically relevant data for texture and stress of copper lines were obtained with a micro X-ray diffractometer (micro-XRD) in reasonable periods of time for process control. The specific test structures consist of arrays of parallel single inlaid copper lines with an array size of 120×120 μm^2. According to [2.28] stress and texture in copper lines are the most sensible microstructure parameters to control deposition and anneal processes in the interconnect technology. For instance, stress changes in the capping layer are reflected also in stress changes in the copper lines. The stress was determined for several directions from lattice parameter shifts based on the Cu (311) Debye ring intensity analysis. The texture was monitored quantitatively along and across the copper lines, based on Cu $\langle 111 \rangle$, $\langle 110 \rangle$ and $\langle 100 \rangle$ pole figures.

Figure 2.6 shows the grain size, texture and stress data for a production period of 8 weeks [2.28]. In this particular case, the microstructure of single inlaid copper lines with a line width of 0.18 μm was monitored. The inlaid copper lines were fabricated in SiO_2 using an oxide patterning and etch technique. PVD-Ta and PE-CVD Si_3N_4 were used as barrier and capping layer, respectively. Overall, the data demonstrate a high stability of interconnect manufacturing.

In terms of the characterization of the microstructure of thin metallic films several, sometimes complementary, techniques are employed. Chapter 3 describes these techniques and their applications in detail. So far we have addressed pure metallic systems as in the case of aluminum or a contaminated system as in the case of electroplated copper. In the latter system contamination even had a beneficial effect (self-anneal). In the next section we will talk about the intentional addition of other elements: alloying.

Grain Size

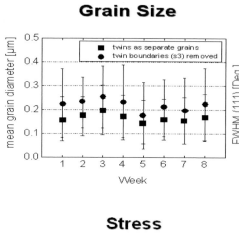

Texture

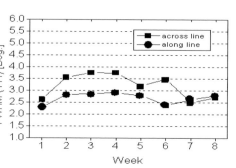

Stress

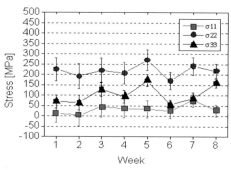

Figure 2.6: Copper line microstructure monitoring: grain size (upper left), texture (upper right), and stress (lower left). The data show high process stability [2.28]

Purpose of Alloying

About twenty years ago aluminum interconnects started to show a failure mode known as electromigration (see Section 4.1). Further miniaturization seemed to be a risk as smaller interconnect cross-sections increased the current densities and thus accelerated the electromigration phenomenon. By chance, minute quantities of Cu were added to a batch of chips and an increase in lifetime of two orders of magnitude was observed in 1970 [2.29]. Since then Al(Cu) alloys (~ 0.5 wt% Cu) are commonly used for interconnects. Details of why Cu has such a beneficial effect will be given at a later point (see Section 4.1). The addition of Cu is just an example; in general trace elements can have the following effects:

- increase yield stress
- prevent interdiffusion
- reduce self diffusivities
- create self passivating layer
- reduce fatigue effects
- promote adhesion
- provide shunt layers
- improve electromigration

Usually, however, beneficial effects come at a cost. As the main task of interconnects is the conduction of electrical current, their resistivity has to be minimized. Any addition of alloying elements will, in general, increase the resistivity. The degree of resistivity increase, however, depends on the kind of alloying element used. The electromigration resistance of aluminum can be improved by various dopants. Cu in Al(Cu) alloys for instance forms intermetallic Al_2Cu precipitates which eventually appear at grain boundaries and interfaces. There, they serve as reservoirs to block fast diffusion paths and reduce the rate of material transport without significantly increasing the resistivity (see Chapter 4). In solid solutions, such as Al(Mg), which also improves the electromigration resistance, the dopant atoms also act as additional scattering centers, and the resistivity is increased significantly.

Let us address some of the effects mentioned above in a little more detail. The yield stress of a thin film is determined by its thickness, the grain size and its texture [2.17–2.19, 2.30]. If the yield stress has to be increased even more, one has to introduce obstacles for dislocations that have a spacing that is significantly smaller than the grain size or thickness. This can be achieved by particles such as Al-Cu precipitates or ceramic yttria, or by single atoms that have a significant size difference to the matrix and pin dislocations caused by their stress field (e.g. Mg).

One of the common alloying elements for Al in addition to Cu is Si. In some circuits the interconnect material is in direct contact with the semiconductor. The resulting undesirable phenomenon is the so-called spiking where Al and Si interdiffuse locally. If the aluminum alloy is already saturated with Si there is no driving force for interdiffusion and the phenomenon is suppressed.

One of the common barrier layers for Al is Ti. On the one hand, this improves the Al texture, on the other hand Ti diffuses into Al and creates an intermetallic phase at the bottom of the interconnect. This usually forms a high resistance conducting path. So if a void is formed by thermal stresses or electromigration, the current path changes from the line down to the intermetallic layer. This causes local heating and in many cases leads to a self-healing of the void.

The relevant issues for inlaid Cu metallization are self-passivation and electromigration. The resistivity of copper alloys increases with the dopant concentration and with the charge difference between copper and the dopant. Mg and Al have been studied as dopant elements in copper since these elements diffuse to the interfaces and form stable oxides. In this way the lines would be self-passivating [2.31, 2.32]. *Hu et al.* [2.33] have shown that the addition of Sn to Cu metallization significantly reduces the electromigration drift velocity. However, the addition of a small percentage of Sn (0.5–2.0 wt%) decreases the average grain size of copper and hence increases the number of fast diffusion paths for material transport. Additionally, the electron scattering rate at grain boundaries increases.

Interconnect Degradation and Reliability

Advanced process technologies and new combinations of materials bring about new reliability challenges: different microstructure of the metallic interconnects, other types of interfaces and new degradation phenomena. Electromigration, stress-induced degradation and mechanical weakness in the case of low-*k* materials are reliability concerns for inlaid copper interconnects. The current generation of highly integrated microprocessors, requiring dense interconnects and increased current densities, has highlighted the electromigration issue. Formation of voids in copper lines induced by electromigration during normal microprocessor operation will cause an interconnect opening or a high resistance, resulting in malfunction or speed degradation, respectively. Stress-induced degradation phenomena and later catastrophic failures are not yet well understood, but they are probably exacerbated by normal stresses at the copper/barrier and copper/nitride interfaces as well as hydrostatic stress components.

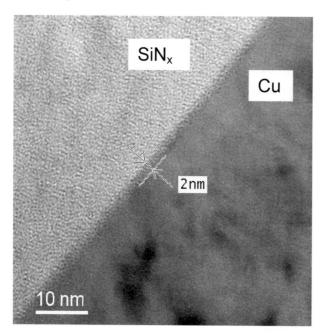

Figure 2.7: TEM brightfield image of a Cu/SiN$_x$ interface [2.49]

In particular, fast diffusion paths have to be identified and failure mechanisms based on directed transport of atoms have to be described, as a basis for process and materials changes that improve interconnect reliability. Considering a high-quality processing technology, degradation of inlaid copper lines is connected with directed material transport. Such a directed transport of copper atoms can be caused by a gradient of the chemical potential, e.g. caused by a concentration gradient (interdiffusion), a gradient of the electrical potential, e.g. caused by a directed current (electromigration), temperature gradients (thermomigration) or stress gradients (stress-induced migration). The atomic transport processes and the resulting degradation mechanisms are discussed in detail in Chapter 4 (see also [2.13, 2.28]).

Although a lot of theoretical and experimental work has been done on electromigration of inlaid copper interconnects, it is still not fully understood how the interconnect degradation takes place. However, numerous experimental studies have indicated that electromigration-induced degradation and eventually interconnect failure depend on interface bonding between copper interconnects and liners or capping layers [2.34–2.39]. In the past, the copper microstructure was a second-order effect since even significant changes in copper stress and texture, e.g. caused by advanced low-k interlayer dielectric materials with extremely reduced Young's modulus compared to silicon oxide, did not or only slightly influence the electromigration lifetime of the test structures. This situation will change as soon as weak interfaces are no longer the fastest pathways for electromigration-induced mass transport [2.40–2.43].

Since the activation energies of atomic transport processes along Cu/(dielectric) capping layer and/or Cu/(metal) liner interfaces are related to the bonding strength, interface strengthening – particularly for the top Cu/capping layer interface – seems to be the consequent way to prolong copper electromigration lifetime [2.44, 2.45]. Recently, several approaches have been published to improve the electromigration lifetime by increasing the bonding strength of the top interface and consequently to reduce the mass transport along the Cu/capping layer interface (compared to the "conventional" Cu/SiN$_x$ or Cu/a-SiC$_x$H$_y$), including a selective electroless coating with CoWP [2.46, 2.47] and the deposition of self-assembled monolayers (SAMs) [2.48] on top of the polished copper line (for reviews see [2.44, 2.45]). In all cases, the electromigration-induced mass transport along the top interface is retarded and interconnect degradation is slowed down significantly. The transmission electron microscopy (TEM) brightfield images of Cu/SiN$_x$ (Fig. 2.7) and Cu/CoWP (Fig. 2.8) interfaces provide the explanation for this observation, i.e., different interface structures. The Cu/SiN$_x$ interface is usually characterized by an "intermixing layer" of about 2 nm thickness with a high degree of disorder. This "intermixing layer" is caused by the PECVD process for SiN$_x$ deposition on top of the polished copper lines. At the Cu/CoWP interface, however, lattice planes are visible in both metal films, sometimes without any interruption at the interface. This kind of interface with regions of coherence (epitaxy-like growth) is much less disordered, and no "intermixing layer" can be found [2.49]. For the Cu/SAM/SiN$_x$ interface, even with TEM imaging it is difficult to identify the influence of the deposited monolayer on the atomic structure or the interfacial strength.

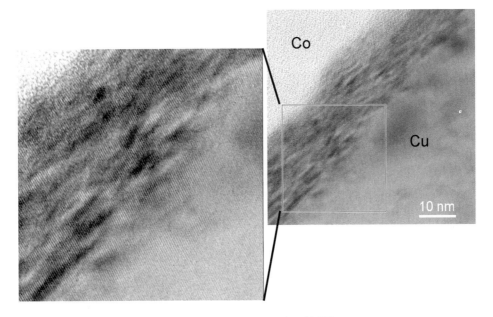

Figure 2.8: TEM brightfield image of a Cu/CoWP interface [2.49]

Numerous experimental and theoretical results have created a basis for understanding the major mechanisms governing the reliability of copper interconnects in the complex dual-inlaid geometry [2.50]. However, most of the experimental data are based on standard electromigration tests that do not reveal degradation processes, or on SEM studies of unpassivated interconnect lines that do not represent the real interfaces and the stress states. In-situ scanning electron microscopy (SEM) studies of electromigration allow for the visualization of the copper mass transport and the time-dependent evolution of voids in inlaid copper interconnects for fully embedded via/line test structures [2.51]. Numerical simulations based on a physical model that incorporates all important migration forces into the mass balance equation [2.52] demonstrate how void nucleation and growth, and eventually interconnect failure depend on interface bonding [2.53, 2.54].

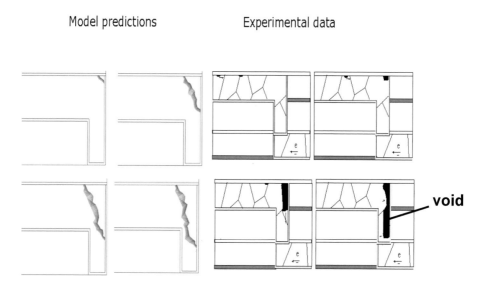

Figure 2.9: Degradation mechanisms in copper via/line structures with standard Cu/SiN$_x$ top interface (experiment and modeling) [2.42]

Figures 2.9 and 2.10 compare the final phase of electromigration-induced void evolution for a dual-inlaid copper interconnect segment – both from experiment and from simulation – for a weak and for a strengthened top interface of the copper interconnects [2.42]. During the electromigration-induced interconnect degradation process, several phases can be distinguished [2.40]: Void formation, void evolution at interfaces, void growth to a volume that causes significant resistance increase and void growth that causes further resistance increase leading to electromigration failure. In the first two phases, agglomerations of vacancies and voids are formed at interfaces and grain boundaries. Depending on the interface bonding energy, voids apparently move along weak interfaces, most of them, eventually, towards the cathode end of the line. Subsequently, the voids grow and merge into a larger void that eventually grows into the via. It can be seen that void evolution depends strongly on both interface bonding of the inlaid copper structures. The increase of the bonding strength of the weakest interface – the top interface – results in a completely changed degradation behavior and failure mechanisms, and in a significantly increased electromigration lifetime.

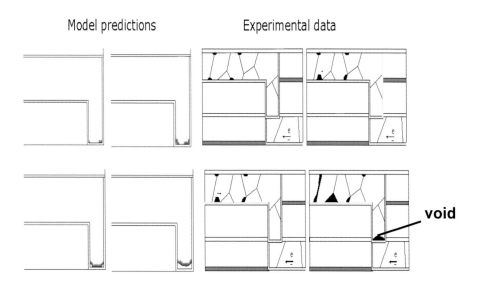

Figure 2.10: Degradation mechanisms in copper via/line structures with strengthened Cu/CoWP/SiN$_x$ top interface (experiment and modeling) [2.42]

Depending on the crystallographic orientation of the grain next to the void, the diffusion rate is expected to be different, and therefore, grains with different orientation will be disintegrated at different speeds. This study gives evidence for theoretical models postulating that voids grow at sites of flux divergences like grain-boundary triple points and interfaces [2.55]. Void formation, growth and movement, and consequently degradation, depend clearly on the microstructure of individual vias. The step-like degradation behavior is in agreement with resistance changes measured in copper dual-inlaid electromigration structures [2.56].

For numerical simulation, which is in excellent agreement with the experimental observations, the electromigration-induced directed atom transport in dual-inlaid copper interconnects is described with an effective diffusivity considering all possible contributions [2.13]. Void nucleation and growth, taking place in the interconnect structures stressed by electrical current, are caused by the momentum transfer from the electrical current carriers to the lattice ions, the so-called "electron wind". The algorithm for simulation of electromigration-induced void evolution developed by *Sukharev* [2.52–2.54] is based on a model that incorporates all important migration forces into the mass balance equation. Its solution – together with the solution of the coupled electromagnetics, heat transfer and elasticity problems – allows simulation of electromigration-induced degradation in dual-inlaid copper interconnect segments [2.52–2.54]. This model makes it possible to explain differences in void dynamics when different interface bonding strengths cause different major channels for atom migration.

The directed mass transport along the Cu/capping layer interface as well as along the copper grain boundaries (and later along the inner surface of a void) are processes with different activation energies. Since their apparent activation energies, as determined by lifetime measurements, depend on interconnect geometry and microstructure as well as on their interfaces, these material-specific data are generally different for each degradation process, and consequently, a multimode failure distribution with strong and weak mode failures may occur. Strong modes relate to high activation energy; weak modes relate to a lower activation energy. Multimodal modeling allows the identification of relative population percentages. The population of early failures is highly geometry and process dependent [2.57]. According to the equation for the effective diffusion coefficient, the interfacial mass transport and consequently the fraction of early failures becomes increasingly important for shrunk interconnect geometries.

The integration of new low-*k* dielectric materials needed for performance enhancement brings about numerous reliability concerns that include thermally or mechanically induced cracking or adhesion loss, poor mechanical strength, low stiffness, moisture absorption, texture effects, and poor thermal conductivity. The mechanical strength of the interlayer dielectrics is considered as a major factor affecting the electromigration lifetime in copper/low-*k* dielectrics structures [2.58]. For dual-inlaid copper structures, out-of-plane stress in vias becomes dominant. Interfacial delamination caused by thermomechanical stress in copper lines is unlikely to take place in single inlaid-copper line structures, but may happen in dual inlaid interconnects [2.59]. This effect is enhanced for Cu/low-*k* dielectric structures because of the very low surface energy of the low-*k*materials.

The typical thermal conductivity of low-k dielectrics is less than one third that of oxide, leading to higher metal wire temperatures and enhanced electromigration. Low-*k* dielectrics with much lower thermal conductivity may be used, which will cause a higher wire temperature and, thus, earlier failure. Furthermore, due to the small Young's modulus of low-*k* materials, mechanical constraints as a counterforce to electromigration will be reduced. For low-*k* dielectrics, therefore, the reliability issues for smaller feature sizes will have to be reconsidered. According to the 2003 ITRS [2.2], bilayer or embedded oxide/low-*k* dielectric schemes may be required to enhance the mechanical strength and heat dissipation of future low-*k* dielectric systems.

Interconnect Challenges According to the ITRS Roadmap [2.2]

The continuing decrease in feature size will pose new issues for Cu processing and Cu/low-*k* integration. Major issues will be increased current densities and an increase in interconnect resistivity due to scaling effects. Eventually, a change in material to, e.g., carbon nanotubes for local interconnects and optical interconnects for the global transport of information seems inevitable. However, a significant research effort is needed to develop them into production technology.

Copper will be the preferred interconnect material for microprocessors for at least the next decade. The ECD process is the dominating process for the void-free filling of inlaid structures. CVD may become competitive as a fill technology, if the same filling behavior and microstructure characteristics of ECD can be achieved. Doping of copper, i.e. copper alloys, seems to improve the reliability of copper interconnects. However, the type and the concentration of the alloying element need to be balanced against the increased electrical resistivity. So far no copper alloy applications have been reported for industrial applications.

The continuous scaling of feature sizes in semiconductors affects not only transistors but also interconnects since resistance and current-carrying capability are determined by the interconnect cross-sectional area. Therefore, the 2003 ITRS has set target values for key parameters of on-chip interconnects (see Table 2.2) [2.2]. In particular, the resistivity target will remain at 2.2 $\mu\Omega$ cm for all future technology nodes down to 22 nm, and the current

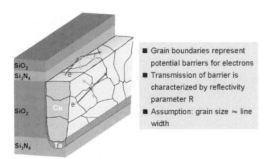

Figure 2.11: Schematic illustration of electron scattering at surfaces and grain boundaries [2.60]

■ Grain boundaries represent potential barriers for electrons

■ Transmission of barrier is characterized by reflectivity parameter R

■ Assumption: grain size ≈ line width

density for IC operation is expected to increase from $1.3 \times 10^6 \, A \, cm^{-2}$ for the 90 nm technology node to $3.9 \times 10^6 \, A \, cm^{-2}$ for the 22 nm technology node.

For small structures, an increase in conductor resistivity due to electron scattering effects will become a limiting factor for the use of copper interconnects. *Steinhögl et al.* [2.61, 2.62] have observed a significant resistivity increase for interconnects with reduced dimensions. This effect was explained by a combined model that considers surface scattering of conducting electrons according to the *Fuchs–Sondheimer* theory [2.63, 2.64] and grain boundary scattering according to a theory provided by *Mayades* and *Shatzkes* [2.65]. Roughness effects will also have to be considered [2.66]. For wire dimensions comparable with the mean free path λ_e of the electrons ($\lambda_e \sim 45$ nm for Cu, $T = 300$ K), the scattering processes of the conduction electrons at the surface have to be taken into account. For model calculations, the exact interconnect geometry (A/R) and the grain-size distribution of the copper grains have to be considered. Figure 2.11 shows schematically the electron scattering at the surface and at grain boundaries [2.60]. Both scattering processes exhibit a certain probability of inelastic processes that contribute to the resistivity increase. Figure 2.12 shows experimental data and theoretical calculations for the linewidth dependence of copper interconnects [2.61]. As a consequence, a reduction in the number of grain boundaries is desirable in order to obtain a lower electrical resistivity. In addition to copper microstructure optimization, interface engineering will be needed to alleviate the impact of this resistivity increase associated with size effects.

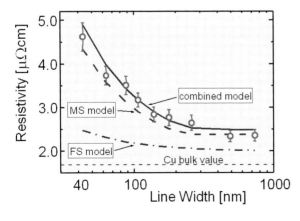

Figure 2.12: Linewidth dependence of the resistivity for copper interconnects, length of metal lines with 40 nm width: 10 cm (2.5×10^8 F) [2.61]

The resistivity of a metallic structure is comprised of several contributions: a temperature-dependent resistivity term originating from scattering of electrons at phonons, and temperature independent resistivity terms originating from electron scattering at crystal defects, such as point defects, dislocations and grain-boundaries, and due to size effects. This picture is an extension of *Matthiesen*'s Rule. *Schindler et al.* [2.60] have shown that the above described temperature independent size effect dominates the resistivity of copper interconnects in small dimensions, and that the resistivity changes only marginally with temperature as the line width decreases. Consequently, cooling of the chip will not be an option to reduce the resistivity of metallic interconnects significantly. For an interconnect dimension of 50 nm and less, cooling will not be able to decrease the resistivity to the roadmap value of 2.2 $\mu\Omega$ cm [2.60].

Schindler et al. [2.60] investigated the linewidth-dependent resistivity of metals other than copper. Interestingly, metals with a small mean free path of the conducting electrons are advantageous as interconnect dimensions decrease, since the size effects described above become important for significantly smaller structures only. Hence, silver has a higher resistivity than copper for linewidths below 50 nm. The reasons are more dominant size effects caused by a larger mean free path of the conducting electrons compared to copper. Aluminum, in contrast, can have an advantage at small dimensions although it has a higher bulk resistivity than copper. The reason is the short mean free path of conducting electrons (14 nm in Al, 45 nm in Cu at $T = 300$ K [2.67]) that reduces the influence of size effects on resistivity. Figure 2.13 shows calculated resistivities vs. linewidth for Al, Cu and Ag, using the same assumptions for grain size and scattering parameters for all materials [2.60]. Possibly, aluminum could be an attractive candidate for low-resistivity interconnects with extremely small dimensions.

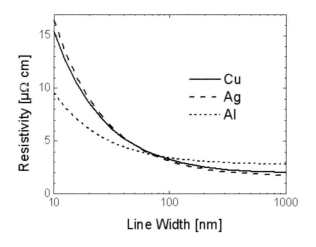

Figure 2.13: Calculated resistivity vs. linewidth for Al, Cu and Ag [2.60]

Since the described feature size effects will become significant contributors to the copper resistivity at the 45 nm technology node, the extendability of the metallic interconnect technology has to be explored. Carbon nanotubes (CNT) are possible candidates for wiring applications in chips, because of their outstanding properties. Since CNTs are ballistic conductors and the electrons within the nanotubes move without resistance, their resistance is length independent at the scale of interest (usually on the order of 10 kΩ). Additional

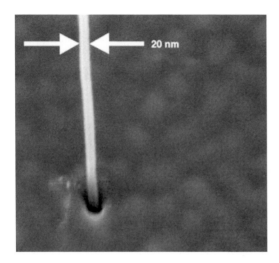

Figure 2.14: Single carbon nanotube acting as a nanoscale vertical connection [2.69].

advantageous properties are the huge thermal conductivity and their high current carrying capacity (current densities of 10^9 A cm^{-2} without heat sinks have been reported) [2.68].

Figures 2.14 and 2.15 show current examples of interconnect vias made out of carbon nanotubes [2.69]. In Fig. 2.14 a single carbon nanotube serves as the vertical connection demonstrating scalability, whereas in the second example (Fig. 2.15) high current carrying capabilities are achieved by vertical arrays of nanotubes. The current issues associated with the implication of nanotubes as interconntects are:

- selection of only metallic nanotubes
- placement of nanotubes
- lateral connections of nanotubes

The first point constitutes an issue but also an opportunity for the use of carbon nanotubes. As their electronic properties depend on the way the carbon sheets are rolled up, carbon nanotubes can be uses for interconnects as well as for transistors. Choosing the right and placing it at the location of interest, however, still remains a major obstacle to be overcome. Currently, for instance, there is no known way to implement horizontal connections that are compatible with today's microelectronics technology.

The introduction of the new low-*k* and ULK dielectrics as the isolating material between the onchip interconnects, CVD barrier/seed layers and additional elements for system-on-chip (SoC) structures provide significant process and process-integration challenges. Interfaces, contamination, adhesion, mechanical stability, electrical parametrics, and thermal budget together with the number of wiring levels for interconnect, ground planes and passive elements, create a high degree of complexity. According to the 2003 ITRS [2.2], copper/low-*k* back-end-of-line structures will continue to find applications in future chip generations. But, the benefit of materials changes alone will not be sufficient to meet overall performance requirements for global wiring. Therefore, alternative interconnect solutions for future generations of ICs will first be needed for global interconnects (see Section 5.1).

Figure 2.15: Bunch of carbon nanotubes to enhance the current-carrying capacity of a vertical interconnect [2.69].

RF and optical interconnects are discussed as technologies to replace conventional wiring and to solve the global interconnect problem. Oxide-based optical interconnects are a logical step, but current deposition technologies are incapable of producing material of sufficient quality for direct implementation. Transition-metal purity and hydroxyl content must improve by several orders of magnitude, in some cases, to meet acceptable loss targets. The challenge for polymeric materials will be the achievement of sufficient refractive index contrast between core and cladding layers to ensure low loss at small radius turns. The conversion efficiencies (electron to photon and photon to electron) still need to be significantly improved for insertion at design frequencies greater than 10 GHz. One advantage of optical interconnects is the application of wavelength and frequency multiplexing of signals simultaneously on the same interconnect. Besides the rapid introduction and integration of new materials and processes into the back-end-of-line process, dimensional control, interconnect reliability and processes with low or no device impact will be the most difficult challenges for interconnects. Considering the mentioned device, materials and integration problems, optical interconnects will probably not be the appropriate solution for onchip interconnects, but for signal transmission over larger distances (> 10 cm).

2.1.4 Function of Barrier and Nucleation Layers and Materials Selection

The function of a barrier layer or layer stack is to prevent copper from diffusing into the isolating dielectrics, to provide the necessary adhesion to the copper and to the dielectric material. Particularly for the copper/low-*k* materials system a barrier layer is required to support the mechanical stability of the on-chip interconnect structure. The fundamental development requirement for the barriers is to meet the mentioned functionality despite further scaling of feature sizes. This task includes the development and the implementation of barrier layers and layer stacks with a high degree of materials compatibility to both the conducting interconnect material and to the isolating dielectrics [2.2] .

The following requirements of the barrier material have been derived from the performance and reliability requirements of the on-chip interconnect system:

- low interdiffusion rate for copper in the barrier material
- high thermal stability of the barrier microstructure
- low resistivity (high electrical conductivity)
- high thermal conductivity
- materials compatibility to the conducting interconnect material and to the isolating dielectric material
- good adhesion to the dielectric material
- mechanical stability (high stiffness and fracture toughness)
- ability to deposit ultrathin films with high conformality
- technology compatibility to the common back-end-of-line process

Until now, Ti/TiN layer stacks have been widely used as liner material for both tungsten plugs and tungsten local interconnects as well as for aluminum interconnects. Several techniques like physical vapor deposition (PVD), chemical vapor deposition (CVD) and atomic layer deposition (ALD) are applicable for the deposition of this layer stack. The Ti reduces the contact resistance significantly, while TiN improves the contact stability and the reliability due to its good adhesion to tungsten. The Ti/TiN layer stack is also used for high aspect ratio (A/R) tungsten contacts, e.g. in dynamic random access memory (DRAM) stacked capacitors.

As a consequence of the use of copper as conducting interconnect material, horizontal lines and vias have to be embedded in ultrathin lining layers (barriers) to prevent copper atoms from diffusing into the isolating dielectric material and to improve the adhesion between interconnect and dielectrics. Ta, W and Ti, and their compounds with N and Si, are widely used as barrier materials, because of their high thermal and chemical stability [2.4]. So far, Ta based barriers have been most commonly used for copper interconnects. The selection of the liner material or material stack is not only based on the interconnect material and the applied deposition technology. Generally, it also depends strongly on the interconnect dimension and on the isolating dielectric material. These barrier layers need to have sufficient thickness and conformality, and a defined microstructure to act reliably as a diffusion barrier. On the other hand, the barrier layers have to be as thin as possible, since the copper interconnect cross-section area decreases and the contact/line resistance increases for thicker barriers. Both effects become more important for smaller interconnect dimensions. For the copper inlaid process, several deposition processes (PVD, CVD, ALD) have been evaluated.

In addition, an ultrathin nucleation layer has to be sputtered on top of the barrier; this acts as a seed layer for the subsequent void-free electrolytic Cu filling (ECD) of trenches and contact holes. A continuous and smooth film is required to carry the current for the ECD process [2.70]. Practically, to date only PVD Cu nucleation layers have been used as starting layers to enable the copper ECD filling of trenches and vias.

The following requirements of the nucleation layer material have been derived from the performance and reliability requirements of the on-chip interconnect system:

- high thermal conductivity
- chemical composition that guarantees a high interconnect reliability (e.g. by alloying)

For both barrier and nucleation layers high conformality requirements exist at sidewalls and at the bottom of trench and via structures. These can be achieved only for very stable deposition processes and using highly sophisticated analytical techniques for barrier/seed step coverage control.

Table 2.3: Selected interconnect technology requirements from the 2003 ITRS [2.2]

Technology node	90 nm	45 nm	22 nm
Local wiring pitch (nm)	214	108	54
Local wiring aspect ratio, Cu	1.7	1.8	2.0
Conductor effective resistivity, Cu intermediate wiring ($\mu\Omega$ cm)	2.2	2.2	2.2
Barrier/cladding thickness, Cu intermediate wiring (nm)	10	7	2.5

Table 2.3 shows some of the 2003 ITRS requirements for barriers used in copper interconnect systems [2.2]. One important parameter is the thickness of the barrier which has a significantly higher resistivity than copper. It has to be as thin as possible for two reasons: For trenches and vias, the cross-sectional area of the highly conductive copper has to be large, and for the via bottom the serial resistance in the via (contact resistance) should be low. According to the roadmap, the minimum trench width for interconnects will decrease from today's > 100 nm to 27 nm in 2016. To achieve these targets it is necessary to scale the barrier thickness to 2.5 nm for the 22 nm technology node.

2.2   Metallization Structures in Acoustoelectronics

2.2.1   Introduction

For a comprehensive treatment of metal-based thin films for electronics we should turn our attention also to types of microstructures which are not in the mainstream of semiconductor microelectronics. The term acoustoelectronics stands for such an area which is associated predominantly with surface acoustic wave (SAW) devices. SAW devices are widely used in consumer electronics, in telecommunications, and they are on the way to entering new markets of wireless control and sensor applications [2.71].

The working principle of SAW devices is based on only a few elementary functions shown schematically in Fig. 2.16. A surface acoustic wave will be launched when applying an ac voltage with appropriate frequency to the electrical input. This is due to the reciprocal piezoelectric effect of the substrate. The surface wave has the strongest amplitude for a wavelength fitting to the period of the interdigital structure of the input electrode. It can be received and converted back into electrical energy by a second interdigital transducer (IDT) in an analogous but inverse manner. Besides the SAW shown, the input transducer launches an additional wave in a backward direction towards the reflector strips arranged nearby. This wave will be reflected completely, so that it can propagate to the output transducer and interfere with the first SAW. A second reflector is shown beyond the output IDT acting similarly to the first reflector. As a result the IDTs together with the reflectors form a cavity named the SAW resonator for standing acoustic waves with maximum acoustic energy at a certain frequency, being the resonance frequency. Besides this type of SAW device, different constructions are also used without reflectors (see Section 5.2). In that case only the SAW traveling from the input to the output IDT is significant for the device function.

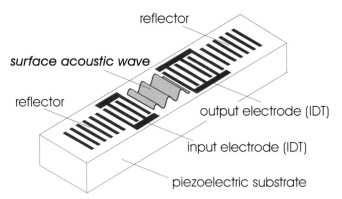

Figure 2.16: Schematic view of a SAW device (resonator)

We learn from inspection of Fig. 2.16, that the metallization structures in acousto-electronics have to fulfil two primary functions: (1) to perform electromechanical trans-duction, (2) to reflect SAWs. Thereby, the simple propagation of SAWs is also affected. Appropriate metallization structures have to meet the following requirements: they must be highly conductive, geometrically well-defined, strongly (in the reflector gratings) or weakly (within the IDTs) reflecting, and long term stable. In practice, non-ideal materials and imperfect technologies set limits to the properties achievable. As a consequence we have to balance the desired device functions and the available materials and technological means. This compromise is based on a thorough knowledge of the relationships between the parameters of SAW excitation, reflection and propagation on the one hand, and the material and geometrical parameters of the metallization structures on the other. The fundamentals of SAW excitation, propagation and reflection presented in the following subsections of Section 2.2 will elucidate this matter.

Within the field of materials science the degradation of a given metallization structure under SAW loading appears to be a challenging question. This becomes most evident when recognizing the technological trends towards higher input power, higher frequencies, or smaller device structures. All these issues result in higher power density levels. It is a well-known fact that finger-shaped interdigital transducers and reflector gratings will change their properties under the influence of the elastic waves. This process is named acoustomigration (see Section 4.3), and becomes more likely at high power density levels. The following sections may support the understanding of acoustomigration and the way in which it affects device performance. For a more detailed insight into the field, the reader is referred to a series of excellent and exhaustive books [2.71–2.78] on different aspects of the fundamentals and the application of SAW.

2.2.2   Fundamentals of Surface Acoustic Waves

Bulk Acoustic Waves

We start with bulk acoustic waves (BAW) in order to simplify the physical problem. They are however closely related to SAW. For a given propagation direction of a plane acoustic wave 3 modes exist in crystals: 1 longitudinal (or pressure) wave and 2 transverse (or shear) waves. The polarization vectors of the particle displacement for different modes are orthogonal to each other. Deviations from the pure longitudinal or shear character, respectively, arise, depending on the strength of the elastic anisotropy resulting in the prefix "quasi-". In addition, the phase velocities can vary drastically with the propagation direction and even the polarization type can change in cases of strong anisotropy. This can be visualized in an appropriate manner by drawing the inverse velocity for a given propagation direction. In such a way, one obtains the so-called slowness surface with 3 sheets: 1 sheet for the longitudinal and 2 sheets for the transverse waves. Slowness surfaces are depicted in Fig. 2.17 for two materials, for glass as an isotropic material and for the anisotropic BaTiO$_3$. Obviously, in both cases the longitudinal bulk wave is the fastest one. In isotropic systems the shear modes are degenerate, i.e. the phase velocities coincide for all directions and form only one slowness sphere.

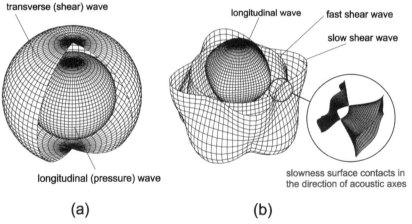

Figure 2.17: Slowness surfaces for bulk acoustic waves. (a) isotropic system: amorphous SiO$_2$, (b) tetragonal crystal: BaTiO$_3$

Acoustic Waves on a Free Surface

A surface acoustic wave (SAW) represents the phenomenon of a periodic motion of particles having an amplitude decaying with the distance from the surface and propagating along the surface with a specific velocity. The simplest SAW is of the Rayleigh type being well-known in nature as being responsible for the transmission of seismic events on the earth's surface. It was mathematically described on the basis of elastodynamics for the first time by *Lord Rayleigh* in 1885, see [2.79].

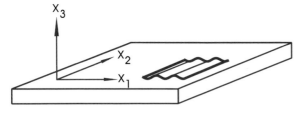

Figure 2.18: Coordinate system for a SAW substrate: x_3–axis is normal to the surface, x_1–axis is parallel to the SAW propagation direction. The x_1x_3–plane is the sagittal plane. The orientation of this coordinate system relative to the crystallographic axes is given by 3 Euler angles resulting from its subsequent rotation around the x_3-, x_1-, and x_3-axes within the crystal

Rayleigh wave depth profiles of particle displacement. Using the coordinate system in Fig. 2.18a Rayleigh wave in the simplest case of an isotropic material has particle motions only in the x_1x_3-plane, also named the sagittal plane. This behavior is depicted in Fig. 2.19 (a). The particle displacement component u_3 has its maximum near the surface and decays considerably within one wavelength. By contrast, the amplitude of u_1 decreases more rapidly with the distance from the surface and touches zero, before it finally decreases exponentially.

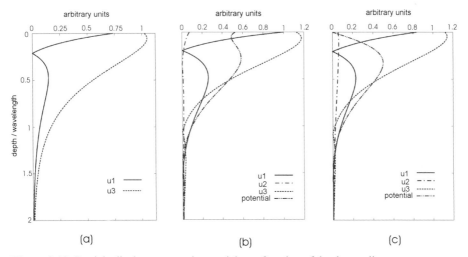

Figure 2.19: Particle displacement and potential as a function of depth coordinate x_3. (a) Amorphous SiO_2, (b) 128°rotYX-LiNbO$_3$ free surface, (c) 128°rotYX-LiNbO$_3$ metallized surface. u_1, u_2, u_3 are the particle displacements in the x_1, x_2, x_3 directions, respectively

Owing to the 90° phase difference between u_1 and u_3, the trajectories of the particles are ellipses. The highest points of the surface wave move in a backwards direction, this is contrary to surface waves in water. The direction of the elliptic motion changes sign, when passing in the depth the $u_1 = 0$ point. Figure 2.19 (b) shows a more general depth profile of particle motion. 128-°rotYX-LiNbO$_3$ denotes a very common single crystal material for SAW devices, which has a symmetry far away from the isotropic case. Nevertheless, the features of u_1 and u_3 discussed above can be recognized, but we observe additionally a small u_2 amplitude near the surface. For this behavior the term quasi-Rayleigh-wave has been introduced. u_3 also has a zero point at a certain depth. From a detailed analysis it turns out that in this case the zero points appear periodically, so we should rather consider this situation as a generalized Rayleigh wave.

Rayleigh wave solution and phase velocity. The phase velocity of Rayleigh waves may be considered in relation to bulk waves. The solution for a SAW obeys the same equations of elastodynamics as those for bulk acoustic waves (BAW). SAW must therefore be comprised of BAW-like partial waves. The coefficients of these partial waves are determined by matching the boundary conditions of a traction free surface. The partial waves contributing to the SAW are, in the simplest case, a longitudinal wave and a sagittally polarized transverse wave, both of the evanescent type (the x_3- component of the wave vector becomes complex-valued in this case) as we have demonstrated by the decaying character of the depth profiles. This is in accordance with the fact that the SAW phase velocity is slower than the velocity of the slowest partial wave with the same wave vector component as that of the SAW (i.e. in the x_1-direction). Otherwise, non evanescent, i. e. sinusoidal, BAW would exist which can transport energy away from the surface to the bulk. In Table 2.4 several examples of crystal cuts used for SAW devices and the corresponding SAW phase velocities are listed. From this it follows that, for wavelengths of 1 µm, frequencies of a few GHz are reached.

SAW types with different polarization. The above described Rayleigh wave with predominantly sagittal polarization is not the only possible type of surface acoustic wave. Because of the crystal anisotropy surface waves with other polarization can also exist.

Table 2.4: Examples of Rayleigh wave parameters

Material	Phase velocity (m s^{-1})	Coupling factor (%)	TCF (ppm K^{-1})
STX quartz	3158	0.11	0
YZ LiNbO$_3$	3488	4.5	-94
128° rotYX LiNbO$_3$	3992	5.3	-75
X112.2°Y LiTaO$_3$	3301	0.9	-18

Corresponding to the 3 bulk acoustic waves (2 transverse, 1 longitudinal), SAWs with different polarization type can be found (see Fig. 2.20). At first glance, this is not surprising because SAWs are governed by the same equations of motion as BAWs. Additionally, however, the surface boundary conditions must be fulfilled. Thus, more than the dominant partial wave is necessary to form the total surface wave solution. For example, shear horizontal (SH-) waves (see Fig. 2.20 (b)) of the so-called *Bleustein–Gulyaev* type [2.80–2.82] are completed by the accompanying wave of electric potential existing as a coupled partial wave in piezoelectrics. In other cases, a radiated bulk wave with small amplitude is needed to form a surface wave which matches the boundary conditions of a free surface. Because of the energy loss due to radiation these waves are named leaky SAW. As a rule of thumb, sagittally polarized SAWs do penetrate to a smaller extent than the other types (b) and (c) because they propagate more slowly than the related BAWs and, consequently, the decaying behaviour is more pronounced. Accordingly, the acoustic energy for the SAW types (b) and (c) has a wider depth distribution. Consequently, its density is reduced near the electrodes, which is important for suppressing acoustomigration.

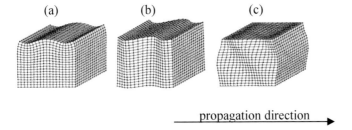

(a) (b) (c)

propagation direction

Figure 2.20: SAW types with different polarization of particle displacement;
(a) sagittal (Rayleigh-type), (b) shear horizontal (SH-type), (c) longitudinal polarization

Dynamic stress fields of SAW. It is a well-known fact that in SAW devices a degradation of the metallization layers, and in turn of the device performance, occurs depending on the level of acoustic power and time of operation (see Section 4.3). Apparently, the finger damage is caused mostly by the dynamic stress fields generated by the SAW. The mechanical stress fields are directly related to the displacement fields (see for example Fig. 2.19) and, because of the piezoelectric properties of the substrate material, to the electric potential.

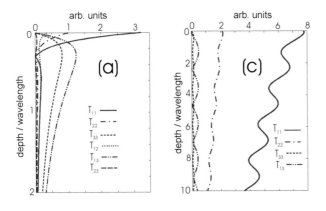

Figure 2.21: Stress tensor components as a function of depth. (a) STX-quartz, (b) $36°\text{rotYX LiTaO}_3$, (c) $\text{Li}_2\text{B}_4\text{O}_7$ with the Euler angles $(0°, 47.3°, 90°)$

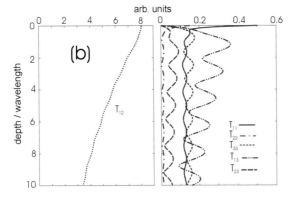

Therefore different stress field distributions will appear in accordance with the different types of displacement polarization. The peculiarities of the stress profiles originating from a metal-coating of the substrates will be considered later. In Fig. 2.21 the non-zero stress tensor elements T_{ij} are depicted as functions of the depth coordinate for 3 crystal cuts relevant for SAW devices. They are representative of the different SAW types shown above in Fig. 2.20: (a) STX-quartz, i.e. with the Euler angles $(0°, 132.75°, 0°)$, (b) $36°\mathrm{rotYX\text{-}LiTaO_3}$, and (c) $\mathrm{Li_2B_4O_7}$ with the Euler angles $(0°, 47.3°, 90°)$. For the definition of Euler angles see Fig. 2.18. It is trivial that the stress components T_{3j} vanish at $x_3 = 0$ because of the traction-free surface. For a Rayleigh wave (Fig. 2.21 (a)) the compressive stresses T_{11} and T_{22} are dominant at the surface. On the contrary, for SAWs with a polarization of the particle displacement in the x_2-direction the shear stress T_{12} is the primary component (Fig. 2.21 (b)). Note also the large penetration of this wave into the bulk, shown on an extended depth coordinate. The same applies for a longitudinally polarized wave (Fig. 2.21 (c)). Similar to the Rayleigh type, the stress components T_{11} and T_{22} are important for the surface forces.

Acoustic Waves on Metallized Surface

Depth profiles of particle displacement and potential. In piezoelectric materials one has to take into account that the deformation and the electric field are generally coupled to each other. Thus the periodic motion of particles of an acoustic wave is connected to a synchronous change of the electrical potential, appearing as a part of the SAW with the same velocity and a similar spatial distribution. Moreover, when changing the electrical boundary conditions at the traction-free surface the SAW can be modified with respect to its parameters. Most important is the covering of the free surface with a conductive layer in an idealized form with neglectable thickness and infinite conductivity. In this case we have a zero surface potential corresponding to the situation already shown above in Fig. 2.19 (c). Apparently, the electric potential curves differ notably at $x_3 = 0$ from the case of Fig. 2.19 (b), but will converge far away from the surface. On the other hand, the curves for the particle displacement components u_1 and u_3 are almost unaffected. This is in contrast to u_2, the penetration depth of which increases considerably for this special case of a conductive surface. However, more likely is an attraction of the SAW fields towards the electrically short-circuiting surface observed for many cases of SH-type SAWs.

Phase velocity and coupling factor. Because piezoelectricity generally acts as a material stiffening phenomenon it is understandable that an infinitesimally thin electrically short-circuiting layer on the surface slows down the SAW velocity. This is because sound velocities increase with the square root of the stiffness constants which are reduced in this case. The extent of this softening effect is a measure of the piezoelectric strength. Therefore the SAW coupling factor K^2 has been introduced given by

$$K^2 = 2 \times (v_{\text{free}} - v_{\text{met}}) / v_{\text{free}} \qquad (2.1)$$

with v_{free} the SAW velocity for a free surface and v_{met} the quantity for a metallized surface, respectively. The coupling factor defined by Eq. (2.1) agrees in many cases with the value determined from electrical admittance measurements. In both Tables 2.4 and 2.6 representative values for the SAW coupling factor are given.

"Mass loading" effect of metallization layer. Mass density and elastic coefficients of a metallization layer of finite thickness will cause a velocity shift of the SAW in addition to that resulting from the electric shorting just discussed. Moreover, a change in the SAW depth profile has to be expected. The directions for the velocity shift and for the change of the localization strength can be qualitatively understood by comparing the SAW velocity of the metallization material with that of the underlying substrate. However, for an exact description of the overall behavior many material constants of both layer and substrate are needed. In addition, the extent of the SAW modification depends on the metallization thickness. Therefore the term "mass loading" widely used to describe the SAW modification by coating has only symbolic meaning. Several examples for a decreasing as well as an increasing of the SAW propagation velocity with the metal layer thickness are depicted in Fig. 2.22. In the limit of small thicknesses the SAW velocity is determined by the substrate. The thicker the metal films become, the more the velocity approaches the SAW velocity of the metal. In the cases of "speeding up" with increasing film thickness the slowest bulk acoustic wave of the substrate sets a velocity limit.

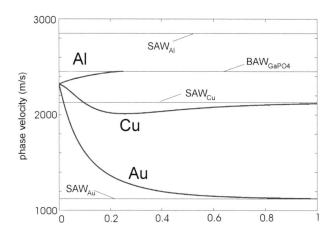

Figure 2.22: SAW velocity as a function of metallization thickness. Al, Cu and Au layers on GaPO$_4$ with Euler angles (0°, 65°, 0°). For comparison, the SAW velocities of all three metals and the limiting BAW velocity of GaPO$_4$ (0°, 65°, 0°) are also given

It turns out that the phase velocity becomes dispersive with respect to the ratio of the metallization thickness to the wavelength. The variation with the frequency can be determined in a straightforward manner from the curves of Fig. 2.22 using the relationship velocity = wavelength × frequency.

Reflection coefficient. The SAW propagation conditions change in a jump-like manner at an edge of the metallization layer. A wave field fulfilling these boundary conditions involves more than one SAW. In such a situation an incident SAW generates BAWs as well as a reflected SAW when passing the edge. Let us focus our interest on the reflected SAW because of its key behavior for the function of reflector structures (see again Fig. 2.16). The complex reflection coefficient appearing as the ratio of reflected and incident SAWs is the sum of two parts ($R = R_E + R_M$): the first part R_E arises from the short-circuiting effect of the electrodes; the second part R_M is of purely mechanical nature and depends on the elastic coefficients, the mass density, and the thickness of the metal layer.

The electric part of the reflection coefficient is proportional to the piezoelectric strength and its absolute value is given by $K^2/4$ (K^2 is the SAW coupling factor according to Eq. (2.1), see values in Tables 2.4 and 2.6). The mechanical part is, to a first order approximation, proportional to the thickness over wavelength ratio ($R_M = a \, h/\lambda$) and dominates in metal structures on weak piezoelectrics like quartz. Typical values for the h/λ ratio in metallization structures of SAW devices are of the order of a few %. It should be mentioned that the reflection coefficient depends strongly on the substrate orientation and can even vanish for special directions. This is due to the complicated dependence of the SAW reflection process on the mixture of the elastic coefficients and mass density of both the metal layer and the crystalline substrate. Additionally, the phase of a reflected SAW (if we consider the SAW potential) can vary between $0°$ and $360°$. Thus the interference between the reflected wave and a SAW excited by the same metal electrode can result in both summing up to a maximum or cancelling each other. The values listed in Table 2.5 show wide variations depending on the material combination. It turns out that accurate predictions about the influence of the electrode material on the reflection coefficient are difficult to make. This can be done only by detailed SAW simulations on the basis of reliable data for the entire material system. On the other hand, the importance of technological and aging stability for the device performance is evident.

Table 2.5: Mechanical reflection coefficients for a metal step. Prefactor a of the thickness/over wavelength ratio h/λ for selected cases

Material	Al	Cu	Au
STX quartz	0.25	0.15	1.0
YZ LiNbO$_3$	0.42	0.16	2.1
128°YX LiNbO$_3$	0.57	0.93	0.12
X112.2°Y LiTaO$_3$	0.27	0.40	0.14

Surface Acoustic Waves in Periodic Structures

The metallization structures of SAW devices are typically arranged as periodic patterns which, together with the bus bars, cover almost the whole chip area (shown schematically in Fig. 2.16). The SAW propagation velocity in such a metallic grating certainly differs from its value for free surfaces as well as for a completely coated surface. In a simple model this value may be assumed to be an average of both cases, additionally taking into account the metallization ratio. The presence of the metal strip edges, however, results in some new features described in the following.

Energy storage effect. As mentioned above, satisfying the boundary conditions at a strip edge requires more than just the solution of the incident SAW. Generated BAWs as well as a reflected SAW have also to be incorporated into the total acoustic field near the edges. First, we will consider the role of BAWs. These contributions to the overall solution are not able to propagate far away from the surface grating, because they arise from many strips and cancel each other due to interference, except when the BAW Bragg condition is fulfilled (see next section). Therefore, instead of being radiated into the bulk, the BAW energy streams back to the surface (symbolized by the ellipses under the strips in Fig. 2.23). This mechanism of energy storage takes time and, consequently, reduces the SAW speed. According to this

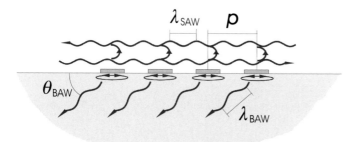

Figure 2.23: Cross section of a metal grating on a SAW substrate. The picture includes the processes of energy storage, *Bragg* reflection of the SAW and the generation of BAWs. For simplicity, the reflections from both edges of a strip are combined to a scattering process at the center

phenomenon the SAW velocity is affected by the metallization thickness and ratio. In Fig. 2.24 different SAW velocities of STX-quartz are depicted as functions of the thickness-wavelength ratio (h/λ) for (a) a surface without metallization layer, (b) with Al layer, and (c) for Al strips with a metallization ratio of 0.5. Note that the effective SAW velocity due to the energy storage effect varies more strongly than linearly with the (h/λ) ratio and becomes considerably slower than for the completely coated surface. It should also be mentioned that the results depicted in Fig. 2.24 (c) are part of an overall numerical treatment of the problem, including the variation of finger-width-to-wavelength ratio that shows the strong influence of all geometrical parameters on SAW behavior.

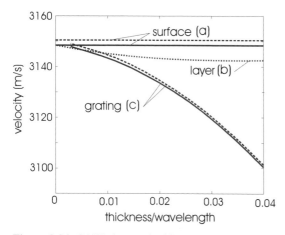

Figure 2.24: SAW phase velocities on quartz (0°, 123°, 0°) for the different cases of a free (dashed) and electrically shorted (solid line) surface (a), continuous layer of Al (b), and periodic Al gratings (c) with a metallization ratio of about 0.5 for the cases of an open (dashed) and electrically shorted (solid) grating [2.83]

SAW Bragg reflection. In analogy to many similar situations in physics, a surface acoustic wave with a wavelength λ passing a periodic structure with period p coinciding with multiples of $\lambda/2$ (Bragg condition) also results in very strong SAW overall reflection. This is caused by the phase matching and summing-up of the single waves reflected from each strip (see Fig. 2.23). The Bragg condition for SAWs can be rewritten using the SAW wavenumber $k_{SAW} = 2\pi / \lambda$:

$$k_{SAW} = n\,\pi / p \qquad (2.2)$$

The working principle of reflector structures shown schematically in Fig. 2.11 is based on this condition.

As discussed above, the energy storage effect is based upon BAWs originating from a scattering of the incident SAW at the strip edges, but they cannot propagate into volume due to destructive interference. This restriction no longer applies, when the Bragg condition is fulfilled by the scattered BAWs. This is possible, whenever the SAW wave number k_{SAW} matches the BAW wave number parallel to the surface $k_{BAW} \cdot \cos\theta_{BAW}$ in such a way that they sum to multiples of the "grating wave number" $2n\pi/p$ (see also Fig. 2.23):

$$k_{SAW} + k_{BAW} \cdot \cos\theta_{BAW} = 2n\,\pi / p \qquad (2.3)$$

We know from the discussion of the Rayleigh wave velocity that BAWs propagate faster than SAWs, so that $k_{SAW} > k_{BAW}$. Consequently, comparing the conditions for SAW Bragg reflection Eq. (2.2) and BAW Bragg reflection Eq. (2.3), the latter case occurs only for frequencies beyond the SAW Bragg frequency (the angular frequency ω is given by $\omega = k_{SAW, BAW} \cdot v_{SAW, BAW}$).

Special Surface Waves: Leaky SAW, SSBW, STW

Leaky SAW. The classical SAW type has been described above: a particle motion, the intensity of which is localized near the surface and which propagates losslessly along the surface. This, however, is an idealized concept. The problems arise if we consider even the simplest experimental configuration needed to verify that behavior. Already in this case small losses are introduced due to the launching of pressure waves from the surface into the ambient air. Usually, this type of leaky SAW is well accepted, because the losses are negligible. A completely analogous situation is the radiation of bulk waves with a small power flow from the surface into the volume. This process arises under the condition that the

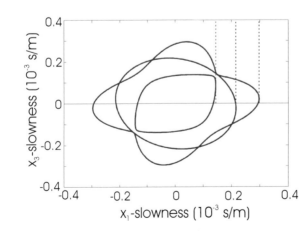

Figure 2.25: Slowness curves in the sagittal plane (x_1x_3–plane) of Li$_2$B$_4$O$_7$ (0°, 47.3°, 90°): The x_1-slowness component, i.e. in the SAW propagation direction, of a considered bulk wave has its maximum for the LBW (limiting bulk wave) direction. The inverse of this x_1-slowness maximum is the LBW velocity. The term LBW corresponds to the name SSBW which has been introduced from a more practical point of view

SAW propagates faster than the limiting bulk wave (LBW) involved as a partial wave in the SAW solution. The term LBW is explained in Fig. 2.25 showing a slowness diagram in the sagittal plane (x_1x_3–plane): The x_1-slowness component, i.e. in the SAW propagation direction, of a considered bulk wave has its maximum for the LBW direction. The inverse of this x_1-slowness maximum is the LBW velocity. Leaky waves (or pseudo surface acoustic waves = PSAW) with small attenuation in piezoelectric crystals are attractive for SAW devices, because they propagate faster than at least one BAW and therefore enable us to achieve higher frequencies without further miniaturization of IDTs and reflector gratings.

Due to the leakage the PSAW velocities become complex-valued. In principle, one can find numerous formal solutions for the same system of equations with the boundary conditions included, also with a complex velocity and a surface localization of particle motion. It is necessary, however, to distinguish between solutions which are excitable by surface sources and several other types without relevance for SAW devices [2.84].

Surface skimming bulk waves (SSBW). This term was introduced after successful experiments [2.85] with horizontally polarized LBWs using interdigital transducers IDTs with periods corresponding to the wavelength of the LBW. One can show that the power flow of bulk waves has a direction normal to the slowness surface. Consequently, LBWs do not transport energy to the bottom of the substrate for a subsequent reflection. Instead, the energy will be transferred parallel to the surface directly to the output transducer. Another name for the same phenomenon has been given after similar observations [2.86]. It is noteworthy to say that this discussion of the SSBW findings in terms of bulk acoustic wave behaviour is only qualitative. One has to keep in mind that limiting bulk waves fulfil the boundary conditions at the surface only in exceptional cases.

Surface transverse waves (STW). A full trapping of SSBWs in the surface region takes place under the influence of a grating [2.87] as is given automatically by IDTs and reflectors according to the device scheme of Fig. 2.16. Thus, in contrast to SSBWs, a genuine eigenmode can exist. One characteristic of STWs is the weak decay of particle motion with distance from the surface. This kind of distribution of the acoustic energy is advantageous for high power devices because of relieving the load on the metallization systems. Another property typical for STW is the dispersive behavior. This is well understood by considering the role of the gratings in the trapping phenomenon. As a consequence, for a technological realization the geometric parameters of the metallic gratings should be well-reproducible in fabrication and stable in time.

Table 2.6: Examples of crystal cuts with high SAW velocities

Material	Phase velocity (m s^{-1})	Coupling factor (%)	TCF (ppm K^{-1})
STZ quartz	4990	1.89	-33
37° LiNbO$_3$	4802	16.7	-70
36° LiTaO$_3$	4211	4.7	-59
Li$_2$B$_4$O$_7$ (0°,47.3°,90°)	6912	>0.8	10

Table 2.6 comprises some examples of crystal cuts relevant for high frequency SAW devices based upon the SAW types discussed in this Section.

2.2.3   Interdigital Transducers (IDTs)

Working Principle and Modeling

Bidirectional IDT.   The excitation of SAWs via the inverse piezoelectric effect by applying
a sinusoidal voltage to an IDT was already mentioned in the introductory part of this section
(Fig. 2.16). The IDT depicted in Fig. 2.26 can be considered as a more advanced type than
the simplest form given in Fig. 2.16, because it has two pecularities. Firstly, all electrode
fingers are split in order to suppress internal reflection of the SAWs within the IDT. This is
possible, because the reflection period is one half of the excitation period.

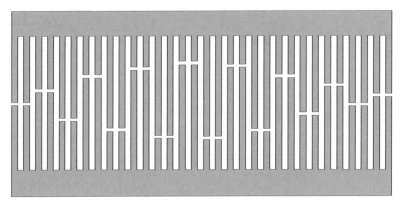

Figure 2.26: Bidirectional IDT with overlapping weighted split fingers

econdly, a weighting function for the individual SAW sources has been introduced via a
variation of the finger lengths. Nevertheless, an IDT according to the construction principle
of Fig. 2.26 launches SAWs of equal amplitudes in both directions. This bidirectionality
without the means to sum both SAW tracks is a drawback of SAW devices, because of the
resulting energy loss. Several solutions to overcome this disadvantage are presented below.

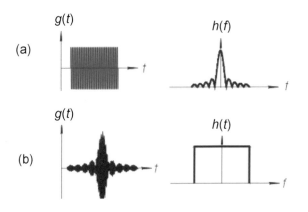

Figure 2.27: Time and
frequency response of an IDT
without (a) and with (b)
apodization. The frequency
response $h(f)$ appears as the
Fourier transform of the time
response $g(t)$

Delta-function model. In a simple model [2.88] to calculate the frequency or time response of an IDT, one assumes that all finger pairs independently generate SAWs which are time-delayed to each other when interfering. For a given frequency the sum over the contributions of all finger pairs to the total SAW has the same form as a Fourier expansion in the time domain with electrode weighting factors as Fourier coefficients. Thus a rectangular shaped IDT weighting ("apodization") function results in a sinc-function (sin(x)/x) for the frequency response and vice versa (Fig. 2.27).

Equivalent circuit model. An excellent theoretical description of the frequency behavior of BAW thickness resonators by equivalent circuit models was achieved by *Mason* [2.89]. The application to IDTs seemed attractive and has been used for a long time [2.90], but also requires a few assumptions for an appropriate approach. There are two main variants to replace finger pair sections of IDTs by BAW resonators (Fig. 2.28). The decision about the variant to be preferred for a selected SAW substrate orientation depends on the polarization of the SAW type used, and the electric field component mainly responsible for the excitation in this case. The inspection of the piezoelectric tensor transformed to the desired substrate orientation will give the required answer.

(a)

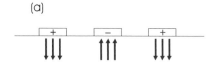

Figure 2.28: Two variants for a BAW transducer approach to IDTs: Crossed-field (a) and in-line (b) model

(b)

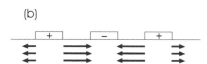

P matrix method. The P matrix method [2.91] describes single sections of the IDT as a three port configuration: the electrodes form one port with the variables I (current) and V (voltage), both SAW propagation directions form the two acoustic ports with the 4 variables for the amplitudes of radiated and incident SAWs on both sides, respectively. The variables are interrelated linearly by a 3×3 matrix containing all parameters which govern the main IDT function: SAW propagation velocity, excitation and reflection coefficients.

Coupling-of-modes (COM) theory. The COM theory [2.92] connects counterpropagating SAWs and the electrical input/output variables in a similar way to the P matrix method, but in a differential manner as a function of the propagation coordinate x_1. This method has been extended and further developed for the simulation of SAW devices also including secondary effects like electrode resistivity. The parameters of the coupling matrix describe the main SAW properties within a grating possessing electrical sources: propagation velocity, reflection behavior, and strength of transduction. They are unknown in the frame of the COM theory and have to be determined independently. This can be achieved in several ways with varying degrees of accuracy. The most modern efforts to calculate COM parameters are based upon finite-element-method (FEM) analyses [2.93].

Electrical Admittance and SAW Parameters

The measurement of the electrical admittance as a function of frequency, for example, using a network analyzer, is a common method of IDT characterization. As it turns out from theoretical treatments, essential information about all SAW parameters can be obtained from the details of the conductance curve $G_d(\omega)$ shown in Fig. 2.29 for a transducer without weighting. As a result, the frequency of the conductance peak is related to the propagation velocity, the height to the coupling factor, and the degree of symmetry to the reflection coefficient per finger. All these parameters depend on the geometry of the fingers, both

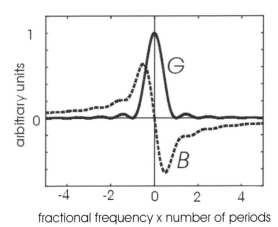

Figure 2.29: Real part G (conductance) and imaginary part B (susceptance) of the electrical admittance of a uniform IDT as a function of frequency

fractional frequency x number of periods

thickness and width, and on their material properties. The dependence of the SAW coupling factor on width cannot be understood in the frame of definition (2.1), in a more general modeling of SAW excitation using equivalent circuits or COM theory, however, the influence of the finger width becomes obvious. Consequently, the analysis of the admittance curve can yield important information about changes within the metallization structures under special influences. Also acoustomigration effects will affect the electrical admittance as a function of frequency (see Section 4.3).

Special Types of IDTs: SPUDT, FEUDT, SFIT

The bidirectional IDT with weighted split fingers shown in Fig. 2.26 fulfils only the simple requirements of an enhanced performance. In many application cases it is desirable to have IDTs which launch SAW only in one direction to ensure an almost lossless energy transfer to the receiving transducer. Currently one mainly uses so-called single-phase unidirectional transducers (SPUDTs) [2.94]. Their functional principle is based on the interference of the waves which are generated by the fields at the electrodes and the waves which are subsequently reflected from the metallization edges. An IDT with $\lambda/4$ fingers will have natural unidirectivity (NSPUDT) [2.95] once the substrate follows a field model being between the situations (a) and (b) in Fig. 2.28. This restrictive condition can be easily circumvented by a special positioning of the edges of the metallization strips yielding the desired loci of SAW generation and reflection, respectively. Following this principle, unidirectional cells with different strip widths can be constructed (EWC = electrode width controlled) [2.96] or with floating electrodes (FEUDTs) [2.28]. Also, the introduction of different thicknesses can cause unidirectivity. Examples are shown in Fig. 2.30. Again, the importance of a constant geometry over long-term operation for the device function is apparent.

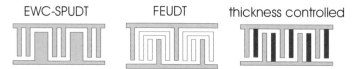

Figure 2.30: Types of single phase unidirectional transducers. EWC-SPUDT: Electrode width controlled single phase unidirectional transducer, FEUDT: floating electrode unidirectional transducer

So far, only IDTs with parallel fingers have been considered. Slanted fingers interdigital transducers (SFIT) [2.97] can be used to attain very high fractional bandwidths without weighting. The working principle is plausible when assuming that SAW transmission from input to output transducer occurs only for a coincidence of the wavelength with the finger period, i.e., within a narrow vertical section of the SFITs (see Fig. 2.31).

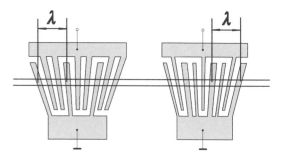

Figure 2.31: Scheme of slanted finger transducers (SFITs)

Secondary Effects: Beam Steering, Diffraction and Scattering

As known from other wave phenomena, in the case of SAW one also has to distinguish between phase and group velocity. Due to the elastic anisotropy the direction of power flow, which is the same as for group velocity, can deviate from the normal to the finger axes (i.e. the direction of the phase velocity). This effect is called beam steering. A transmitting IDT will launch the SAW in the direction of the beam steering angle. Therefore the receiving IDT must be configured just at this angle, but with unchanged finger direction.

Diffraction of SAW also occurs in an analogous manner as we know from light waves. For example, the wave field outside an IDT will diverge like a light beam emerging from a slit. Diffraction has to be taken into account in SAW device simulation. A commonly used approach is the method of the angular spectrum of waves. According to this method, the SAW field is computed as the sum of plane waves with varied wave vectors forming a Fourier series, and with coefficients that are determined from a given amplitude distribution along a line (finger). Contrary to the usual cases in optics, the anisotropic SAW propagation is taken into consideration. Another method using *Green's* functions follows *Huygens'* principle to sum the waves generated by point sources.

The geometry of metallization structures deviates more or less from a perfect pattern. Rough finger edges are responsible for a scattering of SAWs in unwanted directions.

Similarly, irregularities of the flatness of all surfaces or interfaces create disturbing waves. Grain boundaries within thin films or in the substrate also act in such a way. All these mechanisms detract energy and therefore reduce the device performance. In addition, severe device degradation can occur due to SAW scattering. The losses per length increase with the 4…5th power of frequency, or, in terms of distances depend in a similar way on the ratio of defect size to wavelength.

2.2.4   Reflector Gratings

Reflector gratings appear beside IDTs as the key structures in most of the SAW devices. Their main function is to block up the acoustic energy as demonstrated schematically in Fig. 2.16. The basic construction is rather simple and involves a strongly periodic grating with 2 variants of strip connections: open- and short-circuited. All features discussed (surface acoustic waves in periodic structures) also apply to reflectors. The primary application of reflector gratings in SAW devices, i.e. in resonators, will be discussed in Section 5.2.

2.2.5   Waveguides, Energy Trapping

The possibility to slow down the SAW propagation velocity in gratings via the energy storage effect is demonstrated in Fig. 2.24. Following the functional principle of waveguides we can use a periodic structure with such a property to trap the energy of propagating SAWs with respect to adjacent regions, independently of whether these regions are completely metallised or not at all metallized. In other words, the wave field is evanescent outside the grating because of the higher velocity there. Figure 2.32 is a schematic representation of a waveguide which can have 4 modes [2.98]. The common feature of all 4 modes is decaying fields (not shown in Fig. 2.32) above and below the structure according to the waveguide effect. The modes differ with respect to the amplitude distribution across the propagation direction. The special IDT configuration ensures the selective use of these modes. Reflector gratings are added to create resonances along the propagation direction.

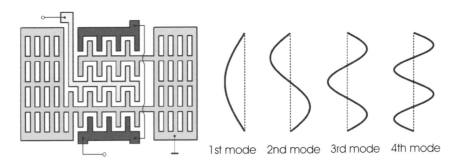

1st mode   2nd mode   3rd mode   4th mode

Figure 2.32: Waveguide with reflector grating and IDTs for selective excitation and detection of 2nd and 4th waveguide modes

2.2.6   Multistrip Couplers

Multistrip couplers (MSC) [2.99] are metallization structures which have been introduced for two reasons. Firstly, to allow the use of two finger length weighted IDTs, and secondly, to suppress unwanted excitation of bulk waves. For this purpose, the transducers are shifted to each other in a sidewards direction (see Fig. 2.33) and a grating (MSC) with a period far away from the SAW Bragg condition is inserted in the centre. The MSC can be considered as a waveguide with two modes of different propagation velocity, a symmetric and an antisymmetric one. The length of the MSC is adjusted in such a way that these modes are in phase on the left and out of phase on the right hand side of the structure. The velocity difference of both modes which is necessary to ensure this condition is proportional to the coupling factor of the substrate material.

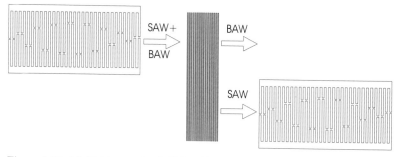

Figure 2.33: Multistrip coupler (MSC) with 2 overlap weighted transducers

2.2.7   Interdigital Transducers with Piezoelectric Layer

The normal construction of an IDT (interdigital transducer) for SAW generation and detection is a metallic pattern on top of a piezoelectric single crystal surface (see Fig. 2.16). In special cases a nonpiezoelectric substrate is desired, e.g. for achieving very high frequencies by use of an acoustically fast substrate material like diamond, or for the reason of technological matching to microelectronics by using silicon wafers as substrates. Under these circumstances a piezoelectric layer has to be added in order to enable the electrodes to convert the electric signals into acoustic ones. ZnO and AlN are the best-suited thin film materials that have to be deposited with crystallographic texture as a pre-condition to exhibit piezoelectric behaviour. Four different configurations of electrodes, piezoelectric thin film layer, and a metal shielding layer are possible (see Fig. 2.34). All these layers cover the surface of the nonpiezoelectric substrate.

a)   b)

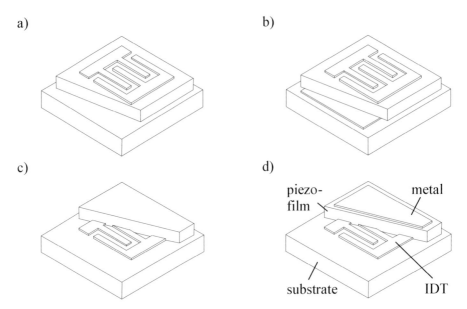

c)   d)

piezo-   metal
film

substrate   IDT

Figure 2.34: Interdigital transducer with a piezoelectric layer. 4 different configurations of IDT electrodes, piezoelectric film, and metal shielding layer.

As has been demonstrated in Fig. 2.22 for the propagation velocity the introduction of a layer results in dispersive behavior of the SAW parameters. In a more complicated manner than the velocity the SAW coupling coefficient depends on the ratio of film thickness and wavelength. Moreover, the 4 configurations of Fig. 2.34 mentioned result in strongly different curves for the coupling coefficient as depicted in Fig. 2.35. A local maximum at small

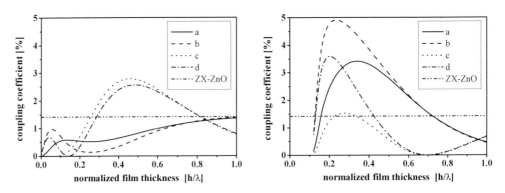

Figure 2.35: Piezoelectric SAW coupling coefficient as a function of the ratio of film thickness h and wavelength λ (ZnO on Si(001)[110]). The curves correspond to the 4 different configurations of Fig. 2.34. The left graph refers to Rayleigh waves with depth profiles similar to Figure 2.19. The right graph corresponds to a surface wave of higher order (so-called first Sezawa mode) that can arise in thin film structures, too.

thicknesses arises when a shielding layer is present (see left graph). Another feature occurring under certain circumstances in layered structures are modes of higher order in the sense that the depth profile of particle displacement is reminiscent of higher harmonics in a cavity (see the coupling coefficient of such a mode on the right hand side of Fig. 2.35).

2.2.8 Metal Strips with Dielectric Coatings

Dielectric layers covering the SAW electrodes and reflectors are used for different reasons. Besides a passivating function and protection against damaging of metal strips caused by surface diffusion, in particular SiO_2 layers serve as temperature-compensating measures, i.e. the overall temperature coefficient of frequency (TCF) of a SAW filter with a SiO_2 overlayer can be almost zero. This is possible due to the fact that single crystal quartz and also amorphous SiO_2 can have positive TCF, by contrast to $LiNbO_3$ and $LiTaO_3$. In such a way the negative TCF of $36°rotYX$ $LiTaO_3$ can be shifted to zero when depositing a SiO_2 layer with a thickness amounting to about 25% of wavelength. In view of a metal thickness of the order of 10% of wavelength, some problems affecting the device function arise.

In particular, for strip-like reflectors (see the function shown in Fig. 2.18) the reflection coefficient (Table 2.5) is changed, and as a more important fact the overall SAW device characteristics is degraded due to enhanced propagation loss. The losses originate from SAW scattering at structures containing grooves and cavities shown in Fig. 2.36 that have a different shape depending on the metallization ratio. To overcome the problem a smoothed SiO_2 surface is demanded that can be achieved by advanced SiO_2 deposition methods [2.100, 2.101].

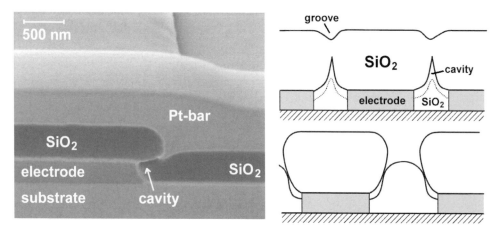

Figure 2.36: Electrodes with SiO_2 overlayers. Left: Cross section with embedded cavity (focused ion beam image, 45° tilt, Pt bar needed for preparation). Right: Effect of metallization ratio on SiO_2 layer on thick electrodes (metallization thickness $h_{met}/\lambda \approx 10\%$, SiO_2 film thickness $h/\lambda \approx 27\%$)

2.3   Silicide Layers for Electronics

2.3.1   Introduction

Silicides belong to the materials group of refractory compounds. They have been attracted scientific and technical interest for more than thirty years. The driving force behind the studies of these compounds is the promise of novel materials with a well-defined combination of physical and chemical properties. These will enable applications under critical environmental conditions, e.g., high temperature, pressure, strong electric and magnetic fields. Silicide compounds are mainly known for their excellent temperature and chemical stability. Among the refractory materials silicides belong to a class with moderate melting points predominantly in the range from 1200 °C to 1800 °C. Nevertheless, there are some silicides, e.g., Hf-, Ta- and Zr-based compounds which possess melting temperatures of about 2000 °C and higher. In many applications, especially in microelectronics, such thermal properties promise a highly stable material behavior. The peculiar properties of certain silicide compounds arise from the specific chemical bonding relations and the chemical stability of silicon dioxide.

The most intensive efforts for introducing silicides into the technology were undertaken in the microelectronics industry. However, the broad spectrum of different properties also allows applications outside the electronics area. Thanks to their chemical and thermal stability some silicide-based alloys have found application as heat resistant materials, in high-temperature soldering, in heat shields for gas turbines and jet engines, as high-temperature electrodes and thermocouples, and as heater materials. In the first prototypes of thermoelectric energy conversion devices Mn and Fe silicides were used as thermocouple materials. In the future these types of silicides are expected to play an important role in advanced thermoelectric materials.

The implementation of silicide-based materials into silicon microelectronics is essentially driven by the remarkable properties of some compounds in this group: (i) the large variability of the electrical transport properties, ranging from metallic to semiconducting behavior, (ii) the high temperature and oxidation stability, and (iii) the existence of direct band gaps in some semiconducting silicides. Silicide applications in microelectronics have begun with the fabrication of ohmic contacts and diode structures - Schottky diodes - on a silicon base material. Further we find on the microelectronics roadmap silicide compounds as complementary interconnection materials. The recently detected photoluminescence on buried silicide layers opens the way for the realization of optoelectronic devices.

Semiconducting silicides also can be prepared as polycrystalline or epitaxial thin films with high optical absorption. This has stimulated a great interest in the potential of transition metal silicides for photovoltaic applications.

The electric transport properties of amorphous and partially crystallized silicide-based resistive materials suggest applications in hybrid electronics as discrete resistors or chip resistors, as well as for integrated passive devices in semiconductor microelectronics.

In the following, the basic physical properties of silicides are discussed in relation to their application potential in microelectronics technology. Due to space constraints we will restrict ourselves to a condensed overview of this extended group within the electronic materials. For detailed studies the reader can be referred to a broad basis of excellent monographs and handbooks in this field [2.102–2.107].

2.3.2 The Basic Chemical and Physical Properties

In this section we summarize the main features of the physical and chemical properties of the large family of metal–silicon compounds. With respect to the chemical bonds silicide compounds may have ionic-covalent (e.g. the compounds of the type Me_xSi_y with Me=Na, K, Rb, Cs), metal-like (e.g., the compounds of the type Me_xSi_y, with Me-transition metal) or covalent binding character (e.g. the compounds of the type A_xSi_y, with A = S, P, C, B). The silicide materials considered in context with microelectronic applications have, preferably, a metallic-like character and are mainly formed by transition metals. In the following considerations we will therefore address the behavior of the transition metal silicides. In addition to the phase formation process and the crystallographic properties, understanding of the electronic structure has a central importance for the development of electronic materials. It serves as a basis for more or less reliable predictions of electrical, thermoelectric, thermal and optical properties of the various silicides.

Silicidation Thermodynamics, Chemical Bonding and Crystallographic Structure of Silicides

The silicidation reaction between metals and silicon is known for more than 180 silicide compounds. They crystallize in a wide variety of different structural types making it very difficult to define consistent structural classification features. Up to an atomic radius ratio of $c_{Si}/c_{Me} \leq 0.85$ we observe silicides with metal-like structures characterized by the substitution of metal atom sites by silicon atoms. Above this ratio radical changes take place in the structure evolution. A further general tendency is found in the accrual of the structural complexity with increasing Si content. Usually the component ratio is used as a quantitative parameter to discriminate the silicide family into metal-rich and silicon-rich compounds. Whereas the silicidation on the metal-rich side proceeds without substantial alterations of the metal lattice structure, on the silicon-rich side we find highly complicated structures characterized by strong changes in the structure of the participating metals. The typical structural elements in these compounds are (i) isolated pairs of Si atoms, (ii) densely packed layers of Me and Si atoms, (iii) chains of Si atoms and networks of silicon and metal atoms (2D layers), and (iv) lattices of silicon atoms (3-dimensional structures).

For all commonly known silicides binary phase diagrams [2.108] and the main thermodynamic parameters, e.g., the heat of formation and melting points are compiled in the mentioned textbooks and monographs [2.102, 2.104, 2.105]. A large fraction of the silicide family is characterized by rather complicated phase diagrams. The properties and the stability of a discrete phase are influenced by the neighboring phases in the diagram. Therefore, as a rule, studies of silicide materials are studies of two- or multi-phase systems.

The melting temperature of a silicide is generally lower than the melting point of the respective metallic component (T_m(silicide) $\approx$ (0.7...0.9) T_m(metal)). This differs significantly from the situation in carbides, borides, and nitrides. Nevertheless, transition metal silicides have a high chemical stability; some of the silicide compounds are particularly stable against oxidation [2.104].

The silicides are characterized by a large variation in their crystal structures which covers almost all known crystal systems. The refractory metal disilicides – as the most interesting materials for electronic applications – typically crystallize in cubic, orthorhombic, hexagonal, or tetragonal structures [2.104–2.107]. Two typical lattice structures for a metallic and a semiconducting silicide are depicted in Fig. 2.37. Some of the metal-silicon systems, e.g.

Mn-Si and Fe-Si, exhibit a particular feature, because they may crystallize in different structures (cubic, hexagonal, orthorhombic, tetragonal) due to the coexistence of several structural phases with different chemical composition. The phase formation of silicides is determined by the fact that the solid solubility of silicon in transition metals proceeds via substitution, whereby the radius of the silicon atom varies with the type of the metal. A typical feature of all silicides is the coexistence of different bonding relations: metal-metal, metal-silicon, and silicon-silicon resulting in highly complex crystal structures. Thus, we find high coordination numbers of about 10–12 in silicides, in contrast to the coordination number of four for pure crystalline silicon.

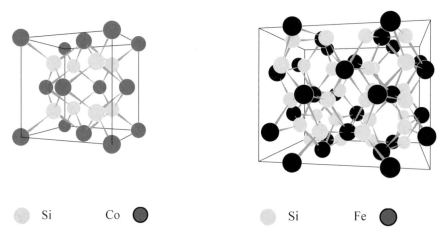

Si   Co ⬤   Si   Fe ⬤

Figure 2.37a: The metallic silicide $CoSi_2$ with a cubic structure (prototype CaF_2, space group *Fm3m*, lattice constant $a = 0.53640$ nm)

Figure 2.37b: The semiconducting silicide β-$FeSi_2$ with an orthorhombic structure (prototype β-leboit, space group *Cmca*, lattice parameter $a = 0.9879$ nm, $b = 0.7799$ nm, $c = 0.7839$ nm)

Depending on the filling of the d- and f-shells in the transition metal atoms we observe a variable intermixing of metallic bonding (between metal and silicon atoms) and covalent bonding (mainly between silicon atoms, but also between metal atoms) in the silicides. The silicide formation may result in several metal-silicon compounds with variable composition. The main phases are the following: Me_3Si, Me_5Si_3, $MeSi$, Me_3Si_5, $MeSi_2$. Due to the variable crystalline structure it may be difficult to assign unambiguously a silicide to a specific compound type. In fact, silicide materials have been recently proposed to hold a hybrid position between interstitial and intermetallic compounds [2.6].

Fundamental Electronic Properties

As already mentioned in the Introduction, transition metal silicides exhibit both metallic and semiconducting behavior depending on the metallic component Me as well as on the Si/Me ratio within the same silicon-metal system. The reason for this interesting feature is found in the electronic structure of these materials. In transition metal silicides we typically observe a gradual change from the silicon sp^3-hybridization to a hybridization between Si p- and metal d-states. This electron-electron interaction results in the formation of bonding and antibonding states which may be separated by an energy gap or a pseudo-gap. The formation of a band gap principally supports the evolution of semiconducting properties. In addition to these Si-Me related states, however, a relatively high density of non-bonding metal d-states exists. This forms, as a rule, distinct bands in the vicinity of the Fermi level. Thus, in all silicides on the metal-rich side up to the MeSi composition we observe a superposition of the gap between bonding and antibonding states and metal-induced d-states resulting in an overall metallic behavior of these compounds.

For the Si-rich transition metal silicides, which are typically characterized by a weakening of the metal-metal interaction, the following trends can be formulated:

(i)   For Si-rich silicides with the metals in groups IV and V of the periodic system the Fermi level is found within the bonding states, whereas for the near-noble metal silicides the Fermi level is shifted into the antibonding states. So, on the silicon-rich side we also observe metallic behaviour (groups IV and V, noble metals) as well as semiconducting behavior (group VI–VIII).

(ii)   The reduction of the metal-metal bonds in silicon-rich silicides results in a decrease in the band width of the non-bonding d-states and in a shift towards higher energies. This process is accompanied by a strong decrease in the density of states at the Fermi level. As a consequence, the crystalline structure proves to be a critical factor in determining the type of electronic transport. The Si-rich compounds $FeSi_2$, WSi_2 and $MoSi_2$ which can crystallize indifferent structural types may serve as an example for the phenomen that, one and the same compound may exhibit both semiconducting and metallic behavior.

(iii)   The increase in the Si content in the $MeSi_x$ phases above $x = 2$ leads to a reappearance of metallic properties. This is caused by the enhancement of the Si p-states near the Fermi level. Thus, the concentration range $1.5 \leq x \leq 2.2$ seems to be the optimum interval for achieving semiconducting behavior in silicides. A nice example is found in the system Ir-Si with the metallic compound $IrSi_3$, whereas the compound $IrSi_{1.75}$ proves to be a typical broad-gap semiconductor.

Electronic band structure calculations have been carried out for a large variety of silicides [2.109]. The accuracy of these theoretical investigations assessed by electrical and optical measurements depends in a larg extent on the system considered. Thus, at present, the size and nature of the band gap have been unambiguously determined only for selected silicides. More efforts in this direction have to be undertaken to shed more light on the correlation between crystalline structure and electronic band structure, in order to arrive at a better understanding of the transport properties.

Transport Properties

As already mentioned above, silicon-rich silicides may exhibit metallic conductivity ($CoSi_2$, $TiSi_2$) as well as semiconducting behavior ($CrSi_2$, $MnSi_{1.73}$, $ReSi_{1.75}$, ß-$FeSi_2$). In some cases silicides may even be low-temperature superconductors (V_3Si). Many transition metal silicides show paramagnetic properties, however, the paramagnetism is less pronounced than in the corresponding metals.

The stimulus for a comprehensive understanding of the electronic transport properties of silicides arises from their application potential. Applications are known in two main areas: (i) metallic silicides as contact and interconnect materials and as *Schottky* barrier components, and (ii) semiconducting silicides as photovoltaic, optoelectronic and thermoelectric materials. In this context, for metallic silicides the most important criteria for an evaluation of the material performance are the resistivity and the barrier heights of silicides on silicon. In semiconducting silicides the band gap properties, the electrical and thermal conductivity and the *Seebeck* coefficient are the most important properties.

The resistivity of metallic silicides covers a broad range from 14 $\mu\Omega$ cm for $TiSi_2$ to 100 $\mu\Omega$ cm for $MoSi_2$. It should be pointed out that the resistivity of thin-film silicides may be higher than the corresponding values obtained from bulk samples. This is due to the well-known influence of structural disorder as well as interface and surface related effects in thin film samples. The resistivity systematically correlates with some general material properties such as crystalline structure and electron configuration of the metal. For example, the resistivity increases within a period of the periodic table if the structure of the silicide changes from orthorhombic over hexagonal and tetragonal to the cubic state. In addition, this variation depends also on the size of the metal atoms: the larger the metal atoms the lower the influence of structural change on the resistivity. Several other effects, however, such as the formation of short metal-silicon bond lengths may partially reverse this tendency. This example reveals that the structural complexity of the silicides results in a complicated resistivity behavior determined by many different factors. These may be the point defect concentration, a deviation from stoichiometry or the nature of the bonding [2.104]. In Table 2.7 the temperature of formation, the resistivity and the Schottky barrier values on n-silicon for some metallic silicides we compile are compiled.

The Schottky barrier heights of metallic silicides on silicon vary over a broad range. For $TiSi_2$ and $TaSi_2$ on *n*-Si relatively low values of about 0.6 eV were found, whereas PtSi and IrSi exhibit the highest barrier values within the silicide compounds (about 0.9 eV). It is still an open question which mechanisms control the barrier height. One approach correlates the barrier height to the electronic configuration of the metal atoms, others to the thermodynamic properties of the silicides and to the silicon/silicide boundary properties. All approaches indicate that the nature of bonding in the silicide plays an important role [2.104]. Recently, a critical re-examination of the Schottky barrier problem has shown that a coherent explanation of the abundant experimental data can be achieved by assuming a realistic relation between bulk and interface contributions to the barrier heights. This leads to a competition between chemical bond formation and the evolution of interface dipoles. In this context, knowledge of the atomic structure at the metal/semiconductor interface is a mandatory requirement for the prediction of interface dipole properties [2.110].

Table 2.7: Properties of metallic silicides

Material	Formation temperature, T_{ph} (°C)	Resistivity, ρ ($\mu\Omega$ cm)	Temperature coefficient of resistivity TCR (K^{-1})	Schottky barrier values, Φ_B (eV)
Co Si$_2$	550	17.25	$+5.0\times10^{-3}$	0.64
Mo Si$_2$	525	100	$+6.4\times10^{-3}$	0.65
Ni Si$_2$	750	50 – 60	–	0.66
Ta Si$_2$	650	50 – 55	$+3.3\times10^{-3}$	0.59
Ti Si$_2$	500	35	$+4.3\times10^{-3}$	0.60
W Si$_2$	650	40 – 70	$+2.9\times10^{-3}$	0.65
Pt Si	225	28 – 35	–	0.88

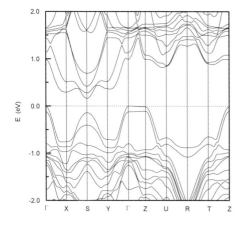

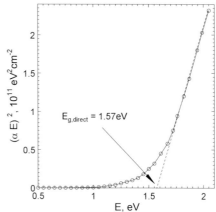

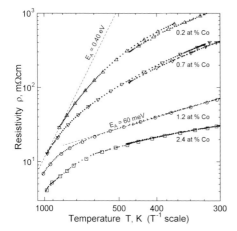

Figure 2.38: Band gap estimation for semiconducting silicides by band structure calculations (upper, left), by optical absorption measurements (upper, right) and by resistivity measurements (lower, left).
Upper, left: Band structure of ReSi$_{1.75}$ indicating a direct band gap at the symmetry point S of the Brillouin zone.
Upper, right: Square of the optical absorption coefficient vs. photon energy for a polycrystalline Ir$_3$Si$_5$ film. Lower, left: Resistivity of a Co-doped poly-crystalline β-FeSi$_2$ films in the intrinsic and extrinsic conductivity range

The band gap parameters of *semiconducting silicides* have been extracted mainly from band structure calculations and from optical and electrical measurements [2.109, 2.111]. The semiconducting silicides show gap values between $E_G = 0.15$eV and $E_G = 2.3$eV. In this way, the compounds CrSi$_2$ ($E_G = 0.35$eV) and ReSi$_{1.75}$ ($E_G = 0.15...0.20$eV) can be classified as narrow-gap semiconductors, the compounds Ir$_3$Si$_5$ ($E_G = 1.6$eV) and Os$_2$Si$_3$ ($E_G = 2.3$ eV) are wide-gap semiconductors, whereas the compounds MnSi$_x$ ($E_G = 0.8$eV), β-FeSi$_2$ ($E_G = 0.8$eV) and Ru$_2$Si$_3$ ($E_G = 0.7$eV) have medium gap values (see Figure 2.38). Concerning the type of the gap there are many theoretical and experimental results illustrating that silicides may be direct gap semiconductors (e.g. β-FeSi$_2$, Ir$_3$Si$_5$) or indirect gap semiconductors (e.g. CrSi$_2$, ReSi$_{1.75}$). For optimizing silicide materials in thermoelectric and optical applications these band gap data are very important, e.g. for the development of doping strategies.

Table 2.8: Representative transport parameters of semiconducting silicides

Material	Melting point, T_m (°C)	Band gap, E_G (eV)	Type of conduc- tivity	Mobility (300 K), μ (cm^2 V^{-1} s^{-1})	Thermal conductivity at 300 K, λ_{ph}(W m^{-1} K^{-1})	Seebeck coefficient at 500 K, S (μV/K)	Resistivity at 500 K, ρ (mΩ cm)
CrSi$_2$ [2.12]	1763	0.35	p	15	6.8	220	7.2
MnSi$_x$ [2.13]	1550	~0.7	p	40	2.9	160	3.1
β-FeSi$_2$ [2.14]	1220	0.8	n	2	4.0	−225	30
Ru$_2$Si$_3$ [2.15]	1710	0.7	n	10	4.0	−100	40
ReSi$_{1.75}$ [2.16]	1980	0.23	n	105	5.5	−110	4.5
Ir$_3$Si$_5$ [2.17]	1400	~1.6	p	–	–	240	75

The transport parameters, such as the electrical resistivity ρ, the thermal conductivity κ, and the thermoelectric power S (*Seebeck* coefficient) of semiconducting silicides depend strongly on structural performance, purity, and doping state, as in all semiconductors. A generalized picture of the electronic transport is difficult to give because of the broad spread of band gap values already mentioned above. Generally, one can state that semiconducting silicides which are not intentionally doped usually exhibit p-type conductivity. However, due to the activation energies of the commonly used dopants between 0.05 eV and 0.14 eV the silicides can be prepared with both distinct p- or n-type conductivity character. Acoustic phonon scattering has been identified as the dominant carrier scattering mechanism in the room temperature range. Depending on the structural state and the degree of doping typical room temperature resistivity values range from 1 mΩ cm to 10 Ω cm. A typical feature of silicide thin films is the large deviation of their properties from those of the bulk samples and single crystals. This is caused by deficiencies in the structural quality of the films. In particular, the carrier mobility data show strongly pronounced differences between film and bulk samples. As a result, typical room temperature mobilities range between 10 and 40 cm^2 V^{-1} s^{-1}, whereas in polycrystalline films values lower than 1 cm^2 V^{-1} s^{-1} are often found. Thermoelectric power and thermal conductivity at room temperature have typical values between 100 and 200 μV K^{-1} and 4.5 W m^{-1} K^{-1} and 8.0 W m^{-1} K^{-1}, respectively [2.112]. A theoretical analysis of the behavior of the Seebeck coefficient is beyond the scope of this section due to the peculiar properties of thermoelectric semiconductors. These are often multivalley semiconductors in which intervalley scattering of carriers dominates. The carrier scattering can be caused by acoustic and optic phonons, neutral and ionized impurities and also by grain boundaries. Thus, for the description of a real material, simple semiconductor models, for example, the single spherical band model, usually yield inadequate results. These materials call for more sophisticated and refined models. The thermal conductivity, which consists of an electronic and a lattice contribution, should be minimized for high thermoelectric performance. This may be achieved by selective scattering of phonons by lattice disorder and by grain boundaries.

2.3.3   Preparation of Silicides

At present, only a limited number of silicide compounds with both metallic conductivity and semiconducting properties are applied in microelectronics technology. Despite this very early stage of application the large variety of silicide materials has attracted the interest of basic research, applied materials science and electronic device technology. The main application areas concern interconnects and contacts, Schottky diode structures, optoelectronic devices and light sources, infrared detectors, and photovoltaic and thermoelectric energy converters for sensor and generator applications. The materials required for these applications are preferably thin films with a large structural and compositional diversity and in a wide range of thickness, but bulk materials in the form of sintered samples and single crystals are also required. In the following a short survey of silicide thin film and bulk material preparation will be given.

Silicide Thin Film Preparation

Depending on the application field, the deposition technology of silicide films has to match a broad spectrum of requirements in order to meet the the desired functional parameters. Some of the main parameters of *conducting silicides* used as contacts and interconnects are the following: (i) low resisitivity, (ii) low mechanical stress, (iii) high surface and interface quality, (iv) high thermal stability against oxidation and structural degradation, (v) technological compatibility with silicon device processing.

The preparation technology of *semiconducting silicide* layers should ensure: (i) the availability of high-purity single-phase material, (ii) a homogeneous and reproducible doping process in a broad concentration range, (iii) a deposition temperature as low as possible, and (iv) a low recombination center density at surfaces and interfaces.

Polycrystalline and amorphous films.   Amorphous and polycrystalline silicide films have been investigated for more than twenty years with very different aims. Due to their high chemical and aging stability amorphous chromium-silicon and rhenium-silicon films have been proved to be good model substances for studying electronic transport phenomena in amorphous thin film systems close to the metal-insulator transition [2.113, 2.114]. Metallic transition metal silicides $CoSi_2$ and $TiSi_2$ are applied in silicon microelectronics as interconnection components [2.115, 2.116]. Semiconducting silicides, mainly $FeSi_2$ and $MnSi_{1.75}$, are investigated as components for thermoelectric sensor and generator devices [2.117, 2.118]. In the following, preparation techniques will be presented in which the silicide compound growth on the substrates proceeds without a substrate reaction, i.e., all film components are provided by the deposition sources. Basically, polycrystalline silicide films can be prepared without major problems. The main deposition techniques are physical (PVD) and chemical vapor deposition (CVD).

Physical vapor deposition (PVD): evaporation and sputtering. Very early studies on chromium silicide based systems were performed by *Glang* [2.119] using electron beam co-evaporation. The aim was to realize a stable resistive thin film material which fulfilled the corresponding technical requirements. For fundamental studies of the electrical conductivity, amorphous chromium silicide films were prepared by a single crucible electron beam technique, based on the comparable vapor pressure data of Cr and Si [2.120, 2.121]. Since the investigated metal-silicon based system was expected to be a very promising material for thin film resistors sputtering technology was also included in these studies [2.122–2.124].

In all these investigations single source magnetron sputtering or multiple source co-sputtering was employed for the film preparation. In comparison to co-deposition technologies the compound target sputtering approach makes the deposition process more practical. The key problem in this case, however, is the target preparation. Some of the silicides may be manufactured by melting procedures, but the majority of the silicide materials can only be prepared by powder metallurgical techniques. For the compound targets preferential sputtering effects have to be taken into account. They arise from the different sputtering yields of the constituents [2.125]. As a rule of thumb, in most systems, a depletion of silicon occurs in the grown films. Counteracting this problem requires an appropriate composition shift in the target in order to get the stoichiometric phase composition. Compound target sputtering is employed preferably for Cr-Si, Fe-Si, and Mn-Si. For investigation of a broader composition range, for example, for the purpose of composition optimization, the co-sputtering technique is a more convenient approach. In this case the composition may be finely adjusted by a source power variation. Thin films of atomic mixtures of Re-Si [2.126] and Ir-Si [2.127] were successfully prepared by co-sputtering.

Chemical vapor deposition (CVD). Besides the classical preparation route of metallic silicides in microelectronics technology – the solid state reaction of metallic deposits on silicon surfaces – chemical vapor deposition is also increasingly utilized for the growth of silicide films. The use of CVD techniques simplifies the silicide preparation process and usually results in films with a higher performance. Furthermore, CVD allows the deposition of films both on oxidized and on non-oxidized silicon wafers. Nevertheless, there are many open questions with respect to the growth process and postgrowth treatments, which still need to be answered. For example, *Bain et al.* [2.128] reported that silicon-rich WSi_x films deposited by liquid phase CVD (LPCVD) in a WF_6-SiH_4 process exhibit a variation of the sheet resistance upon the annealing conditions. The lower sheet resistance of the rapid thermal annealing samples (RTA) in comparison to the furnace-annealed ones is accompanied by an increase in grain size, which in turn results in an increase in the surface roughness. Optimizing this approach needs further insight into the correlation between resistivity and microstructure. With respect to applications, *Sell et al.* [2.129] describe the use of CVD WSi_x films as gate and bottom electrodes in storage capacitors and transistor arrays. In the area of 3d transition metal silicides, *Lee et al.* [2.130] demonstrated the growth of polycrystalline $CoSi_2$ films, which reveal an improved morphological stability as compared to layers prepared via the conventional two-step annealing route in a reactive CVD process. In this CVD process the metallic $CoSi_2$ phase grows directly without the phase transformation $CoSi \rightarrow CoSi_2$. It was shown that the morphological stability is correlated with the grain size of the underlying poly-Si layer. The reason for the improved stability is seen in the oriented (111) growth arising from the poly-Si grain orientation.

Chemical vapor deposition is also used to prepare semiconducting silicides, for example, the frequently investigated compound Fe-Si. In a single source precursor technique involving *cis*-$Fe(SiCl_3)_2(CO)_4$ the formation of FeSi and $FeSi_2$ can be achieved by LPCVD in the temperature range between 350 and 500°C [2.131]. On pyrex glass substrates polycrystalline

FeSi films were obtained, whereas on Si(100) oriented $FeSi_2$ films were grown. The latter growth is promoted by an interfacial stress minimization as a consequence of an appropriate lattice match. β-$FeSi_2$ can also be prepared from a tetramethylsilane/ferrocene or a silane/ferrocene mixture [2.132]. For film growth high temperatures of 1000 K...1373 K are needed at a pressure between 1.3 and 6.7 kPa. For the first precursor mentioned above no single phased films are forming (Fe_5Si_3, β-SiC – mixture). For the second case, however, better results are obtained, if a two-step process is employed in which the CVD starts with a silane reaction and then pure ferrocene is introduced. The thermoelectric properties found in this material may need further optimization (thermoelectric power s = 845 μV K^{-1} and conductivity σ = 1000 S cm^{-1} at 1000 K).

Single-crystalline (epitaxial) films. A highly oriented silicide layer growth on silicon has been observed for both some of the electrically conducting silicides and the semiconducting silicides. Due to the outstanding properties of epitaxial silicide films and silicide/silicon interfaces the deposition techniques and the resulting epilayers are subject to intensive ongoing studies. The interest is stimulated by potential applications in microelectronics, photovoltaics, and optoelectronics. Comprehensive investigations in this field are reported in several monographs [2.133, 2.134] and original publications [2.135–2.137].

The most frequently studied materials comprise the metallic silicides $NiSi_2$ and $CoSi_2$ and the semiconducting compounds $CrSi_2$, $MnSi_{1.73}$ and β-$FeSi_2$. An epitaxial layer growth can be achieved in different ways. In the following we will address basically established techniques for an individual layer growth, that is, neglecting neighboring device structures. Special preparation approaches taking into account the compatibility with the CMOS technology will be discussed in the corresponding sections.

- **Solid phase epitaxy (SPE).** In this technique the silicidation is realized by a solid state reaction between an atomically clean silicon surface and a transition metal layer deposited under UHV conditions by means of evaporation. Usually, the formation of the epitaxial silicide requires an additional heat treatment. For the above mentioned silicides the thermodynamic data (melting temperatures, formation enthalpies, diffusion parameters, phase formation sequences) to a great extent, are known [2.102, 2.104–2.106].

- This abundant data library enables a relatively reliable prediction of the complex reactions in the course of compound formation in metal-silicon systems.

- **Reactive deposition epitaxy (RDE).** In contrast to SPE in reactive deposition epitaxy the metal atoms are deposited onto a hot silicon substrate. Compound formation proceeds simultaneously with metal atom condensation due to the highly activated atomic diffusion. For ternary and more complex systems the RDE can be carried out also by means of coevaporation of more than one chemical component.

- **Molecular beam epitaxy (MBE).** The molecular beam epitaxy is the classical codeposition technique involving a simultaneous condensation of the film components – preferably in their stoichiometric ratio – on substrates at a well-defined temperature. This technique often introduces additional template and buffer layers in order to promote the epitaxial film growth and to extend the growth conditions to higher thickness ranges. Effusion cells as well as electron beam guns may be used as evaporation sources allowing the thin film growth process to be carried out under UHV conditions at pressures between 10^{-9} and 10^{-11} mbar.

- **Ion beam synthesis (IBS).** Depending on the beam energy ion beam synthesis can be realized in two different ways: In the energy range greater than 10 keV this technique is known as ion implantation. It involves a solid state reaction between the matrix material and the implanted species and is typically accompanied by a post-growth treatment to reduce the implantation damages. In the low energy range the ion beam synthesis can be carried out as ion beam sputtering. In this deposition method noble gas atoms are accelerated towards a target surface by means of an ion source. On impact with the target they eject material which is then deposited onto a substrate.
- **Pulsed laser deposition (PLD).** Single or multi-pulse laser irradiation is used to vaporize materials from metal or compound targets. The laser ablation is knownfor a very precise reproduction of the target composition in the growing film. Therefore, using stoichiometric targets PLD may yield epitaxial films of high quality.
- **Magnetron sputter epitaxy (MSE).** More recently, besides evaporation sources magnetron sputter sources have also been employed to prepare epitaxial films. Thanks to the substantial progress in vacuum technology and to the availability of UHV compatible sputtering sources, sputtering deposition requiring noble gas pressures in the 10^{-3} mbar range can be successfully utilized for the growth of epitaxial layers [2.138]. In order to suppress the influence of highly energized plasma particles from the source discharge on the film growth process a special source arrangement according to Fig. 2.39 – the Facing Target Sputtering systems (FTS) – is very advantageous. The main benefit of this sputter configuration consists in the fact that the substrate is placed outside the plasma space of two facing sputter sources equipped with identical targets.

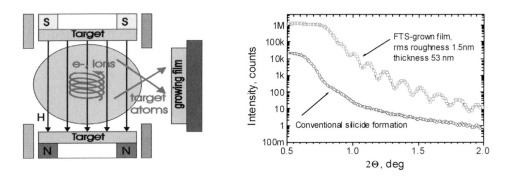

Figure 2.39: Magnetron sputter epitaxy of $ReSi_{1.75}$ by means of FTS yielding an improved interface and surface quality, (left) principle of Facing Target Sputtering; (right) X-ray reflectometry of $ReSi_{1.75}/Si$ (100)

Silicide Bulk Material Preparation

Silicide bulk materials are used up to now for applications in thermoelectric devices, mainly as a polycrystalline sintered modification. High-quality single crystals are still of great interest with respect to fundamental investigations. They offer various possibilities in the study of properties as a function of crystallographic orientation and composition within in the homogeneity range. In particular, spectroscopic measurements for analyzing details of the electronic structure can be carried out only on the basis of single crystals. Furthermore, it is well known that thin film studies often yield properties differing greatly from those found in bulk samples due to grain boundary effects, interface (substrate/film) and surface influences and size effects. In this context, the single crystal information proves to be very helpful in understanding the complex formation processes in compound thin films. Therefore, we will take a short excursion to describe the preparation of poly- and single crystalline bulk material, mainly for the case of semiconducting silicides.

Single crystal growth. The binary phase diagrams show clearly that all metal-silicon systems are characterized by a coexistence of different phases. Congruent and peritectically melting compounds can coexist within the same binary system together with phases which are not in equilibrium with the melt. In this situation each preparation of a well-defined phase clearly requires a specially adapted crystal growth technique. Due to the large variety of phase diagram types, the silicide growth involves all known standard growth techniques from the gaseous to the liquid phase. The growth from the gaseous phase is carried out by chemical vapor transport (CVT) with different transport agents in closed silica ampoules. The growth from the melt is realized either by the flux technique using zinc and tin as solvents or by classical techniques such as the *Bridgman* and *Czochralski* methods, the zone melting or the floating zone method. For some semiconducting silicides which are considered as promising candidates for electronic and thermoelectric applications, further details of the crystal growth are presented.

CrSi$_2$. In the phase diagram of the semiconducting silicide CrSi$_2$ we find a congruent melting compound at 1490 °C. Together with the relatively low partial pressures of the elements this allows the growth of crystals from the melt [2.139–2.141]. However, also chemical vapour transport [2.142, 2.143] and the growth from a solvent flux [2.144–2.150] have also been utilized successfully to grow crystals with size, purity and perfection sufficient for electrical and spectroscopic measurements.

MnSi$_{2-x}$. The Mn-Si system has an exceptionally complicated phase diagram. A whole group of phases exist, especially in the metal-rich regime which is characterized by peritectics formation, non-stability in the equilibrium and polymorphic phase transitions. In the silicon-rich range a whole family of the compound type MnSi$_{1.75 \pm x}$ also exists. The most studied semiconducting phase Mn$_{11}$Si$_{19}$ melts non-congruently at 1155 °C, whereas the peritectic and eutectic temperatures differ by only 5 °C. Due to the deviation of the peritectic concentration from the stoichiometric composition MnSi$_{1.73}$ crystals can be grown from the melt [2.147–2.150]. The crystals obtained in this way, however, are found to have a two-phase composition, i.e. the evaluation of the single-phase properties in such a system needs further effort.

ReSi$_{1.75}$. According to the latest investigations on the Re-Si system the phase ReSi$_{1.75}$ has proved to be the only stable compound in the silicon-rich range with narrow-gap semiconducting properties. This compound melts congruently at about 1940 °C supporting a crystal growth from the melt. However, the high melting point and the reactivity of the molten material require the use of either a cold crucible *Czochralski* technique or a floating zone technique with optical heating. CVT-grown crystals could only be obtained with very small size and inferior purity, whereas the melt growth results in very large crystals (see Fig.

2.40). These yield reliable results in both crystallographic and electrical transport investigations [2.151–2.153].

β-FeSi₂. The orthorhombic semiconducting phase does not coexist with the melt. Therefore, the only possibility for a melt growth involves the preparation of the high-temperature α-phase which is stable at temperatures above 950 °C. This α-phase is transformed via a eutectic reaction according to α-FeSi₂ → β-FeSi₂ + Si by applying an appropriate heat treatment. The crystals obtained in this way are very small and contain secondary phases. The more successful pathway to β-FeSi₂ crystals employs the CVT method [2.154]. Single phase crystals with an orthorhombic structure are obtained from the chemical transport reaction in closed silica ampoules filled with I₂ by using high purity starting materials (silicon with a purity of better than 5N and iron with a purity of better than 4N8). The crystals grow to a considerable size ($5...10 \times 2 \times 0.5$ mm^3) allowing their electrical, magnetic and optical characterization (see Fig. 2.40) [2.155]. For the study of thermoelectric properties the crystals may also be doped with chromium, cobalt and manganese.

Ru₂Si₃. The Ru-Si system is investigated only by a few groups, mainly for thermoelectric applications. The semiconducting phase melts congruently at 1710 °C, but the prediction of a transformation into the stable α-phase below 1690 °C could not be confirmed until now. Single crystals were grown by a *Bridgeman*-like method [2.156] with different crucible materials and by zone melting [2.157, 2.158]. The crystals obtained in this way are large (e.g. diameter 8 mm, length 50 mm). Nevertheless, the transport parameters could not be determined reliably. Recently, new growth experiments by floating zone methods and optical heating [2.159] yielded single phase crystals without secondary phases and inclusions (see Fig. 2.40).

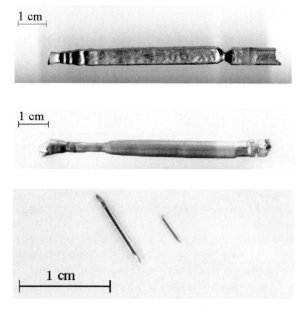

Figure 2.40: Bulk single crystals of semiconducting silicides.
Top: ReSi₁.₇₅ single crystal grown by zone melting.
Center: Ru₂Si₃ single crystal grown by a floating zone technique with radiation heating.
Bottom: β-FeSi₂ grown by chemical transport reaction

Ir₃Si₅.   The system Ir-Si is a silicide compound studied only very recently. Even the equilibrium phase diagram is not yet fully explored, especially not on the metal-rich side. On the Si-rich side three phases have been identified up to now: Ir_3Si_4, Ir_3Si_5, $IrSi_{3\pm x}$. The compound Ir_3Si_5 finds special interest as a semiconductor with a wide band gap ($E_G \approx 1.6eV$) and may be considered in future photovoltaic and thermoelectric device applications. This silicide is formed peritectically from a silicon-rich melt at temperatures of about 1670 °C. Thus, single crystal growth should be possible both from the non-stoichiometric melt and by flux growth. Ir_3Si_5 crystals doped with platinum and osmium were grown from the melt by peritectical transformation in a Bridgeman-like furnace [2.160].

Polycrystalline material preparation.   For thermoelectric applications besides thin films polycrystalline bulk material is also required in order to realize thermoelectric cooling and generator devices in the medium and high power range. The materials mainly used are $FeSi_2$ and $MnSi_{1.75}$, prepared either by sintering techniques, thermal spraying or mechanical alloying. The mentioned approaches have proved to be suitable methods for thermoelectric applications, because compositional and structural properties as well as doping parameters can be controlled by an appropriate process design. Especially, the thermal conductivity can be significantly reduced in comparison to single crystals due to the generation of a high number of grain boundaries.

Sintering techniques.   The typical process steps in this technology are the preparation of the starting ingot, melting, grinding, cold/hot pressing and sintering, final annealing. Special careis required for the choice of the starting materials: in order to achieve a high thermoelectric performance high purity materials should be used and the purity level should be maintained during the whole fabrication process. The additives for doping also have to be of high purity. If the host material contaminants prove to be active dopants, however, a lower grade starting material can also be used. As a rule of thumb, the alloy formation should be performed in a reducing or inert atmosphere (Ar/H_2, Ar, vacuum). The main doping elements for β-$FeSi_2$ are Co and Ni for n-type samples and Cr and Mn for p-type samples. $MnSi_{1.73}$ – higher manganese silicide (HMS) – is typically used as p-leg material whitout intensional doping. In the case of cold pressing typical hydrostatic pressure values are 300–400 MPa for both $FeSi_2$ and $MnSi_{1.73}$. The sintering temperature and time should be optimized for the material and powder consistency to result in a high material density. For Fe and Mn silicides sintering temperatures of 1175 °C and 1155 °C, respectively, are usually applied which yield about 95% of the theoretical density. A faster technique than sintering (requiring normally 6...8 h) is available with hot isostatic pressing. At pressures of about 40 MPa in the temperature range 900 °C $\leq T \leq$ 1020 °C densities of better than 95% were achieved in a relatively short time of 10....20 min for both thermoelectrics. In several cases the stabilization of the semiconducting β-$FeSi_2$ phase (low-temperature phase) needs a long annealing treatment in the temperature range between 600 °C and 800 °C. This treatment also has a positive effect on the long-term stability of the material [2.161, 2.162].

Thermal spraying.   The thermal spraying process is a well-established technique for the preparation of thermal barriers as well as corrosion and wear-resistant coatings. It is also considered as a prospective method for the fabrication of thermoelectric devices. It was found that the microstructure of thermally sprayed $FeSi_2$ differs significantly from the microstructure of hot pressed $FeSi_2$ and is characterized by a strong disorder in the sub-micrometer range [2.163]. By means of a special spray technique, shroud plasma spraying, a significant increase in the thermoelectric performance Z was achieved in comparison to hot pressed samples.

Mechanical alloying.   Whereas most of the $FeSi_2$ preparation techniques need carefully optimized heat treatment and phase formation conditions, mechanical alloying is expected to simplify this complex fabrication procedure. By the preparation of Si-rich Co-doped material *n*-type samples can be obtained in which finely dispersed Si inclusions within the β-$FeSi_2$ matrix play the role of additional phonon scattering centers. An electrical conductivity of 109 $(\Omega\,cm)^{-1}$, a Seebeck coefficient of -183 $\mu V\,K^{-1}$ and a thermal conductivity of 2.1 $W\,m^{-1}\,K^{-1}$ demonstrate the good performance for this material. We find that by optimizing the sintering temperature in the range between 790 °C and 880 °C, the mean size of the Si precipitates increases from 0.19 μm to 0.22 μm and the medium distance between precipitates increases from 0.29 μm to 1.92 μm. The highest performance of the material was observed for the lowest value of interspacing between the precipitates [2.164].

2.3.4   Silicides with Metallic Conductivity

In this section we will address in more detail silicides with metallic conductivity. We thereby focus on $CoSi_2$ representing many other transition metal silicides with metallic conductivity, such as $TiSi_2$, $PtSi$, and WSi_x. $CoSi_2$ is used in integrated circuits (ICs) as interconnecting material in the multilevel metallization technology because of its particular properties. These include a low electrical resistivity, high process temperature stability, chemical compatibility with metals and dielectrics used in IC technology, and good electromigration resistance. Particularly in comparison to doped poly-Si, cobalt silicide films exhibit lower resistivity and a better thermal process stability. The preparation is usually realized via a metal-silicon silicidation reaction or by sputter techniques resulting in a resistivity of about 20 $\mu\Omega$ cm. This yields a sheet resistance level of $(1...3)$ Ω/square. In general, the heat treatment conditions very strongly determine the film performance. Especially in the case of a silicide formation by surface reaction, a two-stage high-temperature treatment has been shown to improve the thickness definition and the transformation of the film into the low-resistance phase. In advanced IC technology not only the resistivity, but also the increasing stress induced in the silicon at decreasing lateral dimensions of the interconnect systems are key parameters. Systematic studies by *Maex and Lauwers* (in [2.107]) have shown that the final stress level is determined by the phase formation process and depends on the structural film features. In addition, the interface quality plays an important role in the case of thermal silicide formation, i.e., using the reaction between Co and Si. It has been shown that by means of appropriate gettering layers the interface reaction can be kept on a high–purity level due to the capturing of interior and exterior contaminants.

Besides the use of silicide films as electrically active components in integrated circuits [2.165] silicides can fulfil a more complex functionality in ICs due to their unique properties. For this reason, in Si-based microelectronic technology the SALICIDE (self-aligned silicide) technique has been established. It describes a combined deposition/patterning method in which source/drain contacts and diffused interconnects can be connected without an additional lithography step. This technique is based on the silicidation reaction proceeding at the interface between Co, the polysilicon gate, and source/drain diffusion regions, but not with silicon dioxide regions. The unreacted metal can be selectively removed by an etching process. The $CoSi_2$ interconnects produced by means of the SALICIDE procedure are characterized by low resistivity, good adhesion and high temperature stability.

The oxidation behavior of silicides can be also used as a non-conventional technological approach to patterning integrated circuits. The oxidation process of a silicide layer on top of a Si wafer involves the formation of a surface SiO_2 film on the silicide, accompanied by a simultaneous progressive growth of the silicide layer into the depth of the silicon wafer. If the oxidation process is carried out locally by masks, buried silicide layers which are laterally separated can be generated. This approach can be used to realize dedicated transistors for the high-frequency range [2.166].

The two transition metal silicides with the fluorite type lattice, $CoSi_2$ and $NiSi_2$, exhibit a sufficient lattice parameter matching ($a_{NiSi2} \approx 0.541$ nm, $a_{CoSi2} \approx 0.537$ nm) with silicon to support epitaxial layer growth. Additionally, these two silicides are known to be materials with a high degree of crystalline perfection within the silicide group. Therefore, with respect to the epitaxial growth conditions cobalt and nickel silicide are the compounds most studied among the silicides. With the progressive reduction of the lateral dimensions in IC technology the requirements on layer thicknesses and interface roughness of interconnects and Schottky barriers are permanently increasing. Therefore, due to their outstanding properties in these respects epitaxial layers will play an important role. Unfortunately, the growth conditions for dislocation-free films are not yet fully optimized. Several sophisticated approaches have been developed for improving the quality, e.g., Ti interlayer mediated epitaxy (TIME) or the oxide mediated epitaxy (OME).

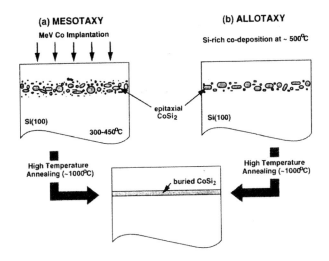

(a) MESOTAXY

MeV Co Implantation

Si(100)

300-450°C

High Temperature Annealing (~1000°C)

(b) ALLOTAXY

Si-rich co-deposition at ~ 500°C

epitaxial CoSi₂

Si(100)

High Temperature Annealing (~1000°C)

buried CoSi₂

Figure 2.41: Schematic diagrams of (a) mesotaxy and (b) allotaxy process (from *Hull et al.* [2.91])

The special feature of silicides – phase formation by a solid state reaction – is reflected also in special techniques for preparing epitaxial silicides. High-quality silicide films can be obtained as buried films by means of the mesotaxy or allotaxy methods (see Fig. 2.41). The characteristic feature common to these two methods is the generation of an embedded epitaxial precipitate network within a silicon matrix by means of metal ion implantation or silicide co-deposition, respectively. The following high-temperature annealing initiates a transformation of the precipitates into a buried single crystalline silicide layer. In the mesotaxy approach introduced by *White et al.* [2.167] the problem consists in finding the critical implantation doses to achieve a continuously buried silicide layer. For the silicide layer itself a lower critical thickness of about 20 nm was found. The allotaxy introduced by *Mantl et al.* [2.168] is characterized by a higher degree of free process parameters determining the coalescence process and the properties of the buried silicide layer. Nevertheless, in analogy to the mesotaxy, the problem of growing layers with a thickness lower than 20 nm is not yet solved. The examples cited show that the application of silicides in silicon-based IC technology makes use not only of the intrinsic material parameters of the compounds, but also takes advantage of specific aspects of the growth and phase formation processes of metal-silicon compounds. The technique of buried layer growth has the outstanding feature of forming very clean interior interfaces with high structural perfection – an essential requirement for the realization of devices based on low-dimensional electronic transport effects.

2.3.5   Semiconducting Silicides

Within the group of semiconducting silicides β-FeSi$_2$ is one of the most studied materials. It possesses a high application potential both in thermoelectric and in optoelectronic devices. During the last 10 years comprehensive investigations covering both theoretical and experimental aspects have been carried out. Remarkable progress has been made in many fields: the preparation of epitaxial films, band structure aspects, electronic, optical, and magnetic properties have been studied to a relatively large extent.

The semiconducting β-FeSi$_2$ has an orthorhombic structure containing 16 molecules with 48 atoms. The lattice parameters are $a = 0.9879$ nm, $b = 0.7799$ nm and $c = 0.7839$ nm, at about 950 °C the β-FeSi$_2$ phase undergoes a phase transformation into the tetragonal metallic α-FeSi$_2$ phase with the parameters $a = 0.269$ nm and $c = 0.5134$ nm. Band structure calculations, optical and electrical measurements result in very different data for the band gap, concerning both the gap size and the type of gap. For the room temperature range as the most probable case a direct band gap of $E_G = 0.87$ eV is generally accepted. Thanks to this electronic structure the β-FeSi$_2$ phase can be effectively doped by metallic dopants to yield *n*- and *p*-type thermoelectrics. Thus, it seems to be a possible material for photoelectronic energy conversion as a less expensive alternative to the Si-based and CuInSe-based photovoltaics. Recently, photo- and electroluminescence measurements on crystalline β-FeSi$_2$ have attracted the attention of research groups working on light emitting materials and devices.

β-FeSi$_2$ is favored as an ecologically friendly semiconductor and in this sense is regarded as a potential material for the next generation energy research and optoelectronics [2.169]. For thermoelectric investigations iron disilicide is studied on samples with different consti-tution: (a) polycrystalline bulk material prepared by powder metallurgical methods, (b) poly-crystalline thin films prepared by electron beam evaporation, magnetron sputtering and plasma ion processing and (c) single crystalline samples prepared by chemical transport reaction. Low-resistivity β-FeSi$_2$ is mainly obtained by doping with transition metals replacing the Fe atoms in the orthorhombic cell of the semiconducting iron disilicide phase. As expected from the behavior of other semiconducting compounds, the transition metals located to the right of Fe in the periodic table act as donors and the metals to the left as acceptors. The best thermoelectric power factor data ($P = S^2/\rho$) were obtained with the 3d transition metals Cr, Mn and Co. This is illustrated in Figs. 2.42 and 2.43.

The transport parameter studies generally suffer from a relatively large spread in the data sets. This is due to the high sensitivity of the electronic properties to stoichiometric deviations, structural defects, and trace-contamination and is related to different preparation methods and process parameters.

In bulk materials the semiconducting phase is formed via the phase transformation α-FeSi$_2$ $\rightarrow$ β-FeSi$_2$, i.e., from the high-temperature side. By contrast, in thin films the phase growth is initiated by heat treatment from the low-temperature side without participation of

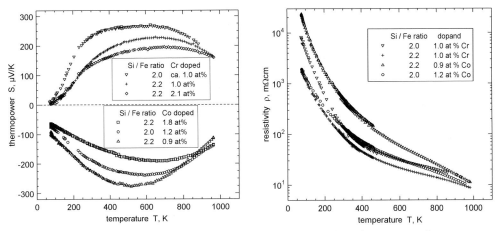

Figure 2.42: Seebeck coefficient (left) and resistivity (right) as a function of temperature for polycrystalline doped iron disilicide films

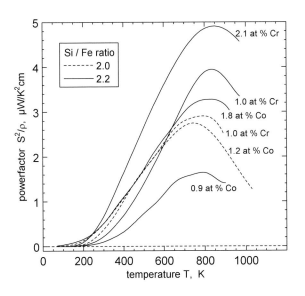

Figure 2.43: Power factor as a function of temperature for polycrystalline doped iron disilicide films

the metallic α-phase. Against this background stressing the highly complex correlation between structure evolution, phase formation and doping process it is understandable that bulk and thin film behavior shows pronounced differences.

In this context single crystal studies contribute valuable information, because their results allow conclusions on the optimization limits in thin film investigations. In Fig. 2.44 the thermoelectric power of polycrystalline β-FeSi$_{2+x}$ samples is compared with single crystal data. As is obvious from this diagram the two polycrystalline samples show an analogous temperature dependence, whereas the single crystal is characterized by quite different behavior.

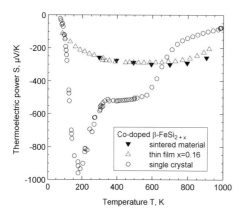 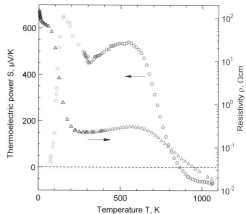

Figure 2.44: Thermoelectric power of *n*-type Co-doped β-FeSi$_{2+x}$ samples with different constitution

Figure 2.45: Thermoelectric power and resistivity of *p*-type Cr-doped β-FeSi$_2$ single crystal

In Fig. 2.45 the Seebeck coefficient and the resistivity of a *p*-type single crystal are depicted. The temperature dependence of the thermoelectric power can be divided into four regions with increasing temperature: impurity band conduction (variable range hopping), extrinsic conduction range, saturation range, intrinsic conduction range. Over the entire extrinsic range we find much larger thermopower values for the single crystals than for the polycrystalline samples. The large thermopower values in the low temperature range result from a pronounced phonon drag effect below 250 K. The higher level of thermoelectric power values is obviously the result of a higher degree of structural and compositional perfection of the crystals. This finding also suggests that the polycrystalline material has potential for performance improvements.

In Table 2.9 we summarize thermoelectric data at room temperature for doped polycrystalline materials. As can be seen quite easily, the conductivity and thermoelectric power values of doped β-FeSi$_2$ allow device applications with emphasis on a high power factor $P = S^2\sigma$ rather than a high thermoelectric figure of merit $ZT = (S^2\sigma/\kappa) T$. Such cases are, e.g., substrate-based thin film sensors in which the thermal conductivity is predominantly determined by the substrates.

The positive effect of the higher crystalline perfection in single crystals was also confirmed by studies on thin films and precipitates grown epitaxially in contact with silicon.

These investigations were preferably carried in the course of electro- and photoluminescence experiments. As reported in [2.170] single crystalline balls embedded in silicon can be generated by means of epitaxial β-FeSi$_2$ films grown by RDE on Si with a subsequent segregation annealing. These buried crystalline 3-dimensional silicide islands exhibit a pronounced photoluminescence signal at about 1.5 μm, if the samples are exposed to an optimized long-time heat treatment at 900 °C. In embedded β-FeSi$_2$–Si diode structures electroluminescence has been found by *Leong et al.* [2.171] at room temperature. These measurements of luminescence effects on samples prepared by different epitaxial growth techniques have stimulated discussion about the nature and origin of the luminescence in β-FeSi$_2$. As the band alignment in β-FeSi$_2$–Si heterostructures has an important influence on the luminescence yield, the strain state of the silicide phase is intensively discussed, in many cases rather controversially. It has been proposed that the intense 1.54 μm signal in the low-temperature range is generated by unstrained iron disilicide [2.172]. The luminescence investigations have also brought about a revival of β-FeSi$_2$ as a direct band gap semiconductor. Time dependent photoluminescence measurements suggest that the light generation arises from indirect transitions [2.173]. Although at present the existence of luminescence in β-FeSi$_2$ has been doubtless demonstrated the question to what extent β-FeSi$_2$ can be used as suitable light emitter for optoelectronic applications has not yet finally been answered [2.174].

Table 2.9: Thermoelectric parameters of doped β-FeSi$_2$ (from [2.175])

Dopant	Sample state	Conductivity	Seebeck coefficient	Power factor	Thermal conductivity	Mobility
		σ, $(\Omega\ cm)^{-1}$	S, $\mu V\ K^{-1}$	$P = S^2\sigma$, $\mu W\ cm^{-1}\ K^{-2}$	κ, $W\ m^{-1}\ K^{-1}$	μ, $cm^2\ V^{-1}\ s^{-1}$
Co	bulk/thin film	40...150	-150...-450	2.2...3.5	4...10	0.3
Co	bulk	140...500	-150	3...11-	5	-
Co+Al	bulk	230	-190	8.3	4.5	-
Ni	thin film	13	-110	0.17	-	-
Pt	thin film	110	-200	4.9	-	-
Mn	bulk/thin film	5...10	280...450	0.4...1.5	6	8
Cr	bulk/thin film	30...35	220...300	1.4...3.0	12	-
Al	bulk	190	170	6	6	1.6...4

The origin of the 1.54 μm luminescence in Si samples with β-FeSi$_2$ precipitates formed by ion implantation has proved to be associated with large, fully relaxed, disk-shaped inclusions generating conditions for indirect bandgap transitions of charge carriers [2.176, 2.177]. This statement is confirmed by TEM and *Mössbauer* spectroscopy on different samples, as well as by optical measurements of photoluminescence (PL), photoreflectance and absorbance. In detail, within the optimum process window at least two different types of precipitates are observed in separate sample regions by TEM: small ball-shaped precipitates in the surface region and large disk-shaped precipitates deeper in the sample. The latter are found to possess very few defects (dangling bonds) at the interface with Si matrix, which is a confirmation of the results from the molecular dynamics simulations. The decrease of the PL signal intensity is detected when progressively removing the region where disk-shaped precipitates are located, while a contribution to the PL signal from small ball-shaped precipitates is marginal. By performing the PL excitation at energies in the gap range of Si and β-FeSi$_2$ a luminescence signal is observed only in the case of samples with β-FeSi$_2$ precipitates and not in the case of mechanically induced dislocations in nonimplanted Si. This issue clearly shows the 1.54 μm luminescence to be originated from β-FeSi$_2$ precipitates. A comparative analysis of photoluminescence, photoreflectance and absorbance measurements in correlation with *ab initio* calculations indicates the nature of the 1.54 μm luminescence as indirect band transition. The model, which considers the disk-shaped precipitates acting as a trapping well for carriers generated in the Si matrix, is proposed to explain both the quenching of the PL signal with temperature and the confusion why PL arises from precipitates and not from good-quality epitaxial films.

Thanks to the technological progress in the film and bulk material preparation of β-FeSi$_2$ another field of application for the semiconducting silicides recently has found again large interest – the photovoltaics. In particular, Japanese groups are trying to place β-FeSi$_2$ as the next material generation of environmentally friendly semiconductors (Kankyo semiconductors) by means of the demonstration of its advantages in real photovoltaic elements [2.178].

Owing to its direct band gap and the high optical absorption coefficient ($\alpha > 10^{-5}$ cm^{-1}) β-FeSi$_2$ is proposed by several groups as a promising alternative photovoltaic material characterized by a high theoretically predicted solar energy conversion efficiency of about 23 % [iv]. The state-of-the-art in the field of β-FeSi$_2$ – based solar cell devices could not yet confirm these optimistic theoretical expectations, however the intensive efforts, especially in the thin film technology, show that the actual low performance level can be ascribed to the low quality of β-FeSi$_2$ films. *Makita et al.* [iv] have shown that flat and homogeneous β-FeSi$_2$ thin films (100 – 300 nm in thickness) with minimized defects can be grown on Si(111) substrates by Fe and Si codeposition and by superlattice formation methods in MBE and sputtering chambers. Typical values of the carrier density and mobility for sputtered n-type material of 3×10^{17} cm^{-3} and 100 – 550 cm^2/V s, respectively, demonstrate the good level of film preparation. Furthermore, the authors state that under an air mass of 1.5, an energy conversion efficiency of 3.7 % can be obtained demonstrating the feasibility of β-FeSi$_2$ as a semiconductor for ultrathin film solar cell applications.

In conclusion, in this section we have addressed the high prospective potential of β-FeSi$_2$ as an energy-conversion material. The feasibility of β-FeSi$_2$ for both solar cell and thermogenerator applications opens up new ways for a unified technology of combined energy conversion modules on the base of photo- and thermoelectricity.

2.3.6   Heterogeneously Disordered Silicide Films

Thermostable electronic materials based on composite thin films with an intrinsic heterogeneity on the length scale of nanometers have found increasing interest. This is related to their high variability of electrical, optical, and other properties. The properties can be adjusted via a defined mixture of metallic and non-metallic components arranged in a microstructure which may vary from amorphous to nanodisperse. In this section we consider nanodispersed thin film materials which have found applications in resistors and thermoelectric sensors.

Silicide-Based Films for Resistive Layers

The materials usually utilized in resistor devices are TaN, Ni-Cr, Ta_2N, TiN_xO_y, Cr-Si and cermets. The key material parameters for resistor applications are the resistivity range ρ, particularly the high-ohmic limit ρ_c, the temperature coefficient of the resistivity, TCR, and the stability $\Delta R/R$. The systems Cr-Si-O and Cr-Si-N with an atomic ratio [Si]/[Cr]>1 were shown to be highly variable systems with the potential for optimization of the electrical transport properties. Their properties permit the fabrication of high-precision thin film resistors.

For heterogeneously disordered thin films such as nc-Cr_nSi_m-$SiO_x(SiN_y)$ films in the nanodisperse state, an estimation of the resistivity ρ_c at the metal-insulator transition leads to $\rho_c = A(\hbar/e2)d$ [2.179] with A being of the order of unity and d denoting the average grain size. This corresponds to a resistance per square area of $R_\square \approx 10$ k$\Omega/\square$ as the high ohmic limit.

The electrical transport behavior of Cr-$SiO_x(SiN_y)$ thin films in the amorphous state turns out to be comparable to amorphous highly doped semiconductors [2.180]. The behavior close to room temperature is characterized by a large negative TCR, which decreases with increasing oxygen content from -700 ppm K^{-1} to about -2000 ppm K^{-1}. Within this range of TCR changes a metal-insulator transition (MIT) is observed in the low-temperature conductivity behaviour. The latter can be divided into three different regions: (i) the TCR range $0 \leq$ TCR $\leq$ -400 ppm K^{-1} can be described by a balance of weak localization effects and a Boltzmann behaviour, (ii) the TCR range -400 ppm K$^{-1} \leq$ TCR $\leq$ -1500 ppm K^{-1} can be solely explained by weak localization, and (iii) the TCR range below -1500 ppm K^{-1} which is usually described by variable range hopping.

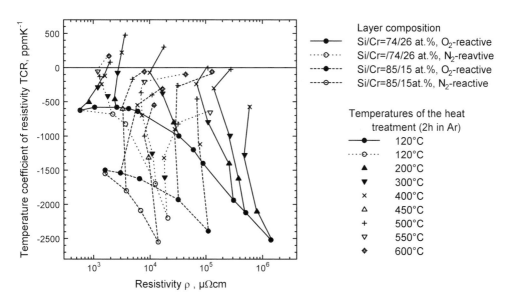

Figure 2.46: Annealing characteristics (TCR$_{20°C}^{120°C}$ as function of ρ [300 K]) of Cr-Si-(O,N) thin films prepared by reactive d.c. sputtering

During an annealing treatment above 250 °C the films undergo a transformation into the nanodisperse state which can be described by the formula nc-Cr_nSi_m-SiO_x(SiN_y). Typically, the TCR of the films in the nanodisperse state changes to positive values over a wide temperature range (see Fig. 2.46). For the application of CrSi(O,N) in thin film resistors it is important to know the resistivity regime in which a TCR close to zero can be realized. In addition, the shape of the function $\rho(T)$ over the whole working temperature range extending usually from –55 °C to + 155 °C (corresponding to the requirements in microelectronics) is of interest. As a quantitative measure for the shape of $\rho(T)$ the so-called parabolicity is commonly used. It is defined as the difference between "high-temperature" and "low-temperature" coefficients

$$\Delta\text{TCR} = \text{TCR}\,^{120°C}_{20°C} - \text{TCR}\,^{+20°C}_{-50°C} \tag{2.4}$$

By careful optimization of the [Si]/[Cr] ratio and the oxygen concentration, by a well-defined addition of Al and by choosing appropriate annealing conditions, resistive films with a resistivity up to 2×10^5 $\mu\Omega$ cm can be prepared. These films are characterized by low TCR and ΔTCR values and a high stability. It was found that the low parabolicity is related to the $CrSi_2$ grain structure. It turns out that the small TCR and ΔTCR values are the result of a competition between weak localization effects and macroscopic silicide properties. Furthermore, the experiments showed that oxide-containing films have the best properties with respect to precision (high nominal value accuracy and low TCR) whereas nitride-containing films have the best thermal stability [2.181].

Partially crystallized silicide films for thermoelectrics.   Nanocrystalline (nc) composites are well-known for providing new opportunities to extend the variability of the structure-property correlation in solid materials. As the morphology of the nanocrystalline state is characterized by ultrafine grains and a large volume fraction of the associated interfaces, many properties of nanocrystalline materials differ significantly from those of their coarse-grained or bulk counterparts. The electronic transport properties of nc-conductors are strongly affected by the interaction of the conduction electrons with the interfaces. In the case of nc-metals the interfaces provide an additional scattering channel for the electrons, resulting in an enhanced resistivity. In the case of nc-semiconductors two effects should be taken into account: (i) the band structure of nanocrystals may differ from the band structure of the bulk semiconductor; (ii) the screening length of the charge carriers trapped at the interface may be comparable to the grain's size. As a consequence, the effective medium approximation (EMA) commonly used for theoretical analysis of composite materials may fail in the case of a nc-composite. Thus, for thermoelectric materials in the nc-state the figure of merit limitation, i.e. the fact that $Z = \sigma S^2/\kappa$ of the composite cannot be higher than the largest component value, can be overcome.

The Re-Si system is known for its high stability in the amorphous state [2.114]. In the thin film state it forms crystalline $ReSi_x$-phases at a temperature of 850 K [2.182]. This is more than 100 K lower than the crystallization temperature of the components Re and Si. Therefore, this system can be used as a model system to study nc-composite materials with respect to their structural evolution as well as their electronic transport behavior. Nc-composite films are obtained by crystallization from the amorphous state via a solid state reaction resulting in very clean internal interfaces [2.183]. The composite film contains only two phases: an amorphous phase and the nanocrystalline $ReSi_{1.75}$ phase with a mean grain size of about 10 nm. However, a large volume fraction of the composite films is occupied by interfaces and grain boundary regions. The crystallization process is characterized by an upper limit of the grain size: after reaching a maximum size only the number of grains is increasing (for more details see Section 4.7.2).

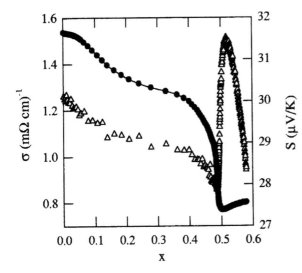

Figure 2.47: Non-monotonic behavior of σ (●) and S ($\triangle$) at 860 K versus the nanocrystalline $ReSi_{1.75}$ volume fraction x

The electrical conductivity and the thermoelectric power of these thin film composites show a non-monotonic dependence on the volume fraction x of the nanocrystalline phase (see Fig. 2.47). In the vicinity of the percolation threshold $x = x_c$ the thermoelectric transport parameters reach extreme values which opens the way for an optimization of the thermoelectric efficiency [2.184].

2.4 Complex Layered Systems for Magnetoelectronics

2.4.1 Introduction

Microelectronics is one of the high-technology areas with enormous economical impact. Products from the microelectronics industry have revolutionized our working environment and have found entrance into many aspects of our everyday life. With the development of a modern information technology society, microelectronics faces many new challenges, for example, with respect to speed. The basic materials for current microelectronic devices are semiconductors, or more specifically, silicon. The fundamental principle of operation of these devices involves the control of the generation and flow of charge, with electrons comprising the charge carriers.

Traditional microelectronics exploits only a fraction of the potential inherent in electron transport processes. Besides the charge the electron has another important property – the electron *spin* – which is entirely of quantum-mechanical nature. The spin becomes manifest in the magnetic moment of an atom and is responsible for the magnetic properties of matter.

In particular, in the classical ferromagnets Fe, Co, and Ni, the electrons generating ferromagnetism are located at the Fermi level and thus also participate in the electrical current. Quite early on this bifunctionality of the electrons – carriers of charge and magnetic moment – raised the question about interferences between magnetism and electrical transport. Such an interference may show up, for example, as a variation of the electric current through the material upon changes of its magnetization state. This ***magnetoresistance*** (MR) may arise from two sources: (*i*) interaction of the charge with the external magnetic field (Lorentz magnetoresistance, Hall effect), and (*ii*) spin-dependent scattering of the charge carriers while they move through the material. In fact, magnetoresistive effects of the latter type, such as the anisotropic magnetoresistance (AMR) were already known for more than a century and have found widespread applications in magnetic sensor technology. The AMR is most pronounced in $3d$ transition-metal ferromagnetic alloys. In the case of Permalloy (a magnetically soft FeNi alloy) the normalized field-induced resistivity change reaches values of $\Delta\rho/\rho_0 \sim 6$ % for bulk samples [2.185] and $\Delta\rho/\rho_0 \sim 2$ % for thin films. The AMR magnitude may be acceptable for sensor purposes, but is very poor in comparison to the on/off ratio in a field effect transistor ($> 10^3$). Active microelectronic devices on the basis of AMR elements are therefore beyond reach.

It was not until 1988 that *Baibich et al.* [2.186] and *Grünberg et al.* [2.187, 2.188] discovered much higher magnetoresistance effects in multilayer structures. This so-called giant magnetoresistance (GMR) may reach a magnitude of $\Delta\rho/\rho_0 > 100$ % in certain systems [2.189]. The GMR effect is based on spin-dependent scattering processes at the interfaces of layered magnet structures, comprising individual thicknesses in the nanometer range. The discovery of GMR and related magnetotransport phenomena in layered structures led to the advent of a new research field, called ***magnetoelectronics*** or ***spintronics*** [2.190]. The first commercial use of spintronic devices came with GMR-based read heads in ultrahigh density hard disks [2.191], introduced by IBM in 1998. It is fair to say that spintronics has a chance to revolutionize the field of microelectronics. Novel non volatile storage concepts – the magnetic random access memory (MRAM) – are emerging as an alternative to FRAM-based memory cards for digital cameras or pocket databases. Currently, prototype MRAM are considered for military and space exploration purposes [2.192, 2.193]. Several international industry consortia are currently working hard to come up with MRAMs for the consumer market soon [2.194, 2.195]. The long-term goal in spintronics is the development of passive and active electronic devices based on spin-dependent transport phenomena, for example, a spin transistor [2.196] or novel devices with completely new functionalities. As the electron spin is a realization of a quantum-mechanical two-level system, spintronics may even pave a pathway towards quantum information technology.

Spintronics is still an emerging technology [2.197]. In this section we will therefore review fundamental aspects rather than applications. The first part gives a very brief introduction to the electronic origin of magnetism. The second part addresses the important issue of magnetic coupling. The fact that neighboring magnetic films may mutually affect their magnetic properties and thus also their transport behavior has no counterpart in conventional microelectronics. The third part of the section is devoted to the basic principles of spin-dependent transport phenomena in layered structures. The section closes with a discussion of the role of the individual components in a complex magnetoelectronic layered structure with respect to their magnetic and electric transport functionalities. A more detailed treatment of the physical mechanisms and the various device concepts may be found in Refs. [2.198–2.200].

2.4.2 Magnetism: A Primer

The microscopic basis of spintronics rests on two main aspects, both of which are closely connected to the electronic structure and in particular to the quantity **spin**. The first one concerns the spontaneous formation of a long-ranged ordered spin structure in the system of interest. In the simplest case, the magnetic order may be of ferromagnetic or antiferromagnetic type, but in many materials also more complex spin structures (ferrimagnets, helical spin order, spin density waves, etc.) occur. The category of spin structure determines the macroscopic and microscopic magnetic properties. The second aspect is related to electronic transport and scattering processes that do not only depend on the charge, but may also be strongly affected by the electron spin. In order to understand the phenomena occurring in magnetoelectronics, it is thus useful to inspect the role of the electronic structure in magnetic – particularly ferromagnetic – systems in more detail.

The macroscopic magnetization M may be seen as the sum over the atomic magnetic moments m_i. In a (single domain) ferromagnet the moments m_i are all aligned parallel. The magnetization defines a spatial quantization axis for the electron spin. The quantum mechanical property spin S has the character of an angular momentum and is directly related to the magnetic moment μ of the electron

$$\mu = -g \cdot \mu_B \cdot S \tag{2.5}$$

with μ_B denoting the Bohr magneton and g the gyromagnetic factor. S can take two orientations with respect to M: parallel ($S = -1/2$, "spin-down", $|\downarrow\rangle$) and antiparallel ($S = +1/2$, "spin-up", $|\uparrow\rangle$). The spin also discriminates the electronic states in a ferromagnetic solid: each state is occupied either by a spin-up or a spin-down electron. This is in marked contrast to a nonmagnetic material, where the spin-up and spin-down states are degenerate, i.e. each state is occupied by both types of electrons. This difference of the spin-up and spin-down states in magnetic materials is also the fundamental origin of spin-dependent charge transport effects.

In a simple atomistic picture, the magnetic moment m_i of a free atom can be determined by counting the number of unpaired electrons when filling the individual shells according to *Hund's* rules. The transition metal ion Fe^{2+} (in this case we don't have to worry about the two $4s$ electrons), for example, has 6 electrons in the $3d$ shell, resulting in a spin magnetic moment of $\mu_S = 4 \mu_B$, an orbital magnetic moment of $\mu_L = 2 \mu_B$, and thus a total moment of $\mu_{Fe} = \mu_S + \mu_L = 6 \mu_B$ for the iron atom [2.201]. The corresponding values for Ni^{2+} are $\mu_S = 2 \mu_B$, $\mu_L = 3 \mu_B$, and $\mu_{Ni} = \mu_S + \mu_L = 5 \mu_B$. The experimentally determined magnetic moments of Fe and Ni solids, however, are significantly smaller, i.e. $m_{Fe} = 2.2 \mu_B$ and $m_{Ni} = 0.6 \mu_B$ [2.202]. respectively. The reason for this discrepancy is the strong interaction of the $3d$ electrons in a solid with each other and with the electrostatic potential of the crystal. The latter gives rise to so-called crystal field effects that lead to an almost complete quenching of the orbital magnetic moment in the $3d$ transition metal ferromagnets Fe, Co, and Ni ($\mu_L < 0.05 \mu_B$) [2.202]. This is a consequence of the $3d$ electrons being weakly bound and participating in the chemical bonding. In rare-earth materials, by contrast, in which the $4f$ electrons are more strongly bound, the magnetic moment is quite close to the atomic value.

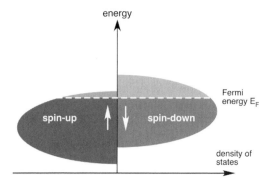

Figure 2.48: Schematic density of states (DOS) picture of a ferromagnet reflecting the contribution of the *d*-like electrons. The spin-down DOS is shifted to higher energies with respect to the spin-up DOS. The difference in the occupied states is responsible for the magnetic moment.

So far, we have only discussed atomic magnetic moments. How does ferromagnetism, i.e. a long-range ordered ferromagnetic state in a solid, arise at all? The underlying quantum-mechanical mechanism is the strong correlation of the electrons in the solid. First, this electron-electron interaction causes the formation of a band structure. Secondly, the Pauli principle results in a spin dependent correlation contribution – the so-called exchange interaction – which favors a parallel spin alignment of neighboring electrons. If the density of occupied states at the Fermi level is high enough, the exchange interaction may force a redistribution of the spin-up and spin-down states, resulting in a lower total energy of the system. The tendency for the formation of this ferromagnetic state is described by the Stoner criterion [2.202]. In the lower energy "ferromagnetic" arrangement, the density of states (DOS) distributions $D(E)$ of the $|\uparrow\rangle$ and $|\downarrow\rangle$ states are energetically shifted with respect to each other (Fig. 2.48). This energy difference is also known as exchange splitting. Since the Fermi energy is the same for both spin states, there are more occupied $|\uparrow\rangle$ ("majority") than $|\downarrow\rangle$ states ("minority"). The excess number of spin-up electrons is directly proportional to the magnetization density m of the material

$$m = \mu_B \int_0^{E_F} \left(D_\uparrow(E) - D_\downarrow(E) \right) dE \tag{2.6}$$

This type of magnetic ordering is also termed *itinerant* ferromagnetism, because the electrons generating the magnetism are delocalized. In the above discussion and in Fig. 2.48, only *d*-states have been considered so far, because the main issue was the magnetic moment. In addition to the *d*-states, however, also *sp*-like states are contributing to the spin-split electronic structure, although their DOS fraction is considerably smaller. These *sp*-electrons are more delocalized than the *d*-states and may therefore play a dominant role in the electrical transport processes. From the imbalance of spin-up and spin-down states at the Fermi level (Fig. 2.48) follows immediately a spin dependence of the electrical transport properties. A more detailed discussion of the spin transport is given in Section 2.4.4.

2.4.3 Magnetic Coupling Phenomena

We will now turn to the interaction between two separated magnetic entities. Two magnetic films brought in close proximity (in the nanometer regime) will influence each other, resulting in an effective magnetic coupling between the two. The interactions responsible for these coupling effects may be separated into two classes according to their physical origin. The magnetic stray field generated by a ferro- or ferrimagnetic film (the stray field emerges at deviations from the ideal infinite film geometry, i.e. film edges, hillocks and kinks at the film surface, etc.) causes a rather long-ranged interaction of dipolar nature. The second class comprises short-ranged interactions of electronic origin, such as the exchange coupling or the interlayer coupling.

The technological application of complex magnetic films requires a careful adjustment of the magnetic behavior of each functional layer and the entire layer stack. The magnetic behavior can be sensitively tuned by an appropriate balance of the various coupling effects. This "magnetic engineering" is an important step in a successful device design and must be based on a thorough understanding of the microscopic mechanisms underlying these coupling phenomena.

"Orange Peel" Coupling

Each magnetic body with a uniform magnetization creates a dipolar stray field at its boundaries. This stray field may be often reduced or minimized by the formation of closure domains [2.203]. In an ideally flat thin film with a macroscopic in-plane magnetization the stray field is confined mainly to the film edges. Along the film surfaces, the magnetization vector lies within the surface plane on microscopic and macroscopic length scales.

In a more realistic scenario, however, a thin film will always grow with a certain surface roughness, due to the kinetic and thermodynamic processes determining its morphology. In polycrystalline films the roughness is mainly determined by the mechanisms of grain formation and lateral/vertical grain growth. When growing a stack of different films from materials that have a tendency to wet each other, the morphological corrugation of the first layer is often transcribed into the subsequent ones ("conformal mapping"). In a thin film system with such a conformal roughness, the surface or interface of a ferromagnetic film is no longer stray-field free on a microscopic scale (Fig. 2.49). The magnetization vector develops components perpendicular to the local surface orientation, the emerging flux thereby creating magnetic charges and a locally confined stray field outside the magnetic material. If two ferromagnetic layers with conformal roughness are separated by an only a few nanometers thick nonmagnetic spacer, the magnetic charges can interact with each other. The result is an effective *ferromagnetic* coupling of the two layers which has first been predicted by *L. Néel* in 1962. It is commonly known as "orange peel" coupling [2.204].

The concept of *Néel* originally considered thick layers and was later extended to describe also a thin film situation [2.205]. Approximating the roughness at the interface by a periodic

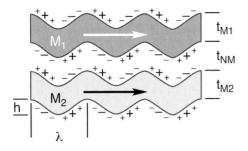

Figure 2.49: Principle of the *Néel* or orange peel coupling due to conformal roughness between two separated ferromagnetic layers of magnetization M_1 and M_2. The roughness amplitude h is not drawn to scale.

function of amplitude h and wavelength λ, the interaction field H_d acting on the softer layer M_1 may be described by the expression

$$H_d = \frac{\pi^2 h^2 M_2}{\sqrt{2}\lambda t_1} e^{-\frac{\pi\sqrt{8}t_{NM}}{\lambda}} \times \left[1 - e^{-\frac{\pi\sqrt{8}t_1}{\lambda}} \right] \times \left[1 - e^{-\frac{\pi\sqrt{8}t_2}{\lambda}} \right] \tag{2.7}$$

The quantities t_1, t_2, and t_{NM} denote the thickness of the first magnetic layer, second magnetic layer, and nonmagnetic interlayer, respectively. The two magnetic layers may have different magnetization M_1 and M_2. The main result is the exponential decay of the Néel coupling strength with the thickness of the nonmagnetic interlayer. As a consequence, the orange peel coupling mechanism becomes important at very thin interlayer thicknesses. This situation occurs, for example, in magnetic tunneling junctions with their very thin ($t_{NM} \leq 1$ nm) oxide barriers. In addition, the coupling increases with the magnetization of the magnetically harder layer M_2. As a consequence of a strong Néel coupling, the two ferromagnetic films can no longer be switched independently. An example of this behavior is discussed in Section 4.5.3.

Interlayer Exchange Coupling

The reduction of the film thickness into the nanometer and subnanometer regime gives rise to novel effects that have their origin in the quantum nature of the electronic system. The appearance of these effects, however, is closely related to the surface or interface quality of the thin films and depends strongly on the preparation conditions and techniques. It was therefore not until 1986 that *Grünberg et al.* [2.187] discovered the so-called ***interlayer exchange coupling*** (IEC) in metallic multilayer stacks consisting of alternating magnetic and nonmagnetic layers. The interlayer coupling was the most important prerequisite for the subsequent discovery of the giant magnetotransport effects in multilayers [2.186, 2.188].

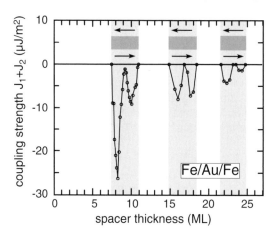

Figure 2.50: Interlayer exchange coupling in epitaxial Fe/Au/Fe(001) trilayers. The regions with antiparallel orientation of the magnetization vectors ("antiferromagnetic coupling") are shaded. They are separated by regions of parallel magnetization orientation ("ferromagnetic coupling"), which are not accessible by the chosen experimental approach. The film thickness is given in monoatomic layers (monolayers, ML). After [2.206, 2.207]

The experimental and theoretical investigations of the interlayer coupling quickly revealed its universal nature. It was found to occur for a wide variety of transition-metal spacer layers [2.208, 2.209], and recently also in semiconducting interlayers [2.210, 2.211]. Although the original discovery was made on single crystalline layer stacks, it became quickly apparent that the effect occurs in a similar form also in polycrystalline film systems. In addition, it was shown that the strength and the sign of the interlayer coupling varies in an oscillatory manner with the thickness of the nonmagnetic spacer [2.212–2.214], i.e., the coupling switches between parallel and antiparallel alignment of the magnetization in the adjacent magnetic layers. This peculiar behavior already points towards the quantum nature of the effect. An example of the interlayer coupling through the noble metal Au is shown in Fig. 2.50 [2.206, 2.207]. The coupling strength $J=J_1+J_2$ is determined from the magnetic field necessary to align the magnetization vectors M_1 and M_2. Consequently, J can be measured reliably only in the regions of the antiparallel coupling state, which is also referred to as *antiferromagnetic* coupling. The distance of these regions is given by an oscillation period $\Lambda_1 \approx 8$ monolayers (ML, monoatomic layers). Interestingly, within these regimes an additional modulation of the coupling strength with a shorter period $\Lambda_2 \approx 2.5$ ML becomes visible [2.215]. In addition, a monotonous decrease of the interlayer coupling strength is superimposed to the oscillations. This short oscillation periods have been reported for a variety of systems, but mostly for single-crystalline layers with sharp interfaces [2.216]. This finding demonstrates the importance of the interfaces in the interlayer coupling.

The oscillatory variation of the interlayer coupling has a close similarity to effects based on RKKY-type interactions. The RKKY model was originally developed to describe the spin polarization of a conduction electron system induced by a magnetic impurity [2.217, 2.218–2.219]. The spin polarization of the electron gas decays in an oscillating manner with increasing distance from the impurity. The oscillation period is determined by the Fermi wave vector k_F. A similar approach has been shown to describe several aspects of the oscillatory variation of the interlayer coupling in the asymptotic limit of thick layers [2.220]. In particular, the failure of most experiments to detect the short-period oscillations was interpreted as an *aliasing* effect. This is related to the fact that the film thickness can only be changed in units of a monatomic layer, i.e. in integer multiples of the lattice plane spacing d. Since $1/d$ and k_F are both of a similar size, a thickness-dependent measurement of the IEC probes a beating pattern between these two periods rather than the short-period oscillation.

A more general picture considers the formation of particular spin-polarized states in the nonmagnetic interlayers [2.221–2.224]. These *quantum-well states* form in bandgaps of the nonmagnetic layer. In the Co/Cu system, for example, the majority spin bands overlap closely with the Cu bands. As a result, majority spin electrons can move freely through the entire layer stack. The minority electrons, however, are confined to the ferromagnetic layers, because electronic states of the appropriate symmetry are not available in the nonmagnetic interlayer. In these bandgaps minority spin quantum-well states are forming that are thought to mediate the interlayer exchange coupling. The presence of these minority spin quantum-well states, particularly in the Co/Cu system, has been proven by numerous experiments [2.225–2.227]. Therefore, the quantum-well origin of the interlayer exchange coupling is nowadays widely accepted.

Phenomenologically, the interlayer coupling exhibits a characteristic angular variation that may be described by the relation

$$E_{IEC} = -J_1 \frac{\vec{M}_1 \cdot \vec{M}_2}{|\vec{M}_1| \cdot |\vec{M}_2|} - J_2 \left(\frac{\vec{M}_1 \cdot \vec{M}_2}{|\vec{M}_1| \cdot |\vec{M}_2|} \right)^2 = -J_1 \cos(\varphi) - J_2 \cos^2(\varphi) \qquad (2.8)$$

with φ denoting the angle between the two magnetization directions.

The first **bilinear** term with the coupling constant J_1 is related to the antiparallel alignment already discussed above. In addition, one also finds a **biquadratic** contribution that favors a $90°$ coupling of the magnetization vectors M_1 and M_2 in the coupled layers. The biquadratic coupling constant J_2 is usually considerably smaller than J_1 and thus becomes important mostly in the case of weak bilinear coupling, i.e. in the transition region between antiparallel and parallel coupling regimes or at rather thick interlayers.

The interlayer coupling is not only important for the observation of GMR phenomena in multilayers, but may also be employed for a successful magnetic engineering in complex layer stacks. The highest values of the interlayer coupling in metallic systems have been found in the Co/Ru or Co/Rh systems. Therefore, antiferromagnetically coupled Co/Ru/Co trilayers are employed as synthetic or artificial antiferromagnets (SAF, AAF) in order to define a magnetic reference direction [2.228].

Exchange Bias

Bringing a ferromagnet into contact with an antiferromagnet can also result in a unique magnetic coupling mechanism. *Meiklejohn and Bean* observed in 1956 that hysteresis loops taken from small Co particles appeared to be shifted along the magnetic field axis, when the sample was cooled below a critical temperature in an external field [2.229]. This critical point turned out to be close to the Néel temperature of the antiferromagnet CoO ($T_N \approx 290$ K) [2.230]. The interpretation of these field-cooling results assumed that the Co particles were covered by a thin CoO layer that takes an antiferromagnetic ordering below T_N, whereby the spin alignment axis of the antiferromagnetic lattice was defined by the external field. The direct spin-spin coupling of ferro- and antiferromagnet at the Co/CoO interface stabilizes the magnetization of the ferromagnet against the external field, causing a so-called **unidirectional anisotropy**. As a consequence, the magnetization reversal occurs no longer at the same absolute field strength for opposite field polarities, and the hysteresis loop is shifted into one field direction – **biased** – with respect to the field-free case.

This exchange biasing concept has been proposed for application to thin film systems by *Dieny et al.* [2.231] in order to create a reliable magnetic reference in so-called spin valve structures. Although spin valves have already found widespread applications in magnetic data storage and sensor technologies, an understanding of the microscopic processes leading to the unidirectional anisotropy is only emerging now. A rather intuitive picture assuming a direct exchange coupling of the spins at the ferromagnet/antiferromagnet interface is described in Fig. 2.51. As long as the sample temperature exceeds the Néel temperature T_N of the antiferromagnet, the spins in the antiferromagnet are disordered. The sample is cooled in the presence of an external field, which orients the spins in the ferromagnetic layer. At the FM/AFM interface the spins are subject to a direct exchange coupling to the ferromagnet.

Very close to T_N the susceptibility in the antiferromagnet is high and the exchange coupling can easily align the AFM spins parallel to the magnetization in the ferromagnet. Starting from this interface the AFM spin order in the film is established when the temperature is lowered further. If the external field is switched off, the orientation of the spins in the ferromagnet is locked by the antiferromagnet. In this exchange biased state, the hysteresis loop is shifted into the opposite field direction. The exchange bias field H_{EB} is conventionally defined by the difference of the switching fields H_A and H_B as

$$H_{EB} = \frac{1}{2}\left(H_B + H_A\right) \tag{2.9}$$

The external magnetic field in the field cooling process is only needed to establish a homogeneous spatial orientation of the magnetization in the ferromagnet. In principle, a local exchange biasing can also be obtained by a thermal treatment without an external magnetic field, exploiting the exchange coupling of the AFM to a ferromagnetic domain. This approach can be used, for example, in magnetic microstructures with a well-defined domain configuration.

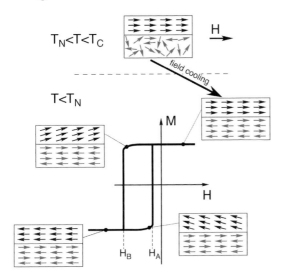

Figure 2.51: Principle of the process leading to exchange biasing in the case of an uncompensated surface of the antiferromagnet. The quantities T_C and T_N denote the Curie temperature of the ferromagnet and the Néel temperature of the antiferromagnet, respectively. After [2.232]

From a phenomenological point of view the exchange biasing is characterized by two important parameters: the biasing field H_{EB} and the blocking temperature T_B. For a given material combination of ferro- and antiferromagnet the biasing strength H_{EB} is often found to vary as

$$H_{EB} = \frac{J_{FAF}}{\mu_0 \cdot M_F \cdot t_F} \tag{2.10}$$

with the film thickness t_F and magnetization M_F of the ferromagnet. The characteristic $1/t_F$ dependence is a clear indication for an interfacial origin of the exchange-biasing mechanism. The microscopic mechanism is related to the exchange coupling of the spins in the ferromagnet to the spin structure at the antiferromagnet/ferromagnet boundary, which is described by the interfacial coupling strength J_{FAF}. This coupling strength is the relevant quantity if the exchange bias in different systems must be compared.

The above picture has three major shortcomings. Firstly, it is based on a so-called ***uncompensated*** surface, i.e. a lattice plane of the antiferromagnet in which the majority of the spins are aligned in the same direction resulting in a net magnetic moment of the lattice plane. This is fulfilled, for example, in NiO{111} planes. Exchange biasing is observed, however, also for ***compensated*** afm surfaces where the local spin arrangement is such that the net magnetic moment of the plane vanishes. Secondly, the magnitude of the interfacial spin-spin coupling J_{FAF} is often assumed to be of the order of the ferromagnetic exchange. The exchange-biasing fields H_{EB} calculated within this picture are usually several orders of magnitude higher than the experimental results. Thirdly, a sizable exchange bias is found even if antiferro- and ferromagnetic layers are not in direct contact, but are separated by a thin nonmagnetic interlayer. In order to explain the first two discrepancies, various microscopic mechanisms have been involved in refined models of the exchange biasing, considering the role of domains in the antiferromagnet, magnetic frustrations, uncompensated surface spins, or non collinearity of the FM-AFM spins, to name but a few (for further information the reader is referred to some concise reviews of exchange biasing in Refs. [2.232, 2.233]). There seems to be a general consensus now that the magnetic coupling is mediated by uncompensated spins of the antiferromagnet at the interface. These uncompensated spins form at morphological defects (step edges or kinks) or as a consequence of antiferromagnetic domain walls. This implies that the strength of the exchange bias can be controlled by tailoring the morphology of the FM/AFM interface. The observation of an exchange biasing through a nonmagnetic interlayer also suggests an electronic component of this phenomenon. This view may be supported by the finding of the exchange-biasing strength to vary oscillatory with the nonmagnetic interlayer thickness, similar to the behavior of the interlayer coupling in multilayers [2.234–2.236].

With increasing sample temperature the exchange coupling between ferro- and antiferromagnet is weakened. Following the simple picture introduced above this coupling should break down with the long-range antiferromagnetic order at the Néel temperature T_N. At $T > T_N$, the spins in the antiferromagnet are randomly oriented. It should be recalled, however, that T_N is an intrinsic property of the antiferromagnet, whereas the exchange biasing is largely determined by the FM-AFM interface and is thus affected by magnetic defects and frustration effects at the interface. In addition, T_N is a well-known quantity only for bulk systems, and may be markedly changed in thin films due to finite size or proximity effects. As a consequence, the exchange biasing usually breaks down at a less well-defined temperature T_B below the Néel temperature, the ***blocking temperature***. Only for idealized systems, for example, single-crystalline systems with perfect interfaces, can $T_B = T_N$ be expected. Although T_B somewhat scales with T_N when comparing different antiferromagnets, within a given system T_B may vary significantly with the ferromagnetic material, interface condition, or even the magnetization direction [2.237]. Table 2.10 compiles some properties of antiferromagnetic materials used for exchange biasing purposes. Note the considerable spread in the blocking temperature and the interfacial coupling strength J_{FAF}.

Table 2.10: Material parameters of some selected antiferromagnets (after [2.232]).

Material	T_N (bulk) [°C]	T_B [°C]	J_{FAF} [μJ/m^2]
NiO	250	180 – 210	10 – 310
FeMn	220	120 – 270	10 – 470
IrMn	420	130 – 250	10 – 190
PtMn	710	130 – 380	20 – 320
NiMn	800	~500	10 – 460

Considering technological applications of the exchange-biasing effect, its stability is determined not only by the blocking temperature, but also the strength of an external magnetic field. As will be discussed in Section 4.4, the unidirectional anisotropy can already be affected at $T \ll T_B$ if a magnetic field is applied perpendicular or antiparallel to the exchange-biasing direction.

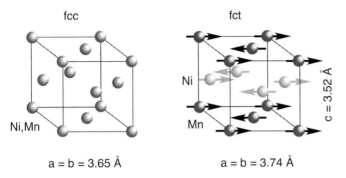

Figure 2.52: Comparison of the unit cells of NiMn: (left) chemically disordered paramagnetic state and (right) chemically ordered antiferromagnetic spin state. The ordering is associated with a tetragonal distortion of the fcc lattice.

The exchange-biasing layer in spin valves or magnetic tunnel junctions usually forms part of the electrode structure. Therefore, the materials currently favored consist mainly of binary metallic compounds, such as $Fe_{50}Mn_{50}$, $Ir_{20}Mn_{80}$, or $Pt_{50}Mn_{50}$. While FeMn has been widely used in read heads, it offers too little chemical stability and blocking temperatures that are too low for other applications, for example, in automotive engineering. IrMn and PtMn have $T_B > 200°C$ and are thus more suitable for these purposes. However, the X-Mn class of materials (X = Ni, Pd, Pt) exhibits an intricate complication. During sputter deposition, the films crystallize in a face centered cubic (fcc) structural modification, which is *paramagnetic*.

The desired antiferromagnetically ordered state is tied to a tetragonally distorted fcc (fct) phase. This fct phase is usually obtained by an extended annealing (several hours) of the

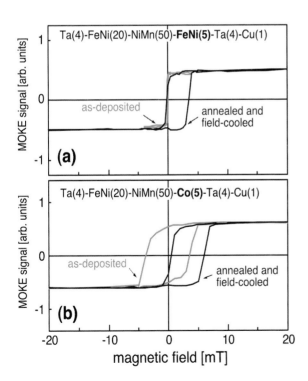

Figure 2.53: Results of a field-cooling process in FeNi/NiMn (a) and Co/NiMn bilayers (b). The hysteresis loops show the magnetization reversal of the ferromagnetic contrast layer (marked boldface in the layer sequence) and indicate the exchange-bias-related loop shift. The individual film thicknesses are given in nanometers. The combination Ta(4)/FeNi(20) serves as a seed layer for optimizing the growth of the NiMn film.

films at about 250°C. The annealing procedure re-establishes a local chemical order, which is necessary for the formation of the spin structure in the afm state. The situation for NiMn is depicted in Fig. 2.52.

The formation of the antiferromagnetic phase can be determined by means of a ferromagnetic "contrast layer". The hysteresis loop of the contrast layer appreciably changes after the phase transformation. If the system before the annealing step is magnetically very soft, i.e. has a small coercive field, the coupling to the antiferromagnet causes a significant increase in the coercive field (Fig. 2.53b). In addition, the magnetization loop may display the loop shift characteristic for an exchange biasing, if the system was field cooled. The field-cooling process establishes the spin quantization axis orientation in the antiferromagnet and thus the unidirectional anisotropy in the ferromagnetic contrast layer. In some cases, the coercive field may be significantly enhanced even before the field-cooling process. This suggests the NiMn layer already assumes a partial antiferromagnetic order with randomly oriented quantization axes (Fig. 2.53b). This behavior is a result of particular growth conditions. A unidirectional alignment of these axes and thus an exchange biasing, however, is obtained only by a subsequent field-cooling procedure.

2.4.4   Electric Transport in Layered Magnetic Systems

Spin-dependent Transport: General Considerations

The basic structure in magnetoelectronic devices comprises two ferromagnetic electrodes, which are connected by a non ferromagnetic material (Fig. 2.54). An important property of this structure must be an individual magnetic switching of the electrodes, i.e. the magnetization direction M_1 of one electrode can be changed without affecting the other one (M_2). The resistivity ρ of this two-terminal device is determined by the scattering of the charge carriers flowing between the electrodes. It contains two contributions. Charge scattering processes from defects and phonons lead to a spin-independent resistivity ρ_0, just as in a nonmagnetic system. In addition, there is a spin-dependent scattering of the charge carriers that depends on the relative orientation of M_1 and M_2.

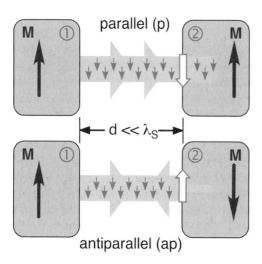

Figure 2.54: Principle of spin-dependent transport phenomena in a two-electrode magnetic systems. The outlined arrow on the right-hand electrode symbolizes the preferred spin transmission channel that depends on the spin character of the empty states and the spatial orientation of the magnetization. The distance d of the electrodes is assumed to be small compared to the spin dephasing length λ_S.

In order to see how this spin-dependent term ρ_S comes about, we take a very idealized approach. We assume that the current consists only of electrons of minority spin (completely filled majority spin bands) and neglect the spatial extension of the orbital part of the wav functions. When the electrons leave electrode ①, they are spin polarized, their quantization axis being defined by M_1. In terms of the spatial coordinates, the spin is pointing downwards. We neglect any spin-dependent scattering events in the intermediate region between the electrodes. This is justified if the spin dephasing length λ_S is much larger than the distance d between the electrodes. Upon arrival at electrode ② (comprising the same material as electrode ①), we must distinguish two cases: M_2 is either parallel or antiparallel to M_1. In the first case, the electrons find enough empty minority spin states, i.e., spin-down states, to occupy. The resulting resistivity ρ_P will be low. If M_2 is oriented antiparallel to M_1, however, the empty states in electrode ② now have their spins pointing upward, because their quantization axis is tied to M_2. With respect to M_1 the states effectively have a spin-up character. As a consequence of the density of states distribution, only very few spin-down states are available, and most of the charge carriers are reflected at the electrode boundary. The resulting conductivity ρ_{AP} is high, i.e. $\rho_{AP} > \rho_P$. We thus observe a magnetoresistance when changing either M_1 or M_2. It is convenient to define the magnetoresistance as a normalized quantity of the type

$$\frac{\Delta\rho}{\rho_0} = \frac{\rho_{AP} - \rho_P}{\rho_P} \tag{2.11}$$

In this definition, the magnetic ground state without external field is an antiparallel orientation between M_1 and M_2. The presence of an external magnetic field causes M_1 and M_2 to orient parallel, being accompanied by the transition $\rho_{AP} \Rightarrow \rho_P$.

Tunneling Magnetoresistance (TMR)

The quantum-mechanical tunneling effect has been extensively used for electronic spectroscopy purposes (elastic and inelastic tunneling spectroscopy). In this case, two extended metallic electrodes – one or both of them superconducting – are separated by an insulating layer, mainly consisting of Al_2O_3. If this insulating barrier is sufficiently thin the electronic wavefunctions of the two metallic electrodes may have some overlap in the barrier region. Via this overlap electrons can tunnel from one electrode to the other. In the presence of an applied electric field this leads to a tunneling current. The tunneling probability depends exponentially on the distance between the two electrodes. In order to obtain tunneling currents of technical use the barrier thickness must be of the order of $1 - 2$ nm.

The magnitude of the tunneling current is determined by the details of the density of states at and above the Fermi level E_F. Usually, in a ferromagnet the majority and minority spin density of states are distinctly different, in particular, in the vicinity of E_F. One may thus expect a spin dependence in the tunneling process, at least, at low temperatures [2.238]. This spin dependence leads to the conductivity of the tunneling junction to depend on the magnetization direction, i.e. to a tunneling magnetoresistance (TMR). The major breakthrough was achieved, however, when improved preparation conditions of the tunneling contacts led to the observation of spin-dependent tunneling phenomena at room temperature [2.239–2.241].

The principle of the spin-dependent tunneling process is sketched in Fig. 2.55 and follows very closely the general considerations given above. For reasons of simplicity we assume the

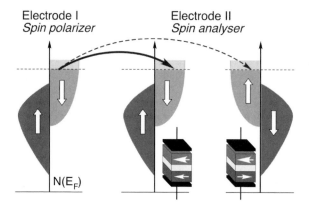

Figure 2.55: Schematic description of the spin-dependent tunneling of minority spin electrons in a thin film junction. One electrode acts as a spin polarizer, the other one as a spin analyzer. A large tunneling probability requires the same spin direction in both electrodes.

electrical current to be carried predominantly by minority spin electrons. The electrons leaving the first electrode (spin polarizer) have a spin-down character with respect to the quantization axis defined by the electrode magnetization M_1. If the magnetization M_2 in the second electrode points into the same direction (parallel configuration), the electrons can easily tunnel via elastic processes into unoccupied states in the second electrode. The respective resistance ρ_P of the tunneling junction is low. Note that the tunneling processes are mainly involving electronic states at the electrode/insulator boundary rather than the bulk states within the electrode. If the magnetization M_2 is reversed, however, also the quantization axis of the electronic states in the second electrode is reversed. They thus become effectively majority spin states. Consequently, the tunneling spin-down electrons find only very few empty states and the charge transport is strongly suppressed as long as spin-flip scattering can be neglected. The resulting resistivity ρ_{AP} for this antiparallel magnetization configuration is high. As a quantitative measure of this transport effect the TMR ratio $\Delta\rho/\rho$ is defined as in Eq. (2.11). In the tunneling junction the current flows in the direction normal to the plane of the layers. This geometry is therefore also referred to a "current perpendicular to plane", CPP.

A sizable tunneling magnetoresistance can be observed only if the switching between parallel and antiparallel magnetization configurations of the two electrodes is well defined. This requires one electrode (e.g., M_1) to be magnetically "harder" than the other one, i.e. the magnetization reversal takes place at a higher external magnetic field H_1 than the switching of the "soft" electrode's magnetization M_2 ($H_2 < H_1$). In practice, these different switching fields are achieved by choosing different magnetic materials for the soft (Ni-based alloys) and hard electrodes (Co-based alloys). In order to further increase H_1, the hard-layer magnetization is often exchange coupled to a synthetic or natural antiferromagnet.

In the materials mentioned above (Ni- and Co-based alloys) the majority spin bands are not fully occupied resulting in a sizable majority spin density of states at the Fermi level $D^\uparrow(E_F)$. A quantitative treatment of the electrical transport must therefore consider conductivity channels for both spin-up and spin-down electrons, the relative weight of which is described by the quantity spin polarization P

$$P(E) = \frac{D^\uparrow(E) - D^\downarrow(E)}{D^\uparrow(E) + D^\downarrow(E)} \qquad\qquad (2.12)$$

Note that in the above example used to introduce the TMR effect the spin polarization is $P(E_F) = -100$ %, a situation approximately realized in halfmetallic ferromagnets. In Co or Ni the majority spin contribution $D^\uparrow(E_F)$ renders the spin polarization $P(E_F) < -50$ %. This leads to an upper limit for the TMR ratio that can be achieved with a given combination of electrode materials. According to *Jullière* [2.240] the TMR ratio can be related to the spin polarization values P_1 and P_2 of the two electrodes by

$$\frac{\Delta \rho}{\rho_p} = \frac{2 \cdot P_1(E_F) \cdot P_2(E_F)}{1 - P_1(E_F) \cdot P_2(E_F)} \qquad (2.13)$$

Using commonly accepted spin polarization values for Co of $P(E_F) \approx -40$ %, one obtains a TMR ratio of $\Delta \rho / \rho_p \approx 38$ % for a magnetic tunnel junction with two Co based electrodes. This corresponds reasonably well to experimental observations [2.242, 2.243].

The Jullière model suggests that very high TMR values may be achieved by using electrode materials with a high spin polarization at the Fermi level. On the one hand, this perspective drives the current materials research in magnetic oxides, Heusler alloys and other halfmetallic ferromagnets. On the other hand, the Jullière approach is based on a very simplified picture of the electronic structure, which deviates more or less strongly from the real situation. Firstly, the tunneling process is governed mainly by the electronic states at the electrode/barrier interfaces. The interfacial electronic structure and thus the spin polarization become very important issues. They may considerably differ from the bulk case. Secondly, not all electronic states at the interface contribute in the same manner to the tunneling current. This is due to the spatial extension of the various states, which is larger for *s*- than for *d*-electrons. The *s*-electrons have a larger wavefunction overlap in the barrier and the corresponding tunneling probability should be higher. Thirdly, electronic states located in the insulating barrier, for example, arising from defects may significantly affect the tunneling process. If these defects carry a magnetic moment, their presence may enhance the spin-flip scattering and thereby reduce the tunneling magnetoresistance. A solid understanding of the tunneling magnetoresistance effect must thus be based on a detailed knowledge of the electronic states in the layers and at the interfaces of the thin film system.

Further improvements of the TMR yield may also be achieved by alternative tunneling barriers. Although a variety of different oxides has been investigated over the years, Al_2O_3 was long considered the optimum choice. Theoretical investigations predicted the system Fe/MgO(001)/Fe to exhibit very large tunneling magnetoresistance, the reason being a particular configuration of the electronic band structure and a strong k-conservation in the tunneling processes leading to coherent tunneling [2.244–2.246]. Since iron reacts strongly with oxygen, however, the formation of an interfacial iron oxide layer was theoretically shown to destroy the spin-dependent tunneling effect almost completely [2.247]. Only very recently, a careful control of the growth conditions for the case of Fe/MgO/Fe(001) resulted in very high tunneling magnetoresistance of more than $\Delta R/R_o > 300$ % at cryogenic temperatures and $\Delta R/R_o \approx 220$ % at 300 K [2.248–2.250]. This opens up a very interesting avenue toward highly efficient TMR elements.

Giant Magnetoresistance (GMR)

The CPP geometry introduced above can also be used in all-metal thin film stacks. In this case the tunneling barrier is replaced by a nonmagnetic conductive interlayer, mostly copper. This configuration raises a problem, however, in that the resistivity of a metal layer stack is much smaller than the barrier-dominated resistivity of a magnetic tunnel junction (MTJ). In order to obtain sizable magnetoresistance effects in CPP-GMR, the conducting element must have a cross section of 100 nm or less, which is difficult to fabricate. Nevertheless, in most systems a somewhat smaller GMR signal can be measured if the current is flowing within the film plane (current-in-plane, CIP). In fact, most of the GMR applications in sensor technology employ the CIP geometry.

The transport of charge carriers through this nonmagnetic interlayer is diffuse in nature. Since the mean free path of electrons in transition metals is in the nanometer regime, the nonmagnetic interlayer must be of the same order of magnitude to observe spin-dependent transport effects. Just as in the TMR case, the magnetoresistance is determined as the resistivity difference between a parallel and antiparallel magnetization configuration of the ferromagnetic layers. In the parallel configuration the spin-polarized electrons leaving the first electrode are traveling through the Cu interlayer and can easily enter the second electrode, whereas for antiparallel alignment they are mainly scattered back at the second electrode.

Giant magnetoresistive effects were first observed in metallic Fe/Cr nanoscale multilayers [2.186, 2.188]. These experiments employed the current-in-plane (CIP) geometry. Subsequent studies proved this phenomenon to be of a general nature occurring for many combinations of layered ferromagnetic and non-ferromagnetic materials and over a wide range of temperatures [2.189, 2.213, 2.251–2.256]. It may even be realized in granular systems [2.257, 2.258]. Several investigations pointed out the importance of the structural and chemical interface quality for the magnitude of the GMR effects [2.259–2.264]. The role of the interfaces was further detailed by theoretical studies of the spin-dependent transport in metallic multilayers that revealed characteristic contributions from the layers and the interfaces to the GMR signal [2.265, 2.266].

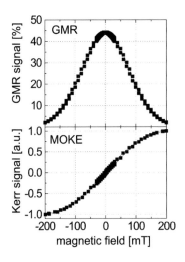

Figure 2.56: Giant magnetoresistance as a function of the external magnetic field in a Co/Cu multilayer. The Cu interlayer thickness corresponds to the first antiferromagnetic interlayer coupling maximum. The GMR measurement refers to a CIP geometry. The Kerr effect measurement shows a typical hard axis magnetization loop indicating the gradual break-up of the antiferromagnetically coupled ground state of the multilayer.

It is important to realize that the appearance of GMR in multilayers is intimately connected to the interlayer exchange coupling discussed in Section 2.4.3. For appropriate interlayer thicknesses, the coupling establishes a strong "antiferromagnetic" ground state, i.e. the magnetization vectors in neighboring ferromagnetic layers assume a mutual antiparallel orientation. According to the simple argument given above, spin-polarized electrons originating from a certain ferromagnetic layer will be able to travel through the adjacent nonmagnetic spacer layers. They will be strongly scattered, however, in the neighboring ferromagnetic layers when the multilayer is in the antiferromagnetic ground state. This magnetization configuration corresponds to a state of high resistance ρ_{AP} of the multilayer. The interlayer coupling may be overcome by a sufficiently high external magnetic field resulting in a parallel orientation of the magnetization in all ferromagnetic layers. In this case, the spin-polarized electrons are only weakly scattered in the neighboring magnetic layers and the respective resistivity ρ_P is smaller. By means of the external field H we can continuously vary the magnetic configuration between the antiparallel and parallel alignment leading to the pronounced "bell"-shaped GMR curve $\Delta\rho(H)/\rho_0$ shown in Fig. 2.56 for Co/Cu multilayers. The "S"-shaped hysteresis loop in Fig. 2.56 documents the strong interlayer coupling in the Co/Cu system that acts as an effective uniaxial anisotropy.

The close connection of interlayer exchange coupling and giant magnetoresistance is also clearly reflected in the variation of the GMR signal as a function of the interlayer thickness. The GMR signal takes its highest values at the interlayer thicknesses corresponding to the subsequent antiferromagnetic maxima and disappears in between (Fig. 2.57). The reduction of the GMR value from the 1st to the 2nd and from the 2nd to the 3rd afm coupling maximum is related to the decrease of the overall resistance of the layer stack and the limited mean free path. These results also show that GMR ratios of $\Delta\rho/\rho_0 >100\%$ may be obtained at low temperatures. This enhancement at low temperatures can be ascribed to the drop of the average resistance due to the reduction of phonon-related scattering processes [2.267].

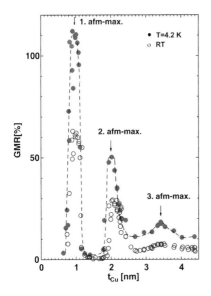

Figure 2.57: Thickness variation of the GMR signal in polycrystalline Co/Cu multilayers prepared by magnetron sputtering.

In the simplistic picture introduced above we have implicitly assumed a totally spin-polarized current. In reality, however, the spin polarization of the charge carriers is usually incomplete, because both majority and minority spin states are available in the vicinity of the Fermi level E_F. As a consequence, spin-up and spin-down electrons contribute to the electrical current and must thus be considered in a refined treatment of the magnetotransport effects. In addition, the charge carriers pass through all (ferro- and nonmagnetic) layers, the individual contributions of which to the total conductivity must be taken into account. In a phenomenological approach this is achieved by the so-called **two-current model** [2.256] that assumes independent spin-up and spin-down current channels, i.e. resistances $\rho_\uparrow$ and $\rho_\downarrow$.

The resistivities of the spin-up and spin-down current channels in a given material are determined by the details of the electronic structure close to the Fermi level (Fermi surfaces), in particular, the quantities $D^\uparrow(E_F)$ and $D^\downarrow(E_F)$, as well as their *sp*- and *d*-type fractions. The ratio of the spin dependent resistivities may be denoted by the asymmetry parameter $\alpha = \rho_\downarrow/\rho_\uparrow$. For the Co/Cu system, for example, the minority spin electrons are more strongly scattered than the majority spin electrons, i.e. $\alpha > 1$. For the Fe/Cr system, the situation is reversed yielding $\alpha < 1$.

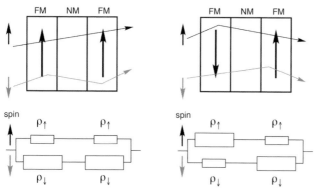

Figure 2.58: Graphical representation of the two-current model for the case of Co/Cu, i.e. minority electrons are scattered more strongly.

For a parallel coupling of the layers a spin-up (spin-down) electron remains spin-up (spin-down) in all ferromagnetic layers (Fig. 2.58). In this case $\rho_\uparrow$ and $\rho_\downarrow$ are behaving as two parallel resistors resulting in a total resistance ρ_P for the parallel state given by

$$\rho_P = \frac{\rho_\uparrow \cdot \rho_\downarrow}{\rho_\uparrow + \rho_\downarrow} \tag{2.14}$$

For the antiferromagnetic configuration of the multilayer (Fig. 2.58), a spin-up electron in one ferromagnetic layer becomes a spin-down electron in the neighboring ones and vice versa. The resistance in each channel may be expressed as the average of the spin-up and spin-down resistances ($\rho_\uparrow/2 + \rho_\downarrow/2$). The total resistance ρ_{AP} becomes then

$$\rho_{AP} = \frac{\rho_\uparrow + \rho_\downarrow}{4} \tag{2.15}$$

Following Eq. (2.11) the giant magnetoresistance $\Delta\rho/\rho$ can be calculated as

$$\frac{\Delta\rho}{\rho_P} = \frac{\rho_{AP} - \rho_P}{\rho_P} = \frac{(\rho_\downarrow - \rho_\uparrow)^2}{4 \cdot \rho_\uparrow \cdot \rho_\downarrow} \tag{2.16}$$

This two-current model holds only, if the layer thickness is much smaller than the inelastic mean free path of the charge carriers. Although in Fig. 2.58 the CPP geometry is chosen, the same argument holds for the CIP configuration. In this case, the individual layers must be thin enough for a spin-polarized electron to be able to probe at least the neighboring ferro-magnetic layer. This requirement is fulfilled at least for the 1st and 2nd antiferromagnetic coupling maximum in most GMR multilayer systems.

In order to understand the different GMR behavior in the various materials systems, the microscopic mechanisms that are behind the quantities $\rho_\uparrow$ and $\rho_\downarrow$ need to be understood. These mechanisms are closely related to the details of the electronic structure of the constituents of the multilayer system. In addition, both interface and bulk related contributions must be considered, which in some cases may be counteracting with respect to the GMR magnitude [2.266]. One example may suffice to demonstrate the particular role of the electronic structure. The Co/Cu system is known to exhibit the highest GMR signals in moderate magnetic fields at room temperature. This is partly due to a peculiar matching of the electronic states of Co and Cu at the Fermi surface, which dominate the electrical transport behavior.

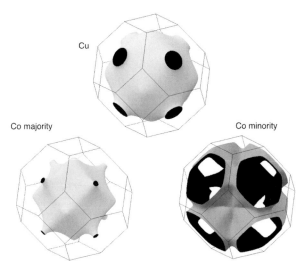

Figure 2.59: Bulk Fermi surfaces of Cu (top), Co majority (bottom left), and Co minority spin states (bottom right). Taken from [2.268]

The Co majority spin and Cu Fermi surfaces have similar shapes and geometries, but differ significantly from the Co minority spin Fermi surface (Fig. 2.59). In the multilayer the Co and Cu layers are coherently strained, i.e. assume locally comparable lattice parameters. As a consequence, the Co majority spin and Cu Fermi surfaces have very little difference and the majority spin electrons can move freely throughout the entire layer stack. On the other hand, there is only a small overlap between the Co minority spin and Cu Fermi surfaces. Therefore, these electrons carry only a small contribution to the overall current in the system.

Spin Injection

In contrast to metallic systems where due to the strong scattering the electrical transport takes place only in the diffusive limit, in semiconductors a ballistic charge transport may be achieved. Due to the high purity of these materials the defect scattering is much reduced. As a consequence, the spin dephasing length in semiconductors may be several orders of magnitude higher than in metals, reaching values in the µm range [2.269, 2.270]. A semiconductor-based spinelectronics approach promises novel types of devices such as, for example, the spin transistor [2.271, 2.272]. The physical realization of these devices, however, requires the solution of major problems in materials science, in magnetism and solid-state physics. This will be illustrated by the following example.

A widely used device in microelectronics is the field effect transistor (FET). In 1990, *Datta* and *Das* proposed a spin-polarized version of the FET by introducing ferromagnetic source and drain contacts [2.271]. This device has two modes of operation. The spin-polarized electrons are injected into the 2-dimensional electron gas forming the conductive channel between source and drain (Fig. 2.60). The ferromagnetic source basically acts as a spin polarizer. Given a sufficiently large spin dephasing length the electrons will reach the drain contact (spin analyzer) without a significant loss of their degree of polarization. In the case of zero gate voltage V_G, the source-drain current depends only on the relative orientation of the source and drain electrode magnetization directions M_S and M_D, respectively. By switching the magnetization between a parallel and antiparallel configuration, two distinct source-drain current levels can be achieved. If a gate voltage $V_G \neq 0$ is applied, a second operational mode becomes accessible. While traveling through the electric field under the gate the electrons experience a gradual rotation of their spins, the amount of which depends on V_G and the channel length. This rotation is caused by the *Rashba* effect [2.273]. As a consequence, the gate voltage can be used to continuously vary the orientation of the spin polarization vector P of the current. The spin analyzing drain contact detects only the component of the spin polarization P parallel to the magnetization direction M_D. In this way the current can be continuously modulated via the gate voltage. In addition to this spin-dependent effect, the electric field acts of course also directly on the charge carriers, as in a conventional FET.

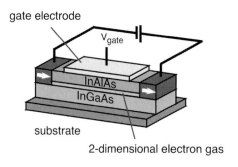

gate electrode

V_{gate}

InAlAs
InGaAs

substrate

2-dimensional electron gas

Figure 2.60: Layout of a spin field effect transistor (Spin FET) as proposed by *Datta* and *Das* [2.271].

As a first step towards a semiconductor-based spinelectronics the feasibility of the spin injection concept must be proven. This turned out to be a formidable challenge. Most of the experimental efforts to realize a spin transistor or similar devices employed a hybrid metal-semiconductor approach. For this purpose, thin ferromagnetic metal films are brought in direct contact with a semiconductor surface, such as Si or GaAs. Apart from the numerous well-known problems with metal/semiconductor interfaces (stability, reactivity, etc.), additional difficulties are introduced by the magnetic properties of the metallic components. The magnetic field emanating from the ferromagnetic electrodes will generate a Hall effect in the semiconductor material. If the spin-injection effect is to be determined from an electrical transport measurement, contributions due to the Hall effect must be carefully taken into account.

An additional problem arises due to the conductivity mismatch between metal and semiconductor at least, as far as diffusive transport is concerned. Whereas the typical conductivity of a metal is of the order of $10^6 - 10^7$ $(\Omega m)^{-1}$, the conductivity of the semiconductor is several orders of magnitude lower. Since the magnetotransport effects are measured as conductivity or resistivity changes in an applied magnetic field with respect to a base value in zero field, the magnetoresistance value is dominated by the semiconductor resistivity. Recent simulations have shown that the $\Delta\rho/\rho_0$ values that can be achieved in this way will be of the order of 10^{-4} and less [2.274]. In addition, the depletion zone resulting from the Schottky-barrier formation at the metal/semiconductor interface may have a detrimental effect on the spin injection [2.275, 2.276]. Two strategies are currently pursued to improve this situation: (i) a proper conductivity matching, or (ii) the use of electrode materials with a high spin polarization. The conductivity matching can be achieved in two ways, at least in principle. In a first approach, *Fert* and *Jaffrès* proposed to insert a tunneling barrier at the interface between the metallic ferromagnet and the semiconductor [2.277, 2.278]. This generates a spin-dependent interface resistance at the metal-semiconductor junction. As a result, a sizable spin injection is predicted for a proper choice of resistivity, spin diffusion length and length of the semiconductor channel in a *Datta-Das* arrangement. Recent experiments seem to indicate that the spin-injection efficiency may indeed be enhanced by the tunneling barrier [2.279].

The second approach to directly avoid the conductivity mismatch uses magnetic semiconductors instead of metals as electrodes. An inherent disadvantage of dilute magnetic semiconductors (DMS) comes from the low Curie temperatures of these materials, which are generally far below 300 K [2.209, 2.280]. There are some recent theoretical predictions for GaN- and ZnO-based magnetic semiconductors with potentially high Curie temperatures [2.281], which started an extensive materials research on this field [2.282]. In addition, a lot of attention is devoted to Mn-doped GaAs, as a model system for DMS. In the ideal case, these materials should exhibit a carrier-induced ferromagnetism, i.e. the magnetism is directly related to the density and imbalance of the spin-up and spin-down charge carriers. As a marked property of this type of magnetism, typical properties such as the Curie temperature or magnetic anisotropy may be varied via the charge density, for example, in a field-effect geometry. In reality, however, carrier-induced ferromagnetism in these materials is extremely difficult to achieve. The magnetic response observed stems most often from ferromagnetic clusters formed by the impurity atoms Mn or Cr and depends very sensitively on the preparation conditions [2.283–2.285]. Moreover, the theoretical predictions in [2.281] were based on a mean-field treatment. A more refined modeling of the exchange mechanisms within the Monte Carlo framework predicts Curie temperatures significantly below the mean-field values, which is more in accordance with recent experiments [2.286, 2.287]. Clearly more materials research is needed to solve this problem. The same holds for the use of highly spin-polarized materials as electrode materials – this issue will be discussed in Section 2.4.5.

The difficulties discussed above affect not only the injection of polarized electrons into a semiconductor, but also their detection with the second ferromagnetic electrode. Establishing an all-electrical spin injection therefore remains one of the formidable challenges in spintronics.

In an effort to circumvent some of the problems discussed above, recently an optical detection scheme for the spin-injection effect has been successfully employed [2.288, 2.289]. It involves the fabrication of a light emitting diode with one of the ferromagnetic electrodes consisting of a dilute magnetic semiconductor. The spin polarization of the injected electrons is proven by detecting the amount of circularly polarized light emitted during the recombination of the excited charge carriers in the spin analyzer. In this way, spin injection was first demonstrated at low temperatures (4.2 K). At present, an extensive experimental activity addresses the details of spin injection for a variety of materials combinations and as a function of temperature [2.276, 2.279, 2.290–2.294].

Spin Angular Momentum Transfer

A more detailed consideration of the microscopic processes taking place during spin transport and injection leads to a very interesting aspect. Let us assume a standard GMR configuration, in which the two ferromagnetic layers are separated by a nonmagnetic spacer. The spin-polarized electrons traversing the spacer are spin filtered upon their penetration into the ferromagnet. If their spin-quantization axis $\underline{P}$ is fully aligned with the electrode magnetization $\underline{M}$, i.e. $\underline{P} \parallel \underline{M}$, we have a situation as sketched in Fig. 2.54: electrons with the "wrong" spin direction will be reflected at the interface, whereas those with the "right" spin will enter the electrode. What will happen, however, if $\underline{P}$ and $\underline{M}$ make a finite angle Θ with each other, that is, the spin polarization is not aligned with the magnetization direction, but has a transversal component? This transversal component must be somehow absorbed by the system. As *Berger* [2.295] and *Slonczewski* [2.296] showed in their seminal papers, the absorption of the spin angular momentum acts as a torque – spin torque – onto the electrode magnetization.

In order to see this more clearly, we turn to the simple sketch in Fig. 2.61, depicting the behavior of a single electron in a semiclassical picture. In the rest frame of the magnetic electrode, the spin of the incident electron has two components. The transmitted fraction of the spin polarization must be aligned with $\underline{M}$ by definition. On the other hand, the reflected contribution must have the spin opposite to $\underline{M}$, assuming a complete spin filtering. As a result, the transversal spin components are transferred to the magnetic system of the electrode exerting a total magnetic torque $\sim \sin \Theta$. A more detailed analysis reveals that spin filtering is not the only mechanism responsible for the absorption of the transversal spin components [2.297]. In particular, dynamic effects such as spin precession and rotation contribute as well.

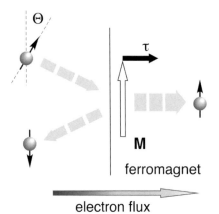

Figure 2.61: Principle of spin-torque generation. The scattering of polarized electrons in the ferromagnet creates a torque, which is absorbed by the magnetization of the system.

In fact, this spin-torque transfer mechanism was first derived by *Berger* and *Slonczewski* in the context of a spin-current induced spin wave excitation. If the spin-polarized current density is sufficiently high, however, the transferred torque may be strong enough to change the magnetization direction in the electrode and thus to initiate a magnetization reversal. This "spin-current induced switching" currently receives much attention as it bears the promise of being a smart alternative to the rather inefficient process of switching by a magnetic field.

2.4.5 Functional Thin Film Systems

For technological applications the various magnetic coupling and transport mechanisms may be combined in a functional magnetic layer stack. In order to obtain an operating device, the magnetic properties, such as magnetization, anisotropies, etc., must be tailored to the specific needs. At the same time the transport properties must be optimized. This situation sometimes results in conflicting requirements in the preparation of the layer stack.

Thin Film Preparation

There are various deposition techniques used to fabricate layered systems (cf. Section 3.3). As we will discuss in more detail below a spinelectronic film stack contains a larger number of magnetically and chemically diverse layers. In order to avoid contaminations at the interfaces it is favorable to deposit the entire layer stack in one run. The method of choice, which also allows the efficient deposition onto large substrates, for example, 6 in silicon wafers, is the magnetron sputtering technique. This approach uses a noble gas discharge between a magnetron source and the substrate. A good command of the sputtering conditions and the deposition parameters allows a precise control of the film thickness down to the sub-nanometer regime. Commercial sputtering deposition chambers may contain up to a dozen individual magnetron sources for the various materials needed. For the deposition of ferromagnetic materials the magnetron sources must be equipped with stronger magnetic fields than those necessary for non- or antiferromagnetic materials. Magnetron sources may also be used for the deposition of insulating materials if the discharge is established by a high frequency (rf) instead of a constant (dc) electric field.

Complex Layer Stacks

Depending on the nature of the spin-dependent transport process one may distinguish spin valves (GMR-based) and magnetic tunneling junctions (TMR-based). The expression "valve" may be a little bit misleading as it implies a large on-off ratio of the electrical current induced by the external magnetic field. With present devices, however, a resistivity modulation of a factor of 2 can be achieved, at best. This is still insufficient for active spintronic devices, e.g. a spin transistor, but can be conveniently exploited in many magnetic sensor applications.

The general layer stacking sequence of spin valves (SV) and magnetic tunneling junctions (MTJ) is very similar with respect to the functional subunits (Fig. 2.62). We will therefore discuss these two type of systems in combination. The main subunits are the ***magnetotransport layer*** *stack* providing the spin-dependent transport effects and the ***reference layer*** *stack*, which defines the reference magnetization. In addition, there may be various seed and cap layers.

The functional subunit providing the magnetotransport effect consists of two ferromagnetic layers, separated by a nonmagnetic metal (e.g. Cu) or insulator (e.g. Al_2O_3). The main difference is found in the direction of current flow that runs parallel (perpendicular) to the film plane in GMR (TMR)-based structures. This puts some constraints to the physical properties of the adjacent subunits and the size of the junctions, as will be discussed below.

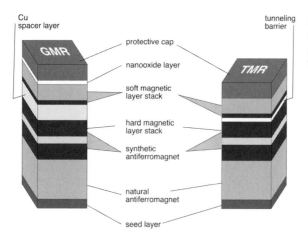

Figure 2.62: Comparison of the layer stacking for a GMR spin valve element and a magnetic tunneling junction (MTJ).

Simple heterostructures. Assuming the two magnetic layers to be the same, we have a situation similar to a multilayer stack. If the two ferromagnetic layers are magnetically decoupled, both will independently switch in an external field resulting in a null GMR signal. In this case a sufficiently strong interlayer coupling is needed in order to establish a well-defined magnetic ground state with antiferromagnetic alignment. The transport curves $\Delta\rho/\rho_0$ will show the same characteristic "bell" shape known from multilayers (cf., Fig. 2.56), but with a comparably smaller GMR effect. Since the GMR signal depends only on the relative angle between the magnetization in the two layers, this type of layer stack can be employed to determine the presence and strength of an external magnetic field, rather than its direction.

Trilayer structures ("pseudospin valves"). Clearly the layer stack described above provides only very limited functionality. This situation can be improved quite significantly

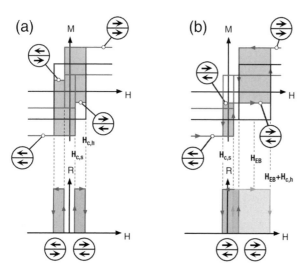

Figure 2.63: Relationship between magnetization curves (top) and GMR characteristics (bottom) in pseudospin valves (a) and exchange biased spin valves (b). The shaded magnetization loops reflect the switching of the entire layer stack, built up from the magnetization reversal of the individual layers (unshaded loops). Note that for the exchange biased spin valve (b) the exchange biased switching characteristic of the hard layer results in an extended plateau on the right-hand side of the resistivity curve. This is favorable for sensor applications.

by making the switching fields (coercivities) of the two electrodes different. As a consequence, one of the electrodes, usually called the (magnetically) "soft" electrode reverses its
magnetization at a much lower field $H_{C,s}$ than the hard layer that switches at $H_{C,h} > H_{C,s}$ (Fig.
2.63). Permalloy ($Ni_{80}Fe_{20}$) is the material of choice for the soft layer, the magnetization
direction of which may be easily oriented in an external field. The $\Delta\rho(H)/\rho_0$ curve splits up
into two symmetric branches with sharp transitions at $H_{C,s}$ due to the magnetization reversal
of the soft layer. The gradual reduction of the GMR signal at higher fields is caused by the
gradual magnetization reversal in the hard layer. The switching characteristics can be directly
measured by magnetometry. In this so-called "pseudospin valve" scheme the GMR signal
may be varied continuously by the external field and serves as a measure of the angle between M_1 and M_2, as long as $H \ll H_{C,h}$ (Fig. 2.64).

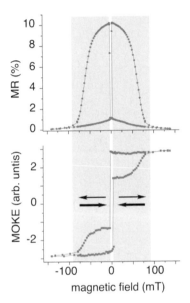

Figure 2.64: Magnetoresistance (top) and magnetization
reversal (bottom) in a Co/Cu/FeNi trilayer stack as a function
of the applied field. The steps in the hysteresis loop mark the
switching of the hard and soft layer. The magnetoresistance is
high only when the two ferromagnetic layers are antiparallel
(shaded area). Whereas the steep onset of the GMR at low
fields is caused by the switching of the soft layer, the gradual
decrease of the GMR at higher fields leading to the bell-
shaped characteristics is related to domain formation and
rotation processes.

The highest GMR signals at room temperature have been found in the Co/Cu system (see
above). Therefore, the magnitude of the GMR response in the trilayer may be increased by
introducing a Co/Cu interface at the soft layer side. For this purpose the surface of the Cu
layer is "dusted" by a very thin (< 1 nm) Co layer before depositing the Fe-Ni soft layer.
This Co dusting does not significantly change the magnetic properties of the FeNi film, but
increases markedly the spin-dependent scattering at the interface and thus the GMR. For this
reason, Co/FeNi or CoFe/NiFe bilayers find widespread use as soft magnetic electrodes. As
an additional advantage in GMR systems, the Co dusting layer increases the thermal stability
by reducing the CuNi interfacial segregation.

Similar arguments hold for the magnetic tunneling junctions. Since the TMR signal can be
increased by using electrode materials with a high spin polarization, many efforts are directed towards the development of half-metallic magnets, such as Heusler alloys, manga-
nates, etc. The magnetic properties of these systems are less known and more difficult to
control. One may therefore envision hybrid electrode structures combining Co or CoFe layers with a highly spin-polarized material on the tunneling barrier side.

Exchange biased spin valves.   For many applications it is mandatory that not only the alignment, but also the orientation of the reference magnetization M_2 is well defined. This is realized by coupling the hard electrode to an antiferromagnet. By adjusting an appropriate exchange bias the switching field of the hard layer can be shifted to the desired values. As a result the $\Delta\rho(H)/\rho_0$ characteristic becomes asymmetric and develops a more or less extended plateau of constant GMR (Fig. 2.63b). This configuration is regularly denoted as spin valve (SV) or exchange biased SV and is widely employed in read heads and magnetic rotation angle sensors. By tuning the magnetic anisotropies of the hard and soft layers the magnetic film properties of the stack can be adjusted to specific needs. The magnetic optimization of MTJ films follows the same guidelines. Various antiferromagnetic materials, mainly Mn-based metals or transition-metal oxides are employed for this purpose. The latter, however, are excluded in the case of MTJs. The $\Delta\rho(H)/\rho_0$ characteristics of several spin valve systems and magnetic tunneling junctions are discussed in detail in Section 4.5.

Each realistic magnetic sensor is characterized by a working regime. A rotation angle sensor, for example, will usually exploit the plateau region of the GMR curve. The highest external field in which the SV system may be safely operated is given by the magnetic stability of the reference electrode. If the external field starts to affect the magnetization direction of the reference electrode, the GMR signal develops unwanted nonlinear contributions. As it turns out this already happens on the high-field side of the plateau region. A modeling of experimental results shows the reliable working regime of such a SV sensor to be limited to $0 < H < H_{EB}$, although the GMR plateau extends far beyond H_{EB} [2.298].

In some applications higher values of H_{EB} than may be obtained with natural antiferromagnets are required. For this purpose use is made of the oscillatory interlayer coupling found in metallic multilayers. An additional trilayer of the type Co/Cu/Co or Co/Ru/Co is inserted between the natural antiferromagnet and the hard magnetic electrode. The interlayer thickness is chosen such that the Co layers are strongly antiferromagnetically coupled (Co: ~ 1 nm; Ru: ~ 0.8 nm). This system is termed artificial or synthetic antiferromagnet, AAF or SAF, respectively. In most cases the top Co layer simultaneously forms the hard magnetic electrode layer. In the resulting $\Delta\rho(H)/\rho_0$ characteristics the plateau is extended to higher field values and the operational window of the film stack is enhanced. The magnetic stability of such a system, however, is still mainly determined by the unidirectional anisotropy and not the strength of the interlayer coupling. This concept of artificial antiferromagnets is also widely exploited in MTJs to achieve increased magnetic stability and a smaller spread of the switching fields. The latter aspect is of great importance of MTJ arrays, which are considered in magnetic random access memories (MRAM).

In the case of SV layer stacks in CIP geometry a further improvement of the GMR signal may be obtained by introducing so-called nanooxide layers (NOL) on one or both sides of the GMR functional unit [2.299]. These NOLs increase the specular scattering of the electrons. Since the electrons are thus more confined to the GMR active layer and undergo a more efficient spin-dependent scattering cascade, the GMR signal can be increased by up to 30 %.

Taking all the above aspects into account the entire functional GMR or TMR thin film stack consists of several subunits, each with 1–3 individual layers. Not considered yet are seed layers (often Ta, FeNi, or Cu) to introduce a peculiar crystallinity and texture in the stack and cap layers (e.g. Ta, Cu, Au) to protect the entire system from oxidation. It is quite obvious that the optimization of such a complex system with respect to magnetic, magneto-transport properties and thermal stability is a challenging task (cf. Section 4.5).

Highly spin-polarized electrodes. The maximum GMR value achievable in metallic spin valves with CIP geometry at room temperature is of the order of $\Delta\rho/\rho_0 \sim 20$ % and may be somewhat increased in the CPP configuration. In the latter case, however, the spin valve elements must be confined to very small lateral dimensions by sophisticated lithography techniques. This is due to the low voltage drop across an extended metallic layer stack, which would otherwise give rise to only minute magnetoresistance effects. Therefore, a significant enhancement of the GMR signal is not easily within reach.

In magnetic tunneling contacts – inherently involving a CPP geometry – the situation is somewhat different. As the electrical transport in a MTJ is mainly determined by the insulating barrier, an MTJ's resistivity per unit area is several orders of magnitude higher than that of a metallic CPP-GMR contact. Thus, MTJs can be made larger, which is an advantage in the application. In fact, for very small MTJs (lateral dimension a < 500 nm) extremely thin tunneling barriers ($t_B \sim 0.7 - 0.9$ nm) are required to keep the junction resistivity in a technologically acceptable regime. The maximum TMR values obtained in dedicated MTJs approach $\Delta\rho/\rho_0 \sim 40 - 50$ % at room temperature. From a theoretical point of view, however, there is still a lot of room for improvement in the TMR signal. A close inspection of Jullière's model leading to Eq. (2.13) shows that the tunneling magnetoresistance may assume values $\Delta\rho/\rho_0 \gg 100$ %, if the product of the spin polarization values of the two electrodes approaches unity, i.e. $P_1(E_F) \cdot P_2(E_F) \sim 1$. A high spin polarization value $P(E_F)$ requires the electronic system to exhibit a gap in one type of spin (usually majority) bands around the Fermi level. Such a behavior may be found in various ferromagnetic oxides and half-metallic systems [2.300]. Classical ferromagnetic oxides, which have been already extensively investigated comprise CrO_2 and Fe_3O_4. More recently also perovskites such as $La_{0.7}Sr_{0.3}MnO_3$, and double perovskites (e.g., Sr_2FeMoO_6) have been considered as ferromagnetic electrode materials. The potential use in spintronics has also strongly revived the interest in the half-metallic Heusler alloys, such as the half-Heuslers NiMnSb and PtMnSb, or the full-Heuslers Co_2CrAl and Co_2MnSi. In order to be useful for applications in spin-dependent tunneling, however, the materials must have $T_C \gg 300$ K and – even more importantly – their half-metallic character in the bulk must be preserved also at the surface or interface. The latter requirement is usually difficult to meet as surface or interface-related electronic states tend to fill the gaps in the bulk band structure and may thus destroy the half-metallicity at the interfaces [2.301]. Considerable efforts in materials science are still needed to overcome these problems.

Spin-torque-induced magnetization dynamics. As we have briefly discussed in Section 2.4.4., a spin-polarized electrical current can generate a spin-transfer torque, which acts onto the magnetization of the magnetic layers in a multilayered structure. In order to obtain a sizable effect, however, the charge density flux must be sufficiently high, of the order of $10^7 - 10^8$ A/cm^2. This implies confined geometries, such as nanocontacts or nanowires. Stimulated by the predictions of *Berger* [2.295] and *Sloncewski* [2.296], spin-torque-induced changes in the magnetization were investigated in such magnetic nanostructures [2.302–2.304]. A first clear signature for a magnetization reversal caused by the spin-torque effect was observed in a nanosized pillar structure [2.304, 2.305].

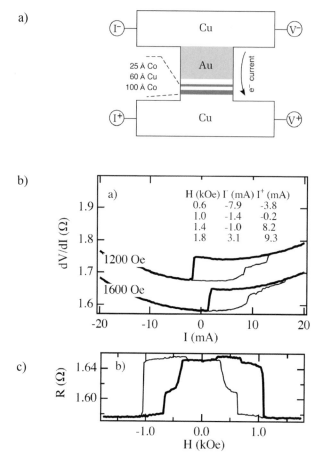

Figure 2.65: Current-induced magnetic switching in Co/Cu/Co nanopillars (sample geometry, see (a)). (b) The dV/dI characteristics of the pillar device exhibits hysteretic jumps as the current is swept, starting from $I = 0$ at constant external field H_{ext}. Light and dark lines indicate increasing and decreasing current, respectively. The traces have been offset vertically for better visibility. The inset table lists the critical currents at which the device begins to depart from the fully parallel configuration (I^+) and begins to return to the fully aligned state (I^-). (c) Zero-bias magnetoresistive hysteresis loop for the same sample. Taken from [2.305].

Some of the results are reproduced in Fig. 2.65. The pillar structure contained a Co/Cu/Co trilayer as a magnetic subunit, which was nanostructured to a diameter of about 130 nm. The Co layers were chosen with different thicknesses, corresponding to different coercive fields (Figure 2.65a). The transport measurements were performed in the 4-point probe geometry at a constant external magnetic field H_{ext}. The pillar structures exhibited a clear GMR signal with well-defined switching behavior between the parallel (P) and antiparallel (AP) state as a function of H_{ext} (Fig. 2.65a). The fact that the system turns into the high-resistivity "AP" state before H_{ext} has reversed sign, suggests the presence of an antiferromagnetic coupling between the Co layers. At a Cu spacer layer thickness of 6 nm, however, the interlayer coupling is rather weak. A stronger coupling can be obtained by the dipolar interaction of the magnetic layers at the edges of the pillar structure (magnetostatic edge coupling). The differential conductivity curves dV/dI shown in Fig. 2.65b have been recorded with a lock-in technique at external bias fields H_{ext} sufficiently high to stabilize the "P" state and bring both Co layers into a single-domain configuration.

The dV/dI curves reveal a characteristic hysteretic behavior depending on the magnitude of the external field. Upon increasing the currrent through the device (light trace), the differential conductivity first follows a smooth rising path, caused by an increasing amount of electron/phonon and electron/magnon scattering events. At a critical current value I^+, however, dV/dI jumps up to a higher resistivity level, before it follows a smooth rising path again. This jump to a higher resistivity indicates that the system leaves the "P" state and starts to switch towards the "AP" state. Apparently, the spin torque exerted by the polarized current is large enough to overcome the stabilizing action of the external bias field and induce a magnetic switching event. If the current is then decreased (heavy trace), dV/dI returns on the curved path, but jumps down to the initial resistivity level of the "P" state at a significantly lower critical current value I^-. Further investigations prove that the critical current values I^+ and I^- depend in a characteristic manner on the magnitude of the bias field: an increase of H_{ext} shifts I^+ and I^- to more positive values (see inset table in Fig. 2.65b). This means that at a higher external field a larger spin torque is needed to reverse the system into the antiparallel state.

These findings prove that a spin-torque-induced switching of a magnetic system is indeed feasible. This opens an avenue for smart alternatives to the established switching modes by external magnetic fields. The main difficulty that has to be overcome is the high current density needed to initiate the magnetization reversal. Therefore, the current experimental and theoretical investigations concentrate on an understanding of the microscopic mechanisms behind the spin-transfer torque effect, in order to be able to develop optimization strategies.

One of the key aspects in this understanding concerns the dynamic response of the magnetization on a microscopic scale. If a magnetic system is subjected to an external field, the magnetization $\underline{M}$ starts to process around an axis defined by an effective field H_{eff}. The precession is damped, usually causing the magnetization to settle in a spiral path along H_{eff} on the nanosecond timescale. The effective field is composed of the external field and the total magnetic anisotropy. The damped precessional motion of $\underline{M}$ is described by the well-known *Landau – Lifshitz - Gilbert* (LLG) equation. In the experiment described in Fig. 2.65, the transfer of spin angular momentum to the magnetic system creates a torque, which deflects the magnetization away from H_{eff}, and increases the precession angle. This large-angle precession is the first step towards magnetization reversal. If the effective field is stronger than the spin-transfer torque, however, a particular interesting condition occurs. The system is kept in a permanent state of large-angle precession – a precessional spin-wave mode [2.306] – because the polarized current delivers a spin torque that balances the damping of the precessional motion.

The response of the magnetic system in such a situation is shown in Fig. 2.66. The sample is again a Co/Cu/Co nanopillar (Co[40 nm]/Cu[10 nm]/Co[3 nm]) similar to the one described above [2.307]. The differential resistivity (Fig. 2.66a) shows the clear signatures of hysteretic switching at low values of H_{ext}. At larger external fields the switching is suppressed. Instead of step-like changes, spike-like features appear in the dV/dI curves up to rather high current values.

These spectral structures are related to the generation of a high-frequency current signal component, as is shown in Fig. 2.66b. This microwave signal is decoupled from the dc background and analyzed with respect to the frequency distribution. The obtained microwave spectra reveal several distinct features in the GHz range, which disperse with the strength of the current flowing through the device. The most pronounced maxima appear in the spectrum taken at I = 3.6 mA (trace 3), whereas no microwave signals can be discerned at I = 7.6 mA (trace 5). These microwave signatures are directly related to the precessional spin-wave modes, which are excited by the spin angular momentum transfer and can be used to determine the damping parameters in the magnetic systems [2.305, 2.308]. In addition to being a possible smart switching alternative, the spin-torque effect also opens up a new approach to magnetization and spin dynamics [2.307, 2.309–2.311].

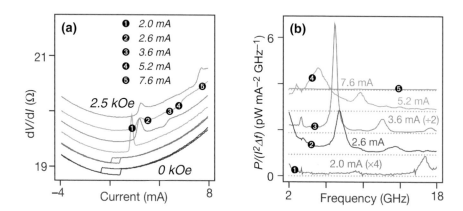

Figure 2.66: Current-induced microwave generation in a Co/Cu/Co nanopillar. (a) Differential resistance vs. current curves for magnetic fields $H_{ext} = 0$ (bottom), 0.5, 1.0, 1.5, 2.0 and 2.5 kOe (top), with current sweeps in both directions. At $H_{ext} = 0$, the switching currents are $I_C^+ = 0.88$ mA and $I_C^- = -0.71$ mA, and $\Delta R_{max} = 0.11$ Ω between the P and AP states. Numbered marks on the 2 kOe curve correspond to spectra shown in (b). (b) Microwave spectra for $H_{ext} = 2.0$ kOe at various current values. Taken from [2.307].

Magnetic Hybrid Structures Involving Semiconductors

On the way towards active spintronic devices two aspects are becoming of importance. Firstly, an electrical control of the spin-dependent transport processes requires a transition from the present two-terminal (sensor, MTJ) to a three-terminal device. Secondly, implementing spintronics functionality into conventional semiconductor microelectronics may in the long run either result in metal-semiconductor hybrid structures, or in the use of magnetic semiconductors. Two approaches being currently pursued will be briefly discussed in the remainder of this section. They may outline the type of problems, which still have to be solved for a future semiconductor-based spintronics.

Spin-valve transistor. In GMR systems with a metallic conduction characteristic the electronic states participating in the electrical transport are located at the Fermi level. The spin-polarized current is mainly carried by equilibrium electrons. In spin polarized tunneling and spin injection phenomena, however, the situation is different. Due to a barrier between the two conducting electrodes, a sizable voltage drop over the barrier (0.1–3 V) can be established. As a consequence, the spin-polarized current will be carried by energetic non equilibrium ("hot") electrons. This holds particularly for metal/semiconductor hybrid structures in which *Schottky* barriers are forming at the interface regions between the metal and semiconductor.

In order to exploit the magnetotransport effects occurring in such systems, one first has to understand the mechanisms governing the spin-dependent transport of these hot electrons. With this aim in mind, a so-called spin-valve transistor (SVT) has been constructed and fabricated [2.196, 2.272]. The SVT adapts the concept of a classical metal-base transistor and comprises a metallic spin valve or magnetic multilayer as a base, which is sandwiched by n-Si layers as emitter and collector electrodes. Its functional principle is sketched in Fig. 2.67a. The unpolarized energetic electrons are injected over the Schottky barrier into the magnetic base, which is effectively acting as a CPP-GMR stack. The hot electrons undergo spin-dependent and spin-independent scattering processes in this GMR layer and the amount of current passing through the GMR stack depends on the relative orientation of the magnetization directions of the two ferromagnetic layers in Fig. 2.67a. For the antiparallel ground state the current will be low. The collector Schottky barrier is operated in reverse bias voltage. This feature may be used to select quasi-ballistically transferred hot electrons and reject those electrons that have suffered large energy losses. The measured quantity is thus the collector current I_C as a function of the collector bias voltage and the applied magnetic field.

The first experiments used a Co/Cu multilayer as a base and showed a magnetization-induced change of the collector current of $I_C/I_{C0} \sim 215$ % at cryogenic temperatures of 77 K [2.272].

(a)

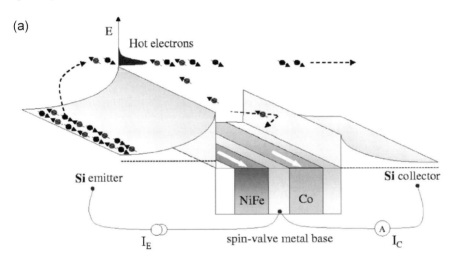

(b)

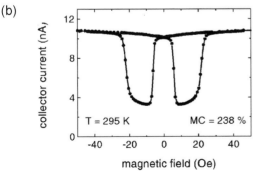

Figure 2.67: (a) Functional principle of the spin-valve transistor. (b) Example of the magnetocurrent data obtained from a SVT with Si(001) emitter and collector structures and a Pt(2)/NiFe(3)/Au(3.5)/Co(3)/Au(4) basis (thicknesses in nanometers). The collector current is highest for the parallel orientation of the NiFe and Co layers. Taken from [2.196].

By improving the fabrication procedure and employing dedicated spin-valve systems as a base, collector magnetocurrent variations of $I_C/I_{C0} \sim 240$ % at room temperature have been achieved (Fig. 2.67b) [2.196]. Note that the magnetic field dependence of the collector current reflects the characteristic symmetric switching behavior of a pseudo-spin valve. Several other features of these latest SVT devices include large magnetocurrent values at small applied fields (H < 4 mT) and a linear variation of I_C with the emitter current I_E. These studies have considerably improved the understanding of the spin polarized hot electron transport in magnetic layers. It must be noted, however, that the SVT is not operating as a real transistor as the gains are very small. In the above example in Fig. 2.67b the transfer ratio I_C/I_B is only of the order of 6×10^{-6}. The large magnetocurrent variations may still be successfully employed, for example, in sensor applications.

Spin-polarized LED. The operation of a true spintronics three-terminal device in the *Datta* and *Das* framework [2.271] involves three crucial steps: the injection of *spin-polarized* electrons, their ballistic transport, and their detection. As has been discussed in Section 2.4.4, the conductivity mismatch between ferromagnetic metal and semiconductor renders spin injection effects in such a device very small, if the spin polarization of the injected electrons is well below $P = 100$ % (as is the case for classical ferromagnetic materials) [2.274]. In order to reduce the complexity of the problem, it is thus useful to study each of the three transport steps individually. Of particular interest is the spin injection step as it determines the spin polarization of the electrons in the subsequent ballistic transport channel. A considerable amount of work has been devoted to this issue. It took until 1999, however, before the existence of the electrical spin injection mechanism could be reliably proven [2.288, 2.312].

The characterization of the spin injection process involves a peculiar problem. How can the spin polarization of the electrons injected into the nonmagnetic semiconductor be detected? This can be achieved by an optical process, exploiting the radiative recombination (electroluminescence) of the electrons in the nonmagnetic semiconductor layer. If the electrons are spin polarized along a defined quantization axis, the electroluminescence light will be circularly polarized, too. This is the inversion of the well-known process of optical spin orientation [2.313]. Consequently, the device to investigate the spin injection mechanism is constructed as a light emitting diode (Fig. 2.68a), comprising a AlGaAs/GaAs based layer stack. The spin alignment is obtained by means of a layer of a dilute magnetic semiconductor n–Be$_x$Mn$_y$Zn$_{1-x-y}$Se. The spin-polarized electrons are injected into a n-AlGaAs layer and recombine in a thin layer (15 nm) of intrinsic GaAs. The second electrode is formed by a p–AlGaAs/p-GaAs layer combination. The circular polarization of the electroluminescence light depends on the external magnetic field used to saturate the spin aligner and reaches maximum values of up to $P_c = 43$ % at a temperature of 4.2 K. From this value a spin polarization of the injected electrons of $P_e \sim 90$ % can be estimated. A certain disadvantage of the dilute magnetic semiconductors is their low Curie temperature that limits the experiments to cryogenic temperatures. In addition, they require large magnetic fields ($B \sim 3$ T) to saturate. In a similar experiment the spin alignment was provided by a (Ga,Mn)As layer that had a Curie temperature around 40 – 90 K [2.312]. In this case, the measurements at 6 K revealed light polarization changes of $\Delta P_c \sim 2$ % in small magnetic fields of $B \sim 5$ mT. These results stimulated an extensive ongoing search for novel magnetic semiconductors with higher magnetic ordering temperatures. Some theoretical predictions suggest room temperature ferromagnetic semiconductors on the basis of ZnO and GaN [2.314].

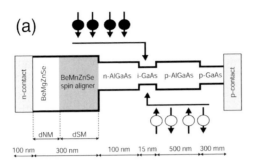

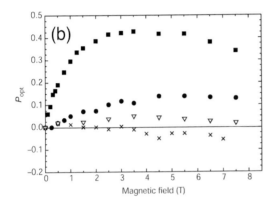

Figure 2.68: (a) Functional principle of a spin-polarized light emitting diode. (b) Circular polarization of the electroluminescence after the injection of the spin-polarized electrons in the i-GaAs layer. The electrons are spin aligned by a n-Be$_x$Mn$_y$Zn$_{1-x-y}$Se layer of varying thickness (d_{SM} = 300 nm, squares; d_{SM} = 3 nm, circles). For comparison, data for a non-polarizing n-BeMgZnSe electrode (d_{NM} = 300 nm, d_{SM} = 0 nm, triangles) and the polarization degree of the intrinsic GaAs layer (crosses) are also included. After [2.288]

Recently, spin injection has been shown to occur at room temperature in a LED device involving an Fe electrode on GaAs [2.315]. The spin injection efficiency takes values of about 2 % at magnetic fields in excess of 2 T. This result suggests that indeed efficient spin injection devices working at room temperature may be realized. However, there is plenty of room for improvements. Further optimization strategies will be needed to reduce the required magnetic fields and push the spin injection efficiency to higher levels [2.316].

2.5 Multilayer and Single-Surface Reflectors for X-Ray Optics

2.5.1 Introduction

The manipulation of X-rays with optical elements or combinations thereof in complex optical instruments is the topic of X-ray optics. The basic principles of that field are analogous to other regions of the electromagnetic spectrum – e.g. that of visible or ultraviolet light. Compared to other spectral regions there are similarities and differences in the interaction of X-rays with matter and thus in the possibility to manipulate X-rays by optical instruments or even their components. Thus some peculiarities of X-rays in contrast to other wavelength intervals have to be considered in the following paragraphs.

The optical systems use elements such as mirrors to deflect, to collimate or to focus the X-rays or even to form images from extended objects; zone plates to form images from microscopic objects and natural crystals, artificially synthesized multilayers (layered synthetic microstructures **LSM**) and diffraction gratings to separate the X-rays into their individual spectral components.

As there is some arbitrariness in the literature [2.317–2.322] about standardization of boundaries and subdivision into partial spectral intervals of the X-ray region, a definition will first be suggested that may be a compromise between the collection of definitions:

X-rays are a particular region of the electromagnetic spectrum (Fig. 2.69) adjoining at its high energy boundary (typically $\lambda \approx 0.0001\ldots 0.001$ nm) the γ-rays and at low energies ($\lambda \approx 30$ nm) the region of Vacuum Ultra Violet (VUV). Over this X-ray region (except for some very few exceptions) the refractive index n ($n = c/v$) in any material is less than unity and for the phase velocity v the relation $v > c$ is valid. A distinction between hard and soft X-rays has already been given by *R. W. Pohl* in 1940 [2.320] by taking the Ag K-edge as reference. Recently the spectral interval from typically $10 \leq \lambda_X \leq 30$ nm has been defined as the extreme ultraviolet (EUV) region to distinguish a newly developing lithography technique (EUVL) with working wavelengths $\lambda_X = 11.4$ and $\lambda_X = 13.4$ nm, respectively, from the already conventional proximity X-ray lithography ($0.4 < \lambda_X < 5$ nm). The water window region, where low X-ray absorption from water occurs, is indicated by "H_2O".

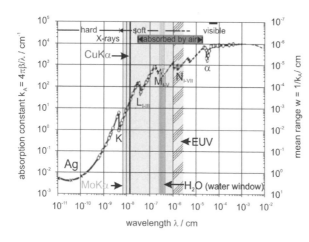

Figure 2.69: Typical absorption spectrum of a metal (Ag) for wavelength region $\lambda_X = 10^{-11}$ through $\lambda_X = 10^{-2}$ cm (after [2.320])

A wide field of X-ray applications has emerged in various areas of science and technology. Besides the recently started development of EUVL, there are many – during the last century already well established – research disciplines utilizing X-ray optics. Outstanding fields worth mentioning are: medical research, biology, astronomy, plasma diagnostics, microscopy and various spectroscopic techniques for basic research and materials characterization (e.g. Electron probe micro-analysis, X-ray fluorescence analysis, X-ray and neutron diffraction, beam lines of synchrotrons and particle storage rings).

Already in the early days after the discovery and the first application of X-rays it was realized as a serious drawback that there were no possibilities for beams of this new type of radiation [2.323] to be deflected or refracted by means of conventional lenses or mirrors, as can be done with visible or UV light [2.320–2.323]. This phenomenon was understood when it was discovered that X-rays are electromagnetic waves of very short wavelength and that this spectral region is strongly absorbed but only weakly refracted by any material. Alternative proposals for optical elements led to the application of curved crystal planes (e.g., bent mica or NaCl). Such elements, however, showed rather low acceptance angles and severe image defects (e.g. astigmatism).

When X-rays interact with the highly polished surface of matter[1] over the entire X-ray region a remarkable reflectance is observed only for high angles of beam incidence α on that surface. Hence, for an angle Θ (glancing angle between beam axis and deflector surface, i.e. $\Theta = 90° - \alpha$) below a certain critical angle Θ_c[2] the reflected beam propagates along the surface and according to *Snell's* law, total external reflection (i.e. reflectance $R \approx 1$) must occur.

2.5.2 Refraction and Reflection at Single Boundaries

Inspection of Fig. 2.69 shows that there is a wide range of penetration depth for the various X-ray wavelengths observed in the interaction with a metal. Whereas hard X-rays with wavelengths 0.0001 to 0.05 nm will penetrate hundreds of µm, soft X-rays with wavelengths in the range 0.05 to 10 nm are already absorbed by air and have only ranges of the order of tenths of 1 µm in Ag. This means that, in contrast to the behavior of dielectrics in the visible range with vanishing absorption and a high transparency, similar materials are not available for the X-ray regions. Most materials are only partially transparent. They absorb energy from the transmitted waves.

Thus, the interaction of such electromagnetic waves with matter will be controlled by refraction *and* absorption. The refractivity coefficients in or at the boundary between media then have to be described in a complex form:

$$\tilde{n} = n + i\beta = 1 - \delta + i\beta \tag{2.17}$$

where δ is the refractive index decrement and β the absorption index and the representation of the amplitude $E(r,t)$ of a plane wave propagating in homogeneous matter with an incident amplitude E_0 is given by

$$E(r, t) = E_0 \times e^{-i(\omega t - kr)} \tag{2.18}$$

involving the complex dispersion relation

---

[1] Wave vector **k**- and angles α-, Θ-definitions see Fig. 2.70, inset

[2] For the interface vacuum/arbitrary material from *Snell's* law ($\sin \alpha' = \sin \alpha/(1 - \delta)$) the critical angle $\Theta_c \approx \sqrt{2\delta}$ is a rough estimate for the absorption free case. It is to be pointed out that the curve shape $R = R(\Theta)$ is strongly influenced by the size of the ratio δ/β and the definition of Θ_c is not always unambiguous. Even for one and the same material the approximation $\delta/\beta \gg 1$ is not fulfilled throughout the total wavelength region of the X-rays – as it varies between values <1 and >1.

$$\frac{\omega}{k} = \frac{c}{\tilde{n}} = \frac{c}{(1 - \delta + i\beta)}$$

where ω is the radiation frequency, k the wave vector, k the scalar wavenumber $(k = |k| = \frac{2\pi}{\lambda})$ and c the phase velocity in vacuum.

Then the individual amplitude factors may be separated as

$$E(r,t) = E_0 \times e^{\{-(2\pi\beta/\lambda)r\} + \{-i(2\pi\delta/\lambda)r\} + \{-i\omega(t-r/c)\}} \tag{2.19}$$

and the components of the exponential function represent "decay", "phase shift" and "vacuum propagation" of the wave in the given order.

When X-rays propagating in a high refractive index material (vacuum) illuminate a surface of low refractive index material (any matter) the partial processes of reflection and transmission and thus the magnitudes of the amplitudes E_r and E_t of the electric fields of the reflected and transmitted components of the incident wave are characterized by the reflection and transmission coefficients $r = E_r/E_0$ and $t = E_t/E_0$ of these amplitudes. The *Fresnel* equations describe these coefficients as the relation between the optical constants of a material and the angle of incidence α of the incident wave for the relevant wavelengths. These relations are given for the perpendicular or "s"-polarized components of electromagnetic radiation in Eqs. (2.20) and (2.21):

$$r_{\perp} = \frac{\tilde{n}_1 \times \cos\alpha_1 - \tilde{n}_2 \times \cos\alpha_2}{\tilde{n}_1 \times \cos\alpha_1 + \tilde{n}_2 \times \cos\alpha_2} \tag{2.20}$$

$$t_{\perp} = \frac{2\tilde{n}_1 \times \cos\alpha_1}{\tilde{n}_1 \times \cos\alpha_1 + \tilde{n}_2 \times \cos\alpha_2} \tag{2.21}$$

where α_1 is the angle of incidence, α_2 the angle of refraction and $\tilde{n}_1$, $\tilde{n}_2$ the refractive indices of the adjacent materials.

The reflectance $R_{\perp}$ of such a boundary as the fraction of the reflected intensity I_r and the incident intensity I_0 is then obtained for normal incidence ($\alpha_1 = \alpha_0$ and δ, $\beta \ll 1$, compare with Fig. 2.71) from Eq. (2.20) as:

$$R_{\perp}(\alpha_0) = |r_{\perp}(\alpha_0)|^2 \approx \frac{\delta^2 + \beta^2}{4} \tag{2.22}$$

Figure 2.70 shows, as a typical example, the specular s-reflectance of an idealized molybdenum surface (surface roughness $\sigma \approx 0$) within a wavelength interval $0.01 < \lambda_X < 30$ nm. For an angle of incidence $\alpha = 5°$ the reflectance R always stays below 1%. The increase in R from an extremely low level for hard X-rays towards larger wavelengths is rather monotonic except for the L–edge ($\lambda \approx 0.433 - 0.492$ nm) and M–edge ($\lambda \approx 2.449 - 5.44$ nm) regions.

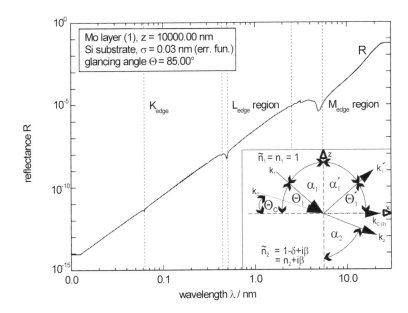

Figure 2.70: Spectral reflectance of the Mo surface in the X-ray region $0.01 \leq \lambda_X < 30$ nm (from [2.324]), inset shows wave vector and angle definitions for X-ray incidence on matter: k_1, k'_1, k_2, wave vectors of incident, reflected and transmitted wave, respectively; α_i the appropriate angles with reference to the surface normal z; Θ. the glancing angle, subscript c refers to critical ray path

The coefficients of refractivity n and extinction β are plotted in Fig. 2.71 for the same material. Whereas n shows almost no remarkable deviation from unity (i.e. $0.9953 < n < 1$) for $\lambda_X < 5$ nm, a substantial change is observed for $5 < \lambda_X < 30$ nm. The extinction coefficient β shows an almost monotonic increase for $\lambda_X < 5$ nm except at the absorption edges.

Above the M-absorption edges a moderate slope followed by a steep rise is found for $5.5 < \lambda_X < 13$ nm and a relative maximum is approached at, typically, 30 nm, after which another decrease is caused by the N-edges. However, the refractive index distinctly decreases to below 1 for this spectral band.

For nearly all useful materials the decrements δ of the refractivity coefficient n and the extinction coefficient β show qualitatively similar behavior, as demonstrated for Mo in Fig. 2.71 and, as can be deduced from Eq. (2.22), in any case the reflectance $R_\perp$ remains very low.

Thus for the entire X-ray range no conventional refractive optical elements are imaginable. Therefore new mirror concepts have to be considered in order to be able to design appropriate X-ray optical elements and to combine powerful systems. There are two alternatives that are presently under development and in application for the various demands of X-ray beam handling. Their basic functional principles involve either

- Glancing incidence reflection (GIR) for $\Theta \leq \Theta_c$ or
- Normal incidence reflection (NIR) for $\alpha \approx 0°$

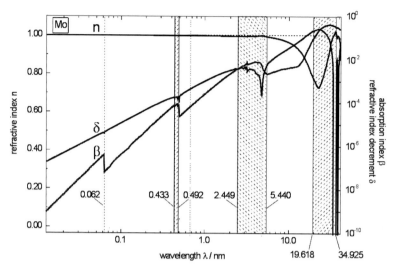

Figure 2.71: Optical constants (n,β) of Mo in the wavelength region of $0.01 \leq \lambda_X \leq 40$ nm (from [2.324])

GIR

Optics of these types, using only a curved or highly polished plane surface[3] as a reflector, are used to collimate or to concentrate (focus) X-ray beams of the shorter wavelength region in particular experiments in astronomy, plasma diagnostics and in synchrotron radiation beamlines. Imaging with glancing incidence elements (utilized e.g. in GI microscopes for laser-fusion studies [2.325] involves certain drawbacks as there are severe image defects (e.g. astigmatism, spherical aberration, coma) and rather limited collection angles.

In particular, concave spherical mirrors proved to result in poor imaging optics. For grazing incidence the formation of two line foci from the same object point by the meridional and sagittal rays is caused due to astigmatism. Both types of rays differ in location and orientation [2.326]. An image defect correction for an identical focal length $f = f_s = f_m$ in both sagittal and meridional planes leads to a toroidal shape of the mirror with distinct radii of curvature $P_s \neq P_m$. The relation ([2.322], p. 530)

$$P_s = P_m \times \sin^2 \Theta \tag{2.23}$$

demonstrates that there are extreme demands on the flexibility of the optical shaping and polishing techniques to fabricate mirrors with very different curvatures and superior figure strength and surface smoothness (compare appropriate tolerances in Table 11). For the Mo-mirror shape using CuKα-radiation with $\Theta_c \approx 0.444°$ a ratio of the radii $P_m/P_s \approx 6 \times 10^{-5}$ is found! This becomes slightly more tolerable for wavelengths in the soft X-ray region: $\lambda_X = 13.4$ nm, $\Theta_c \approx 22.4°$ and the input $P_m = 100$ cm requires $P_s \cong 14.5$ cm. Unfortunately, spherical aberration and coma play a negative role in this case.

---

[3] The surface is usually coated with a substantial Au film for surface optimization, stabilization and conservation, thus introducing our phenomenon of single boundary interaction to the general topic "metal based films ...".

Meanwhile multilayer coated GIR optics have a wide field of applications in X-ray diffractometry etc. They will be discussed together with the NIR elements in the following. Therefore, refined solutions allowing image defect correction by X-ray optical systems comprising more than one element have been developed as a consequence. In the middle of the last century *Kirkpatrick* and *Baez* [2.327] proposed an astigmatism-corrected combination of two crossed spherical mirrors and thus realized a successfully operating reflection microscope for X-rays.

Only shortly afterwards toroidal systems usable as telescopes were proposed by *Wolter* [2.328]. In this approach, paraboloid or ellipsoid and hyperboloid elements are combined and consequently a coma correction is obtained. On this basis, improved X-ray telescopes[4] (angular resolution < 5" for white X-rays) could be constructed [2.329].

NIR

Single-surface reflectors show, as explained, a very low X-ray reflection efficiency at NIR conditions. For small angles of incidence (e.g. $\alpha < 20$ °) and $\lambda_X < 13$ nm the reflectance of a single Mo surface always remains below 0.15 %. Furthermore, at GIR, a rather poor acceptance and very limited image fields can be realized. Thus, for beam deflection, shaping and analysis or monochromatization as well as for imaging, new solutions in comparison to visible light or UV optics had to be found. This situation has been substantially improved by upscaling the decisive features of X-ray diffraction in natural crystals to diffractor dimensions which can be handled by the techniques of precision coating.

Moreover, normal-incidence optics for imaging in the soft X-ray region require that the figure and smoothness of the deflector elements (or substrates) are manufactured to a precision much better than that usually required for visible optics. This is done in order to achieve an optimum reflectance and the improved spatial resolution intrinsically possible with the shorter-wavelength radiation. The surface quality characterized by the tolerances given in Table 2.11 represents, in conventional terms, quality levels for EUVL applications of typically $\lambda_X/130$ and $\lambda_X/65$, respectively.

2.5.3 Bragg Reflection at 1D Lattice Systems

Analogies to Conventional Optics and X-Ray Techniques

The conventional mirror – consisting of the mirror body or deflector with its extremely polished and thus highly reflecting surface or comprising one or a few additional metallic or dielectric thin films [2.330] on its surface – has to be replaced by a new optical element (see Fig. 2.72), consisting of a combination of the optical deflector body (providing the mirror figure deduced from ray tracing calculations) and a multilayer system of high reflectance (providing effective throughput of the beam intensity).

As one of the features of prime importance of a layered system, its microscopic tolerances must not deteriorate the imaging quality of the deflector; i.e. both layer stack and the completed mirror must conserve the shape (uniformity) and surface (smoothness) quality of the deflector surface with the high precision reasoned in the preceding paragraph. This requires the utmost regularity of the layer stack itself.

---

[4] The X-ray space telescopes of ROSAT mission and others (e.g. CHANDRA, XMM, EXOSAT) [2.329] are based on W*olter's* systems.

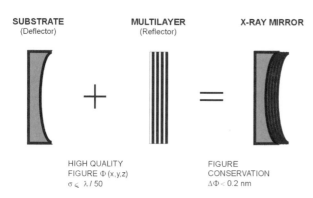

The figure of real mirror surfaces, on the one hand, has to meet the ideal shape of the optical element within tolerances of typically 0.1 nm to reduce wave front distortions in the ray path to an acceptable level. On the other hand, the roughness of these surfaces has been reduced by super-polishing techniques to such an extent that an optimum specular reflectance R of the imaging rays is achieved. It follows from Eq. (2.25) that roughness levels of typically 0.1 nm must be obtained if a loss by non-specular reflection is to be kept within a few percent.

The principles of *Bragg* reflection known from X-ray diffraction experiments on natural crystals, and thus the "constructive" interference of a large number of partial waves reflected from millions of lattice planes, offer a realistic chance to generate a substantially higher reflectance than a mirror's single surface can deliver. The lattice planes of a crystal are formed by the regular arrangement of its atoms in the respective lattice type. Thus a 3D space is filled by a sequence of alternating absorbing and non-absorbing spatial elements. The absorber ranges of this "Bragg lattice" are formed by the atoms with their electron density distributions, whereas the vacuum of the interatomic spacing acts as an ideal non-absorber. In X-ray optical multilayers a 1D approximation to this structure is synthesized by alternating deposition of many individual nanometer films (of two different materials) on top of each other. In principle, X-rays incident on such structures show similar interference pattern as those from natural crystals (Fig. 2.73).

Specular reflection from the lattice for a particular wavelength is obtained when lattice spacing and glancing angle meet the well-known *Bragg* equation which also treats refraction within the stack:

$$m \times \lambda_x = 2\Lambda \times \sin\Theta \times \left(1 - \frac{\delta - \delta^2}{\sin^2\Theta}\right) \tag{2.24}$$

where: m is the spectral diffraction order, λ_x the wavelength of the X-rays, Λ the period thickness and Θ the glancing angle (complement of angle of incidence).

From Eq. (2.24) it is easily seen that the maximum wavelength $\lambda_{\max}$ which can be diffracted by a particular crystal is determined by a value of typically two times its appropriate lattice spacing. Therefore, a series of different crystals or diffracting elements is necessary to cover an extended spectral range of interest.

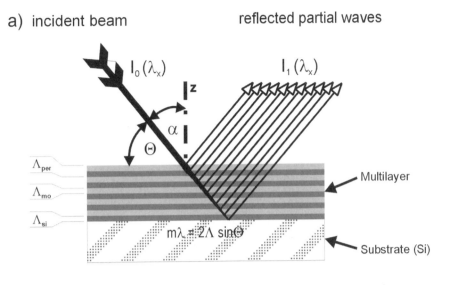

a) incident beam   reflected partial waves

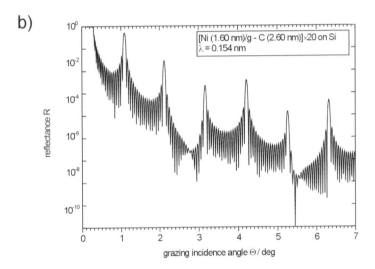

Figure 2.73: Basic principle of Bragg reflection. (a) Scheme of multilayer/beam interaction; (b) X-ray reflectograph for CuKα radiation

In the past, particular solutions for X-ray spectrometers and monochromators in the short wavelength regime have been developed. Fully- or semi-focusing systems have been designed comprising curved crystals (e.g. *Johansson*-type spectrometer [2.331]) and are applied in X-ray reflectometry (XRR), X-ray diffractometry (XRD), X-ray fluorescence analysis (XRFA) and electron probe micro analysis (EPMA). From Fig. 2.74 we can deduce that for chemical analyses by EPMA or XRFA various LiF and SiO_2 crystals of different orientation ($d \approx 0.08$ through 0.35 nm) will allow the determination of all elements of the periodic table with $Z \geq 14$. For the light element range there are no useful natural inorganic crystals having d-spacings > 0.6 nm. Therefore, crystallized compounds of organic acids (e.g. phthalates like TAP or KAP, $d \approx 1.3$ nm) have to be applied for element determination in the range $13 > Z > 8$.

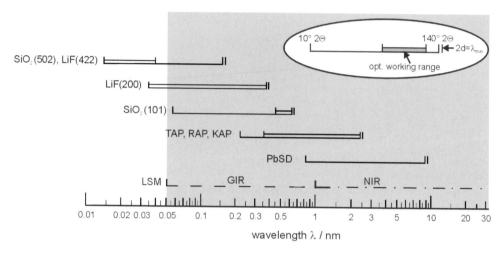

Figure 2.74: Working ranges of diffractive X-ray optical elements in EPMA applications (after [2.332])

The investigation of a few, but very important light elements in materials science like O, N, C, B and Be needs components with still larger lattice spacings to realize *Bragg* reflection under appropriate measuring conditions.

The forerunners of LSM are the so-called *Langmuir Blodgett* films (LB films), i.e. artificial organic multilayers [2.333] (made from the lead soaps of e.g. lauric, stearic – "PbSD" – , or melissic acid) that have interplanar spacings $d \approx 3.5$, 5 or 8 nm, respectively (determined by the appropriate length of the organic molecules) and will allow one to detect also these light elements.

Unfortunately, as for organic crystals, their spectral reflectance is rather poor (Fig. 2.75). Since the absorber layers of the stack are formed merely by one or two atomic layers of Pb localized at the end of the molecular chains of these compounds, a rather insufficient absorber/spacer ratio is established for the selected d-spacing.

In contrast to the multilayers prepared by physical coating techniques, where this ratio can be chosen in a useful working range, for the various lead soap compounds the d-spacing can be produced only for discrete values due to their fixed molecular chain lengths. Furthermore, the structural stability is very sensitive to mechanical and thermal load, since all these exotic acids have a melting point below 100 °C.

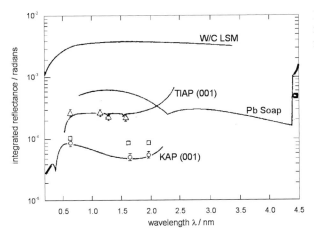

Figure 2.75: Comparison of spectral reflectance for various types of Bragg reflectors (after [2.334])

Artificial Multilayers with Arbitrarily Adjustable *d*-Spacings

For $\lambda_{max} > 2.5$ nm neither natural crystals with adequate *d*-spacings nor LB films with sufficient reflectance are available. Therefore high precision coating techniques are required to synthesize multilayer stacks on super-polished substrates and to tailor their stack structure for optimum peak reflectance of the 1st order Bragg reflex, desired attenuation of higher order reflexes, and working range by an appropriate choice of layer materials and layer thicknesses.

Structure of the Ideal Layered Synthetic Microstructure (LSM). A periodic layer stack that comprises a larger number of alternating thin solid films of a high regularity from two chemical elements or compounds is called LSM. The period of this structure is the bi-layer thickness Λ formed by two adjacent films of distinguished features. If the alternating films have different optical properties, e.g. refractive indices associated with high and low absorption indices β, a period is formed by an absorber and a spacer layer, respectively. The stack may then be used to deflect electromagnetic radiation having wavelengths of the order of the individual layer thickness for normal incidence reflection and even shorter wavelengths for grazing incidence reflection.

This one-dimensional analog of a natural crystal is penetrated by the incident X-rays within a thickness of N periods. It permits constructive interference under *Bragg* conditions by coherent superposition of the partial waves being reflected at the interfaces of the stack and having equal phase (Eq. (2.24)). The efficiency of the reflection at an ideal stack relative to that of a single-surface reflection will increase proportional to N^2.

The reflectance R of a real LSM is influenced by various factors including

- stack regularity,
- period design,
- choice of layer materials (optical constants),
- degree of their mutual physical and chemical interaction during deposition,
- (layer growth, interface formation), storage and thermal or radiant,
- exposure in use (diffusion, reaction).

Therefore, deviations ($\Delta\Lambda$) of the real stack structure from the ideal layer system ($\Delta\Lambda = 0$) have to be kept small in comparison to the wavelength λ_X of the incident beam. A substantial reflection of X-rays can be obtained with nanometer layer stacks provided that they have a very good lateral uniformity and reproducibility of the layer thickness throughout the total stack established during the coating process. In addition, the reflecting interfaces must exhibit a well defined and steep gradient of the electron density (equivalents are: mass density, atomic number) with minimum deterioration due to interface roughness or diffusivity.

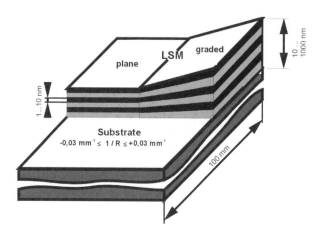

Figure 2.76: Scheme of an idealized multilayer mirror and typical tolerances to be achieved for real optical elements

The schematic representation of a typical X-ray mirror is given in Fig. 2.76. A super-polished plane substrate of dm^2 dimension is taken as a prototype for the various configurations of optical elements (plane or curved surfaces of different kinds). Its ultimate surface smoothness (roughness typically $\sigma \approx 0.1 - 0.2$ nm) is the requirement for a vanishing contribution to the final roughness of the entire mirror. For applications in imaging systems also the utmost planarity (for a high quality a figure of $\Delta z \leq 0.1$ nm) should result in minimized wave front distortions. The structure of the multilayer coating is represented by a three period layer stack, in which light films, or spacer layers, mark low density materials (e.g. C, Si, B$_4$C) and dark films, or absorber layers, stand for high density materials (e.g. Ni, Mo, W).

Typical layer thicknesses Λ_a and Λ_s or period thicknesses Λ ($\Lambda = \Lambda_a + \Lambda_s$, combination of one spacer and one absorber layer) take values in the range 1 to 10 nm. For period numbers N between 20 and 500 a total thickness of the coating within a range of $D_{tot} = N \times \Lambda \approx 10 - 1000$ nm will be obtained. The tolerances to be met by the coating techniques are so stringent that neither figure nor surface quality of the completed mirror should be substantially deteriorated in comparison to those of the as-received substrate.

Thus, typical numbers that can be taken as prerequisites of an acceptable optical element or system are compiled in Table 2.11.

Table 2.11: Critical tolerances of X-ray multilayer mirrors

Substrate quality	
• Figure preservation	$\Delta r \leq 0.1$ nm
• Surface roughness	$\sigma \leq 0.2$ nm
Multilayer quality	
• Interface roughness	$\sigma \leq 0.2$ nm
• Reproducibility of layer thickness on stack	$\Delta\Lambda(z) \leq 0.01$ nm
• Lateral uniformity in optical cross section	$\Delta\Lambda(x,y) \leq 0.1$ %
• Reproducibility run-to-run	$\Delta\lambda_X \leq 0.01$ nm

Non-planar layer stacks (e.g. graded LSM, see right hand side of Fig. 2.76) (i.e. for a correction of the glancing angle or λ_X shift) are employed for the compensation of imaging errors or the enlargement of the acceptance angle of an optical element or system. Therefore, graded systems must be prepared having 1D or 2D period gradients over the entire mirror cross section with their lower limit error typically around $(\Lambda_1 - \Lambda_2)/(x_1 - x_2) \approx 5 \times 10^{-9}$.

Since even the spacer layers of low density materials will always show a finite absorption, the reflectance of the multilayers remains distinctly below $R = 1$. However, with a careful selection of suitable material combinations, X-ray mirrors can be prepared that have a substantial reflectance either in the GI region for sub-nm wavelengths [2.335–2.337] or in the NI region, where typically $\lambda_X > 1$ nm is used [2.338, 2.339]. The highest reflectances achieved until now are about 90 % for Ni/C *Goebel* mirrors with a glancing incidence ray path, and 70 % for Mo/Be as well as for Mo/Si mirrors at nearly normal beam incidence.

For GI operation of multilayer mirrors the loss of reflectance compared to $\Theta \leq \Theta_c$ geometries is balanced by a series of advantages. As the intensity of the first order Bragg peak appears at an angle Θ substantially larger than Θ_c, an increased numerical aperture, an enhanced collection solid angle, and reduced aberrations are achieved. For mirrors operated at near normal incidence ($\Theta \rightarrow 0$) of the X-rays, from Eq. (2.24) follows for constructive interference a multilayer period width of approximately $\Lambda \approx \lambda_X/2$. Consequently, for high quality reflector stacks in an optimum range of working wavelengths, sharp interfaces with $\sigma \leq 0.1\Lambda$ must be provided to approach maximum reflectance. On the other hand, a lower limit for the NI working wavelength of $\lambda_X > 1$ nm results from the atomic nature of matter. The smallest imaginable Λ_a or Λ_s values are given by an atomic monolayer.

In practice, both the growth of uniform thin solid films and the influence of the layer morphology on the resulting reflectance R are responsible for a substantial increase in this wavelength boundary. First attempts for the Ni/C system with the rather small $\Lambda_{Ni}/\Lambda \approx 0.026$ showed a good regularity of the stack, but it was confirmed that Ni does not form continuous layers [2.340]. For C on Mo continuous barrier layers of $d = 0.2$ nm in Mo/Si stacks have been previously obtained by magnetron sputter deposition [2.339]. *Windt, D.L.* reported about W/B$_4$C multilayers optimized in the range $1.4 < \lambda_X < 2.4$ nm, where average period thicknesses between 0.8 and 1.19 nm could be realized and peak reflectance ranges from $R \approx 0.08$ % at $\lambda_X = 1.4$ nm to 1.3 % at $\lambda_X = 2.4$ nm. For such structures an interface width $\sigma \approx 0.29$ nm is determined from NI reflectometry [2.341]. Other thin period multilayers of C/C [2.342] and a variety of material combinations (Fe/Sc, Cr/Sc, W/Si, W/Sb, W/B$_4$C) showed for typical Λ-spacings of 1.3–2.4 nm reflectances $R \approx 1$–2% at $\lambda_X \approx 1$–4.5 nm

[2.343]. W/Si [2.344] layer combinations were characterized by CuKα radiation to $\Lambda = 1.11$ nm and $\Lambda = 1.5$ nm, respectively.

From purely geometric considerations a value of $\lambda_X(min) \approx 2–3$ nm (water window range, see Fig. 2.69) may be assumed. Even for this limit, the influence of layer morphology will cause an appropriate decrease in reflectance, unless a highly stable coating procedure is available to prepare stacks of some hundreds of periods with utmost thickness precision.

Particular Interface Properties

The 0.1 nm order of magnitude tolerances of the entire X-ray mirror are critical not only for the general geometry of the individual component and the regularity of its multilayer design but also for the realization of a period design deduced from the idealized theoretical layer combination supporting an optimized constructive interference of the reflected partial waves. The interaction process (reflection, penetration) of the X-rays with the layer surface or interface is unambiguously characterized by the electron density or the chemical profiles at the boundary between absorber and spacer. This transition region is formed by a superposition of the intrinsic boundary roughness giving a heterogeneous interlocking of adjacent layers on an atomic through µm scale, and a near-surface region of continuous interdiffusion and reaction of both major chemical components of the period. Both contributions can cause an equivalent reflectance deterioration, either by generating a substantial amount of non-specularly reflected photons, or by weakening the desired sharp electron density gradient, thus forming an additional sub-layer of the period which comprises a material composition with unfavorable optical constants.

The influence of this interface layer of width Σ, caused by roughness *and* diffusivity, on the specular reflectance R_0 of an infinitely sharp interface is taken into account by an analog of the *Debye–Waller factor* (DW).

Originally this was used in X-ray physics to include the effects of thermally stimulated oscillations of lattice atoms. One obtains:

$$\frac{R}{R_0} = \exp\left\{-\left(\frac{2\pi m \Sigma}{\Lambda}\right)^2\right\} \tag{2.25}$$

It is easily seen that decreasing the period width Λ or increasing Σ and the spectral order m will have a tremendous effect on the optical output of a reflector. Assuming an interface width of $\Sigma \approx 0.1 – 0.3$ nm for Λ in the range of 5 to 1 nm, the resulting loss of reflected energy amounts to $\Delta R(0.1$ nm$) = 1 – R/R_0 \approx 1.6 – 32.5$ % and $\Delta R(0.3$ nm$) \approx 13.2 – 97$ %. These considerations demonstrate that the preparation of sharp and narrow interfaces is an aim of vital importance for high-quality X-ray mirror production.

Design of Layered Synthetic Microstructures

The reflectance R of multilayer mirrors is calculated by a recursive application of the *Fresnel* equations (for a detailed explanation of the calculational model see e.g. [2.345]) to determine the amplitude reflection coefficient r of the s- or p-component of the light for each boundary. Starting with a single layer model the reflected and transmitted amplitudes (r and t, respectively) are determined and a further layer is added on top of this structure. This new system is calculated again by inserting the former results r and t as the new amplitudes of the bottom boundary. This procedure is continued step by step until the top of the structure is approached. The reflected amplitude coefficient of the appropriate interim solution f is obtained from:

$$r_f = \frac{r_t + r_b \times \exp(2i\varphi)}{1 + r_t r_b \times \exp(2i\varphi)} \tag{2.26}$$

where subscripts t and b indicate "top" and "bottom" parts of the layered system and φ is the phase shift caused by wave propagation through a film of thickness d. It is obtained as a function of the angle of incidence α by the formula $\varphi = 2\pi \tilde{n} d \times \cos\alpha / \lambda$.

For *Bragg* conditions the optimum reflectance of a mirror is achieved when λ_X experiences a phase shift of 2π during a propagation of period thickness Λ.

The reflectance of a multilayer mirror is not only determined by the factors discussed in the preceding section. Major properties of their "substructure" will greatly influence the optical quality characterized by reflectance, spectral resolution, and higher order suppression.

Compared to visible light optics, where quarter wave stacks of dielectric materials give optimum reflectance due to n appreciably > 1 and $\beta \approx 0$ for both multilayer components and the optimum phase shift $\varphi = 2\pi$, there are no similar materials available for the X-ray regions. Here all materials absorb energy from the X-rays while they pass through the layers. Thus, for the standing wave model [2.345] a compromise must be found to account for these absorption losses. As there is no spacer material with $\beta \approx 0$ available, an approximation to the ideal *Bragg* crystal with its very thin absorber layers and high number of periods N cannot be realized in practice.

For real X-ray multilayer stacks for the desired λ_X a spacer material with minimum absorption index and an absorber material with highest reflection coefficient (and for that matter with smallest possible absorption index) should be selected. This choice would allow a design closest to a quarter wave stack. The symmetric stacking has to be dropped for an optimum reflectance of the X-ray mirrors in the first order *Bragg* reflex. This is achieved by a reduction of the absorber thickness Λ_a resulting in practice in a relative absorber thickness $\Gamma = \Lambda_a/\Lambda \approx 0.4$. A general relation for Γ_{opt} has been worked out [2.30] as:

$$\tan(\pi\Gamma_{opt}) = \pi[\Gamma_{opt} + \frac{\beta_L}{\beta_H - \beta_L}] \tag{2.27}$$

For practical multilayer mirrors it turns out that the relative absorber thickness $\Gamma = \Lambda_a / \Lambda$ must be kept in the range $0.3 \leq \Gamma \leq 0.5$ in order to achieve the highest reflectivity.

A further optimization of the stack can be obtained by the determination of the period number N that supports the reflection process. The minimum number N_{min} that is necessary to approach $R \Rightarrow 1$ is given by:

$$N_{min} = \frac{1}{2|r|} \tag{2.28}$$

This quantity, however, is only of value when the stack absorption is so low that for a particular angle of incidence the entire stack is penetrated and thus reflection of the incoming radiation is obtained also from the period closest to the substrate.

The maximum number of periods N_{max} taking part in a constructive interference is determined by the amount of absorption in the multilayer and is given by:

$$N_{max} = \frac{\cos^2 \alpha}{2\pi\beta_s} \qquad (2.29)$$

This quantity gives the optimum λ_X-resolution ($\lambda / \Delta\lambda = m \times N_{max}$) of the mirror obtainable for a selected spectral order m. Period numbers larger than N_{max} cannot give a further advantage, neither for an additional reflectivity increase nor from the point of view of surface quality. From each additional period, due to columnar growth, an increase in the total surface roughness has to be expected (see also p. 145 in [2.345]).

Order Selection by a Particular Ratio Λ_a / Λ_s

The application of multilayer mirrors, e.g., in XRFA or EPMA can offer two advantages: First, a substantial signal increase is achieved, as already demonstrated in Fig. 2.75, by the enhanced reflectivity of the LSM compared to organic crystals (e.g. TAP, RAP) and soap multilayers (e.g. PbSD). Secondly, a proper selection of Γ, even with non-optimized R, allows the complete extinction of particular interference orders, thus yielding an elimination of coincidence of spectral lines from different elements and being reflected in different diffraction orders.

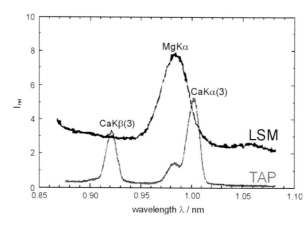

Figure 2.77: Application of a LSM in chemical analysis: determination of Mg contents in cement. Enhanced detection limits and blind value reduction are obtained by a tailored period structure of the multilayer ($\Gamma = 1/3$), after [2.346]

The particular ratio of Γ causes a decrease in R almost to zero as explained by Eq. (2.30):

$$R_m = \exp\left\{-\left(\frac{4\pi^2\sigma^2 m^2}{\Lambda^2}\right)\right\} \times th^2 A_m \qquad (2.30)$$

where $A_m \sim \dfrac{\Lambda^2}{m^2(1 - \cos 2\pi m\Gamma)^{1/2}}$

R_m is the reflectivity of the mth diffraction order peak, σ the interface roughness, Λ the period width and Γ the relative absorber thickness Λ_a / Λ.

In Fig. 2.77 a typical example of order extinction is illustrated for the determination of Mg in a cement matrix. Using TAP as the dispersive element the interference of a tiny 1st order MgKα and an intense 3rd order CaKα line is obtained, whereas the same spectral region observed in LSM dispersion, with $\Gamma = 1/3$, does not reveal any 3rd order interference, but an enhanced MgKα line.

2.5.4 Multilayer Preparation

As already explained, the quality of the reflector stack should be similar to that of the deflector surface, when powerful X-ray optical components are fabricated. High precision and uniformity, long term stability and the utmost run-to-run reproducibility of the coating process are essential prerequisites for the vacuum deposition technologies to be used for the synthesis of multilayer interference coatings. The critical layer parameters (layer thickness, thickness error) to be achieved are typically one or two orders of magnitude smaller and the number of layers needed per stack are at least a factor of 10 larger than conventional optical coatings. Besides the appropriate choice and purity of the layer materials for optimum optical constants, morphological and structural parameters of the layer stack (regularity and interface sharpness) will finally determine the practical output of the mirror.

Preferred technologies for multilayer synthesis employed up to now are:

- Magnetron sputter deposition (MSD)
- Ion beam sputter deposition (IBSD)
- Electron beam vapor deposition (EBVD)
- Pulsed laser deposition (PLD)

The basic principles of these techniques have already been described in Section 3.1. Some additional remarks will be given from the point of view of their application to multilayer synthesis and its particular demands of uniformity, precision and reproducibility.

In order to precisely control the average coating thickness in MSD and PLD, a careful calibration of the effective coating time taking into account the stabilization of the plasma properties is developed to a very high standard. Thus, thickness errors well below 0.01 nm for a coating uniformity better than 99.9 % have routinely been achieved. For EBVD, since an appropriate stabilization as for plasma sources can not yet be obtained, in-situ monitoring of the X-ray reflectance of the growing multilayers is needed to avoid disturbing thickness errors in the period width.

Typical coating devices fulfilling the stringent demands of X-ray multilayer preparation are described in the specialist literature [2.347–2.349]. Figure 2.78 shows schematically typical equipment for MSD (a) and PLD (b) for X-ray coatings. Blank deflector substrates are placed face down over the appropriate plasma particle sources. The sputter technology is characterized by separate magnetrons for each of the layer materials and the substrate is rotated on a turntable, passing above the discharges with predetermined rotation speeds, thus producing one layer period for each revolution. To improve the coating uniformity an additional substrate spin with a higher rotation frequency is applied.

The laser coating technology combines a linear substrate motion across the region of the plasma plume with an oscillation of the plume in a perpendicular direction. The change in the spatial orientation of the plume axis is induced by scanning the laser beam waist across the cylindrical target surface. The layer material atomized by the laser ablation from the appropriate solid state targets produces an individual layer for each substrate run. For alternating deposition of the two period materials the targets are synchronously changed after each run by a revolving mechanism.

a)   b)

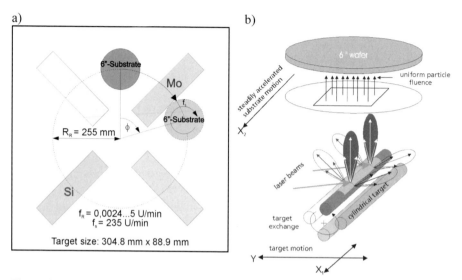

Figure 2.78: Representation of the principles of multilayer coating devices. (a) magnetron sputter deposition (MSD), (b) pulsed laser deposition (PLD)

Besides process stability particular efforts are directed to the synthesis of an optimum stack structure involving an optimum material composition and a high stack regularity with optimum morphology of the layers and their interfaces. The real multilayer, in order to provide an efficient constructive interference, has to approach its ideal model as closely as possible from the points of view of optical constants, identical period width throughout the entire stack, and interface layers as thin and as smooth as possible. The final substructure of the layered periods (absorber, spacer and interface layers) is determined by the growth regime that proceeds during particle condensation. The intrinsic energy distribution of the mixture of condensing atoms and ions plays a crucial role for the achieveable layer quality.

Structural zone models (SZM) have been developed for various coating techniques to explain the dependence of the resulting layer structures on the parameters of the deposition process [2.350, 2.351]. It was found that for conventional low energy coating regimes, where low temperature substrate conditions and low energy particle energies prevail (i.e. normalized substrate temperature $T_S/T_M \ll 1$, where T_S is the actual substrate temperature, T_M the melting point of the bulk coating material) columnar films containing appropriate void concentrations are grown. Multilayers prepared by EBVD thus show a substantial roughness of their interfaces, which accumulates towards the mirror surface with increasing period number. To obtain stacks of high X-ray optical quality by this technique an additional process step, ion polishing,has to be applied after deposition of each layer to achieve the desired interface smoothness [2.352]).

"... formation of a TRANSIENT LIQUID MONOLAYER, which gets IMMEDIATELY QUENCHED into the solid state, because the actual substrate temperature ... is cold with respect to the melting temperature. The quenching preserves the AMORPHOUS STRUCTURE of the liquid solidified film."

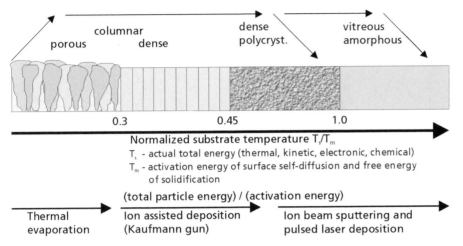

Figure 2.79: Extended structure zone model comprising additional zone for amorphous state of thin solid films

Latest improvements of the SZM [2.353] take into account the far from thermal equilibrium conditions of the ion-assisted coating technologies. These peculiar conditions during layer growth are caused by the high energy tail of the energy distribution of the plasma emitted particles. The conventionally considered region of the normalized substrate temperature is therefore supplemented by a fourth zone beyond $T_S/T_M = 1$ (Fig. 2.79). There the state variable T, defined for Maxwellian velocity distributions. is replaced by an equivalent of the total particle energy/activation energy of surface self-diffusion. The conditions in zone 4, during layer growth, involve the formation of transient liquid monolayers getting rapidly quenched into the solid state due to an actual $T_S \leq 400$ K. Quenching is accompanied by a conservation of the vitreous-amorphous structure with a rather smooth surface of the solidified film. A typical energy distribution supporting this "Guenther-model" is found in PLD (Fig. 2.80). This technology, providing a high degree of ionization in the emitted plasma particle flux, creates surfaces and interfaces with lowest roughnesses (see e.g. [2.349]), thus delivering one prerequisite for well organized and smooth interfaces. The high energy tail, however, causes a substantial inter-diffusion of the condensing film material which results in an increase in the interface width. Following Eq. (2.25) this will also result in a deterioration of the X-ray optical properties of the multilayer. As roughness zone and intermixing zone give a similar contribution to the "Debye–Waller" factor a reduced reflectance is obtained in any case. In order to achieve optimum structures having a minimum DW, a compromise for the amounts of roughness and diffusivity contributing to the entire interface width must be aimed at.

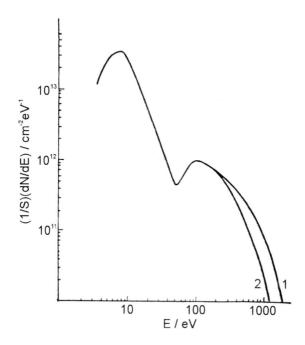

Figure 2.80: Kinetic energy distributions of particles emitted from a PbTe-pulsed laser plasma. Note the different high energy tails due to a power density change:
(1) – $dq/dt = 10^9$ W cm^{-2},
(2) – $dq/dt = 4 \times 10^8$ W cm^2
(after [2.354])

In practice this is realized for MSD coatings by reducing the discharge pressure of the magnetrons to a level as low as compatible with a stable discharge regime (e.g. $p_{Ar} \approx 1$ mbar) thus achieving a reduced roughness contribution at moderate intermixing. For PLD a decrease of the laser power density in the beam waist and/or a change to a shorter laser wavelength results in a decreased mean particle energy accompanied by reduced intermixing. The conservation of a good interface smoothness thus delivers also a minimum width for the interface region.

2.6   Metallic Layers for Photovoltaics

2.6.1   Introduction

Solar energy plays a central role in creating a sustainable future and developing renewable energy sources. Many experts believe it will be the main base of our global energy supply if fossil sources became exhausted. Conventional thermal solar collectors, solar thermal power plants and photovoltaics (PV) are the three possibilities to open up solar energy for human usage.

The photovoltaics industry and the environment in which it operates are developing at exceptional speed. In 2003 global solar cell production expanded by 34 %. From 2003 to 2004 we observed a 67 % growth. This trend will flatten in the next few years (Fig. 2.81) before rising again.

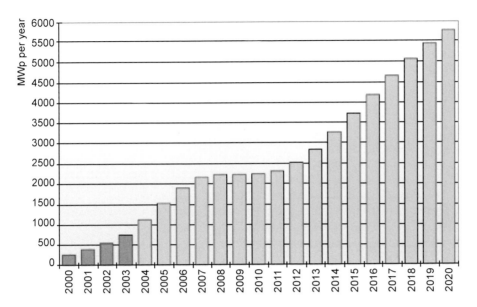

Figure 2.81: Sarasin forecast for global solar cell production in MW up to 2020 [2.355]

The reasons for the forecasted temporary growth pause are as follows [2.355]:
1. Temporary bottlenecks regarding solar-grade silicon over the next few years.
2. Run out respectively shortening of subsidy programs in Japan (2006) and probably in Germany.
3. Limited cost reduction potential of wafer silicon based PV:
   In principle there are two ways for cutting costs in PV cell and module production. In crystalline silicon technology, the focus here is on the economy of scale effects provided by mass production, larger and thinner wafers and improved efficiency. Thereby the price per module has dropped by around 40% in the past five years. However, many experts doubt that this is a feasible way of braking the 1-EUR/Wp "sonic wall". If the material costs and the cell production costs are reduced systematically, the lower cost limit is given by the modules, which must be compact and stable to protect the brittle wafers.

The second strategy that the manufacturers favor, is to research innovative materials and technologies and test them in pilot production. Thin-film technologies – one option is thin-film silicon - will gain more momentum as solar-grade silicon becomes more expensive and in short supply. This requires intensive research and is risky. The breakthrough of a profitable mass production will not be achieved before 2012, in the opinion of the experts.

2.6.2   Solar Cells

The physics of photovoltaic cells is described in detail, e.g., in the book of Würfel [2.356]. In Fig. 2.82 the principle structure of a photovoltaic cell is shown.

The main components are

- the absorber made from one or more semiconductor materials or from a metal-organic dye,
- the transparent front contact,
- the rear contact.

The front window of crystalline silicon wafer based cells is an antireflective passive-coating from amorphous a-SiN:H and a "fired-through" Al contact grid made by thick film technology. The rear side is built out of passive oxide and a "fired through" Al contact grid. The aluminum reacts with the silicon and generates a heavily Al doped p^+ zone.

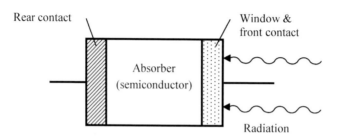

Rear contact   Window & front contact

Absorber (semiconductor)

Radiation

Figure 2.82: Main components of solar cells

Thin film photovoltaic cells – this is the long-term future of PV as we have seen – are more sophisticated in their structure compared to silicon wafer cells. Specific thin film technologies – evaporation, sputtering, electrodeposition etc. play a significant roll and metallic layers are important components.

In most cases the front window is from a transparent conductive oxide (TCO) - indium tin oxide (ITO = SnO_2:In), fluorine doped tin oxide (SnO_2:F) or aluminum doped zinc oxide (ZnO:Al).

Metallic films are used on the rear side. Besides their primary function, to form an electrical contact, these films have to fulfil in some cases several secondary functions:

- back reflection of the transmitted part of the light ("Light trapping"),
- source of dopants,
- barrier coating as well as
- solderability.

2.6.3   Functionalities of Thin Metallic Films in PV Cells

Metal-Semiconductor Contact

The work function of a metal Φ_m or semiconductor Φ_s is the energy required to raise an electron from the Fermi level E_F to a state outside the surface at infinity (Fig. 2.83). The work function is equal to the electrochemical potential η of the electrons. In general, the work function of a given metal and the semiconductor are different, caused by the different electronegativity of both materials. Metals, particularly alkali metals, have low work functions.

Assume we have an n-type semiconductor with gap energy ε_g and a metal with a high work function Φ_m (Fig. 2.83 a). After connecting the materials by a wire, the Fermi levels are forced to coincide (Fig. 2.83 b). Electrons pass from the semiconductor to the metal, generating a step in the potential energy of height

$$\Delta_\varphi = \Phi_m - \Phi_s \tag{2.31}$$

In the metal, the extra charge is screened within the Fermi screening length of ~ 0.5 Å, but the missing electrons in the semiconductor leave the uncompensated donors behind them. Because of the low density of the donors, a charged depletion layer of finite width w will be formed. Integrating the Poisson equation we get an extra part of the potential energy, which superimposes onto the energy levels of the valence band as well as of the conduction band: The bands will be bent (Fig. 2.83 b, c).

In the depletion layer the electrons will feel a potential barrier of width w and height

$$\Phi_b = \Phi_m - \chi_s \tag{2.32}$$

which is the distance between the Fermi level of the metal and the bottom of the conduction band immediately behind the interface. χ_s is the electron affinity of the semiconductor. This barrier is called a Schottky barrier.

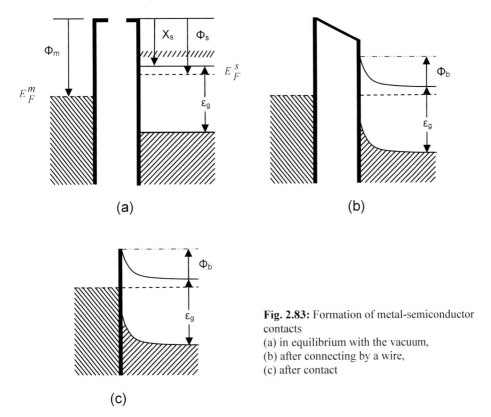

Fig. 2.83: Formation of metal-semiconductor contacts
(a) in equilibrium with the vacuum,
(b) after connecting by a wire,
(c) after contact

If a voltage V is applied to the metal-semiconductor contact, caused by the barrier, we have a low rate of charge exchange across the interface. This results in different thermodynamic equilibria and different Fermi energies at both sides of the interface. The voltage V drops within the very thin interface. At positive bias of the semiconductor the barrier may hardly be tunneled by electrons from the metal side and emitted into the conduction band of the (n-type) semiconductor. On the other hand, if the bias is negative, the Fermi energy of the semiconductor will be raised relative to the metal and to the barrier at the interface. The probability for thermionic emission over the top of the barrier into the metal increases exponentially and a measurable current will flow.

As a result, the Schottky barrier works as a rectifier.

Schottky barriers will form if we contact an n-type semiconductor with a high work function metal or a p-type semiconductor with a low work function metal (Fig. 2.84). If we combine an n-type semiconductor with a low work function metal or a p-type semiconductor with a high work function metal we get good conducting ohmic contacts.

More details regarding metal-semiconductor contacts can be found in [2.357] and [2.358].

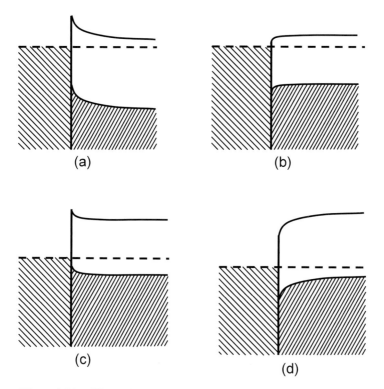

Figure 2.84: Different forms of metal-semiconductor contacts.
(a) Metal with high work function – n-type semiconductor (Schottky)
(b) Metal with low work function – n-type semiconductor (ohmic)
(c) Metal with high work function – p-type semiconductor (ohmic)
(d) Metal with low work function – p-type semiconductor (Schottky)

Rectification of the current is necessary in photovoltaic cells to force the photocurrent into a preferred direction. In principle, a cell can be built by a homogeneously doped semiconductor with two contacts, one ohmic and one rectifying Schottky type. However, according to the voltage drop across the interface a Schottky barrier generates power loss (heat) in the pass direction. This would mean an efficiency loss of the cell. Consequently, all metals in contact with the semiconducting absorber should be chosen in this way that ohmic contacts are formed. The rectification of the photocurrent will be carried out by p-n transitions within the semiconductor's structure itself.

In most practical cases Eq. (2.32) does not hold. Caused by many effects we have a high density of electronic interface states on the semiconductor's side of the interface, which may easily absorb extra electrons or holes. The Fermi level of the semiconductor will be "pinned" to the neutral level of the interface states. Therefore, in designing a photovoltaic cell the selection of adequate metals for the contacts is made much more empirically than starting from first principles.

A disadvantage of interface states is the high recombination rate if both electrons and holes are present. To avoid this e.g. p-type silicon will be heavily p^+ doped near the contact to minimize the electron concentration at the interface. A second way is to passivate the semiconductor by a thin oxide interlayer; which may easily be tunneled by electrons (<u>M</u>etal-<u>I</u>solator-<u>S</u>ilicon / MIS structure [2.356]).

Light Trapping

At present silicon is the most used but not the ideal material for photovoltaic cells because it is an indirect semiconductor. When exciting an electron by a photon from the top of the valence band to the bottom of the conduction band momentum is not conserved and an additional phonon is needed. Consequently, the absorption coefficient of silicon is relatively low. We need thick wafers to absorb all incoming photons and this in turn means low material efficiency.

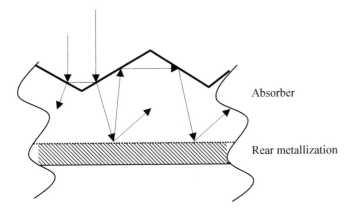

Absorber

Rear metallization

Figure 2.85: Light trapping in thin film cells.

How can we avoid this disadvantage in our effort to utilize thinner wafers with thickness $\leq$ 100 μm or thin-film silicon of a few micrometers?

The solution is "light trapping". Two measures are combined for this purpose:
- Rear metallization with a highly reflecting metal like silver or aluminum to reflect back the transmitted photons; a fine roughness makes for diffuse reflection,
- Rough front-side structure of the absorber; most suitable are etched or otherwise- made pyramidal structure are used (Fig. 2.85).

Figure 2.85 illustrates that by combination of refraction, backreflection, total reflection, etc. the possible photon path length could be very large. Calculations have shown that the mean path length without absorption would be increased by a factor of 50, large enough for the absorption of most of the photons, also in the case of some micrometer film thickness [2.356].

Unfortunately, the metals with the highest reflectivity are not the best ones from the electrical point of view (ohmic contact and low recombination rate) and vice versa. This may be solved by coating a highly reflecting metal (silver) with a transparent conductive oxide, e.g. ZnO:Al [2.359].

Reactivity

Whether low or high reactivity is demanded depends on the design:
- High reactivity is demanded if a special reaction layer is part of the design, e.g. a silicide, or if the material may serve as a dopants source, e.g. aluminum as acceptor source for in silicon.
  In some industrial processes compound semiconductors like, e.g., copper indium diselenide (CIS) are made by sequential sputtering the components onto a substrate and a following solid-state reaction at high temperatures [2.360].
- If low reactivity gives the best performance: molybdenum for instance is the best choice for CIS or CdTe type solar cells.

Solderability

To ensure a good solderability in some designs a top coat of a NiV-alloy is used.

Table 2.12: Typical cell structures.
gl: glass substrate, PET: plastic PET substrate, pr: printed (thick film technology); MeO-TPD: N,N,N',N''-Tetrakis(4-methoxyphenyl)benzidine, ZnPc: Zn-Phtalocyanin

Cell type	Rear contact	Absorber	Front contact	Ref.
c-Si-wafer	pr-Al/SiO$_2$	c-Si	a-SiN:H/pr-Al	[2.356]
CIGS	gl/Mo	Cu(In,Ga)Se$_2$	CdS/i-ZnO/ZnO:Al/gl	[2.361]
CdTe	Mo/Sb$_2$Te$_3$/(Te)	CdTe	CdS/i-SnO$_2$/SnO$_2$:In	[2.361]
a-Si (p-i-n)	NiV/Ag(100 nm)/ ZnO:Al(100 nm)	a-SiGe:H/a-Si:H/μc-Si	ZnO:Al (0,5-1.5 μm)/gl	[2.359]
a-Si (n-i-p)	gl/Ag/ ZnO:Al	a-SiGe:H/a-Si:H/μc-Si	ZnO:Al (80 nm)	[2.359]
a-Si Flexcell®	PET/Al/ SnO$_2$:In	a-Si:H	SnO$_2$:In/Ag-TF	[2.362]
Organic (experimental)	gl/ SnO$_2$:In	p-MeOTPD(55 nm)/ ZnPc:C$_{60}$(32 nm)/ C$_{60}$(10 nm)/n-C$_{60}$(30 nm)	Al (50 nm)	[2.363]

2.6.4   Examples

In Table 2.12 some designs are compiled. In thin film cells, two basic designs are in use:
- substrate technology, where the coating starts with a rear contact on the substrate,
- superstrate technology, where one starts with a front contact and window.

In the silicon thin-film cells and in the cited organic cell backreflection is ensured by metal/oxide structures (metal= Al, Ag or Au). At present, the organic cell is only an experimental study.

There are some cell types that are designed without any metals. The fabrication of crystalline silicon thin film cells, e.g., includes a recrystallization step at high temperatures after evaporating the silicon. Ceramic substrates from graphite or silicon nitride are required and metals should be avoided because of their reactivity with silicon at higher temperatures.

2.7   References

[2.1]   E. Zschech, H.-J. Engelmann, H. Stegmann, H. Saage, Q. de Robillard, *Future Fab. Int.* **14**, 127 (2003).

[2.2]   *The International Technology Roadmap for Semiconductors (ITRS)*, 2003 edition.

[2.3]   A. Diebold, R. Goodall, in *Proceedings of the International Interconnect Technology Confeence*, San Francisco, 1999, p. 77.

[2.4]   D. Edelstein, J. Heidenreich, R. Goldblatt, W. Cote, C. Uzoh, N. Lustig, P. Roper, T. McDevitt, W. Motsiff, A. Simon, J. Dukovic, R. Wachnik, H. Rathore, R. Schulz, L. Su, S. Luce, J. Slattery, in *Proc. IEDM*, Piscataway, New Jersey, 1997, p. 773.

[2.5]   P. R. Besser, A. Marathe, L. Zhao, M. Herrick, C. Capasso, H. Kawasaki, in *Proc. IEDM*, San Francisco, 2000, p. 119.

[2.6]   E. T. Ogawa, J. W. McPherson, J. A. Rosal, K. J. Dickerson, T.-C. Chiu, L. Y. Tsung, M. K. Jain, T. D. Bonifield, J. C. Ondrusek, W. R. McKee, in *Proc. IEEE IRPS Conf.*, Dallas, 2002.

[2.7]   S. Murarka, *Solid State Technol.* **83** (1996).

[2.8]   W. Lee and P. Ho, *MRS Bull.* **19** (1997).

[2.9]   W. W. Mullins, *Acta Metall.* **6**, 414 (1958).

[2.10] C. Lingk, M. E. Gross, W. L. Brown, *J. Appl. Phys.* **87**, 2232 (2000).

[2.11] A. Gangulee, *J. Appl. Phys.* **43**, 867 (1972).

[2.12] R. Spolenak, C. A. Volkert, K. Takahashi, S. Fiorillo, J. Miner, W. L. Brown, *Mater. Res. Soc. Proc.* **594**, 63 (2000).

[2.13] E. Zschech, W. Blum, I. Zienert, P. R. Besser, *Z. Metallkd.* **92**, 803 (2001).

[2.14] P. R. Besser, E. Zschech, W. Blum, D. Winter, R. Ortega, S. Rose, M. Herrick, M. Gall, S. Thrasher, M. Tiner, B. Baker, G. Braeckelmann, L. Zhao, C. Simpson, C. Capasso, H. Kawasaki, E. Weitzman, *J. Electr. Mater.* **30**, 320 (2001).

[2.15] C. V. Thompson and R. Carel, *Mater. Sci. Eng.* **B32**, 211 (1995).

[2.16] R. Venkatraman and J. C. Bravman, *J. Mater. Res.* **7**, 2040 (1992).

[2.17] W. D. Nix, *Metall. Trans. A* **20**, 2217 (1989).

[2.18] L. B. Freund, *J. Appl. Mech.* **54**, 553 (1987).

[2.19] C. V. Thompson, *J. Mater. Res.* **8**, 237 (1993).

[2.20] R. Spolenak, N. Tamura, B. Valek, R. S. Celestre, A.A.MacDowell, T.Marieb, H. Fujimoto, W. L. Brown, J. C. Bravman, H. A. Padmore, B. W. Batterman, J. R. Patel, *Phys. Rev. Lett.* **90** (2003).

[2.21]   P. R. Besser, Y. C. Joo, D. Winter, M. Ngo, R. Ortega, *Mater. Res. Soc. Symp. Proc.* **563**, 189 (1999).

[2.22]   Y. Du, P. S. Ho, J. Kasthurirangan, C. Capasso, M. Gall, M. Jawarani, H. Kawasaki, *Mater. Res. Soc. Symp. Proc.*, 563 (1999).

[2.23]   H. Prinz, I. Zienert, J. Rinderknecht, H. Geisler, E. Zschech, P. Besser, in *Stress-Induced Phenomena in Metallization,* ed. P. S. Ho et al, AIP Proc. **741**, 85 (2004).

[2.24]   I. Zienert, in *Materials for Information Technology,* eds. E. Zschech, C. Whelan, T. Miko-lajick, Springer London, in press (2005).

[2.25]   R. P. Vinci, E. M. Zielinski, J. C. Bravman, *Mater. Res. Soc. Symp. Proc.* **308**, 297 (1992).

[2.26]   R. P. Vinci, T. N. Marieb, J. C. Bravman, *Mater. Res. Soc. Symp. Proc.* **338**, 289 (1994).

[2.27]   T. M. Shaw, L. Gignac, X.-H. Liu, R. R. Rosenberg, E. Levine, P. Mclaughlin, P.-C. Wang, S. Greco, G. Biery, in *Proceedings of the 6th International Workshop "Stress Induced Phenomena",* Ithaca, (AIP 612, Melville), 2001, p. 177.

[2.28]   E. Zschech, H. Geisler, I. Zienert, H. Prinz, E. Langer, A. M. Meyer, G. Schneider, in *Proceedings of the Advanced Metallization Conference,* San Diego, 2002, p. 305.

[2.29]   I. Ames, F. M. d'Heurle, R. E. Horstmann, *IBM J. Res. Dev.* **14**, 461 (1970).

[2.30]   J. E. Sanchez E. Arzt, *Scr. Metall. Mater.* **27**, 285 (1992).

[2.31]   D. Weiss, O. Kraft, E. Arzt, *J. Mater. Res.* **17**, 1363 (2002).

[2.32]   T. Takewaki, R. Kaihara, T. Ohmi, in *Technical Digest IEEE.* (1995), p. 253.

[2.33]   K. L. Lee, C. K. Hu, K. N. Tu, *J. Appl. Phys.* **78**, 4428 (1995).

[2.34]   D. Edelstein, C. Uzoh, C. Cabral Jr., P. Dehaven, P. Buchwalter, A. Simon, E. Cooney, S. Malhotra, D. Klaus, H. Rathore, B. Agarwala, D. Nguyen, *Proc. of the International Inter-connect Technology Conference (IITC),* (Piscataway, NJ, 2001) 9.

[2.35]   C. S. Hau-Riege, C. V. Thompson, *Appl. Phys. Lett.* **78**, 3451 (2001).

[2.36]   K. D. Lee, E. T. Ogawa, H. Matsuhashi, P. R. Justison, K. S. Ko, P. S. Ho, V. A. Blaschke, *Appl. Phys. Lett.* **79**, 3236 (2001).

[2.37]   C. K. Hu, L. Gignac, S. G. Malhotra, R. Rosenberg, *Appl. Phys. Lett.* (2001) 904.

[2.38]   E. Zschech, H. Geisler, I. Zienert, H. Prinz, E. Langer, M. A. Meyer, G. Schneider, *Proc. of the Advanced Metallization Conference (AMC),* San Diego (2002) 305.

[2.39]   A. V. Vairagar, A. Krishnamoorthy, K. N. Tu, S. G. Mhaisalkar, A. M. Gusak, M. A. Meyer, E. Zschech, *Appl. Phys. Lett.* **85**, 2502 (2004).

[2.40]   E. Zschech, M. A. Meyer, E. Langer, *Proc. MRS Spring,* **812**, 361 (2004).

[2.41]   E. Zschech, M. A. Meyer, H. Prinz, I. Zienert, M. Grafe, E. Langer, in *Stress-Induced Phenomena in Metal Interconnects,* eds. P. S. Ho et al., AIP Proc. **741**, 196 (2004).

[2.42]   E. Zschech, V. Sukharev, *Proc. Materials of Advanced Metallization (MAM),* Dresden, in press (2005).

[2.43]   E. Zschech, M. A. Meyer, I. Zienert, E. Langer, H. Geisler, A. Preusse, P. Huebler., *Proc. IPFA, Singapore,* submitted (2005).

[2.44]   C. K. Hu, D. Canaperi, S. T. Chen, L. M. Gignac, B. Herbst, S. Kaldor, M. Krishnan, E. Liniger, D. L. Rath, D. Restaino, R. Rosenberg, J. Rubino, S. C. Seo, A. Simon, S. Smith, W. T. Tseng, *2004 IEEE Int. Integrated Reliability Workshop, Final Report* (Lake Tahoe, CA) (2004) 2.

[2.45]   E. Zschech, H. J. Engelmann, M. A. Meyer, V. Kahlert, A. V. Vairagar, S. G. Mhaisalkar, A. Krishnamoorthy, M. Yan, K. N. Tu, V. Sukharev, *Z. Metallkde.,* submitted (2005).

[2.46]   C. K. Hu, L. Gignac, R. Rosenberg, E. Liniger, J. Rubino, C. Sambucetti, A. Domenicucci, X. Chen, A. K. Stamper, *Appl. Phys. Lett.* **81**, 1782 (2002).

[2.47]   C. K. Hu, L. M. Gignac, E. Liniger, B. Herbst, D. L. Rath, S. T. Chen, S. Kaldor, D. A. Simon, W. T. Tseng, *Appl. Phys. Lett.* **83**, 869 (2003).

[2.48]   G. Ramanath, G. Cui, P. G. Ganesan, X. Guo, A. V. Ellis, M. Stukowski, K. Vijayamoha-nan, P. Doppelt, M. Lane, *Appl. Phys. Lett.* **83**, 383 (2003).

[2.49]   E. Zschech, M. A. Meyer, I. Zienert, H. Prinz, E. Langer, H. Geisler, H. J. Engelmann, *Proc. of the 3rd Int. Conf. on Semiconductor Technology (ISCT),* Shanghai, (2004) 386.

[2.50]   C.-K. Hu, L. Gignac, E. Liniger, R. Rosenberg, *J. Electrohem. Soc.* **149**, G408 (2002).
[2.51]   M. A. Meyer, M. Herrmann, E. Langer, E. Zschech, *Microelectron. Eng.* **64**, 375 (2002).
[2.52]   V. Sukharev, in *Stress-Induced Phenomena in Metal Interconnects*, eds. P. S. Ho et al., AIP Proc. **741**, 85 (2004).
[2.53]   V. Sukharev, E. Zschech, 2004 IEEE Int. Integrated Reliability Workshop, Final Report (LakeTahoe, CA) 79 (2004).
[2.54]   R. Spolanak, PhD Thesis, Universität Stuttgart, 1999.
[2.55]   Q. Guo, K. Nyunt, A. Krishnamoorthy, M. M. San, in *Proceedings of the Advanced Metallization Conference*, San Diego, 2002.
[2.56]   V. Sukharev, E. Zschech, *J. Appl. Phys.* **96**, 6337 (2004).
[2.57]   P. S. Ho, personal communication.
[2.58]   P.S. Ho, in *Proceedings of the Characterization and Metrology for ULSI Technology Conference,* Austin, 2003.
[2.59]   Y. Du, PhD Thesis, University of Texas, Austin, 2001.
[2.60]   G. Schindler, W. Steinhögl, G. Steinlesberger, M. Traving, M. Engelhardt, in *Proceedings of the Advanced Metallization Conference*, San Diego, 2002, p. 13.
[2.61]   W. Steinhögl, G. Schindler, G. Steinlesberger, M. Engelhardt, *Phys. Rev. B* **66**, 075414 (2002).
[2.62]   G. Steinlesberger, M. Engelhardt, G. Schindler, W. Steinhögl, A. v. Glasow, K. Mosig, E. Bertagnolli, *Microelectron. Eng.* **64**, 409 (2002).
[2.63]   K. Fuchs, *Proc. Cambridge Philos. Soc.* **34**, 100 (1938).
[2.64]   E. H. Sondheimer, *Adv. Phys.* **1**, 1 (1952).
[2.65]   A. F. Mayadas M. Shatzkes, *Phys. Rev. B* **1**, 1382 (1970).
[2.66]   S. M. Rossnagel H. Kim, in *Proceedings of the International Interconnect Technology Conference*, San Francisco, 2001.
[2.67]   C. Kittel, *Introduction to Solid State Physics* John Wiley and Sons, New York, 1971.
[2.68]   F. Kreupl, A. P. Graham, G. S. Duesberg, W. Steinhögl, M. Liebau, E. Unger, W. Hönlein, *Microelectron. Eng.* **64**, 399 (2002).
[2.69]   A. P. Graham, G. S. Duesberrg, R. V. Seidel, M. Liebau, E. Unger, W. Pamler, F. Kreupl, W. Hoenlein, *small* **1**, 382 (2005).
[2.70]   B. L. Chin, *Semiconductor International*, May 2001.
[2.71]   F. S. Hickernell, in *Physical Acoustics*; *Vol. XXIV*, ed. E. P. Papadakis, Academic Press, San Diego, 1999, p. 135.
[2.72]   B. A. Auld, *Acoustic Waves and Fields in Solids,* Wiley, New York, 1973.
[2.73]   A. A. Oliner, *Acoustic Surface Waves*, Vol. 24 ,Springer-Verlag, Berlin, 1978.
[2.74]   D. P. Morgan, *Surface-Wave Devices for Signal Processing*, Vol. 19, Elsevier, Amsterdam, 1985.
[2.75]   K. Hashimoto, *Surface Acoustic Wave Devices in Telecommunications,* Springer-Verlag, Berlin, 2000.
[2.76]   S. V. Biryukov, Y. V. Gulyaev, V. V. Krylov, V. P. Plessky, *Surface Acoustic Wave Devices in Inhomogeneous Media,* Springer-Verlag, Berlin, 1995.
[2.77]   C. K. Campbell, Surface Acoustic Wave Devices for Mobile and Wireless Communications, Academic Press, New York, 1998.
[2.78]   C. C. W. Ruppel, T. A. Fjeldy, *Advances in Surface Acoustic Wave Technology, Systems and Applications,* World Scientific, Singapore, 2000.
[2.79]   L. Rayleigh, *Proc. London Math. Soc.* **7**, 4 (1885).
[2.80]   J. L. Bleustein, *Appl. Phys. Lett.* **13**, 412 (1968).
[2.81]   Y. V. Gulyaev, *Sov. Phys. JETP Lett.* **9**, 7 (1969).
[2.82]   Y. Ohta, K. Nakamura, H. Shimizu, *Appl. Phys. Lett.* **13**, 412 (1968).
[2.83]   S. V. Biryukov, personal communication.
[2.84]   S. V. Biryukov, M. Weihnacht, *J. Appl. Phys.* **83**, 3276 (1998).

[2.85]   M. Lewis, in Proceedings of the 1977 IEEE Ultrasonics Symp., Phoenix, IEEE, Piscataway, 1977, p. 744.

[2.86]   K. F. Lau, K. H. Yen, R. S. Kagiwada, K. L. Gong, in *Proceedings of the IEEE Ultrasonics Symposium, Phoenix* (IEEE, Piscataway, 1977), p. 996.

[2.87]   B. A. Auld, J. J. Gagnepain, M. Tan, *Electron. Lett.* **12,** 650 (1976).

[2.88]   R. H. Tancrell, M. G. Holland, *Proc. IEEE* **59,** 393 (1971). [2.70] W. P. Mason, *Physical Acoustics,* Vol. 1 A, Academic Press, New York, 1964.

[2.89]   W. P. Mason, *Physical Acoustics,* Vol. 1 A, Academic Press, New York, 1964.

[2.90]   W. R. Smith, H. M. Gerard, J. H. Collins, T. M. Reeder, H. J. Show, *IEEE Trans. MTT* **17,** 856 (1969).

[2.91]   G. Tobolka, *IEEE Trans. SU* **26,** 426 (1979).

[2.92]   C. S. Hartmann, P. V. Wright, R. J. Kansy, E. M. Garber, in *Proceedings of the IEEE Ultrasonics Symposium, San Diego* (IEEE, Piscataway, 1982), p. 40.

[2.93]   P. Ventura, J. M. Hodé, M. Solal, in *Proceedings of the IEEE Ultrasonics Symposium, Seattle,*IEEE, Piscataway, 1995, p. 263.

[2.94]   T. Kodama, H. Kawabata, Y. Yasuhara, H. Sato, in *Proceedings of the IEEE Ultrasonics Symposium, Willliamsburg* (IEEE, Piscataway, 1986), p. 313.

[2.95]   P. V. Wright, in Proceedings of theIEEE Ultrasonics Symposium, San Francisco (IEEE, Piscataway, 1985), p. 58.

[2.96]   C. S. Hartmann, B. P. Abbott, in *Proceedings of the IEEE Ultrasonics Symposium, Montreal,* IEEE, Piscataway, 1989, p. 79.

[2.97]   H. Yatsuda, *IEEE Trans. UFFC* **44,** 453 (1997).

[2.98]   G. Martin, in Proceedings of the IEEE Ultrasonics Symposium, Caesars Tahoe, IEEE, Piscataway, 1999, p. 15.

[2.99]   F. G. Marshall, E. G. S. Page, *Electron. Lett.* **7,** 460 (1971).

[2.100]   R. Takayama, H. Nakanishi, Y. Iwasaki, T. Kawasaki, in *Proceedings of the IEEE Ultrasonics Symposium, Montreal,* (IEEE, Piscataway, 2004), p. 959.

[2.101]   G. Kovacs, W. Ruile, M. Jakob, U. Rösler, E. Maier, U. Knauer, H. Zottl, in *Proceedings of the IEEE Ultrasonics Symposium, Montreal,* (IEEE, Piscataway, 2004), p. 974.

[2.102]   G. V. Samsonov, L. A. Dvorina, B. M. Rud, *Silicides,* Metallurgia, Moscow, 1979 (in Russian).

[2.103]   G. V. Samsonov, L. M. Vinitskii, *Handbook of Refractory Compounds,* IFI/Plenum, New York, 1980.

[2.104]   S. P. Murarka, *Silicides for VLSI Applications,* Academic Press, New York, 1983.

[2.105]   K. Maex, M. v. Rossum, *Properties of Metal Silicides,* INSPEC, IEE, London, 1995.

[2.106]   V. E. Borisenko, *Semiconducting Silicides,* Springer-Verlag, Berlin, 1999.

[2.107]   *Silicides-Fundamentals and Application,* ed. L. Miglio and F. d'Heurle, World Scientific, Singapore, 2000.

[2.108]   T. B. Massalski, H. Okamoto, P. R. Subramanian, L. Kasprzak, *Binary Alloy Phase Diagrams,* ASM International, Materials Park, OH, 1990.

[2.109]   H. Lange, Phys. Status Solidi B **201,** 3 (1997).

[2.110]   R. T. Tung, Mater. Sci. Eng. **R35,** 1 (2001).

[2.111]   H. Lange, S. Brehme, W. Henrion, A. Heinrich, G. Behr, H. Griessmann, in *Proceedings of the 16th International Conference on Thermoelectrics,* Dresden, 1997, p. 1267.

[2.112]   C. B. Vining, *CRC Handbook of Thermoelectricity,* CRC Press, Boca Raton, 1995.

[2.113]   A. Heinrich, J. Schumann, H. Vinzelberg, U. Bruestel, C. Gladun, *Thin Solid Films* **223,** 311 (1993).

[2.114]   A. Heinrich, H. Vinzelberg, C. Metz, J. Schumann, A. Zyuzin, *Phys. Rev. B* **67,** 035302 (2003).

[2.115]   G. Queirolo, in *Silicides-Fundamentals and Application,* ed. L. Miglio, F. d'Heurle, World Scientific, Singapore, 2000, p. 276.

[2.116] R. Lindsay, A. Lauwers, M. d. Potter, N. Roelandts, C. Vrancken, K. Maex, *Microelectron. Eng.* **55**, 157 (2001).

[2.117] S. Teichert, R. Kilper, J. Erben, D. Franke, B. Gebhard, T. Franke, P. Häussler, W. Henrion, H. Lange, *Appl. Surf. Sci.* **104/105**, 679 (1996).

[2.118] S. Tsunoda, M. Mukaida, Y. Imai, *Thin Solid Films* **381**, 296 (2001).

[2.119] R. Glang, R. A. Holmwood, S. R. Herd, *J, Vac. Sci. Technol.* **4**, 1963 (1967).

[2.120] A. Moebius, D. Elefant, A. Heinrich, R. Mueller, J. Schumann, H. Vinzelberg, G. Zies, *J. Phys. C: Solid State Phys.* **16**, 6491 (1983).

[2.121] A. Moebius, H. Vinzelberg, C. Gladun, A. Heinrich, D. Elefant, J. Schumann, G. Zies, *Phys. C: Solid State Phys.* **18**, 3337 (1985).

[2.122] C. Gladun, A. Heinrich, F. Lange, J. Schumann, H. Vinzelberg, *Thin Solid Films* **125**, 101 (1985).

[2.123] A. Heinrich, H. Vinzelberg, G. Gladun, J. Schumann, G. Sobe, J. Sonntag, G. Zies, *Materi. Sci.* **8**, 111 (1987).

[2.124] A. Heinrich, C. Gladun, H. Schreiber, J. Schumann, H. Vinzelberg, *Vacuum* **41**, 1408 (1990).

[2.125] V. S. Smentkowski, *Prog. Surf. Sci.* **64**, 1 (2000).

[2.126] A. T. Burkov, C. Gladun, A. Heinrich, W. Pitschke, J. Schumann, *J. Non-Cryst. Solids* **205-207**, 737 (1996).

[2.127] J. Schumann, D. Elefant, C. Gladun, A. Heinrich, W. Pitschke, H. Lange, W. Henrion, G. Groetzschel, *Phys. Status Solidi A* **145**, 429 (1994).

[2.128] M. F. Bain, B. M. Armstrong, H. S. Gamble, *Vacuum* **64**, 227 (2002).

[2.129] B. Sell, J. Willer, K. Pomplun, A. Saenger, D. Schumann, W. Krautschneider, Microelectron. Eng. **55**, 197 (2001).

[2.130] H. S. Lee, H. S. Rhee, B. T. Ahn, *J. Electrochem. Soc.* **149** (2002).

[2.131] L. Luo, C. E. Zybill, H. G. Ang, S. F. Lim, D. H. C. Chua, J. Lim, A. T. S. Wee, K. L. Tan, *Thin Solid Films* **325**, 87 (1998).

[2.132] M. Mukaida, I. Hiyama, T. Tsunoda, Y. Imai, in *Proceedings of the 17th International Conference on Thermoelectrics*, Nagoya, 1998 (IEEE, Piscataway).

[2.133] T. T. Tung, K. Maex, P. W. Pellegrini, L. H. Allen, *Silicide Thin Films-Fabrication, Properties and Applications*, Vol. 402, Materials Research Society, Pittsburgh, 1996.

[2.134] *Handbook of Semiconductor Technology; Vol. 1*, ed. K. A. Jackson, W. Schroeter Wiley-VCH, Weinheim, 2000.

[2.135] A. Cros, P. Muret, *Mater. Sci. Rep.* **8**, 271 (1992).

[2.136] H. v. Kaenel, *Mater. Sci. Rep.* **8**, 193 (1992).

[2.137] L. Miglio, V. Meregalli, *J. Vac. Sci. Technol. B* **16**, 1604 (1998).

[2.138] C. A. Kleint, A. Heinrich, H. Grießmann, D. Hofmann, H. Vinzelberg, J. Schumann, D. Schlaefer, G. Behr, L. Ivanenko, Mater. Res. Soc. Symp. Proc. **545**, 165 (1999).

[2.139] D. Shinoda, S. Atanabe, and Y. Sasaki, *J. Phys. Soc. Jpn.* **19(3)**, 269 (1964).

[2.140] B. K. Voronov, L. D. Dudkin, N. N. Trusova, *Kristallografija* **12**, 519 (1967).

[2.141] I. J. Ohsugi, T. Kojima, I. A. Nishida, *Phys. Rev. B* **42**, 10761 (1990).

[2.142] J. J. Nickl, J. D. Koukoussas, *J. Less-Common Met.* **23 (1)**, 73 (1971).

[2.143] R. Krausze, M. Khristov, P.Peshev, G. Krabbes, *Z. Anorg. Allg. Chem.* **579**, 231 (1989).

[2.144] W. N. Gurin, A. P. Obukhov, M. M. Kursukowa, Z. P. Terentjewa, I. R. Kozlowa, *Planseeberichte für Pulvermetallurgie* **19**, 86 (1971).

[2.145] S. Okada, T. Atoda, *Nippon Kagaku Kaishi* **5**, 746 (1983).

[2.146] P. Peshev, M. Khristov, G. Gyurov, *J. Less-Common Met.* **153**, 15 (1989).

[2.147] L. M. Levinson, Report No. 72CRD111 (1972).

[2.148] L. M. Levinson, *J. Solid State Chem.* **6**, 126 (1973).

[2.149] M. A. Morokhovets, E. I. Elagina, *Izv. Acad. Nauk SSSR, Neorgan. Mater.* **2**, 650 (1966).

[2.150] L. D. Ivanova, N. K. Abrikosov, E. I. Elagina, V. D. Khvostikova, *Izv. Acad. Nauk SSSR, Neorgan. Mater.* **5** (1933).

[2.151]   T. Siegrist, F. Hulliger, G. Travaglini, *J. Less-Common Met.* **92**, 119 (1983).
[2.152]   U. Gottlieb, B. Lambert-Andron, F. Nava, M. Affronte, O. Laborde, A. Rouault, M. Madar, *J. Appl. Phys.* **78**, 3902 (1995).
[2.153]   L. Ivanenko, V. L. Shaposhnikov, A. B. Filinov, D. B. Migas, G. Behr, J. Schumann, H. Vinzelberg, V. E. Borisenko, *Microelectron. Eng.* **64**, 225 (2002).
[2.154]   C. Kloc, E. Arushanov, M. Wendl, H. Hohl, U. Malang, E. Bucher, *J. Alloys Comp.* **219**, 93 (1995).
[2.155]   G. Behr, J. Werner, G. Weise, A. Heinrich, A. Burkov, C. Gladun, *Phys. Status Solidi A* **160**, 549 (1997).
[2.156]   C. B. Vining, C. E. Alevato, in *Proceedings of the 10th International Conference on Thermoelectrics*, Cardiff, 1991 (Babrow Press, Cardiff), p. 167.
[2.157]   U. Gottlieb, O. Laborde, A. Rouault, M. Madar, *Appl. Surf. Sci.* **73**, 243 (1993).
[2.158]   F. Nava, K. N. Tu, O. Thomas, J. P. Senateur, R. Madar, A. Borghesi, G. Guizetti, U. Gottlieb, O. Laborde, O. Bisi, *Mater. Sci. Rep.* **9**, 141 (1993).
[2.159]   D. Souptel, G. Behr, L. Ivanenko, H. Vinzelberg, J. Schumann, *J. Cryst. Growth* **244**, 296 (2002).
[2.160]   C. E. Allevato, C. B. Vining, in *Proceedings of the 28th Intersociety Energy Conversion Engineering Conference*, Washington, 1993, American Chemical Society, Atlanta, p. 1239.
[2.161]   E. Gross, *Int. J. Powder Metall.* **31**, 239 (1995).
[2.162]   M. Riffel, E. Gross, U. Stoehrer, *J. Mater. Sci.: Materials in Electronics* **6**, 182 (1995).
[2.163]   E. Mueller, K. Schackenberg, J. Schilz, in *Proceedings of the 19th International Conference on Thermoelectrics*, Cardiff, 2000, BABROW Press, Wales, p. 121.
[2.164]   B.-G. Min, J.-D. Shim, D.-H. Lee, in *Proceedings of the 17th International Conference on Thermoelectrics*, Nagoya, 1998 (IEEEE), p. 386.
[2.165]   G. D. J. Smit, S. Rogge, T. M. Klapwijk, in *Proc. MAM 2002*, Vaals, 2002, Philips Research, Eindhoven, p. 21.
[2.166]   S. Mantl, *Phys. Bl.* **51**, 951 (1995).
[2.167]   R. Hull, A. Ourmazd, W. D. Rau, P. Schwander, M. L. Green, R. T. Tung, in *Handbook of Semiconductor Technology; Vol. 1*, ed. K. A. Jackson W. Schröter, Wiley-VCH, Weinheim, 2000, p. 453.
[2.168]   S. Mantl, H. L. Bay, *Appl. Phys. Lett.* **61**, 267 (1992).
[2.169]   K. Miyake, Y. Makita, Y. Maeda, T. Suemasu, *Thin Solid Films* **381**, 171 (2001).
[2.170]   T. Suemasu, Y. Iikura, T. Fujii, K. Takakura, N. Hiroi, F. Hasegawa, *Jpn. J. Appl. Phys.* **38**, L620 (1999).
[2.171]   D. Leong, M. Harry, K. J. Reeson, K. P. Homewood, *Nature* **387**, 686 (1997).
[2.172]   M. G. Grimaldi, S. Coffa, C. Spinella, in *Silicides-Fundamentals and Application*, ed. L. Miglio, F. d'Heurle, World Scientific, Singapore, 2000, p. 93.
[2.173]   B. Schuller, R. Carius, S. Lenk, S. Mantl, *Microelectron. Eng.* **60**, 205 (2002).
[2.174]   L.I. Ivanenko, V.L. Shaposhnikov, A.B. Filonov, A.V. Krivosheeva, V.E. Borisenko, D.B. Migas, L. Miglio, G. Behr, J. Schumann, Thin Solid Films **461**, 141 (2004).
[2.175]   A. Heinrich, in *Silicides-Fundamentals and Application*, ed. L. Miglio, F. d'Heurle, World Scientific, Singapore, 2000, p. 126.
[2.176]   M.G. Grimaldi, C. Buongiorno, C. Spinella, E. Grilli, L. Martinelli, M. Gemelli, D.B. Migas, L. Miglio, M. Fanciulli, *Phys. Rev.*, **B 66**, 085319 (2002).
[2.177]   L. Martinelli, E. Grilli, D.B. Migas, L. Miglio, F. Marabelli, C. Soci, M. Geddo, M.G. Grimaldi, C. Spinella, *Phys. Rev.*, **B 66**, 085320 (2002).
[2.178]   Y. Makita, Y. Nakayama, Y. Fukuzawa, S.N. Wang, N. Otogawa, Y. Suzuki, Z.X. Liu, M. Osamura, T. Ootsuka, T. Mise, and H. Tanoue, *Thin Solid Films* **461**, 202 (2004).
[2.179]   Y. Imry and M.Strongin, *Phys. Rev. B* **24**, 6353 (1981).
[2.180]   H. Vinzelberg, A. Heinrich, C. Gladun, D. Elefant, *Philos. Mag. B* **65**, 651 (1992).
[2.181]   A. Heinrich, J. Schumann, H. Vinzelberg, U. Brüstel, C. Gladun, *Thin Solid Films* **223**, 311 (1993).

[2.182]   A. Heinrich, C. Kleint, H. Grießmann, G. Behr, L. Ivanenko, V. Shaposhnikov, J. Schumann, in *Proceedings of the 18th International Conference on Thermoelectrics*, Baltimore, 1999 (IEEE, Piscataway), p. 161.

[2.183]   A. T. Burkov, A. Heinrich, C. Gladun, W. Pitschke, J. Schumann, *Phys. Rev. B* **58**, 9644 (1998).

[2.184]   W. Pitschke, D. Hofman, J. Schumann, C. A. Kleint, A. Heinrich, *J. Appl. Phys.* **89**, 3229 (2001).

[2.185]   I. A. Campbell, A. Fert, in *Ferromagnetic Materials*; Vol. 3, ed. E. P. Wohlfarth, North-Holland, Amsterdam, 1982, p. 747.

[2.186]   M. N. Baibich, J. M. Broto, A. Fert, F. N. V. Dau, F. Petroff, P. Etienne, G. Creuzet, A. Friederich, J. Chazelas, *Phys. Rev. Lett.* **61**, 2472 (1988).

[2.187]   P. Grünberg, R. Schreiber, Y. Pang, M. B. Brodsky, H. Sowers, *Phys. Rev. Lett.* **57**, 2442 (1986).

[2.188]   G. Binasch, P. Grünberg, F. Saurenbach, W. Zinn, *Phys. Rev. B* **39**, 4828 (1989).

[2.189]   S. S. P. Parkin, Z. G. Li, D. J. Smith, *Appl. Phys. Lett.* **58**, 2710 (1991).

[2.190]   G. A. Prinz, *Phys. Today* **48**, 58 (1995).

[2.191]   C. H. Tsang, R. E. Fontana, T. Lin, D. E. Heim, B. A. Gurney, M. L. Williams, *IBM J. Res. Develop.* **42**, 103 (1998).

[2.192]   S. Tehrani, J. M. Slaughter, E. Chen, M. Durlam, J. Shi, M. DeHerrera, *IEEE Trans. Magn.* **35**, 2814 (1999).

[2.193]   R. R. Katti, *J. Appl. Phys.* **91**, 7245 (2002).

[2.194]   J.-G. Zhu, Y. Zheng, G. A. Prinz, *J. Appl. Phys.* **87**, 6668 (2000).

[2.195]   J. Zhu, G. A. Prinz, *Data Storage* **7**, 40 (2000).

[2.196]   R. Jansen, O. M. J. v. t. Erve, S. D. Kim, R. Vlutters, P. S. Anil Kumar, J. C. Lodder, *J. Appl. Phys.* **89**, 7431 (2001).

[2.197]   S. A. Wolf, D. D. Awschalom, R. A. Buhrman, J. M. Daughton, S. v. Molnár, M. L. Roukes, A. Y. Chtchelkanova, D. M. Treger, *Science* **294**, 1488 (2001).

[2.198]   I. Zutic, J. Fabian, S. D. Sarma, *Rev. Mod. Phys.* **76**, 323 (2004).

[2.199]   *Spin Electronics;* Vol., eds. M. Ziese and M. J. Thornton (Springer-Verlag, Berlin, 2001).

[2.200]   *Semiconductor Spintronics and Quantum Computation;* Vol., eds. D. D. Awschalom, D. Loss, N. Samarth (Springer-Verlag, Berlin, 2002).

[2.201]   N. W. Ashcroft, N. D. Mermin, *Solid State Physics,* Saunders College, Philadelphia, 1976.

[2.202]   E. P. Wohlfarth, in *Ferromagnetic Materials*; Vol. 1, ed. E. P. Wohlfarth (North-Holland Publishing Company, Amsterdam, 1980), p. 3.

[2.203]   A. Hubert and R. Schäfer, *Magnetic Domains* (Springer-Verlag, Berlin, 1998).

[2.204]   L. Néel, *C. R. Acad. Sci.* **255**, 1545 (1962).

[2.205]   J. C. S. Kools, *IEEE Trans. Magn.* **32**, 3165 (1996).

[2.206]   Q. Leng, V. Cross, R. Schäfer, A. Fuss, P. Grünberg, W. Zinn, *J. Magn. Magn. Mater.* **126**, 367 (1993).

[2.207]   A. Yoshihara, J. T. Wang, K. Takanashi, K. Himi, Y. Kawazoe, H. Fujimori, P. Grünberg, *Phys. Rev. B* **63**, 100405(R) (2001).

[2.208]   S. S. P. Parkin, *Phys. Rev. Lett.* **67**, 3598 (1991).

[2.209]   W. Dobrowolski, J. Kossut, T. Story, in *Handbook of Magnetic Materials; Vol. 14*, ed.K. H. J. Buschow (Elsevier Science Publishers, Amsterdam, 2003), p. 289.

[2.210]   S. Toscano, B. Briner, H. Hopster, M. Landolt, *J. Magn. Magn. Mater.* **114**, L6 (1992).

[2.211]   S.-S. Yan, R. Schreiber, F. Voges, C. Osthöver, P. Grünberg, *Phys. Rev. B* **59**, R11641 (1999).

[2.212]   C. Carbone, S. F. Alvarado, *Phys. Rev. B* **36**, 2433 (1987).

[2.213]   S. S. P. Parkin, N. More, K. P. Roche, *Phys. Rev. Lett.* **64**, 2304 (1990).

[2.214]   S. S. P. Parkin, R. Bahdra, K. P. Roche, *Phys. Rev. Lett.* **66**, 2151 (1991).

[2.215]   J. Unguris, R. J. Celotta, D. T. Pierce, *Phys. Rev. Lett.* **79**, 2734 (1997).

[2.216]   D. E. Bürgler, P. Grünberg, S. Demokritov, M. Johnson, in *Handbook of Magnetic Materials*; *Vol. 13*, ed. K. H. J. Buschow, Elsevier, Amsterdam, 2000.
[2.217]   M. A. Ruderman, C. Kittel, *Phys. Rev.* **96,** 99 (1954).
[2.218]   T. Kasuya, *Prog. Theor. Phys.* **16,** 45 (1956).
[2.219]   K. Yosida, *Phys. Rev.* **106,** 893 (1957).
[2.220]   P. Bruno, C. Chappert, *Phys. Rev. Lett.* **67,** 1602 (1991).
[2.221]   P. Bruno, *J. Phys.: Condens. Matter* **11,** 9403 (1999).
[2.222]   P. Bruno, *J. Magn. Magn. Mater.* **164,** 27 (1996).
[2.223]   P. Bruno, *J. Magn. Magn. Mater.* **121,** 248 (1993).
[2.224]   P. Bruno, *Phys. Rev. B* **52,** 411 (1995).
[2.225]   K. Garrison, Y. Chang, P. D. Johnson, *Phys. Rev. Lett.* **71,** 2801 (1993).
[2.226]   R. Kläsges, D. Schmitz, C. Carbone, W. Eberhardt, P. Lang, R. Zeller, P. H. Dederichs, *Phys. Rev. B* **57,** R696 (1998).
[2.227]   C. Carbone, E. Vescovo, O. Rader, W. Gudat, W. Eberhardt, *Phys. Rev. Lett.* **71,** 2805 (1993).
[2.228]   J. L. Leal, M. H. Kryder, *J. Appl. Phys.* **83,** 3720 (1998).
[2.229]   W. Meiklejohn, C. Bean, *Phys. Rev.* **102,** 1413 (1956).
[2.230]   *Magnetic Properties of Metals*; ed. H. P. J. Wijn, Springer-Verlag, Berlin, 1991.
[2.231]   B. Dieny, V. Speriosu, S. S. P. Parkin, P. Baumgart, D. Wilhoit, *J. Appl. Phys.*, **69,** 4774 (1991).
[2.232]   J. Nogués, I. K. Schuller, *J. Magn. Magn. Mater.* **192,** 203 (1999).
[2.233]   A. E. Berkowitz, K. Takano, *J. Magn. Magn. Mater.* **200,** 552 (1999).
[2.234]   M.-T. Lin, C. H. Ho, C.-R. Chang, Y. D. Yao, *Phys. Rev. B* **63,** 100404(R) (2001).
[2.235]   Y.-J. Lee, C.-R. Chang, T.-M. Hong, C. H. Ho, M.-T. Lin, *J. Magn. Magn. Mater.* **239,** 57 (2002).
[2.236]   Y.-J. Lee, C.-R. Chang, T.-M. Hong, C. H. Ho, M.-T. Lin, *J. Magn. Magn. Mater.* **240,** 264 (2002).
[2.237]   O. d. Haas, R. Schäfer, L. Schultz, C. M. Schneider, Y. M. Chang, M.-T. Lin, *Phys. Rev. B* **67,** 054405 (2003).
[2.238]   R. Meservey, P. M. Tedrow, *Phys. Rep.* **238,** 173 (1994).
[2.239]   P. M. Tedrow, R. Meservey, *Phys. Rev. Lett.* **26,** 192 (1971).
[2.240]   M. Jullière, *Phys. Lett.* **54A,** 225 (1975).
[2.241]   J. S. Moodera, L. R. Kinder, T. M. Wong, R. Meservey, *Phys. Rev. Lett.* **74,** 3273 (1995).
[2.242]   J. S. Moodera and G. Mathon, *J. Magn. Magn. Mater.* **200,** 248 (1999).
[2.243]   S. S. P. Parkin, K. P. Roche, M. G. Samant, P. M. Rice, R. B. Beyers, R. E. Scheuerlein, E. J. O'Sullivan, S. L. Brown, J. Bucchigano, D. W. Abraham, Y. Lu, M. Rooks, P. L. Trouilloud, R. A. Wanner, W. J. Gallagher, *J. Appl. Phys.* **85,** 5828 (1999).
[2.244]   W. H. Butler, X.-G. Zhang, T. C. Schultheiss, J. M. MacLaren, *Phys. Rev. B* **63,** 054416 (2001).
[2.245]   J. Mathon, A. Umerski, *Phys. Rev. B* **63,** 220403 (R) (2001).
[2.246]   D. Wortmann, G. Bihlmayer, S. Blügel, *J. Phys.: Condens. Matter* **16,** S5819 (2004).
[2.247]   X.-G. Zhang, W. H. Butler, A. Bandyophadhyay, *Phys. Rev. B* **68,** 092402 (2003).
[2.248]   S. Yuasa, A. Fukushima, T. Nagahama, K. Ando, Y. Suzuki, *Jpn. J. Appl. Phys.* **43,** L588 (2004).
[2.249]   S. S. P. Parkin, C. Kaiser, A. Panchula, P. M. Rice, B. Hughes, M. Samant, S.-H. Yang, *Nature Materials* **3,** 862 (2004).
[2.250]   S. Yuasa, T. Nagahama, A. Fukushima, Y. Suzuki, K. Ando, *Nature Materials* **3,** 868 (2004).
[2.251]   B. Dieny, V. S. Speriosu, S. S. P. Parkin, B. A. Gurney, D. R. Wilhoit, D. Mauri, *Phys. Rev. B* **43,** 1297 (1991).
[2.252]   F. Petroff, A. Barthelemy, D. H. Mosca, D. K. Lottis, A. Fert, P. A. Schroeder, W. P. Pratt, Jr., R. Loloee, S. Lequien, *Phys. Rev. B* **44,** 5355 (1991).

[2.253]   S. S. P. Parkin, R. F. Marks, R. F. C. Farrow, G. R. Harp, Q. H. Lam, R. J. Savoy, *Phys. Rev. B* **46**, 9262 (1992).

[2.254]   S. S. P. Parkin, in *Ultrathin Magnetic Structures*; *Vol. 2*, eds. B. Heinrich, J. A. C. Bland, Springer Verlag, Berlin, 1994, p. 132.

[2.255]   S. S. P. Parkin, R. F. C. Farrow, R. F. Marks, A. Cebollada, G. R. Harp, R. J. Savoy, *Phys. Rev. Lett.* **72**, 3718 (1994).

[2.256]   A. Barthélémy, A. Fert, F. Petroff, in *Handbook of Magnetic Materials*; *Vol. 12*, ed. K. H. J. Buschow, (Elsevier , Amsterdam, 1999, p. 3.

[2.257]   L. Dimesso, H. Hahn, *J. Appl. Phys.* **84**, 953 (1998).

[2.258]   Y. G. Pogorelov, M. M. P. d. Azevedo, J. B. Sousa, *Phys. Rev. B* **58**, 425 (1998).

[2.259]   R. Schad, C. D. Potter, P. Beliën, G. Verbanck, V. V. Moshchalkov, Y. Bruynseraede, *Appl. Phys. Lett.* **64**, 3500 (1994).

[2.260]   R. Schad, P. Belien, G.Verbanck, *Phys. Rev. B* **57**, 13692 (1998).

[2.261]   D. Olligs, D. E. Bürgler, Y. G. Wang, E. Kentzinger, U. Rücker, R. Schreiber, T. Brückel, P. Grünberg, *Europhys. Lett.* **59**, 458 (2002).

[2.262]   J. Unguris, R. J. Celotta, D. T. Pierce, *Phys. Rev. Lett.* **67**, 140 (1991).

[2.263]   D. T. Pierce, J. Unguris, R. J. Celotta, M. D. Stiles, *J. Magn. Magn. Mater.* **200**, 290 (1999).

[2.264]   S. S. P. Parkin, *Appl. Phys. Lett.* **61**, 1358 (1991).

[2.265]   J. Barnas, A. Fuss, R. E. Camley, P. Grünberg, W. Zinn, *Phys. Rev. B* **42**, 8110 (1990).

[2.266]   P. Zahn, I. Mertig, M. Richter, H. Eschrig, *Phys. Rev. Lett.* **75**, 2996 (1995).

[2.267]   D. Elefant, D. Tietjen, L. van Loyen, J.-I. Mönch, C. M. Schneider, *J. Appl. Phys.* **91**, 8590 (2001).

[2.268]   see, http://www.physik.tu-dresden.de/~fermisur/

[2.269]   J. M. Kikkawa, I. P. Smorchkova, N. Samarth, D. D. Awschalom, *Science* **277**, 1284 (1997).

[2.270]   J. M. Kikkawa, D. D. Awschalom, *Nature* **397**, 139 (1999).

[2.271]   S. Datta and B. Das, *Appl. Phys. Lett.* **50**, 665 (1990).

[2.272]   D. J. Monsma, J. C. Lodder, T. J. A. Popma, B. Dieny, *Phys. Rev. Lett.* **74**, 5260 (1995).

[2.273]   E. I. Rashba, *Sov. Phys. Solid State*, **2**, 1109 (1960).

[2.274]   G. Schmidt, D. Ferrand, L. W. Molenkamp, A. T. Filip, B. J. v. Wees, *Phys. Rev. B,* **62**, R4790 (2000).

[2.275]   J. D. Albrecht, D. L. Smith, *Phys. Rev. B* **68**, 035340 (2003).

[2.276]   C. Adelmann, X. Lou, J. Strand, C. J. Palmstrom, P. A. Crowell, *Phys. Rev. B* **71**, 12130 (R) (2005).

[2.277]   A. Fert, H. Jaffrés, *Phys. Rev. B* **64**, 184420 (2001).

[2.278]   H. Jaffrés, A. Fert, *J. Appl. Phys.* **91**, 8111 (2002).

[2.279]   X. Jiang, R. Wang, R. M. Shelby, R. M. Macfarlane, S. R. Bank, J. S. Haris, and S. S. P. Parkin, *Phys. Rev. Lett.* **94**, 056601 (2005).

[2.280]   F. matsakura, H. Ohno, T. Dietl, in *Handbook of Magnetic Materials; Vol. 14*, ed. K. H. J. Buschow (Elsevier Science Publishers, Amsterdam, 2002), p. 1.

[2.281]   T. Dietl, H. Ohno, F. Matsukura, J. Cibert, D. Ferrand, *Science* **287**, 1019 (2000).

[2.282]   A. H. MacDonald, P. Schiffer, S. Samarth, *Nature Mater.* **4**, 195 (2005).

[2.283]   K. W. Edmonds, P. Boguslawski, K. Y. Wang, R. P. Campion, S. N. Novikov, N. R. S. Farley, B. L. Gallagher, C. T. Foxon, M. Sawicki, T. Dietl, M. B. Nardelli, J. Bernhole, *Phys. Rev. Lett.* **92**, 037201 (2004).

[2.284]   J. H. Park, M. G. Kim, H. M. Jang, S. Ryu, Y. M. Kim, *Appl. Phys. Lett.* **84**, 1338 (2004).

[2.285]   K. Hamaya, T. Taniyama, Y. Kitamoto, T. Fujii, and Y. Yamazaki, *Phys. Rev. Lett.* **94**, 147203 (2005).

[2.286]   L. Bergqvist, O. Eriksson, J. Kudrnovsky, V. Drchal, P. Korzhavyi, and I. Turek, *Phys. Rev. Lett.* **93**, 137202 (2004).

[2.287]   K. Sato, W. Schweika, P. H. Dederichs, and H. Katayama-Yoshida, *Phys. Rev.* B **70**, 201202(R) (2004).

[2.288]   R. Fiederling, M. Keim, G. Reuscher, W. Ossau, G. Schmidt, A. Waag, L. Molenkamp, *Nature* **402**, 787 (1999).

[2.289]   M. Oestreich, J. Hübner, D. Hägele, P. J. Klar, W. Heimbrodt, W. W. Rühle, D. E. Ashenford, B. Lunn, *Appl. Phys. Lett.* **74**, 1251 (1999).

[2.290]   K. H. Ploog, *J. Appl. Phys.* **91**, 7256 (2002).

[2.291]   M. Ramsteiner, H. Y. Hao, A. Kawaharazuka, H. J. Zhu, M. Kästner, R. Hey, L. Däweritz, H. T. Grahn, K. H. Ploog, *Phys. Rev.* B **66**, 081304(R) (2002).

[2.292]   X. Jiang, R. Wang, S. v. Dijken, R. Shelby, R. Macfarlane, G. S. Solomon, J. Harris, S. S. P. Parkin, *Phys. Rev. Lett.* **90**, 256603 (2003).

[2.293]   I. A. Buyanova, M. Izadifard, W. M. Chen, J. Kim, F. Ren, G. Thaler, C. R. Abernathy, S. J. Pearton, C. C. Pan, G.-T. Chen, J.-I. Chyi, J. M. Zavada, *Appl. Phys. Lett.* **84**, 2599 (2004).

[2.294]   R. Wang, X. Jiang, R. M. Shelby, R. M. Macfarlane, S. S. P. Parkin, *Appl. Phys. Lett.* **86**, 052901 (2005).

[2.295]   L. Berger, *Phys. Rev.* B **54**, 9353 (1996).

[2.296]   J. C. Slonczewski, *J. Magn. Magn. Mater.* **150**, L1 (1996).

[2.297]   M. D. Stiles, A. Zangwill, *Phys. Rev.* B **66**, 014407 (2002).

[2.298]   D. Tietjen, D. Elefant, C. M. Schneider, *J. Appl. Phys.* **91**, 5951 (2002).

[2.299]   H. J. M. Swagten, G. J. Strijkers, P. J. H. Bloemen, M. M. H. Willekens, W. J. M. d. Jonge, *Phys. Rev.* B **53**, 9108 (1996).

[2.300]   J. M. D. Coey, in *Spin Electronics*, eds. M. Ziese, M. J. Thornton, Springer-Verlag, Berlin, 2001, p. 277.

[2.301]   J. Galanakis, *J Phys.: Condens. Matter* **14**, 6329 (2002).

[2.302]   M. Tsoi, A. G. M. Jansen, J. Bass, W.-C. Chiang, M. Seck, V. Tsoi, P. Wyder, *Phys. Rev. Lett.* **80**, 4281 (1998).

[2.303]   J.-E. Wegrowe, D. Kelly, Y. Jaccard, P. Guittienne, J.-P. Ansermet, *Europhys. Lett.* **45**, 626 (1999).

[2.304]   E. B. Myers, D. C. Ralph, J. A. Katine, R. N. Louie, R. A. Buhrman, *Science* **285**, 867 (1999).

[2.305]   J. A. Katine, F. J. Albert, R. A. Buhrman, E. B. Myers, D. C. Ralph, *Phys. Rev. Lett.* **84**, 3149 (2000).

[2.306]   J. C. Slonczewski, *J. Magn. Magn. Mater.* **195**, L261 (1999).

[2.307]   S. I. Kiselev, J. C. Sankey, I. N. Krivorotov, N. C. Emley, R. J. Schoelkopf, R. A. Buhrman, D. C. Ralph, *Nature* **425**, 380 (2003).

[2.308]   R. H. Koch, J. A. Katine, J. Z. Sun, *Phys. Rev. Lett.* **92**, 088301 (2004).

[2.309]   W. H. Rippard, M. R. Pufall, S. Kaka, S. E. Russek, T. J. Silva, *Phys. Rev. Lett.* **92**, 027201 (2004).

[2.310]   M. R. Pufall, W. H. Rippard, S. Kaka, S. E. Russek, T. J. Silva, J. Katine, M. Carey, *Phys. Rev.* B **69**, 214409 (2004).

[2.311]   I. N. Krivorotov, N. C. Emley, J. C. Sankey, S. I. Kiselev, D. C. Ralph, R. A. Buhrman, *Science* **307**, 228 (2005).

[2.312]   Y. Ohno, D. K. Young, B. Beschoten, F. Matsukura, H. Ohno, D. D. Awschalom, *Nature* **402**, 790 (1999).

[2.313]   F. Meier, D. Pescia, in *Optical Orientation*, eds. F. Meier B. P. Zakharchenya, North-Holland, Amsterdam, 1984, p. 296.

[2.314]   T. Dietl, *J. Appl. Phys.* **89**, 7437 (2001).

[2.315]   H. J. Zhu, M. Ramsteiner, H. Kostial, M. Wassermeier, H.-P. Schönherr, K. H. Ploog, *Phys. Rev. Lett.* **87**, 016601 (2001).

[2.316]   V. F. Motsnyi, P. V. Dorpe, W. V. Roy, E. Goovaerts, V. I. Safarov, G. Borghs, J. D. Boeck, *Phys. Rev.* B **68**, 245319 (2003).

[2.317]   *Lexikon d. Optik II,* Spektrum Akademischer Verlag, Heidelberg/Berlin, 1999.
[2.318]   *Lexikon d. Physik IV* , Spektrum Akademischer Verlag, Heidelberg/Berlin, 2000.
[2.319]   L. Golub, J. M. Pasachoff, *The Solar Corona,* Cambridge University Press, Cambridge, 1997.
[2.320]   R. W. Pohl, *Einführung in die Physik,* Bd. III, Springer Verlag, Berlin, 1954.
[2.321]   W. Finkelnburg, *Einführung in die Atomphysik,* Springer Verlag, Berlin, 1964.
[2.322]   J. H. Underwood, in *Encyclopedia of Applied Physics*; *Vol. 23,* ed. G. L. Trigg, Wiley-VCH Verlag, Weinheim, 1998, p. 525.
[2.323]   C. W. R. Röntgen, *Sitz. Ber. Würzbg. Mediz. Gesellsch.,* 28. Dez. 1895.
[2.324]   D. L. Windt, *Comput. Phys.* **12,** 360 (1998).
[2.325]   F. Seward, *Rev. Sci. Instrum.* **47,** 464 (1976).
[2.326]   E. Hecht, *Optik,* Oldenbourg Verlag, München/Wien, 2001.
[2.327]   P. Kirkpatrick, A. V. Baez, *J. Opt. Soc. Am.* **38,** 766 (1948).
[2.328]   H. Wolter, *Ann. Phys.* **10,** 94 (1952).
[2.329]   see: http://imagine.gsfc.nasa.gov/index.html
[2.330]   N. Kaiser, *Vakuum in Forsch. u. Praxis* **6,** 347 (2001).
[2.331]   T. Johansson, *Z. Phys.* **82,** 507 (1933).
[2.332]   I. C. E. Turcu, J. B. Dance, *X-rays from Laser Plasmas,* J. Wiley & Sons, Chichester, 1998.
[2.333]   M. W. Charles, *J. Appl. Phys.* **42,** 3329 (1971).
[2.334]   J. H. Underwood, T. W. Barbee, *Appl. Opt.* **20,** 3027 (1981).
[2.335]   P. Pianetta, T. W. Barbee, *Nucl. Instrum. Methods A* **266,** 441 (1988).
[2.336]   H. Mai, *Mater. World* **7,** 616 (1999).
[2.337]   M. Schuster, H. Goebel, L. Bruegemann, *Proc. SPIE* **3767,** 183 (1999).
[2.338]   K. M. Skulina, C. S. Alford, R. M. Bionta, *Appl. Opt.* **34,** 3727 (1995).
[2.339]   S. Braun, H. Mai, M. Moss, Jpn. *J. Appl. Phys.* **41,** 4074 (2002).
[2.340]   H. Mai, R. Dietsch, T. Holz, *Proc. SPIE,* 1994, p. 268.
[2.341]   D. L. Windt, E. M. Gullikson, C. C. Walton, *Opt. Lett.* **27,** 2212 (2002).
[2.342]   A. Baranov, R. Dietsch, T. Holz, M. Menzel, in *Proc. SPIE,* 2003.
[2.343]   Y. Platonov, S. Andreev, N. Salashchenko, *Inst. Phys. Conf.* **130,** 583 (1992).
[2.344]   S. Braun, BMBF-Report FKZ 13N7886, (2001).
[2.345]   E. Spiller, *Soft X-ray Optics,* (Optical Engineering Press, Bellingham, 1994.
[2.346]   J. A. Nicolosi, J. P. Groven, D. Merio, R. Jenkins, *Opt. Eng.* **25,** 964 (1986).
[2.347]   D. L. Windt, W. K. Waskiewicz, *J. Vac. Sci. Technol. B* **12,** 3826 (1994).
[2.348]   M. P. Bruijn, P. Chakraborty, H. W. van Essen, *Proc. SPIE,* 1985, p. 36.
[2.349]   R. Dietsch, T. Holz, H. Mai, *Opt. Quant. Electron.* **27,** 1385 (1995).
[2.350]   J. A. Thornton, *J. Vac. Sci. Technol.* **11,** 666 (1974).
[2.351]   B. A. Movchan, A. V. Demchishin, *Fiz. Met. Metalloved.* **28,** 653 (1969).
[2.352]   A. Kloidt, PhD Thesis, Universität Bielefeld, 1993.
[2.353]   K. H. Günther, *Proc. SPIE,* 1990, p. 1324.
[2.354]   A. D. Achsachaljan, J. A. Bitjurin, S. W. Gaponov, A. A. Gudkov, W. I. Lutschin, *J. Technol. Phys.* **52,** 1584 (1982).
[2.355]   M. Fawer-Wasser, in *Current status and outlook for photovoltaics and solar thermal power; Sarasin Basic Report* (Sarasin, Basel, 2004), p. 27.
[2.356]   P. Würfel, *Physik der Solarzellen,* Spektrum Akademischer Verlag, Heidelberg, Berlin, Oxford, 1995.
[2.357]   E. H. Rhoderick, R. H. Williams, *Metal-Semiconductor Contacts,* Clarendon Press, Oxford, 1988.
[2.358]   W. Mönch, *Semiconductor Surfaces and Interfaces,* Springer-Verlag, Berlin, Heidelberg, 2001.
[2.359]   B. Rech, H. Wagner, *Appl. Phys. A* **69,** 155 (1999).
[2.360]   F. H. Karg, in *Solar Energy Mater. Solar Cells* **166,** 645 (2001).

[2.361]   M. Köntges, *Beleuchtungsabhängiger Ladungstransprot durch tiefe kompensierte Störstel-len in CdTe- und Cu(In,Ga)Se$_2$-Solarzellen*, Ph. D. Thesis, Universität Oldenburg, 2002.

[2.362]   B. Maennig, J. Drechsel, D. Gebeyehu, P. Simon, F. Kozlowski, A. Werner, F. Li, S. Grundmann, S. Sonntag, M. Koch, K. Leo, H. Hoppe, D. Messner, N.S. Sacrifici, I. Riedel, V. Dyakonov, J. Parisi, *Appl. Phys. A* **79**, 1 (2004).

[2.363]   D. Fischer, Y. Ziegler, P. Torres, E. Tagliaferi, S. Dubail, A. Closset, in *Proc. Workshop "Materialien und Verfahren zur Herstellung von Solarzellen"*, Frankfurt a.M., EFDS, Dresden, 2004.

3 Thin Film Preparation and Characterization Techniques

3.1 Thin Film Preparation Methods

3.1.1 Introduction

The aim of this section is to give an overview of several main thin metal film deposition methods for fabrication of functional layers in microsystems and electronic or microelectronic devices. It cannot give a full overview of all deposition processes but rather concentrates on such preparation methods, that are able to create thin metallic films functionally, in high quality and reproducibility, onto a large area and with high efficiency. Further information is given in [3.1–3.15].

For classification of deposition processes in electronic fabrication one may distinguish between vacuum-based and non vacuum-based methods. In the category of the vacuum based methods a division can be made between physical and chemical vapor deposition (PVD, CVD). In either case plasma-assisted processing is the favored procedure for thin metal film deposition. Non-vacuum based deposition can be divided into dry and wet chemical and physical processes. In particular, electroless and electrochemical deposition are rediscovered for electronic and microelectronic applications especially. Table 3.1 shows the process classification schematically.

Table 3.1: Process classification for metal-based thin films

Vacuum-Based Deposition	Non Vacuum-Based Deposition
Physical Vapor Deposition (plasma assisted or non-plasma assisted)	Dry Chemical Deposition
	Wet Chemical Deposition
Chemical Vapor Deposition (plasma assisted or non-plasma assisted)	Physical Deposition

The functionality of thin films is strongly influenced by their composition and microstructure. The composition and microstructure of thin films can be changed on a large scale depending on the deposition process and the corresponding parameter set-up. Exact knowledge of this interaction is one of the main conditions for setting the thin film functionality needed.

Metal Based Thin Films for Electronics, Second Edition. Klaus Wetzig and Claus M. Schneider (Eds.)
Copyright © 2006 WILEY-VCH Verlag GmbH & Co. KGaA, Weinheim
ISBN: 3-527-40650-6

In general, the deposition process can be divided into three steps:
1. Generation of the depositing species.
2. Transport of the species from the generation area (source) to the deposition area (substrate).
3. Condensation of the species and thin film growth.

These steps are completely independent from each other, if there is no interaction between their spheres of activity. In contrast, the geometry of the deposition equipment can influence the film growth and the microstructure sensitively.

The energy and the reactivity of the depositing species can be varied in relation to the deposition method and the individual process parameters. Both these main physical and chemical properties determine the course of condensation and the thin film growth. For example, the low kinetic energy of the species in physical evaporation techniques (~ 0.1 eV) can lead to a significantly lower adhesion between film and substrate in comparison with sputter or vacuum arc techniques (~ 1 – 100 eV). However, due to this higher kinetic energy thin films grow in a higher non-equilibrium state. So, for example, the microstructure changes to more polycrystalline with a high defect density. Such changes could be so extreme, that the thin film properties differ from the general material properties very strongly. On the other hand, the generation of new film properties is practicable.

The microstructure of thin films is also strongly dependent on the kind of substrate (e.g. roughness, crystalline structure) and the chemical state of the surface. Surface roughness and geometrical shadowing could lead to a preferential growth, e.g. giving a columnar structure of the thin film grains. Smoothing and cleaning processes immediately before depositing the species influence the adhesion strength and the microstructure of the thin film. The importance of this interaction increases as the films become thinner and thinner. Amorphous thin films are of high interest in microelectronic application, where the film thickness shrinks to a handful of monolayers. A new understanding of how to produce ultra-thin films with high purity is growing.

The aging of thin films plays an important role in their present application in microelectronics. Due to the reduction in dimensions diffusion processes can change the functionality of thin films in a shorter time. The diffusion of impurities and the interdiffusion of immiscible materials via defects increasingly occurs and becomes the main reason for change in the properties and the functionality of thin films in general. An important question to answer is how the stability can be improved for thin films of ever decreasing size and thickness.

In general, the choice of a thin film deposition process is based on a set of material properties to realize the required purpose. These parameters are, for metal-based thin films:
- electrical parameters (conductivity, resistivity, electromigration stability, radiation hardness)
- thermal parameters (expansion coefficient, thermal conductivity, melting point, recrystallization temperature)
- mechanical parameters (intrinsic stress, hardness, elasticity, ductility, adhesion)
- optical parameters (absorption, reflexion, spectral characteristics)
- magnetic parameters (permeability, saturation flux density, coercive force)
- chemical parameters (composition, reactivity, chemical resistance, toxicity, hygroscopicity)
- morphology (crystalline structure, crystallite orientation, kind of defects and defect density, surface topography, structure conformality, interface state)

Any property can be influenced by the deposition technique and its parameters. However, each deposition technique has its limitations but, due to the wide variety of processes, it is probable that a deposition process can be found or developed for a specific set of thin film properties in most cases.

Beside the choice of thin film material parameters there are two other aspects to which attention should be paid. First, the deposition process is a planar one, i.e., for smooth and planar substrate surfaces it is normal to get homogeneous thin films properties. In principle, for large substrates problems exist only in expense. In reality, often a non-planar structured and inhomogeneous surface has to be deposited. So, in microelectronic devices side walls and deep small bottoms have to be deposited with thin films conformal in thickness and properties. Second, the thin film deposition equipment has to meet a set of requirements regarding productivity, deposition reproducibility, reliability and flexibility. The last will become more and more important in future. Online equipment and process control is necessary to increase the system throughput and to decrease the maintenance and repair time, respectively.

In general, two kinds of deposition equipment must be distinguished, the batch processing type and the single substrate facility. In the case of short time deposition the single substrate facility seems to be favored if high productivity is required. This kind of deposition offers additional advantages regarding the process stability, the incorporation of pre-deposition cleaning as well as the post-deposition treatment, such as thermal annealing or surface covering.

The process automation and control is becoming standard. The present deposition equipment provides automatic loading and unloading, process sequencing, parameter measuring and control, failure diagnostics, remote process recipe generation as well as product status information. The coupling of various deposition chambers to a unique system, so-called cluster tools, improves the quality and functionality of the thin films. The sequential deposition of thin films without breaking the vacuum leads to more defined interfaces and a distinctly higher quality of the thin film system. Last, but not least such cluster tools represent a further step towards a completely automated production line.

3.1.2 Physical Vapor Deposition

Physical vapor deposition (PVD) can be divided into three groups: sputtering, evaporation and ion-plating. Among the high number of PVD methods (magnetron) sputtering is the most utilized deposition method for thin metal films. This plasma-assisted method provides deposition of a large number of various materials: metals, alloys, semiconductors, insulators, organic blends etc. The relevance of this technique is increasing as the film thickness becomes thinner and thinner. The thicknesses of sputtered films are usually from 1 nm (a few monolayers) up to a few micrometers. The sputtering process itself uses the interaction of accelerated ions with atoms of a so-called target, containing the material to be deposited. The ions can be generated by a cold plasma or an ion gun. The energy of the ions has to be high enough for multiple scattering in the surface region of the target. The energy and momentum transfer lead to structural changes in the near surface region of the target and to a partial emission of usually neutral atoms and/or clusters of atoms. Fig. 3.1 shows the sputter process schematically.

Due to a considerable kinetic energy of some eV the sputtered atoms and ions move from the target and condense on a substrate, located opposite to the target at distances between 50 and 500 mm. The target surface erodes in dependence on the ion energy, their density and the angle of incidence. The condensation of atoms cannot be completely prevented on surfaces outside the substrate so, a loss of material flux and particle generation has to be considered for fabrication of thin films, especially.

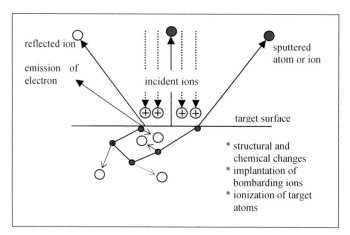

Figure 3.1: The sputtering process shown schematically

In contrast, there are other PVD vacuum-based deposition techniques, which do not use an ion source. Such methods are called classical PVD methods and include all evaporation processes to generate a metal vapor, which should be condensed onto a substrate surface inside the vacuum chamber. The first deposition processes were performed by Joule by heating thin wires in a vacuum chamber. For fabrication of high quality thin films a collisionless flight of particles has to take place from the vapor source to the substrate so, a vacuum of better than 10^{-5} Pa has to be realized. Besides the Joule heating, electron beam or laser induced heating, arcing or eddy currents are the other favored methods for vapor generation. The particle energy is of the order of 0.1 eV, which is much lower than the energy of sputtered particles. If ionization of the vapor atoms is successful the kinetic energy of the ions can be increased substantially. Laser induced and arcing processes especially have a high potential for manipulation of the particle energy. A second main difference is that the evaporation source can be estimated as a point source in a first approximation. Sputtering sources are more planar and thus more suitable for deposition of large area substrates.

The third group of PVD processes is represented by ion-plating techniques. Here, the evaporation process is assisted by the interaction of the evaporated atoms with ions. The ions generated by a cold plasma partially ionize the particles of the vapor flux. Biasing the substrate to a negative potential all positive ions (ionized vapor atoms, ions of the plasma) will accelerate to the substrate and influence the condensation process. This influence should be a constant sputtering of the present substrate surface. The removal of surface impurities and the knock-on effect of condensed atoms lead to a better thin film adhesion, a modification of the microstructure and a lower impurity content. On the other hand, the deposition rate decreases due to the scattering of the vapor particle and the surface sputtering.

Movchan and *Demchishan* (Fig. 3.2) and *Thornton* (Fig. 3.3) have schematically described the microstructure of thin PVD films for evaporation and sputtering, respectively, as a function of the melting point of the condensate, the substrate temperature and the argon pressure in sputtering. Although many physical and reactor parameters are not considered these models give a satisfactory overview of the microstructure which could be expected from PVD processes, generally. Predominantly, a polycrystalline microstructure will be created. A high amount of defects like grain boundaries, pores and dislocations, vacancies and impurities will move the material properties far from those of compact materials. A columnar growth of the grains will very often be observed. Often, it is necessary to improve the thin film microstructure by developing modified deposition processes to create thin films with the material parameters needed for the specific application.

All three groups of PVD techniques have advantages and limitations, which cannot be discussed here in detail. In the following several PVD methods will be described with regard to their application capability for thin metal film deposition. Examples for application will be given.

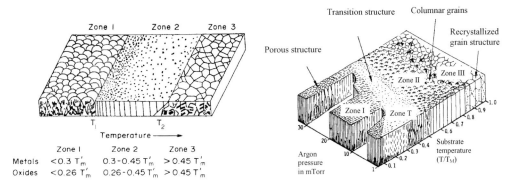

Figure 3.2: Structural model of PVD condensates (after *Movchan* and *Demchishan* [3.16])

Figure 3.3: Structural model of PVD condensates (after *Thornton* [3.17])

Magnetron Sputtering

To understand the advantages and limitations in sputtering it is necessary to focus on the mechanism of sputtering and the sputter yield, both qualitatively and quantitatively. Relating to Fig. 3.1 the interaction between the incoming ion and the target atoms can be treated as a series of binary collisions. A comparison of this process with the opening event in billiard is not too unrealistic. The creating collision cascade can lead to an emission of target atoms (backward scattering) or more likely to dissipation of the ion energy into the interior of the target (forward scattering). Only about 1% of the incident energy is transformed into kinetic energy of the sputtered atoms. From this it follows that the sputter yield is not high or that a high ion density bombardment of the target is necessary to reach deposition rates sufficient for industrial application. High density plasma sources have to be used.

For sputtering of multicomponent materials the non-stoichiometric removal of surface atoms must be considered. Following the mass dependence in sputtering only the lighter component will be sputtered preferentially but this effect is weak in comparison with the role of the surface binding energy. The prediction of the real surface composition demands detailed information about the surface chemistry and the binding forces. It is interesting to remark that preferential sputtering does not influence the composition of the thin film deposited. At the beginning of sputtering there is an imbalance in removal conditions. After a short time at constant sputter parameters the changed surface concentration becomes constant. The element with the higher sputter yield has a reduced surface concentration so, the absolute amount of those sputtered atoms remains constant and equal to the target atomic concentration. Following this, the thin film composition is equal to the target composition. This advantage is disturbed by scattering of sputtered atoms with each other and with the plasma ions (e.g. Ar^+). Lighter atoms are scattered more strongly than heavier ones. This leads to an increase of the heavier element in the thin film. For instance, due only to this effect the concentration of titanium sputtered from a target of 70 at% W and 30 at% Ti decreases to about 22 at% Ti. This loss of the lighter element can be reduced by optimization of the working gas pressure (in the range between 1 and 0.1 Pa) in the plasma and/or the distance between the target and the substrate.

As already mentioned, high density plasma sources have to be used to generate a sufficiently high ion current (e.g. Ar^+). The aim must be to employ all possibilities for getting a high effective ionization of the working gas. This can be done by using an additional magnetic field to exploit the major part of the electron energy in a plasma discharge for impact ionization. So-called magnetron sputter sources have revolutionized the sputtering deposition process. They can be defined as diode devices in which the electrons move on closed spiral trajectories. Magnetrons can be configured in various forms. The most important are planar magnetrons for large area and high rate deposition. In Fig. 3.4 an example is given for two types of planar magnetron sputter sources.

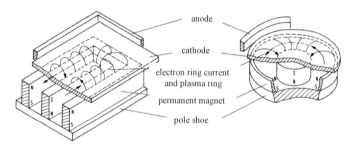

Figure 3.4: Planar magnetron sputter sources (after *Haefer* [3.3])

The magnetic field is perpendicular to the electric field and the electrons have to move in a closed ring-shaped trajectory. The target serves as the cathode and the substrate (not shown here) has to be located face down above the target. From Fig. 3.4 it is clearly seen that the plasma density should not be uniform over the target area. This leads to a non-uniform emission of sputtered atoms from the target. Consequently, in the case of a planar substrate there will be a deposition rate profile as a function of the distance between target and substrate. These differences disappear at higher distances only. Magnetron sputter sources are under development to achieve thin films more uniform in thickness over larger areas.

The plasma discharge can be realized by a dc or a rf electrical field. The advantage of rf sputter sources is that they can deposit nonconducting and semiconducting materials. In the case of reactive deposition, where a material compound is formed after atomic emission from the target more homogeneous thin films can be produced. However, the deposition rate is reduced significantly and cannot be compensated by higher electrical power supplies. Dc power supplies are able to provide more plasma energy.

Although the magnetron deposition process requires an expensive high vacuum chamber this method has found wide application. An important application area is the production of functional thin films in microelectronics (see also Sections 2.1 and 4.2). For instance, in the fabrication of silicon-based integrated circuits the magnetron sputtering of Al wires has been used world-wide. Due to the demand for submicron and nanoscaled thin film structures the deposition rate plays an increasingly minor part. The deposition of high quality ultra-thin films onto patterned surfaces becomes more important. Film thicknesses of less than 10 nm are the order of the day now. Additionally, the deposition of three-dimensional structures is a big challenge for planar techniques. In the present copper chip metallization magnetron sputtering is used for the deposition of Ta and TaN diffusion barriers and a copper seed layer. Here, trenches and contact vias with high aspect ratios (width/height) have to be deposited nearly conformal. Only modified magnetron sputter methods are able to guarantee a satisfactory bottom and side wall deposition. Vertical walls and deep bottom edges are very critical areas for thin film defects. So, the deposition methods have been modified for a more parallel incidence of the atoms and an increase in their kinetic energy. The last can be done by ionization of the sputtered atoms on their way from the target to the substrate. Techniques like long through sputtering (LTS) and ionized metal plasma deposition (IMP) are applied to minimize the risk for barrier and seed layer failure. Figure 3.5 characterizes the nature of the problem. On the left side the critical areas are illustrated for a dual damascene structure.

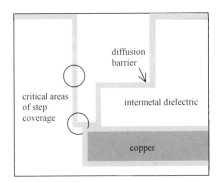

Figure 3.5: Trench and via deposition with diffusion barrier [3.18]

The right side shows the reality in IMP. Due to a more parallel incidence of the deposited atoms the bottom coverage is improved significantly. The side walls in the near bottom region also possess a good thin film coverage. This is caused by the acceleration of a considerable number of ionized sputtered atoms and their partial re-sputtering from the bottom of the trench in the side wall direction. Other deposition techniques which are more suitable for conformal deposition will be discussed below.

Due to the higher kinetic energy of the ionized atoms a modified growth mechanism is expected. The thin films are denser and the adhesion strength is improved. They become a more amorphous-like microstructure with less defect density. This change in the thin film property improves the barrier behavior even if the thickness is further decreased. Similar effects can be achieved by an additional plasma treatment or an ion implantation after the thin film deposition.

Thermal Evaporation

In the evaporation process a material vapor is produced in vacuum by heating the material to be deposited to a sufficiently high temperature. The temperature needed is a function of the vapor pressure of the material. The evaporation process has to be carried out in a high vacuum environment so, the transport from the evaporation point prior to the condensation area should occur without collisions between the vapor atoms and the molecules of the residual gas. The substrate is always located face down above the evaporation point at distances from a few to some ten centimeters depending on the deposition area and the required film thickness homogeneity. In Fig. 3.6 a scheme illustrates the deposition process.

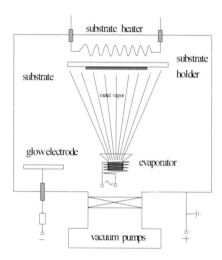

Figure 3.6: Scheme of the evaporation process

In a first approximation the evaporation source can be looked at as a point source. Assuming a spherical, or better a cosine-like, emission function the film thickness will vary over planar substrates. The differences in thickness can be limited by using a spherical substrate holder. An additional rotation will further improve the thickness homogeneity. In dependence on the material to be evaporated a choice has to be made as to which evaporation technique can be used. The main techniques of creating the vapor are direct heating by high electric current (resistant heated source), sublimation source, induction heated source, electron beam heated source, laser beam heated source and arc source.

There is a large range of materials which can be deposited. This variety, however, is restricted by serious problems in evaporation of high melting materials and of compounds or alloys. A technical problem is the reloading procedure due to a limited stock of deposition material. In general, the reload is combined with a break in the vacuum. The general deposition time increases as high vacuum restoration takes an appreciable time in spite of high efficiency pumping systems. So, a multitude of solutions has been developed for reloading the evaporation tank. The composition of the vapor, and consequently of the thin film, is, among other things, a function of the vapor pressure ratios of the components. Composition changes take place in the source. A special feature are so-called azeotropic blends. They possess a specific boiling point independent of the concentration of the components in the source. Thus, a constant thin film composition is possible. Due to the thermal activation of the components reactions may occur during the transport from the source to the substrate. Depending on the type of evaporation source a possible interaction of the evaporant with the source material must also be considered. The higher the evaporation temperature the higher the risk for compounding or alloying. A high degree of competence has been accumulated and subsequently, a deposition procedure can be evolved for most materials. There are compendia for choosing the best method for evaporation.

Remarkable progress has been made in large area deposition. Flat panel displays, large area solar cells and the deposition of windows and foils for industrial or domestic application are excellent examples of the introduction of evaporation techniques into an effective and high output production. This progress was achieved by the development of highly productive equipment, like cluster tools and multilayer deposition chambers without breaking vacuum. The substrates move continuously above broad evaporation sources and the chambers are selected by pumped gate valves. Such equipment combines high rate deposition with continuous-flow surface cleaning and deposition.

Electron and laser beam heated sources as well as arc sources are other important evaporation sources and are further under intensive development. In the following the discussion will be focused on these techniques.

Electron Beam Evaporation

Heating with an electron beam has two main advantages. Firstly, a high power density provides a wide range of evaporation rates and secondly, there is practically no interaction of the evaporant with the source material. The crucible consists of a water-cooled container and the molten area is located inside the evaporation material. Specific examples for evaporation material are tantalum, titanium, tungsten, metal oxides and other dielectrics. Figure 3.7 shows an example of an electron beam evaporator. A W-cathode provides the electron emission. After acceleration of the electrons up to a kinetic energy of tens of keV they are deflected by a magnetic field to reach the evaporant surface perpendicularly.

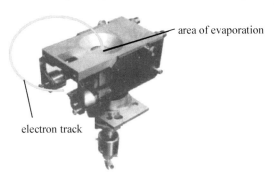
area of evaporation

electron track

Figure 3.7: Electron beam heated evaporation source

The electron current amounts to about 1 A at an area of the beam focus of less than 1 cm^2. The evaporation rate can be varied by the applied power. Besides the hot cathode electron gun as in Fig. 3.7 which works in high vacuum there is another type of electron beam gun. In the plasma electron beam gun the electrons are produced by a plasma discharge. Such a gun can operate at higher pressures as is important for reactive and ion plating evaporation.

Laser Beam Evaporation

If the laser radiation wavelength lies in the absorption band of the target material the interaction of a laser beam with the target leads to a rapid heating of the surface (on a ns time-scale) and an ablation (evaporation) of the first atomic layer(s). This ablation process is often called pulsed laser deposition (PLD) because pulsed irradiation sources are used. For metal surfaces the convenient wavelength is in the UV region. Excimer lasers have to be used with wavelengths less than 400 nm. Here, the light absorption increases very sharply. Laser ablation occurs at an energy density input of a few J cm^{-2} for most metals. The material removal from the surface can take place in the form of vapor, atomic clusters, droplets and debris. The vapor of the erosion plume consists of mainly excited atoms. The laser ablation is accompanied by a plasma formation in the atomized target vapor and the surrounding gas. Further, the incoming laser irradiation is partially absorbed by the plasma making it hot and pressurized. Photoionization and acceleration of the vapor atoms take place. The kinetic energy (erosion plasma temperature) reached is of the order from a few eV up to some hundred eV depending, among others, on the wavelength of the laser irradiation. It is interesting that dealing with atoms and ions with sufficiently higher kinetic energy than in thermal evaporation or sputtering opens new opportunities for managing thin film properties. This will be discussed further later.

The excited or ionized state of the vapor atoms in the erosion plume can be exploited for their additional acceleration. Up to now, there have been investigations to get additional ionization effectively. For instance, a second laser irradiation perpendicular to the spreading direction of the erosion plume heats up the electrons in the plasma. Due to their acceleration the electrons drag along the ions with themselves and thus a significant increase in the kinetic energy can be achieved. In [3.19] a doubling of the median energy was measured due to an additional CO_2 laser irradiation. Furthermore, the concentration of atomic clusters and droplets in the plume can be decreased in favor of the atomic vapor.

The interaction of two synchronized laser beams and two erosion plumes is applied to get droplet free thin films [3.6].

In general, the targets are formed cylindrically and not discoidally like in sputter sources. This is based on the necessity for a uniform material removal. Cylinder rotation is a simple and reliable solution. The most effective exploitation of the target material is reached by additional movement of the cylinder along the axis.

As already mentioned, the ablation of the target is of the order of a few monolayers or less if an excimer laser irradiation is used. Regarding the particle loss during the transport of the vapor atoms and ions from the target to the substrate a fraction of a monolayer is condensed only but this thin film material condenses during a short laser impulse of the order of 10 μs. So, the mean value of the deposition rate is relatively high and is dependent on the possible laser shot rate. A precise amount of thin film material can be deposited. The thin film composition corresponds to the target composition. Together with the high particle energy these three features make PLD an excellent method for preparation of thin films with exceptional properties. Monolayer and submonolayer deposition are a precondition for preparation of metastable phases, super lattices, thin film systems and alloys. The high quenching rate of the deposited material leads to a freezing of the nonequilibrium state immediately after condensation of atoms and ions. X-ray mirrors are a commercially far evolved product produced by PLD. Here, a sandwich structure of very smooth and ultra-thin films with an interface as perfect as possible provides a very high coefficient of reflectance. More detailed information is given in Section 2.5. Sandwich structures with alternating ferromagnetic and non-ferromagnetic layers to get a giant magnetoresistance (GMR) are another example of successful use of PLD.

These very promising properties of PLD are damped by the problem of depositing large area substrates. Due to the necessary smallness of the laser focus a large area deposition can be realized only under precise conditions. If the low absolute amount of the particle flux per laser shot is added there is no effective way of depositing substrates of much more than 10 x 10 cm^2.

Arc Evaporation

Vacuum arc deposition is now widely used for the deposition of hard and decorative coatings. This kind of evaporation is really the first and oldest method of material deposition. The interaction of lightning with matter is always accompanied by a material evaporation. The only problem to be solved is controlling the arc in energy and location. Advanced power supplies are able to provide a high current arc discharge. By generating a high current arc a materials plasma forms and evaporation takes place. It can be distinguished between a direct current arc (dc arc) and pulsed current arc. The current of a dc arc discharge is limited by the repelling forces between the single spots. Using a pulsed arc current of several 10^3 A an instantaneous high deposition rate can be achieved. Values of about 100 μm s^{-1} are attainable. Nevertheless, pulsed arc discharges can produce small and well defined portions of ions.

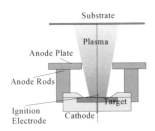

Figure 3.8: Schematic of a pulsed HCA source (left) with magnetic droplet filter (right) (source: FhG IWS Dresden)

In arc evaporation droplets are produced comparable to laser beam deposition. The lower the melting point of the target the higher the droplet size and density. Thus, the particle beam has to be filtered. The filtering can be realized by a magnetic field due to a hundred percent ionization of the evaporated atoms. The ions can be deflected by a bent cylinder provided with an electromagnetic coil. Inside the cylinder the electrically neutral droplets move straight to the wall while the ions move along the bent axis. In Fig. 3.8 the evaporation source, schematically, and the filter unit are shown for a pulsed high current arc (HCA) equipment.

The discharge is ignited by a high-voltage pulse over the ignition electrode. The arc moves radially from the ignition electrode to the target edge and erodes the target material homogeneously. The energy of the arc decreases and arcing can be stopped by an optimized pulse clipping.

The kinetic energy of the ions is of the order of a few tens up to hundreds of eV. An additional substrate dc bias can increase the kinetic energy significantly.

Thin film growth from heavy atoms is not understood completely, based mainly on the difficulty in producing and handling so-called hyperthermal particles (10 eV $< E_{kinetic} <$ 1000 eV). Extensive experimental results are absent in comparison with thermal evaporation or sputtering but the knowledge has been growing intensively in the last few years. The main features of interest are:

1. Surface cleaning
2. Increased surface mobility of adatoms
3. Re-sputtering
4. Generation of structural defects in the layer and the surface
5. Atom mixing and ion penetration beneath the film surface
6. Implantation

Returning to arc evaporation the complex influence of ion energy and charge can be used to modify thin film growth and thin film properties. The adhesion of thin films can be improved. For hard and decorative coatings an in-situ cleaning before deposition is not imperative. Thin films like copper on silicon or on polymers which have, in principle, a low adhesion strength stick very well. Due to the enhanced surface mobility a more layer-by-layer film growth is observed replacing the island-growth typical for thermal evaporation and/or sputtering. Up to now, not nearly all the opportunities have been used to characterize the interrelations to optimize thin film properties in arc evaporation.

Molecular Beam Epitaxy

Molecular beam epitaxy (MBE) was developed in the early 1970s as a means of growing high-purity epitaxial layers of compound semiconductors. MBE can produce high-quality layers with very abrupt interfaces and good control of thickness, doping and composition. It is a valuable tool in the development of sophisticated electronic and optoelectronic devices, especially single crystal thin films. The constituent are deposited onto a heated crystalline substrate. The molecular beams are typically from thermally evaporated elemental sources, but other sources include metal–organic group III precursors (MOMBE), gaseous group V hydride or organic precursors (gas-source MBE), or some combination.

To obtain high-purity layers, the material sources have to be extremely pure and the entire process must be done in an ultra-high vacuum environment. The growth rate is typically of the order of less than 1 nm s^{-1} and the beams can be shuttered in a fraction of a second, allowing for nearly atomically abrupt transitions from one material to another. Due to the complex vacuum environment and to the point-like evaporation sources there is no application for large area deposition.

An important part of the chamber consists of some analytical tools for in situ deposition and thin film control. For instance, reflection of high energy electron diffraction (RHEED) is used to control the crystal structure during deposition. A quadrupole mass spectrometer serves as a highly sensitive detector for fluctuation of the residual gas composition. Surface sensitive elemental analyzing techniques like AES or XPS are applied in spite of the high cost. The arrangement of two ore more evaporation sources in one chamber can cause cross-contamination effects. Thus, multi-chamber tools with one source per chamber become more and more the rule at present. The evaporation sources represent another important technical detail. In general, we can distinguish between thermal evaporation sources (Knudsen cells, direct heating cells, ionization cells and others), electron beam heated sources, implantation sources and gas sources. System controll and automation are inalienable.

The surface preparation is significant for the growth of determined monolayers. A multisequence technique of sputtering and heating (cleaning and crystal restoration) under ultra high vacuum conditions seems to be the best way for successful homoepitaxy or heteroepitaxy. In any case a wet chemical pretreatment is necessary to eliminate strong contamination and to create homogeneous surface conditions.

Ion Plating

The importance of ion plating techniques is high. In general, these techniques can be viewed as an additional tool to overcome the problems of conventional evaporation regarding the adhesion of thin films, their morphology, density and internal stress. Ion plating is typically done in an inert gas discharge. The substrate must be at a negative potential relative to the plasma potential. Argon ions bombard the surface and influence the interface between substrate and thin film, the growth process in general and film density and stress in particular. It must be mentioned that success is not so substantial as in sputtering. This is mainly due to the problems in generating high density plasma. The ion density is not high enough and the kinetic energy of the atoms remains unchanged and low.

3.1.3 Chemical Vapor Deposition

In general, chemical vapor deposition (CVD) is the formation of thin films via chemical reactions from gaseous precursors. The activation of the chemical reaction is initiated by thermal or electric discharge plasma treatment. Decomposition or reduction of compounds like fluorides, chlorides, bromides, organometallics, hydrocarbons, phosphorus trifluoride and ammonia complexes provides the deposition of the metallic component. After dissociation the element or compound to be deposited condenses(reacts) on the substrate surface while the volatile component leaves the reaction chamber. Typical reactions for metal deposition are:

$$WF_6 + 3 H_2 \longrightarrow W + 6 HF \tag{3.1}$$

$$WCl_6 + 3 H_2 \longrightarrow W + 6 HCl \tag{3.2}$$

$$TaCl_5 + 5/2 H_2 \longrightarrow Ta + 5 HCl \tag{3.3}$$

$$TiCl_4 + CH_4 \longrightarrow TiC + 4 HCl \tag{3.4}$$

The temperature has to be in the range between about 350 °C and 1100 °C. Chemical reactions already in the gas phase lead to particle generation (powder formation). The thin film quality is dependent on the reaction kinetics, the temperature, the surface preparation, the purity of the precursor, the gas flow and the chamber conditions. The deposition can be carried out in vacuum, under normal or high pressure conditions. The term "in vacuum" includes all pressures lower than atmospheric. Plasma activation possesses the advantage of significantly lower reaction temperatures. The particle energy is very low. In general, the deposition process is reaction controlled. That means, among others, that the dependence of the film growth on the angle distribution of the incoming reactants is very low. In contrast to the PVD techniques a more conformal deposition is possible. Regarding the demands of the advanced electronics CVD seems to be favored. However, CVD methods have not replaced sputtering or others. Differences in microstructure, film density and defects are some possible reasons. In metal deposition for electronic application low pressure CVD and plasma enhanced CVD are the favored methods.

Low Pressure CVD (LPCVD)

The improvement of thin film quality (uniformity, conformality, purity) and the reduction of processing costs (precursor consumption) are the driving forces for the development of LPCVD techniques. Early reactors were modified diffusion furnaces. Now, there are four types of reactors: horizontal, vertical, bell jar and single wafer reactors. The LPCVD process can be divided into two main categories: film deposition via pyrolysis and via reaction of two or more components. Examples of the first group are the deposition of TEOS (silicon dioxide) or of tungsten (reactions in (3.1) and (3.2)). The deposition of silicon oxide by the reaction of dichlorosilane and nitrous oxide is an example of the second group.

In microelectronics the tungsten LPCVD deposition process on silicon at low temperature has the advantage of selective growth on Si but not on silicon oxide, for instance. The process is not simple and a lot of problems had to be solved. The process operates in two steps: 1. reduction of WF_6 by silicon substrate in an inert gas atmosphere and 2. reduction of WF_6 by H_2. The deposition of tungsten silicide is a process similar to that for tungsten but silane (SiH_4) is used instead of hydrogen. The growing selectivity disappears here.

Another example of LPCVD is aluminum and aluminum silicon alloys. These films can be deposited through pyrolytic decomposition of triisobutyl aluminum. The deposition temperature is between 250 and 300 °C.

Plasma Enhanced CVD (PECVD)

Thin films can be produced at a lower temperature than the corresponding thermochemical reaction claims. This can be done by applying a cold plasma to enhance the chemical reactions. Since the cold plasma is not in a thermal equilibrium and the reaction cross sections are not well known the deposition process is difficult to describe theoretically. In addition, the plasma interacts with the substrate surface and the reactor walls. The gas molecules, atoms and/or radicals are excited, ionized or dissociated, mainly by electron impact. They diffuse to the substrate surface, migrate on it and find corresponding adsorption sites. Reaction between the atoms can take place and the film grows. The thin film growth can be enhanced or disturbed by the excited species and the plasma interaction. A general problem is the deposition rate. This parameter can be increased by higher particle density of active species only. And this can be done by using a higher gas flow and an ultrahigh plasma density. Thus, the development of high density plasma sources (parallel plate reactor, microwave or ECR reactor, ICP reactor) is one of the main tasks in advanced PECVD processing. Deposition rates between 10 and 1000 nm min^{-1} can be reached. The deposition temperature is reduced to $400 - 100$ °C dependent on the thin film material and the corresponding film properties.

PECVD is widely used in electronics. Especially in microelectronics, the deposition of tungsten, tungsten silicide, silicon oxide, silicon nitride, barrier materials like Ti, TiN, WN and low-k or high-k insulators is executed with PECVD processes. The technological demands from industry have moved the process techniques far ahead. The deposition area has been increased from some square centimeters to substrates with a diameter of 300 mm and more. Large area deposition has became a fact.

The problem already mentioned in Section 3.1.2 of the deposition of a thin film with deep trenches and vias conformal in thickness and property is reduced significantly. In Fig. 3.9 a trench is seen in silicon oxide which is deposited with a PECVD TiN barrier layer. The conformality is nearly 100%. In general, the PECVD thin film grows polycrystalline with larger grains than with PVD techniques. This can be explained by a film growth near to the thermodynamic equilibrium. The thin film properties of course differ from those in PVD. Layer adhesion and defect density can differ very strongly due to the absence of kinetic energy of the deposits.

Figure 3.9: PECVD TiN deposited into a deep trench of silicon oxide (source: ZfM TU Chemnitz)

Atomic Layer Deposition (ALD)

Atomic layer deposition is a strongly modified CVD technique developed more than 20 years ago. Chemical reactants are introduced into a vacuum chamber. But in contrast to CVD, the reactants are supplied step by step, where the steps are separated by an additional purge gas stream. The reactants provide a chemical reaction with the substrate surface in a precise monolayer mode. If ALD is used to deposit a metal oxide or nitride film (e.g. Al_2O_3, TaN) the first reactant contains the metal and the second the non-metal oxygen or nitrogen. The second reactant forms a compound with the metal just as a monolayer. Each reaction is self-limiting. That means that a wholly conformal deposition is possible. The reaction temperature is typically in the $200 - 400$ °C range. The deposition rate is very low at about 0.1 nm/cycle. Because of this low deposition rate ALD can be used for ultra-thin films only. The importance of such ultra-thin films is rapidly increasing and so the interest in ALD is now very high.

From the technical point of view, in ALD high efficiency pump and valve systems have to be applied. A speedy switching is necessary. The chamber walls also acquire deposits and an effective cleaning process has to be developed. CVD chambers can be considered as a basic tool but it needs more development to establish ALD as a highly reliable production tool. Another problem to be resolved is the precursor chemistry. Any material needs a specific chemistry and deposition technology. It is no wonder that only a few processes are under development now. There are three possible areas for ALD use in microelectronics: DRAM deep trench dielectrics, gate dielectrics and diffusion barrier/seed layer for interconnects in the back end. These application fields are the main critical ones for the further increase in the transistor density. A structure size of less than 70 nm is the next aim in microelectronic engineering. Figure 3.10 shows the capability of ALD in the case of high-k materials (here hafnium oxide with a thickness of 30 nm in a 6 µm deep and 0.17 µm wide trench). The increase in the dielectric constant and the more conformal deposition inside the trench compensate for the lack of capacity due to the smaller structures. The demand for more effective gate dielectrics can be realized, among others, by a higher dielectric constant of the insulator material. Dielectric ALD with an alternative insulator has a good chance of success in replacing CVD and thermal SiO_2 dielectric deposition.

Another successful application is improving the deposition of tungsten. ALD can provide a more conformal tungsten deposition in via filling. The ALD tungsten film serves as a seed layer. There is even a "production-ready" equipment for void-free via filling with tungsten.

Figure 3.10: Deep trench with 30 nm hafnium oxide [3.20]

The disadvantages of PVD techniques for depositing vias and trenches with a diffusion barrier and a seed layer conformally could be eliminated by ALD. In the back end processing of copper interconnects a TaN barrier and a copper seed layer have to be deposited. In particular the barrier thickness is changing to values below 10 nm now. PVD techniques are less and less able to guarantee regular barrier properties with any possible barrier material. ALD could bring significant progress and there is a chance of reducing the barrier thickness to a few monolayers. The usual seed layer thickness is now of the order of 100 nm but Cu ALD could decrease the required thickness to a few tens of nanometers if a conformal copper deposition becomes reality.

3.1.4   Non Vacuum-Based Deposition

The importance of non-vacuum based deposition processes of metals for electronic application is concentrated on atmospheric CVD and deposition from aqueous solutions. The main process technique for the latter is electrodeposition, also called electroplating. Independent of the wide application of electroplating processes in jewelry, household articles and other decorative coatings, electroplating has found a high application level in the fabrication of printed board assemblies, metallization of ceramic substrates and protective coatings. A second deposition technique from aqueous solutions is electroless or chemical deposition. Like in CVD, a selective layer deposition is possible. A third deposition technique is anodic oxidation. Metal oxide, nitride and sulfide thin films can be formed.

Electroplating

Electroplating works by ions being generated in an aqueous solution (electrolyte). The electrical transfer and the ion generation is realized by a dc source and two electrodes (anode, cathode) dipped into the solution. Positive ions travel to the cathode and the negative charges (negative ions, electrons) move to the anode. The electrolyte contains the elements to be deposited and additives influencing the condensation process. The deposition reaction at the cathode is characterized as a reduction reaction since electrons are consumed to neutralize the metal ion. The following equation describes the situation

$$M^{n+} + n \cdot e \rightarrow M \tag{3.5}$$

The anodic reaction is an oxidation reaction. The consumption of electrons in the cathode reaction must be equal to the value of liberated electrons in the anodic reaction. The process is maintained by the dc source and is limited by a condition of material balance.

In the cathode region a dynamic balance forms between the substrate surface and the solution. A potential difference is caused between the solution and the substrate surface because a small amount of metal is dissolved. This difference, called electrode potential, is dependent on the kind of metal, the ion concentration in the solution and the temperature. The electrode potentials of both electrodes determine the lowest voltage to short the electrical circuit. Actually, the voltage needed is higher because of the diffusion process of the metal ions to the substrate, the discharge of the ions and their incorporation into the lattice of the growing film.

In general, the electrolyte is an alkaline, alkaline-cyanidic or acid solution. The metal to be deposited is contained as a salt. Ni sulfate, Ni chloride or Ni sulfamate are the typical salts for nickel deposition, for instance. Organic additives in low concentration influence the film roughness, the grain size and the intrinsic stress. In the case of alloy deposition the electrode potentials of the components have to be similar or else complex compounds have to be used and/or a decrease in the concentration of the more noble metal in the solution is necessary.

Usually, the anodes are made from the material to be deposited but the use of insoluble anodes is also possible (deposition of chromium, gold or platinum). The results of deposition are mainly dependent on the temperature, pH-value and the current density. Features like the kind and form of the anode, an additional stream and shielding screens in the electrolyte have secondary significance. The current density and its distribution determine the microstructure and the thickness of the thin films. This means that there is a correlation between the surface geometry of the substrate and the deposition result, in particular the film thickness. In electronics the surface design is manifold. Very smooth surfaces alternate with very rough or three-dimensional ones. Provided that current density fluctuations are not too high the deposition result is comparable with that in CVD. The negligible low kinetic energy of the ions provides a regular thin film growth due to the reaction controlled process. A highly conformal thin film can be produced over a live surface (conductive substrate or thin film - seed layer). If the surface is covered with an insulator the film grows only over that area where charge exchange is possible. This effect can be used for the so-called pattern plating. Here, an insulator (usually a photoresin) is patterned over a metallic surface. In that area where the electrolyte is in contact with the metallic surface a film forms and grows. Figure 3.11 shows a copper deposition into openings of a photosensitive patterned resin. If the resin is removed a copper structure is left at positions determined by the design.

In the last years electroplating has conquered the back end processing of highly integrated circuits. In this field trenches with a high aspect ratio have to be filled with copper mainly anisotropically, even though there is a closed live surface. Figure 3.12 illustrates this.

The aim of the deposition is to get quite a lot of copper inside the trench. After electroplating chemical mechanical polishing (CMP) has to smooth the surface and remove all copper from the top of the interconnect dielectric. An optimization of the electroplating process to get a planar surface already after the deposition would be helpful and is the target of development today.

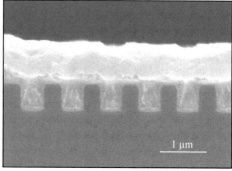

Figure 3.11: Copper pattern plating (source: IHM TU Dresden)

Figure 3.12: Electroplating of trenches with copper (source: IHM TU Dresden)

Moreover, the use of dc current power supplies with periodic reverse current or with pulsed dc can provide a modification of the deposition process and the thin film properties. The following advantages are noticeable:

- a higher deposition rate
- a thinner seed layer
- lower concentration of additives
- lower thin film defect density
- lower intrinsic stress and higher purity

The microstructure of electroplated thin films is generally comparable with that of PVD or CVD deposited layers. Polycrystalline structure and relatively big grains are typical. In general, additives decrease the grain size, but also the film density. A higher deposition rate and a higher metal ion concentration in the solution lead to purer layers and films with a lower intrinsic stress. The plating equipment ranges from baths of different size and form to a special high productivity deposition tool in microelectronics.

Electroless Deposition

In contrast to electroplating the electrons for reduction of the metal ions are provided by a special agent. The reduction is only performed at catalytic metal surfaces. This reaction is accompanied by oxidation of the anode. The deposition material must be catalytic by itself otherwise the deposition process will be interrupted. Contamination layers have to be removed if they prevent the catalytic reaction. Deposition on insulators is possible if a sufficiently dense seed layer of autocatalytic material is deposited initially. Palladium is a widely used seed layer material. The treatment of the substrate surface with a PdCl$_2$ solution enables the creation of catalytic seeds.

For metal deposition from their salts the following reducing agents are used:

- sodiumhypophosphite (Ni, Co)
- sodiumborohydride (Ni, Au)
- dimethylaminoborane (Ni, Co, Au, Cu, Ag)
- formaldehyde (Cu)

As in electroplating, additives can influence the deposition process. There are complexing agents, buffers and accelerators. Complexing agents act like buffering agents and hold the pH value constant. Further, they reduce the free metal ion concentration and prevent the precipitation of metal salts. Accelerators should increase the deposition rate without causing bath instabilities. They are anions such as CN$^-$ and should make the anodic oxidation process easier. Examples are imidazole, pyrimidine and pyridine.

Deposition of metal alloys can be achieved by codeposition. An important example for electronic application is NiP. Phosphorus provides an optimum of hardness, density, corrosion resistance and thickness of Ni thin films. The main composition of a NiP bath consists of NiCl$_2$, NaH$_2$P$_2$, glycine and water.

Electroless deposition is an inexpensive technique and power supplies are not necessary, but process controlling is not easy due to the problems of bath analysis and dosage. In electronic applications electroless deposition is used for making structural microelements. LIGA or LIGA-like processes would be unthinkable without such deposition processes. Other areas of application are hybrid circuits, different versions of printed circuits and multichip modules (MCM). In all cases the advantage of selective deposition makes the electroless process favored for the fabrication of metal lines, electrical contacts and other functional elements.

Physical Deposition

The existing physical deposition processes have a lesser importance for electronic applications because of the continuous miniaturization of electronic products and the use of thinner and thinner films. The physical techniques produce high quality films, but the layer thicknesses cannot be down-scaled to a few monolayers. The physical deposition processes can be divided roughly into:

- thermal spraying
- layer welding
- liquid quenching
- surface and dispersion varnishing (printing of thick-film layers)

The basis of all thermal spraying techniques is the generation of a fine droplet stream from a molten powder or wire. The droplets spread over the surface and cool down. The adhesion is mechanical, a subsequent thermal treatment can lead to interdiffusion and improved layer adhesion. Layer welding is connected with a melting of the deposit. The welding can be carried out by flame, arc, Joule heating, laser beam or electrons. A sufficient interdiffusion takes place and can be further improved if the process is run under pressure. Liquid quenching is used for the production of metallic glasses. This group of amorphous metals is generated by rapid cooling ($10^5 - 10^6$ K s^{-1}) from a molten mass. Typical examples are Co–P– and Fe–Ni–P–B– compounds. Their exceptional properties permit application in magnetic heads, chokes, magnetic shields, sensors and mechanical stressed parts.

Printing techniques play an exceptional part in electronics. With the development of microdosage and microprinting tools (e.g. screen printing) the deposition of metallic or insulating layers is realized from disperse organic materials in a suitable solvent. After printing, the disperse organic materials are converted into a stable film. The film thickness is in the range from a few to hundreds of micrometers. In general, there is a wide variety of so-called thick film pastes. For metal lines Ag–Pd, Au–Pd or Cu-based pastes are used. The actual film conversion occurs in a continuous furnace and a special atmosphere. The formation temperature is between 500 and 800 °C. Another application example is printing solder bumps or contact adhesives for electrical connection of devices and circuits. In the case of printed solder bumps the solvent can also serve as a flux if soldering takes place. The importance of this deposition technique is growing rapidly. The mass production of chip cards and other electronic devices for consumer goods needs inexpensive deposition techniques.

3.1.5   Outlook

A wide variety of deposition techniques for thin metallic films exists. Vacuum-based physical and chemical processes are favored. This is the case especially if thin metallic films have to be deposited with the following properties:
- ultra-thin (atomic scale thickness)
- homogeneous in property and thickness over a structured surface
- homogeneous in property and thickness over a large area
- low intrinsic stress
- low defect density (amorphous, mono-crystalline)
- low temperature stress
- selective deposition
- high adhesion strength etc.

However it does not mean that other techniques like wet chemical deposition or other selected physical techniques are not able to satisfy defined demands of electronic products better and less expensively.

3.2   Electron Microscopy and Diffraction

3.2.1   Transmission Electron Microscopy (TEM)-Imaging

The TEM is an instrument for the highly-magnified imaging of micro- and nanostructures in thin films. With enhancement to an analytical TEM it is possible to investigate the morphology, the crystallographic structure up to the arrangement of the atoms at interfaces and the chemical composition in the nanometer range. In principle the construction of the TEM is similar to that of the transmission light microscope: Radiation is transmitted through the specimen, an objective lens creates a real image which is magnified by additional projective lenses.

Resolution and Magnification

The most essential parameter of a microscope is its resolution. For the light microscope it is limited by the wave aberration to about a half of the wavelength λ of the visible light, i.e. to about 0.3 µm. Improvement of the resolution requires shorter wavelengths such as given by electron waves. The electrons can be deflected by electric and magnetic fields, thus the construction of lenses for electron waves is possible.

For an accelerating voltage of 200 kV, as usual in modern microscopes for materials science, the electron wavelength is $\lambda = 2.5 \cdot 10^{-12}$ m = 0.0025 nm. This can be calculated by means of the energy theorem and the de Broglie relation. Simple extrapolation of the basics known from light microscopy would lead to a resolution limit better than 0.002 nm. In fact, commercial high-end instruments currently reach values of $0.1 - 0.2$ nm. The reason for the difference is the discrepancy in the spherical aberration correction of the objective lens. The correction is possible for glass lenses but impossible for the standard electromagnetic lenses currently used in commercial electron microscopes (rotationally symmetric, free of space charges, time-invariant). Commonly, the calculation of the resolution limit of the TEM requires consideration of the spherical aberration constant C_S. In a simplified approach (sum of wave aberration and spherical aberration disks, Gaussian image plane) the resolution limit can be calculated to be

$$\delta_{min} = 0.3 \sqrt[4]{C_S \cdot \lambda^3} \tag{3.6}$$

with an optimal aperture

$$\alpha_{opt} = 0.67 \sqrt[4]{\frac{\lambda}{C_S}} \tag{3.7}$$

For 200 kV with a typical C_s of 2 mm this results in α_{opt} = 13 mrad and δ_{min} = 0.13 nm .

The distance δ_{min} has to be magnified by the microscope up to the resolution limit of the recording device, e.g. a camera or the human eye (0.1 mm). The useful magnification necessary for the human eye amounts to

$$M_{useful} = \frac{0.1\,\text{mm}}{\delta_{min}} \approx 10^6 \tag{3.8}$$

Currently, modern transmission microscopes are developed which are equipped with a special multipole system. This system allows the correction of the spherical aberration of the rotationally symmetric lenses [3.21, 3.22].

Specimen Preparation

A precondition for the TEM investigation is an electron-transparent foil. For metal-based thin films and a 200 kV microscope the foil thickness should be less than 0.1 µm which can only be reached by special preparation methods and techniques. Commonly, this electron microscopic specimen preparation is the most time-consuming step of the TEM investigation. An overview of the possibilities for thin film preparation is given in Fig. 3.13.

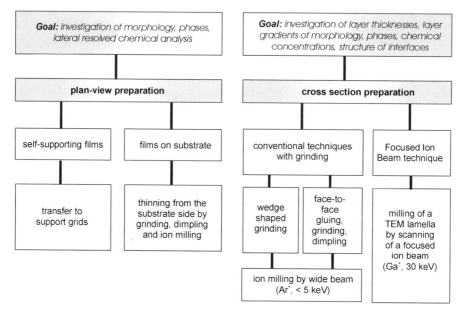

Figure 3.13: Transmission electron microscopic specimen preparation of metal-based thin films

Producing self-supporting films needs deposition in an electron-transparent thickness on a special substrate, e.g. NaCl. In this case the substrate can be partly dissolved in distilled water at the interface to the film and the film can be transferred to a support grid. That is advantageous for in situ investigations, e.g. for heating experiments. The focused ion beam (FIB) method (see Section 3.2.3) allows a comparatively fast cross section preparation with high precision to meet the specimen area of interest. On the other hand, the large energy of the Ga^+ ions can lead to an implantation and film amorphization which can disturb especially the high resolution electron microscopy.

Image Contrast

A main point for TEM, especially for the image interpretation, is the understanding of the image contrast formation.

While the beam electrons are penetrating the specimen they are elastically and inelastically scattered by interaction with the atoms. For the imaging the elastic part is essential, whereas the inelastic part worsens the contrast and is unwanted for imaging. The scattering angle depends on the atomic number and the number of atoms interacted with. An aperture in the back focal plane of the objective lens removes the strongly scattered electrons from the beam (see Fig. 3.14). Specimen details responsible for the strong scattering appear dark in the electron microscopic image ("bright field imaging"). The aperture size determines the magnitude of this scattering absorption contrast (also mass thickness contrast). Therefore, this aperture is also called "contrast aperture". If this aperture is moved in such a way that instead of the unscattered beam the more deflected rays go through, all of the sample areas which scatter weakly appear dark, whereas those responsible for a strong deflection are bright ("dark field imaging").

In the case of crystalline samples the wave character of the electrons has to be considered. The magnitude of scattering is influenced by the orientation between the crystal lattice and the direction of the wave propagation (direction of the electron beam, see Section 3.2.2). Areas with crystal orientations fulfilling Braggs law scatter strongly and these areas appear dark in the image. Because of the dependence of this diffraction contrast on the specimen orientation the electron microscopic image contrast changes drastically if the specimen is tilted. That allows one to distinguish between mass thickness and diffraction contrast. The crystal lattices are often deformed and at each specimen tilt some areas of a single crystal achieve the Bragg condition ("bend contours"). The diffraction contrast is important for the distinction of crystals, e.g. for the measurement of grain sizes in thin films.

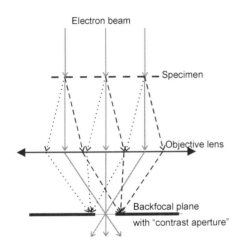

Figure 3.14: Optical path near to the objective lens (bright field imaging)

A further contrast is based also on the wave character of the electrons. On penetrating the specimen the electron waves experience a phase shift because of the inner potential of the sample. After amplification up to values of about π and superposition of the origin wave a visible image contrast can be obtained. In light microscopy this amplification is brought about by a phase plate. In the objective lens of the electron microscope a sufficient amplification can also be reached by the phase shift between near axis and axis distant electron bundles. The shift depends on the spherical aberration and is influenced by the distance to the focus point. In electron microscopy it is possible to amplify this phase contrast by a suitable setting of the lens current of the objective ("defocusing"). This kind of contrast is important, especially in high resolution TEM.

Analytical TEM

The addition of spectrometers for the detection of characteristic X-rays (EDXS) created by inelastic scattering of the primary electrons at the specimen atoms and for the measurement of the electron energy losses (EELS) occurring in this process (see Section 3.4: Spectroscopic Techniques) leads to the analytical TEM (see Fig. 3.15).

The electron optics of an analytical TEM allow the formation of a small electron probe with a diameter < 1 nm to confine the excited sample area. To obtain a sufficient signal-to-

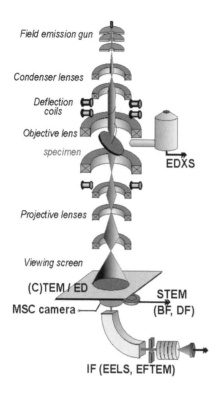

Figure 3.15: Scheme of an Analytical TEM

noise ratio a beam current up to 1 nA in this probe is desirable. That can only be reached by using a field emission gun which provides high current density with small divergence.

Additionally to the fixed beam technique a unit with deflection coils enables the scan of the electron probe over the specimen and the choice of an interesting area for analysis. In contrast to the usual scanning electron microscope the specimen is electron-transparent and therefore this method is named "scanning transmission electron microscopy" (STEM). The use of the very thin specimen avoids the formation of an exciting area with sizes up to the micrometer range within the sample as known from bulk materials. In this way, a very high spatial resolution up to the atomic level can be reached even in the STEM mode. Two detectors acquire the intensity of the weakly (bright field - BF) and the strongly (dark field - DF) scattered electrons. In contrast to fixed beam microscopy the BF and DF images can be recorded synchronously. Conventional (C)TEM images as well as electron diffraction (ED) patterns can be observed on the viewing screen or by use of a multiscan camera (MSC). The imaging filter (IF) allows recording of electron energy loss spectra (EELS) as well as of filtered images (EFTEM, see the respective paragraph). The imaging filter can be attached below the viewing screen as shown in Fig. 3.15 ("post-column filter") as well as between the projective lenses ("in column filter").

High-Resolution TEM (HRTEM)

In HRTEM the phase contrast plays the crucial role. This method needs another approach to the contrast interpretation. Working close to the resolution limit, it is necessary to consider the electron wave modulation not only by the sample but also by the objective lens. For excellent high resolution images specimen foils thinner than 30 nm are necessary, that means a higher demand on the specimen preparation.

The features of the high resolution TEM can be plausibly understood in two ways:

(i)   The wave modulation by the periodically arranged sample atoms depends on the specimen thickness. In a simplified approximation it can be assumed that the beam intensity oscillates between the differently scattered waves with the the foil thickness and therefore the interference result after the sample also oscillates with respect to the foil thickness.

(ii) Close to the resolution limit the objective lens does not transfer the contrast monotonically. In Fig. 3.16 two phase contrast transfer functions (CTF) are shown for a modern TEM. The CTF shows the kind of contrast transfer versus the spatial frequency, i.e. a reciprocal length. It depends on the quality of the objective lens (spherical and chromatic aberration constants C_S, C_C), on the angle of incidence, on the defocusing Δf and on the coherence of the electrons determined by the electron gun (energy spread ΔE). The resolution limit amounts to 0.2 nm in this case ("point resolution"). It coincides with the first zero-crossing of the CTF using the *Scherzer* focus (best focusing). By comparison of Figs. 3.16 a and b the variation of the CTF in dependence on the defocus can be seen. The point resolution is shifted to lower spatial frequencies (i.e. larger distances). On the other hand, some contrast is also transferred at higher spatial frequencies than that of the point resolution but with oscillations. Different distances are transferred with different contrasts which complicates the image interpretation, especially at interfaces with relaxed atoms of alternating distances. Beside special methods for the consideration of the CTF such as holography [3.23] or defocus series reconstruction [3.24] the way out is mainly the image simulation.

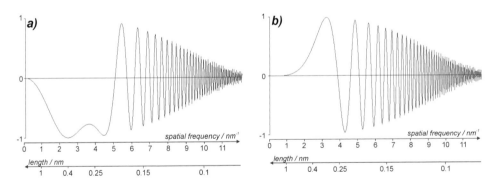

Figure 3.16: Phase contrast transfer functions of a TEM with $C_S = 1.2$ mm, $C_C = 1.4$ mm, $\Delta E = 1$ eV and different defocuses: a) $\Delta f = 58$ nm (*Scherzer* focus), b) $\Delta f = 0$

For this simulation an atomic arrangement is assumed and by variation of the specimen foil thickness and defocusing of the objective lens an array of different images is calculated using suitable software (e.g. EMS [3.25], see Fig. 3.17). All images in Fig. 3.17 are based on the same face centred cubic (fcc) unit cell of Cu. The variation of foil thickness and defocusing leads to completely different patterns and the task is to find the image which represents the best coincidence with the measured image. If coincidence cannot be reached the atomic arrangement has to be changed and the calculation must be repeated. The use of a corrector for the spherical aberration mentioned above can lead to an easier interpretation of the high resolution TEM images.

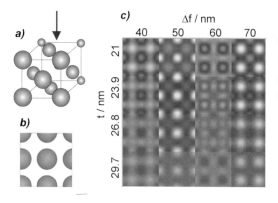

Figure 3.17: High resolution TEM:
a) unit cell of fcc Cu with view direction [001]
b) expected image of one unit cell
c) array of calculated images for different specimen foil thicknesses t and defocuses Δf

Energy filtered TEM (EFTEM)

The analytical TEM equipped with an imaging electron energy filter allows the recording of chemical information (elemental distribution and chemical binding) during an imaging process. The advantage of this method is the visibility of correlations between structural and chemical features of thin films on the nanometer scale.

The tool is based on an energy dispersive element (magnetic sector field) combined with electron optics which create an energy dispersive plane within the optical path. By positioning a small slit in this plane electrons of a chosen energy can be extracted and pass the slit contributing to the image brightness. As a result of the inelastic interaction between primary electrons and specimen atoms the electrons get energy losses characteristic for the elements (see Section 3.4: Spectroscopic Techniques). Combining these two features it is possible to generate an element specific image.

The procedure is demonstrated in Figs. 3.18 to 3.20 using an example of nanocrystallites from nickel. The crystallites (10 – 30 nm in size) lie within a thin NiO layer and the question is: do they also consist of NiO and are visible only by diffraction contrast or are they distinguished by another chemical composition? To analyse the problem an energy window of 855 – 895 eV (Ni-L edge at 855 eV) is chosen in the filter for imaging with electrons having this energy loss that is typical for Ni.

Figure 3.19a shows the region of interest of an energy loss spectrum with the "post-edge" window and the associated image (Fig. 3.19d). To correct the background two additional windows (pre-edge 1 and 2) are set. By means of these pre-edge windows (Figs. 3.19b and 3.19c) a background is approximated for each pixel in the same way as for the quantification of the EEL spectra (see Section 3.4: Spectroscopic Techniques). After background subtraction the image in Fig. 3.20 is obtained which clearly shows the accumulation of Ni in the crystallites mentioned above. That means the crystallites have got another chemical composition. The "three-windows-method" is the most common technique for the recording of energy filtered images. Another possibility is to divide the post-edge by one of the pre-edge images pixel by pixel ("jump ratio").

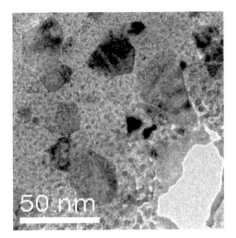

Figure 3.18: TEM brightfield micrograph of Ni crystallites within a NiO layer

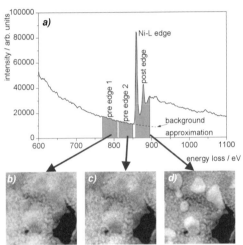

Figure 3.19: Energy loss spectrum in the region of the Ni-L edge (a) with the settings of three windows for energy specific images (b – d)

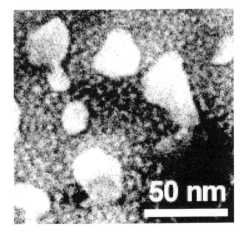

Figure 3.20: EFTEM image for Ni after background subtraction

Concerning the TEM resolution distinction has to be made between the energy and the spatial resolution. At first, the energy resolution is limited by the spectrometer and the energy spread of the electron gun, for the Schottky field emission to about 0.5 – 0.8 eV. On the other hand, practical work needs a sufficient signal intensity to reach the required signal-to-noise ratio within the EFTEM images. Often, the most important criterion for the parameter choice is a high beam intensity. In this sense the width of the energy windows should be as large as possible. Contrary to that, for mapping of chemical bindings details of the near-edge structure are essential and small energy windows are required. Because of the chromatic aberration of the objective lens the energy window width influences not only the energy resolution but also the spatial resolution. The size of the chromatic aberration disk is given by

$$\delta_C = C_C \cdot \alpha \cdot \frac{\Delta E}{E}$$

(3.9)

with the chromatic aberration constant C_C, objective aperture α, energy window width ΔE and primary electron energy E. For an analytical TEM with an acceleration voltage of 300 kV, C_C = 1.4 mm and α = 10 mrad the energy window has to be smaller than 22 eV if structures of 1 nm in size are to be analytically resolved.

3.2.2   TEM-Selected Area Electron Diffraction

Electron diffraction allows crystallographic phase determination in thin films. The samples are often self-supporting films or prepared in plan-view. Commonly, the selected area diffraction is used in this case.

For the diffraction mode the intermediate and projective lenses in the TEM are excited in such a way that the diffraction pattern generated in the back-focal plane of the objective lens is imaged on the viewing screen or a recording device below the screen (see Figs. 3.14 and 3.15). An aperture in the real image plane of the optical path allows the selection of the sample area of interest for diffraction ("selected area aperture"). The easy switching between imaging and diffraction enables correlation of morphological and structural materials features.

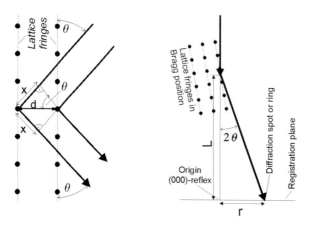

Figure 3.21 (left): Sketch for demonstration of Bragg´s equation

Figure 3.22 (right): Explanation of the term "camera length" L

The diffraction diagrams are interference patterns which reflect the (periodic) arrangement of atoms or groups of atoms in the crystal lattice. A basic knowledge of crystallography is necessary for the interpretation of electron diffraction patterns [3.26]. The crystalline specimen is the precondition for diffraction patterns in the sense described here. Essentially, the appearance of the pattern is influenced by

(i) The relation between the size of the specimen area included in the diffraction pattern and the mean size of the crystallites, i.e. the number N of crystallites which contribute to the pattern. If this number is large ($N \gg 10$) and all of the crystal orientations have the same probability, the pattern is a ring diagram with closed circles. The rings are sharper the larger the crystallites. Decreasing N the circles are no longer closed, they partly dissolve into single points. For $1 < N < 10$ several point diagrams are overlaid and identification can be impossible.

(ii) The crystallographic structure of the sample which is characterized by the unit cell. The complete lattice is generated by stringing together these unit cells. For point diagrams the pattern is additionally influenced by the electron beam direction with respect to the axes of the unit cell.

For interference maxima as the result of wave superposition the retardation Δs between two waves must be a whole-numbered multiple of the wavelength λ:

$$\Delta s = n \cdot \lambda \quad (n = 0, 1, 2, 3, ...) \tag{3.10}$$

Assuming the atoms within the crystal lattice are arranged in planes (lattice fringes) looking like semipermeable mirrors, then the diffraction maxima can be understood as a result of the superposition of electron waves partly scattered on the lattice fringes with the retardation $n \cdot \lambda$ (see Fig. 3.21). The retardation $\Delta s = 2 \cdot x$ depends on the angle θ between the propagation direction of the electron wave and the "reflecting" lattice fringes with spacing d:

$$\Delta s = 2 \cdot x = 2 \cdot d \cdot \sin \theta \tag{3.11}$$

With the condition for interference maxima and using the assumption $\theta \ll 1$ the Bragg equation for the electron diffraction follows

$$2\,d \cdot \theta = n \cdot \lambda \tag{3.12}$$

With the distance r between the origin and the diffraction spot and the camera length L the angle θ is given by

$$\tan(2\,\theta) = \left(\frac{r}{L}\right) \approx 2\,\theta \tag{3.13}$$

(see Fig. 3.22) and the basic equation for the first maximum ($n = 1$) by

$$d \cdot r = \lambda \cdot L \tag{3.14}$$

This equation allows the computation of the lattice spacing d_{hkl} pertaining to a chosen diffraction spot or ring with the distances r_{hkl} to the origin. The importance of the "apparatus constant" $\lambda \cdot L$ for the identification of diffraction patterns can be seen, too. In principle, it is possible to multiply the camera length L by the electron wavelength λ calculated from the acceleration voltage using the energy theorem and the de Broglie relation. Because of the variation of the camera length by the projective lenses it is better to measure the apparatus constant by means of a ring diagram of a known substance.

The analysis of point diagrams is more complicated. A diffraction point arises if the considered set of lattice fringes is orientated to the occurring electron wave in such a way that the retardation of the waves partly reflected at the lattice fringes is a whole-number multiple of the wave length ("Bragg position"). In addition to the distances of the spots from the origin the angles between the spots have to be considered. These angles are equivalent to the angles between the lattice fringes responsible for the spots. The cutting line of all the lattice fringes being in the Bragg position is called the zone axis. It is in keeping with the direction of the electron incidence. The indexing of the diffraction spots is equivalent to the Miller indices of the corresponding lattice fringes and their whole-number multiples (see Fig. 3.23).

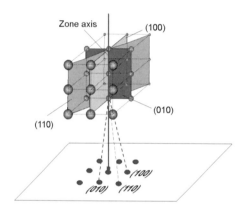

Figure 3.23: Indexing of diffraction point diagrams

For the mathematical handling the implementation of the "reciprocal lattice" has proved to be convenient. To understand this convenience, the Bragg equation will be transformed into the Laue equations: Firstly, the propagation directions of the incident and the diffracted wave are described by the wave vectors k_0 and k (see Fig. 3.24).

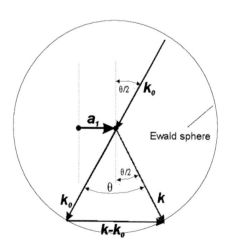

Figure 3.24: Sketch for demonstration of the Laue equations and the Ewald sphere

Using the assumption $\theta \ll 1$ it follows from Fig. 3.24 and the Bragg equation:

$$\left| k - k_0 \right| = \theta \cdot \left| k \right| = \frac{n \cdot \lambda}{d} \left| k \right| \tag{3.15}$$

The absolute values of the wave-vectors are (elastic scattering)

$$\left| k \right| = \left| k_0 \right| = \frac{1}{\lambda} \quad \text{and therefore} \quad d \cdot \left| k - k_0 \right| = n \tag{3.16}$$

After substitution of d by a_1 and n by the whole number h as well as consideration of

$$a_1 \parallel (k - k_0) \qquad \text{it follows:} \qquad a_1 \cdot (k - k_0) = h \tag{3.17}$$

Analogously, it can be written for the two other directions of the three-dimensional lattice:

$$a_2 \cdot (k - k_0) = k$$
$$a_3 \cdot (k - k_0) = 1$$

To construct the diffraction pattern, the reciprocal lattice with the basis vectors b_1, b_2 and b_3 is introduced according to the definition

$$a_i \cdot b_j = \begin{cases} 1 \\ 0 \end{cases} \text{if} \begin{array}{c} i = j \\ i \neq j \end{array} \qquad (i, j = 1,2,3) \tag{3.18}$$

This leads to

$$a_1 \cdot (h \cdot b_1 + k \cdot b_2 + 1 \cdot b_3) = h$$
$$a_2 \cdot (h \cdot b_1 + k \cdot b_2 + 1 \cdot b_3) = k \tag{3.19}$$
$$a_3 \cdot (h \cdot b_1 + k \cdot b_2 + 1 \cdot b_3) = 1$$

and after comparison with (3.17) to

$$k - k_0 = h \cdot b_1 + k \cdot b_2 + 1 \cdot b_3$$

i.e. the vector $k - k_0$ is a vector of the reciprocal lattice. This consideration leads to the Ewald construction: A diffraction spot arises if a sphere with radius $1/\lambda$ constructed around the end of the vector k_0 of the electron incidence touches a point of the reciprocal lattice.

In the reciprocal lattice each point represents one set of lattice fringes. The vector from the origin to the reciprocal lattice point is perpendicular to the corresponding lattice fringes, its length is reciprocal to the lattice spacing.

For the interpretation of electron diffraction patterns different PC software is available. Exercises using this software give an impression of the diversity of possible diffraction patterns if the crystal system and/or the zone axis are changed.

Precondition for reaching sharp diffraction diagrams in the selected area mode is the parallel illumination of the specimen (see Fig. 3.14). Using this method the size of the selected area is limited to about 50 – 100 nm. For smaller areas of interest the focusing of the electron beam on the specimen is necessary (micro- or nanodiffraction). It makes sense only for point diagrams of small single crystallites. In this case the illumination is convergent and the spots become small disks. The increase of the angle of convergence ("condensor aperture") leads to Kikuchi patterns and to patterns within the diffraction disks. They respond sensitively to the tilting of the specimen (Kikuchi) and allow the precise measurement of the lattice constants (Convergent Beam Electron Diffraction – CBED) [3.27].

3.2.3   In situ-SEM Methods

Conventional Scanning Electron Microscopy (SEM)

The most usual method for direct, high resolution imaging and microanalysis of solid surfaces is now scanning electron microscopy (SEM), combined with analytical attachments. A schematic view of the principal operation mode of a SEM is shown in Fig. 3.25.

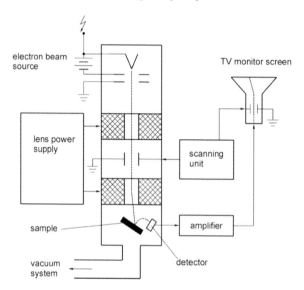

Figure 3.25: SEM operation mode shown schematically

A primary electron beam is emitted from an electron beam source and focused by different magnetic lenses to form a very fine probe of only few nanometers in size. The electron probe is moved over the sample by means of deflection coils, scanning it point by point. Simultaneously, a second electron beam runs over a TV monitor screen in a synchronized mode. Because of the interaction between the focused primary electron beam and the sample surface, different signals are emitted from each scanned surface point. These include secondary and backscattered electrons, Auger electrons, cathodoluminescence and characteristic X-ray radiation. The radiation emitted from each individual sample point may be sequentially collected by a detector. After boosting by an amplifier it controls the brightness of individual positions on the TV monitor screen.

The basic principle of the SEM was laid down by *Knoll* [3.28] and *v. Ardenne* [3.29] as early as in the 1930s, however practical and commercial implementation began only 30 years later, because progress in electronics and video techniques was indispensable for the construction of a practical instrument.

Though the scanning image is produced without any focusing lenses, the aberrations of which limit TEM resolution, the lateral resolution for SEM is ultimately limited by the size of the focused probe. In the conventional secondary electron image mode with a tungsten filament source resolution of about 10 nm is available. From this, a useful magnification of about 50000 times follows. Higher resolution can be obtained by the application of a field emission gun, with its higher source brightness. With it *Crewe* could reach a lateral resolution of 2 nm [3.30] and current performance parameters claimed by SEM producing companies amount to $\delta = (1 - 2)$ nm. A special advantage of the SEM is its high depth of sharpness, resulting in a spatial, almost three-dimensional image. This is caused by the low irradiation aperture, which also allows the investigation of rough material surfaces.

The information content of SEM investigations can be considerably increased by in situ sample treatment [3.31]. As a result a "microlab" exists inside the SEM by different in situ sample loading, based on mechanical, thermal, irradiation or electrical effects. Electrical in situ experiments can principally be divided into experiments with additional contrast by microfields and experiments with pure secondary electron or backscattered electron contrast. In the latter case sample behaviour under electrical loading is the focus of intention. Contrast variations due to microfields can be subdivided into a number of main groups with contrast changes due to local differences of the electrical potential, ferroelectric domains and electron-beam-induced variations of the current (EBIC mode), allowing the study of p–n junction behaviour (see Table 3.2).

Table 3.2: Electrical in situ sample treatments in the SEM

Method	Reaction	Quantity
Potential contrast	Voltage barriers and breakdown	Localized voltage distribution $\Delta U(x,y)$
Ferroelectric domains	Domains size and orientation	Domains distribution
EBIC	p–n junction behaviour	Current distribution $I(x,y)$
Electromigration, acoustomigration	Growth of voids and hillocks	Lifetime of interconnects and SAW structures

Electromigration and Acoustomigration Experiments

The second group of electrical in situ techniques includes both electromigration and acoustomigration experiments with loading simulation inside the SEM. Electromigration is caused by the so-called electron wind, which develops in the interconnecting lines of microelectronic circuits at very high current densities ($j > 10^5$ A cm^{-2}). The electron wind causes a mass transport of metal ions, which leads to the formation, growth, and movement of defects (hillocks and voids) and can result in an electrical failure [3.32]. The experimental set-up for in situ SEM investigations of electromigration of chip interconnect lines was developed by *Wetzig, Wendrock, Buerke* and *Kötter* [3.33] and is given schematically in Fig. 3.26.

The test chip is mounted on a heating stage, which allows constant chip temperatures in the range from 350 K to 750 K. The operation loading of the test chip is achieved by a DC source. The combined heating and circuit contact unit allows electromigration tests in SEM under speed-up conditions with regard to temperature and current density. Typical values are a current density $j = (2 - 7) \, 10^6$ A cm^{-2}, a chip temperature $T_{Al} = (450 - 470)$ K for Al interconnects and $T_{Cu} = (540 - 620)$ K for Cu interconnects respectively. Because of Joule's selfheating, the temperature at the interconnect is $(10 - 20)$ K higher than the chip

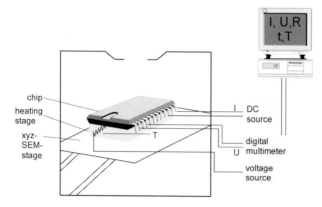

Figure 3.26: Experimental set-up for in situ SEM electromigration tests shown schematically

temperature, which is measured via interconnnect resistance by a digital multimeter. The in situ current exertion takes from a few hours to some days.

The spatial arrangement of the heating stage in the SEM specimen chamber is shown in Fig. 3.27. A copper heating stage is mounted on the specimen holder via a thermal isolated coupling. Onto the heating stage a 24 pin test chip can be fixed which allows the control and measurement of the relevant parameters for electromigration. Not only electromigration but also the progress of acoustomigration can be studied in situ in a SEM. The principal experimental set-up is shown in Fig. 3.28. Surface acoustic waves (SAW) are generated on the surface of a SAW couple by input of an electrical signal from the network analyser. The input signal is amplified by 33 dB in a special amplifier unit, then it passes a directional coupler and reaches the SAW. One part of the wave goes from the directional coupler onto the network analyser as reference signal. The reflected wave is given onto the network analyser as input signal. After symmetrical attenuation by 20 dB both the input and the reference signal are compared with regard to their amplitudes and phase displacement. From these values the admittance can be calculated in the connected computer.

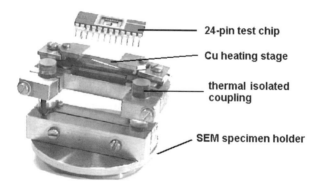

Figure 3.27: Heating stage for in situ SEM electro-migration tests

The admittance amplitude versus frequency gives resonance curves as a characteristic signal of SAW units. The simultaneous SEM image allows the study of acoustomigration as a damage process on the SAW sample surface. Correlations are possible between the electrical behavior (frequency shift and dumping of the admittance curve) and the visible stripe damaging on the surface, which shows nucleation, growth and coalescence of voids and hillocks. The fixing of the SAW sample on the sample holder with the heat current stage can be seen from the view in the left lower corner of Fig. 3.28.

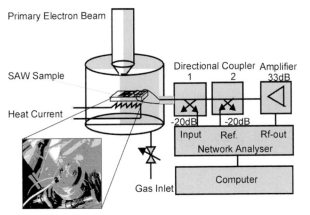

Figure 3.28: Experimental set-up for in situ SEM acoustomigration tests shown schematically

Focused Ion Beam (FIB) Technique

A further in situ SEM method is focused ion beam (FIB) microscopy. In recent years the FIB technique has become widely accepted in materials research for microelectronics, especially for SEM imaging, cross section preparation, failure analysis and device modification in the semiconductor industry. In FIB microscopy a focused ion beam is precisely scanned over the sample surface producing secondary electrons which give a high quality contrast SEM image as well as depositing in situ material on a submicron scale. The operating principle is the following: a strong electric field applied to a liquid metal ion source extracts positively charged Ga^+ ions, which are focused into a beam using electrostatic lenses. The focused ion beam strikes the specimen surface thereby removing material mainly through the physical sputtering mechanism and generating secondary electrons (SE) and ions (SI). The SE form a SEM image of the area of interest that was scanned by the ion beam. Unlike the conventional SEM the FIB shows a strong orientation contrast.

The most frequent application of the FIB microscope is in situ cross section preparation (Fig. 3.29).

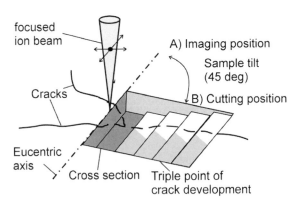

Figure 3.29: In situ cross section preparation inside a focused ion beam microscope

This technique allows one to prepare micro-cross sections precisely at places of interest. In the cutting position of the sample surface the focused ion beam removes material the quantity of which depends on the local position, forming a step pattern. After cutting the sample can be tilted 45° around the eucentric axis to reach its imaging position. The cutstep pattern allows one to produce SEM images from the cross section. As an example Fig. 3.30 shows cuts through two hillocks after an electromigration test of an electroplated Cu interconnect line, tested at $T = 260$ °C and $j = 3 \cdot 10^6$ A cm^{-2}. The images are taken with ion induced secondary electrons and show distinct orientation contrast.

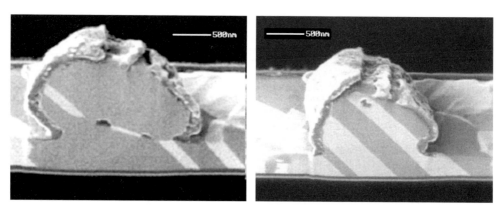

Figure 3.30: FIB cuts through two hillocks after electromigration test of an electroplated Cu interconnect line, SE image

Furthermore, the hillock of the right image has grown epitaxially through the interconnect line. The cap layer has probably been formed by carbon contamination produced in the SEM during electron backscatter diffraction process.

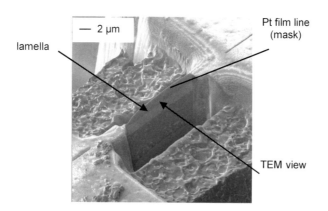

Figure 3.31: SEM image of a FIB prepared TEM lamella in TiN/Al$_2$O$_3$ multilayer system

Besides cross section preparation, the most frequent application of the FIB microscope is TEM lamella preparation. This can be achieved by a material removal according to Fig. 3.29, but from both sides. Using a metal organic vapor injection system some ion-assisted CVD processes can be performed which cause a defined heavy metal film deposition as a mask on the lamella. Figure 3.31 shows the lamella preparation from a TiN/Al$_2$O$_3$ multilayer on a hard metal substrate. The whole sample can be tilted 90° for direct TEM transmission of the lamella.

Disadvantages of the in situ FIB microscopy are sample damaging effects due to the Ga ion bombardment. Furthermore, Ga ion implantation takes places at depths up to some 10 nm depending on the ion energy and the material of the sample. However, these effects are negligible for the "cut and view" study of bulk materials [3.34, 3.35].

3.2.4 Electron Backscatter Diffraction

Electron backscatter diffraction (EBSD) is a technique to obtain local crystallographic information as microtexture or preferred crystal orientation. This can be done by orientation imaging microscopy or by texture diagrams in orientation space. The method is frequently used in combination with scanning or transmission electron microscopy. The spatial resolution covers a wide field from < 50 nm for modern field emission SEM up to some mm with special computer controlled stages. EBSD is also known as backscatter *Kikuchi* diffraction or electron backscatter pattern technique.

The EBSD technique is based on the work of *Kikuchi* on observations of electron diffraction in thin crystals [3.36] and was further developed by *Dingley* in the 1980s (e.g. [3.37]). The first commercially available system was introduced in 1994, and nowadays there exist modern systems with fully automated indexing of EBSD patterns and pseudocoloured orientation imaging.

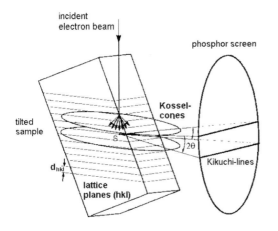

Figure 3.32: Formation of EBSD *Kikuchi* patterns shown schematically

The formation of an EBSD pattern can be understood from the interaction of a primary electron beam with the crystalline solid. This is shown schematically in Fig. 3.32. The incident electron beam is diffusely and quasielastically scattered in the subsurface region of the tilted sample. Some of the scattered primary electrons arrive at the Bragg angle θ_B at every set of lattice planes (*hkl*) and are then elastically scattered at these planes. From this interaction two cones of diffraction result. With typical values for electron wavelength λ and lattice interplanar spacing d_{hkl} Bragg's equation (cf. (3.12))

$$\lambda = 2d_{hkl} \cdot \sin\theta_B \approx 2d_{hkl} \cdot \theta_B \tag{3.20}$$

gives a Bragg angle θ_B of about 0.5°. Consequently, the diffraction cone angle is close to 180°, i.e. the cones are almost flat. At the position of the phosphor screen or of the EBSD camera respectively a pair of parallel conic sections exists. These are the so-called *Kikuchi* lines, with an angular distance of 2 θ_B which is reciprocal to the interplanar spacing d_{hkl}. From the *Kikuchi* lines both the spatial orientation of crystals, and in some case their phase allocation can be determined [3.38].

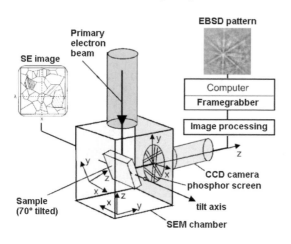

Figure 3.33: SEM with EBSD attachment and schematic illustration of information output

EBSD equipment can be attached easily to a SEM as shown in Fig. 3.33 [4.40]. The primary electron beam is focused on a highly tilted bulk sample in the SEM chamber (70° is used as a standard tilt angle and 20 mm as standard working distance). The electron backscattered diffraction images are displayed on a phosphor screen of a CCD camera which is connected with an image processing unit, and finally the EBSD patterns are evaluated in a computer.

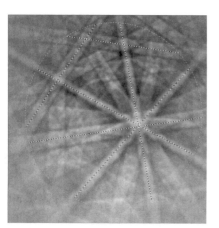
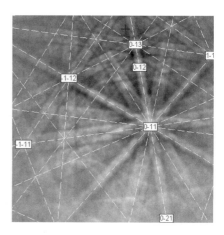

Figure 3.34: Example of EBSD indexing of a Ni sheet after rolling and recrystallization. Left: pattern captured by detector unit, right: image after overlay of Miller's indices of zone axes [3.38]

Using a Hough-transformation, the computer at first determines the position of the clearest *Kikuchi* bands. Using the previously determined possible crystal types, the respective phases and crystal orientations best corresponding in position and angle of intersection to each band are then determined. This is the so-called indexing process. The phase allocation and 3 orientation angles can be stored in a data file, together with the other parameters of the indexing process. As an example Fig. 3.34 shows the indexing process by EBSD at a Ni sheet after rolling and recrystallization. In the left image the pattern captured by the detector unit is visible, the right image shows the overlay of Miller's indices of zone axes [3.38].

The orientation accuracy for EBSD is better than 1°. Because of the information depth of (20 – 50) nm, adsorption layers on the surface do not disturb the EBSD measurement. On the other hand this method requires crystal surfaces free of mechanical damage. For this reason, for the final surface preparation only electropolishing or careful mechanical polishing are suitable.

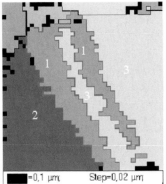
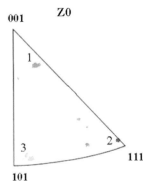

Figure 3.35: EBSD orientation mapping of a twin lamella of electroplated Cu [3.39].
Left: image with 3 different orientations.
Right: reduced pole figure

From orientation mapping one can obtain information on materials texture, microstructure and grain boundary classification. At high magnifications the image quality and thus the pattern recognition is limited due to different disturbances such as image drift caused by charging, inconstancies in instrumentation stability, contamination and effects of surface topography. A compromise must thus be made between the beam current, scanning step width, sensitivity and reliability of recognition.

An example for best attainable spatial resolution of EBSD is given in Fig. 3.35. It shows an orientation mapping at a twin lamella of electroplated Cu [3.39]. The image is 80000 times magnified, and it was taken with field emission (U_B = 20 kV, I_B = 200 nA) and with a primary electron beam step width of 20 nm. From the reproducibility of measured lamella width a spatial resolution of (1–2) steps, that means (20–40) nm is guaranteed.

3.3 X-Ray Scattering Techniques

X-ray scattering is a widely-used tool for material characterization due to the variety of structural information which can be obtained by it in a non-destructive way. Characteristic wavelengths of X-rays are of the order of interatomic distances in typical crystal lattices.

Thus, a variety of X-ray scattering methods exists for the characterization of such lattices by interference phenomena. The penetration depths of X-rays of typically several μm in metals make them also adequate for film investigations. However, advanced X-ray techniques are required to meet the challenges for characterization of very thin layers. Various peculiarities of X-ray scattering from thin films are related to the following points:

(i)   The interaction of X-rays with thin films is weak due to the small scattering volume. Absorption in additional cap layers further reduces the scattered signal.

(ii)   The influence of the surface leads to anisotropy and gradient effects and to deviations in the physical properties of films from bulk behavior.

(iii) The substrates can influence the scattering in an undesired manner (background, overlapping substrate peaks, interface roughness). On the other hand, substrate reflections can be used as references.

(iv)   Peak shift and broadening effects are usually stronger than in bulk materials (small crystallite size, stresses, lattice defects).

(v)   The frequently strong texture of the crystallites in thin films leads to limitations for measuring directions and to anisotropic film properties.

Apart from the interference at the crystal lattice as utilized in X-ray powder diffraction, the film dimension itself, which typically differs from the crystal lattice scale by some orders of magnitude, can also be exploited for scattering effects. For example, the film thickness can be precisely determined both by wide-angle diffraction and by X-ray reflectometry (XRR), making use of the interference between scattered waves from the upper and lower interfaces of a film. Also interface and surface properties like roughness and interface correlations are accessible by XRR. Information about the film composition can be obtained via wide-angle diffraction (lattice constants) and XRR (electron density). Furthermore, conventional techniques of wide-angle diffraction for texture, strain and defect analysis have to be adapted to the peculiarities of thin film scattering. For instance, the grazing-incidence diffraction at lattice planes perpendicular to the film surface (GID) has a strong surface sensitivity and allows determination of depth profiles of composition, strain etc.

Instrumental adaptations to thin film scattering include the utilization of parallel beam optics like mirrors and collimators in diffractometers equipped with conventional X-ray tubes, which are typically used in the line focus mode. The application of high brilliance synchrotron sources allows one to perform also microdiffraction investigations with typical beam cross sections of a few μm, whereas in conventional diffraction cross-sectional areas of about 1 mm² are used. Moreover, the high brilliance, monochromacy and coherency, adjustable wavelength and polarization, and the time structure are unique properties of synchrotron radiation, making it well suited for investigation of thin film properties.

The techniques employed for thin film scattering have to be adapted to the physical properties of the films. In the following section, some aspects of X-ray reflectometry and wide-angle scattering from poly- and nanocrystalline metallic thin films are discussed.

3.3.1   Wide Angle Diffraction

The designation of various methods as "wide angle" techniques is related to the measurement of Bragg reflections of the crystal lattice, corresponding to measuring angles 2θ in a range between ~15° and ~170° for the characteristic radiation of X-ray tubes. This designation does not necessitate large angles between incident or scattered rays to the film surface; one or even both of them may be small. Wide angle diffraction is sensitive to information on a lattice scale level of films, e.g. lattice constants, texture and strains. It utilizes basically the relation between the angular positions θ of interference maxima and the spacing d of corresponding lattice planes (Bragg's law, cf. (3.12), (3.20))

$$2\,d\,\sin\theta = m\,\lambda \tag{3.21}$$

where m is an integer designating the order of reflection, and λ is the wavelength. The below described techniques are based on angle-dispersive set-ups for 2θ measurement in reflection (Bragg) geometry and require a fixed wavelength, i.e. monochromatic radiation. For high precision measurements of peak positions, strongly refracting films or small measuring angles, Eq. (3.21) has to be corrected as derived independently by *Darwin* and *Ewald* [3.40, 3.41]

$$2\,d\,(\,1 - \frac{\delta}{\sin^2\theta}\,)\,\sin\theta = m\,\lambda \tag{3.22}$$

where δ is the dispersive correction of the refraction index n

$$n(\lambda) = 1 - \delta(\lambda) + i\beta(\lambda) \tag{3.23}$$

and β is its absorptive part. For characteristic X-ray radiation, δ and β are small quantities of typically $\sim 10^{-6}$. E.g. for Cu-Kα radiation ($E_{K\alpha 1} = 8.048$ keV) and scattering in copper, they amount to $\delta = 24.3 \cdot 10^{-6}$ and $\beta = 0.55 \cdot 10^{-6}$. Both parameters are related to the material properties ρ_e (electron density) and μ (absorption coefficient) by

$$\delta = \frac{1}{2\pi}\,\lambda^2\,r_0\,\rho_e\,, \qquad \beta = \frac{\lambda\,\mu}{4\pi} \tag{3.24}$$

(r_0 electron radius). For X-ray wavelengths, the dispersive correction δ is usually a *positive* quantity. Therefore, there occurs total reflection of X-rays striking the sample surface below a critical angle θ_c, which is in contrast to the region of visible light. For incidence angles α_i smaller than θ_c, a strong increase in the surface sensitivity of X-ray scattering is reached. A measure of the surface sensitivity is the depth T from which 63% of the intensity is backscattered, i.e. half of the usual penetration depth. T is of the order of ~ 5 nm for total reflection from metals, independent of the X-ray wavelength:

$$T = \frac{\lambda}{4\pi\,\theta_c} = \frac{1}{4\sqrt{\pi\,r_0\,\rho_e}} \tag{3.25}$$

For larger incidence angles $\alpha_i \gg \theta_c$, the intensity decrease of X-rays with increasing depth in a sample is governed by the absorption coefficient according to

$$I(z) = I_0\,\exp(-\mu\,z) \tag{3.26}$$

yielding for symmetrical beam conditions the relation

$$T = \sin\alpha_i\,/\,2\mu \tag{3.27}$$

Further correction factors have to be considered for thin films and non-symmetrical geometries [3.42]. The enhanced surface sensitivity for shallow incidence angles α_i often favors measurement geometries with fixed incidence angle in comparison to symmetrical beam conditions for thin film diffraction (Fig. 3.36), if no other reasons, e.g. of texture, are opposed.

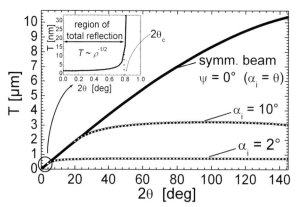

Figure 3.36: T as a measure of the penetration depth for Cu Kα radiation in Cu for symmetrical beam ($\psi = 0$), and measurements with fixed grazing incidence angle α_i

The two measurement geometries compared in Fig. 3.36 are illustrated in Fig. 3.37. As also in the usual *Bragg–Brentano*-geometry applied in powder diffraction, measurements with symmetrical beam (Fig. 3.37a) characterize lattice planes parallel to the sample surface. In this geometry the diffraction vector $\vec{q}$ is always parallel to the surface normal $\vec{n}$.

For grazing incidence measurements (Fig. 3.37b), the direction of the diffraction vector depends on 2θ. Therefore the different diffraction peaks measured in such a configuration correspond to different ψ angles, which can be utilized e.g. for stress measurements applying the $\sin^2\psi$ technique (cf. Section 3.5).

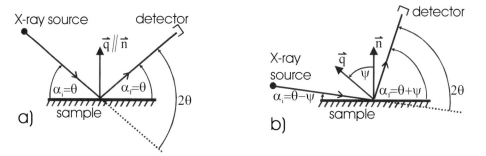

Figure 3.37: Two measuring geometries for back scattering of X-rays (Bragg geometry) a) symmetrical beam, b) grazing incidence

Phase Analysis

Phase analysis of materials is based on the relation between the angular *position* of Bragg peaks and the corresponding lattice spacing according to Eq. (3.21). Apart from the position, the intensity of the peaks also has to be included for quantitative phase determination. However, for thin layers the peak intensity is strongly influenced by film properties like thickness, texture and defects, complicating the evaluation. Furthermore, the lattice constants can also deviate from the bulk values, and non-equilibrium phases are frequently observed in thin films.

Figure 3.38 represents an example of phase analysis in a 400 nm thick Co film. Due to their magnetic features, Co films are of interest for applications in sensors, data storage and other fields of magnetoelectronics. Bulk Co shows a martensitic h.c.p. to f.c.c. transition at about 420 °C sensitively influencing the magnetic properties. Grain growth, texture changes and other structural peculiarities in films can influence the transformation and corresponding changes of magnetic properties [3.43]. In contrast to the clear h.c.p. pattern of the symmetrical beam measurement for the as-deposited state, after annealing at 450 °C only one single reflection is visible, thus indicating a strong texture (Fig. 3.38, lower curves). Since the corresponding reflections (111 f.c.c. and 002 h.c.p.) differ only slightly in angular position, it is difficult to distinguish between both phases.

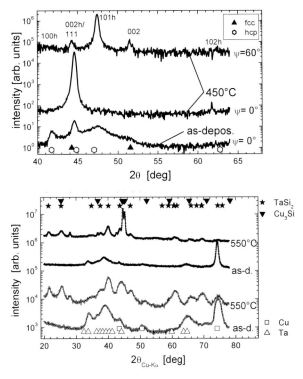

Figure 3.38: Measurement of a 400 nm PVD Co layer before and after a 450 °C/1 h anneal with Cu Kα radiation. Clear detection of the h.c.p. pattern after annealing requires inclination of the sample by an appropriate angle ψ

Figure 3.39: Measurement of as-deposited and annealed Si/Ta/Cu stacks at rotating anode (upper curves) and at synchrotron (lower curves). During annealing at 550 °C, Ta- and Cu-silicides form. Correspondingly, the barrier property of the Ta film degrades. For comparison, the synchrotron measurement curves measured close to the Cu K absorption edge ($E = 8.98$ keV) have been converted to $E = 8.04$ keV

From the additional measurements with the sample surface normal inclined out of the scattering plane it can be inferred, that the main effect of annealing is a transition from a weak to a sharp h.c.p. <001> texture [3.44]. Corresponding to this texture, several h.c.p. reflections occur at special angles $\psi > 0$ (Fig. 3.38, upper curve). Furthermore, a small f.c.c. 002 reflection also appears at $\psi = 60°$, indicating the presence of a minor fraction of additional f.c.c. grains.

As an example of a polycrystalline film with additional absorption in a cap layer, measurements of a 10 nm thick Ta film through a 50 nm Cu cap are displayed in Fig. 3.39. A diffractometer with rotating anode (CuKα radiation, $E = 8.04$ keV), secondary mono-chromator and collimator was optimized for angular resolution with grazing incidence

($\alpha = 2°$). The curve clearly shows the as-deposited nanocrystalline structure of β-Ta and strong Cu reflections, and the annealed state with Cu_3Si and $TaSi_2$ reflections (Fig. 3.39, upper curves). The same sample was measured at the synchrotron source ESRF with a photon energy close below the Cu K absorption edge ($E = 8.97$ keV) in order to minimize the absorption in the cap layer. Due to the small divergence of the synchrotron beam, the incidence angle could be reduced to $\alpha = 0.5°$ thus enhancing the selectivity for the Ta film. Thus, the peak to background ratio of the reflections was significantly improved. A series of such investigations showed that the thermal stability of such diffusion barriers can be improved by modifying their microstructure via nitrogen incorporation [3.45].

Size–Strain Analysis

Intensity and breadth of Bragg peaks depend on the number of reflecting lattice planes. The peak broadening can be utilized to determine a mean size D for the scattering particles (grains) according to the *Scherrer* equation

$$b_p \cos \theta = \lambda / D \tag{3.28}$$

where the breadth of the Bragg peaks b can be measured by the FWHM (full width at half maximum) of a Bragg peak, corrected for the instrumental broadening. The index p stands for the contribution of the particle size (i.e., the finite number of reflecting lattice planes) to the peak width. In the case of a polycrystalline material, the particles correspond to the grains, since the coherent reflection of the incident beam is limited to the aligned lattice planes of those grains complying with the Bragg condition. Note that D is a measure of the mean crystallite extension only in the direction of the measuring vector, i.e. for symmetrical beam conditions it is the direction perpendicular to the film surface. For the crystallite shape an additional correction factor K (approximately equal to unity) is required.

In general, different scale levels can be distinguished for the film strain:

(i)   The strain of the film as a whole occurs due to macroscopic reasons (border-core strains in a solid, e.g. resulting from surface treatment, thermal strains) or due to the film growth. As a special type of such strains coherency strains can arise in epitaxial films.

(ii)   On a crystallite scale mean strains appear in each grain and differ between the grains due to constraints at the grain boundaries and the variation of the anisotropy from grain to grain.

(iii)   Microscopic strains arise on a lattice scale due to defects (e.g. dislocations, point defects) in films.

Film strains of type (i) are discussed in more detail in Section 3.5. The crystallite and microscopic scale is related to fluctuations of the lattice spacing in the film, i.e. to a strain variation $\Delta\varepsilon$, yielding a strain related Bragg peak broadening b_s according to

$$b_s \cos \theta = 4 \sin \theta \, \Delta\varepsilon \tag{3.29}$$

Both the contributions of grain size (3.28) and of strain (3.29) to the peak breadth can be distinguished by their different dependence on the scattering angle θ. This can be done either by superposing both contributions directly according to Williamson and Hall [3.46], using additional assumptions about the shape of the broadening functions, or with fewer presuppositions in the Fourier space after *Warren* and *Averbach* [3.47]. For very thin films, however, it can be difficult to measure enough reflections with sufficient intensity for clear evaluation of the broadening versus θ. Therefore, frequently, additional assumptions are needed, e.g. about the dominating broadening contribution.

For the 222 reflection of a thin Cu film, profile shape changes due to effects of grain size and defect content are represented in Fig. 3.40. The film deposited by electroplating shows the so-called "self-annealing" effect, i.e. grain growth and other structural changes occurring in the first hours/days after deposition at room temperature. Due to the grain growth, the initially common 222 reflection for the Kα radiation begins to split into its Kα$_1$ and Kα$_2$ contribution about 1 day after the film deposition. A high density of lattice defects in the initial film, in particular at the interface to the subjacent thin PVD Cu seed layer, is suggested as a reason for the preferential growth of <001> grains (cf. also [3.48]). According to the shape changes shown in Fig. 3.40, the r.m.s. strain variation and correspondingly, the defect density, reduce during the "self-annealing".

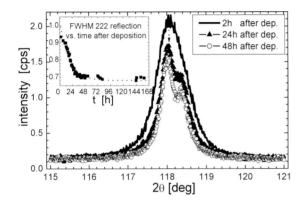

Figure 3.40: Measurement of an electroplated 1 μm Cu film at different times after film deposition (222 reflection, CoKα radiation). The inset shows the decrease in the half width versus time after deposition, roughly following a temporal decay according

to $\exp\left(\dfrac{-t[h]}{17.5}\right)$ (dotted line)

Texture Analysis

The film growth conditions and the violation of the spatial isotropy by the presence of the free surface and neighboring layers or the substrate regularly lead to deviations from a randomly crystallite orientation distribution in thin films, i.e. to a texture. Due to the anisotropy of sample properties with respect to the crystal orientation, the texture of a film significantly influences its physical properties such as elastic, transport or magnetic features. The investigation of the texture by X-ray diffraction is based on the intensity measurement of one or several Bragg reflections for different inclination and rotation angles ψ and φ of the sample (pole figure measurement), or of different Bragg reflections for special angles ψ and φ (cf. Fig. 3.41). Usually, the detector position is fixed during the pole figure measurements at the corresponding Bragg angle 2θ. From measurements with symmetrical beam ($\psi = 0$) or fixed ψ partial texture information can also be derived (cf. Fig. 3.37).

Since strong stresses are a very common feature of films, the corresponding shift of Bragg peak positions, increasing with increasing ψ, can require additional 2θ scans or measurements with a position-sensitive detector to prevent a peak shift out of the detector window. Owing to the low intensity scattered from thin films requiring very long measurement times, partial texture measurements are frequently performed, e.g. pole figure cuts, completed by symmetry considerations for the film texture.

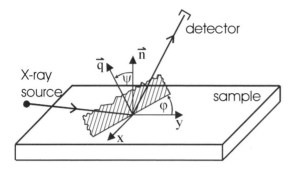

Figure 3.41: Inclination angle ψ and azimuth φ defining the relative orientation between surface normal $\vec{n}$ and diffraction vector $\vec{q}$. The scattering plane containing the incident and reflected beam is marked by the hatched area

As an example Fig. 3.42 represents cuts through the Cu <111> pole figure for two different layer sequences of a Cu/Ta–N layer stack produced by PVD. For the 10 nm thick Ta–N layer acting as a diffusion barrier between a 50 nm Cu film and the Si substrate (open circles), a broad Cu <111> texture results, whereas for a 100 nm Cu film deposited directly onto Si and capped by the Ta–N film (closed squares) a very sharp <111> texture of the Cu film is obtained. Corresponding to the symmetry of the film and the deposition conditions,

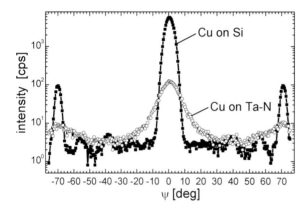

Figure 3.42: Cuts through the <111> pole figure of PVD Cu films in the stacking Si/Cu/Ta–N (closed symbols) and Si/Ta–N/Cu (open symbols). The latter curve shows a broad <111> texture, whereas the copper deposited directly onto Si is sharply textured. Apart from the <111> texture component, it possesses also a <511> component of grains in twin orientation

roughly a <111> fibre texture, i.e. a rotational symmetry of crystallite orientations, all with {111} lattice planes parallel to the surface, can be assumed. Additional peaks in the pole figure cut can be predicted from the angles between the lattice planes in a f.c.c. crystal. Apart from the main peak at $\psi = 0°$, at about $|\psi| = 70°$ a second peak appears due to the reflections from {111} lattice planes having an inclination angle of 70.5° to those parallel to the surface. Furthermore, smaller peaks appear for the film deposited directly onto Si at about $|\psi| = 39°$ and $|\psi| = 56°$. They match to crystallites with {511} planes parallel to the surface according to the angles of 38.9°, 56.2° and 70.5° between {111} and {511} lattice planes. Indeed, for Cu the <511> directions are known to result from a rotation over 60° around the <111> directions, i.e. from a twin relation. The presence of such twins has been also derived from electron microscopy (cf. Figure 3.35).

In the case of laterally structured films, as for parallel Cu interconnect lines of 0.4 μm line width, the measurement of complete pole figures is essential (Fig. 3.43a). Again, the strong intensity in the center of the pole figure marks the predominating <111> texture component. However, the outer ring at $\psi = 70°$ is structured and contains 6 distinct maxima, which correspond according to the 3-fold symmetry of the {111} planes to two preferred crystallite orientations within the film plane. Indeed, for small line widths the crystallite orients preferentially with the {211} planes parallel to the line side walls. This structure is also reflected by the <511> twins (additional rings at 39° and 56°). Furthermore, vertical stripes are resolved corresponding to side-wall oriented grains inclined roughly 90° with respect to the main texture component (cf. Fig. 3.43b). The two stripes from the side walls show a splitting of about 10° into sub-stripes arising from a slight deviation of the side walls from the exact vertical direction of 5°. These structural features depend on the line geometry and on the deposition conditions (e.g. PVD or EP) and can significantly influence the performance of the interconnect lines in integrated circuits.

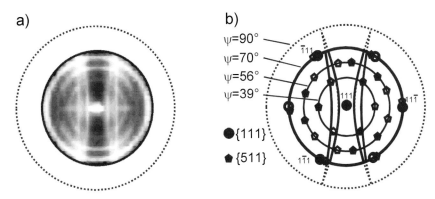

Figure 3.43: {111} pole figure measurement of parallel Cu lines (a), and designation of main texture components of the Cu lines in the schematic {111} pole figure (b). The dotted ring corresponding to $\psi = 90°$ is out of the range of the measurement. Due to the symmetry of the Cu lines running in vertical direction, apart from the regular poles also their mirror images appear, yielding e.g. six 111 maxima instead of the three on a regular $\psi = 70°$ ring. This property is reflected by open and closed symbols, corresponding to the alignment of crystallite orientations at the left and right line side wall, respectively

Diffraction from Multilayers

Periodic layer stacks of nanoscale metallic layers can show outstanding physical properties not observed in the bulk, e.g. giant magnetoresistance (GMR). Due to the periodicity of such stacks, X-ray scattering effects can be enhanced, allowing one to obtain additional structural information compared to scattering from single films. For example, Fig. 3.44 represents diffraction patterns from Co/Cu multilayers recorded after annealing at 450 °C. In the as-deposited state, the samples possess broad <111> fibre textures according to the PVD preparation. The measurements performed at $\psi = 0$ due to the texture and the relatively large grains, show the tendency for a transition from the appearance of separated Co/Cu reflections for the samples with larger thickness of the individual layers, to the formation of common 111 Co/Cu multilayer reflections (at about $2\theta = 44°$), being more pronounced for the very thin individual layers (Fig. 3.44, from top to bottom).

A phenomenon typical for multilayers is the appearance of additional satellite reflections apart from the zero order reflections, as seen in the two lower curves in Fig. 3.44.

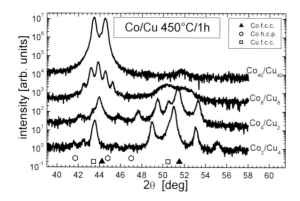

Figure 3.44: θ–2θ-scans of annealed Co/Cu multilayers deposited by PVD with a total stack thickness between 120 nm and 240 nm and different thicknesses of the individual layers (from top to bottom:
[Co$_{40\ nm}$/Cu$_{40\ nm}$] $\cdot$ 3
[Co$_{8\ nm}$/Cu$_{8\ nm}$] $\cdot$ 15
[Co$_{4\ nm}$/Cu$_{2\ nm}$] $\cdot$ 20
[Co$_{2\ nm}$/Cu$_{4\ nm}$] $\cdot$ 20)
The two upper curves show an enhanced h.c.p. <001> / f.c.c. <111> texture component, whereas the two lower curves comply with a texture transition to the <100> component (strong f.c.c. 002 reflection at about $2\theta = 51°$)

Furthermore, in correlation with the decrease in the individual layer thickness, at annealing a texture change occurs in Co/Cu multilayers for small layer thicknesses (Fig. 3.44). This behavior can be related to layer strains and the corresponding elastic energy stored in the multilayer. For a given layer strain, due to the elastic anisotropy of Cu and of f.c.c. Co, the f.c.c. <001> orientation is energetically favored with respect to other orientations, whereas a strain in the <111> directions is connected with a high amount of stored energy. For appropriate thicknesses of the individual layers, the energy release due to a texture alteration can stabilize the multilayer structure. An enhanced stability of GMR properties upon annealing results from such structural changes.

Layer strains originate for instance from the different thermal expansion coefficients of layer stack and substrate. However, such strains also arise in the cases of other material combinations (e.g. NiFe/Cu), where no texture alteration is observed. As follows from an evaluation of the layer coherency, the presence of corresponding coherency strains oscillating within the multilayer stack is a more likely reason for the texture alteration. For the growth of a Cu layer onto the Co and vice versa, two extreme cases can be considered: In the case of a relaxed growth, the mismatch between the larger lattice parameter of Cu and the smaller one of Co is completely accommodated by misfit dislocations. In contrast, for a fully coherent growth of the layers, both the Co and Cu lattice planes perpendicular to the surface would match perfectly, whereas the lattice parameters of planes parallel to the surface are enlarged for Cu and decreased for Co due to the transverse contraction (Fig. 3.45a), giving rise to out-of-plane strain components ε_{op} in the Co and Cu layers. From a determination of ε_{op} via an evaluation of the satellites, the degree of coherency can be inferred. Whereas the angular distance of the satellite peaks is a measure of the layer thickness, which can be calculated from Bragg's equation according to

$$2 <d> \sin\theta_m = (1 \pm m \frac{<d>}{D})\lambda, \quad m = 0, 1, 2, \ldots \tag{3.30}$$

($<d>$, average lattice parameter, D, multilayer periodicity length, cf. [3.49]), the intensity of the reflections is strongly sensitive to the contrast between the Co and Cu layers. This contrast consists of both a chemical component, which is weak between Co and Cu due to their similar electron density, and a component arising from the difference in the lattice parameters between the Co and Cu layers (Fig. 3.45b). The latter contribution is related to the coherency of the lattices in the multilayer. From comparison of the measured and calculated curves, a fully coherent strain state can be assumed (Fig. 3.45b) [3.44]. The reduction of the considerable coherency stresses attained in the case of a <100> texture formation can be taken as a likely driving force for the texture alteration.

Microdiffraction

Apart from integral structural investigations by X-ray scattering on a laterally macroscopic scale, local methods are required for patterned film structures, if the macroscopic area cannot be patterned in a homogeneous way. In the following, examples on interconnect lines and via structures required for Cu metallization in microelectronics are briefly considered.

a)   b)

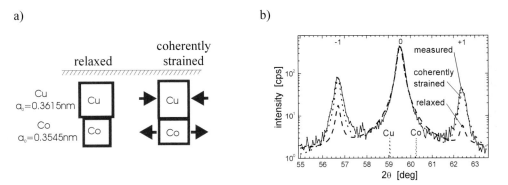

Figure 3.45: Multilayer strain states (a), and measurement (b) of a Co/Cu 002 reflection (magnetron-sputtered $[Co_{2nm}/Cu_{2nm}]$·30 multilayer, in situ annealing at 380 °C) together with curves calculated for the relaxed state (dashed line) and the fully coherent strain state (dotted line)

An important issue in interconnect failure is electromigration resulting in the formation of voids and hillocks during current conduction (cf. Section 4.1). Since these failure mechanisms strongly depend on mechanical stresses and temperature, they should be investigated at intact layer systems including barrier and passivation layers. The investigation of such systems by X-rays can be divided into the following areas: microscopy, averaging techniques with a microbeam, tomography and Laue microdiffraction.

Whereas optical microscopy suffers from insufficient spatial resolution for structures in the sub-100 nm range, conventional scanning electron microscopy is limited by electron scattering in passivation layers. Another possibility is direct microscopic observation using synchrotron radiation. For example, in the experiments of *Schneider et al.* [3.50] the formation of voids in passivated Cu lines has been imaged with a spatial resolution of 40 nm by an X-ray transmission microscope operating at 1.8 keV photon energy equipped with a microzone plate. It was concluded that voids nucleate at grain boundaries within the lines and then migrate to their side walls where they agglomerate.

To measure directly the local strain at such interconnects, microdiffraction analysis is required. For example, *Zhang et al.* found large compressive strains during electromigration test of interconnects with 2 μm linewidth around a hillock by measuring the Cu 111 reflection peak shift at $E = 9.5$ keV in step sizes of 1 μm (resolution 0.2 μm) [3.51]. Using white X-ray radiation confined by a 10 μm x 10 μm pinhole in symmetric reflection geometry and an energy dispersive detector, *Wang et al.* [3.52] proved the presence of electromigration-induced stress gradients in Al interconnects. By simultaneous analysis of strain and fluorescence the mass transport and local changes of the element distributions in Al–Cu alloy lines could also be detected [3.51, 3.53]. An equivalent technique is the measurement of the intensity of the 400 peak of the silicon substrate [3.54] by a microbeam, called microtomography. This technique is extremely sensitive to shear stress induced in the silicon substrate. Tomography maps showed delamination in thin Ni films as well as electromigration induced stress built up in Al conductor lines.

As another promising technique to obtain the full strain tensor in single grains of interconnect lines, the combination of white and monochromatic synchrotron radiation can be utilized [3.55–3.57]. Whereas the deviatoric part of the strain tensor can be derived from analysis of Laue patterns, the hydrostatic component is obtained by energy scans of selected reflections with monochromatic beam. This technique allows a fast (10 s) determination of deviatoric strain with submicron local resolution. By analyzing patterns as in Fig. 3.46 for different locations on the sample, stress maps can be obtained. The measurement of large distributions in inter- and intragranular stresses as well as the observation of electromigration induced plasticity (prior to macroscopic damage) can be attributed to advances in this technique. A further advance of Laue microdiffraction is the so-called 3D X-ray microscopy [3.58], utilizing a thin wire to mask the diffracted beam. By movement of the wire a 3D strain map can be obtained. Another advantage of the Laue technique is that the sample does not have to be rotated. Even with state of the art goniometers limiting sample movement to about 1 micron upon rotation is a difficult task. When focusing devices approach the sub-micron range, sample rotation will have to be avoided completely.

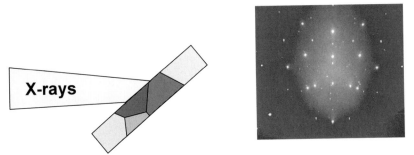

Figure 3.46: Schematic of X-ray beam size vs. crystallite size (left) and Laue pattern of Cu on Si substrate (bright spots, right). The grain size is about 2 μm, and the beam diameter is below 1 μm ([3.55, 3.56]). From the pattern, information can be extracted about grain orientation, 3D deviatoric strain tensor of the Cu grains, and dislocation density

3.3.2 Reflectometry

Whereas the wide-angle diffraction methods are sensitive to a scale level corresponding to the lattice planes and interatomic distances, reflectometry methods require smaller diffraction angles sensitive to a larger scale level in direct space, ranging from about 1 nm to 1 μm. Therefore X-ray reflectometry (XRR) is well suited for investigation of films with thicknesses in this range. In the following section some coplanar scattering geometries are briefly discussed, i.e. with the incident, the reflected beam and the surface normal-vector lying within one plane (the scattering plane). If this condition is fulfilled, the scattered intensity can be described in terms of the scattering vector $\vec{q}$ only, whereas elsewhere it depends on the wave vectors of the incoming and reflected wave independently (cf. [3.59]).

Specular Reflectometry
Specular reflectometry measurements are performed under symmetrical beam conditions for shallow incident and reflected beams. They scan the reciprocal space perpendicular to the sample surface with varying magnitude, but constant direction of the scattering vector $\vec{q}$ (Fig. 3.47a). Presupposing knowledge of the atomic scattering factors (or, equivalently, of the refraction indices) of the investigated layers, in general three parameters for the layers can be derived from such measurements, e.g. by comparison of the measured curves with calculated ones. These parameters are the film thickness D, the density ρ of films or surface layers, and the surface or interface r.m.s. roughness σ. They influence the angular dependence of the reflectivity R in a characteristic way (cf. Fig. 3.47b).

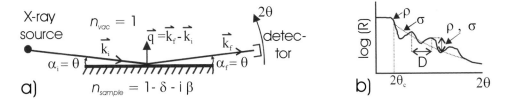

Figure 3.47: Measurement set-up for specular reflectometry (a) and schematic reflectometry curve of a thin film (b)

If the coherence length of the incoming radiation is sufficiently large, interference between the waves reflected from the upper and lower interfaces of films is possible. For constructive interference a series of maxima occurs in the reflected intensity. The maxima positions can be determined in analogy to the Bragg peak positions in lattice diffraction. Due to the small diffraction angles, the dispersive correction of the Bragg equation (3.22) should be considered, yielding in a linear approximation for the maxima positions θ_m

$$\sin^2 \theta_m = (\frac{m\,\lambda}{2\,D})^2 + 2\,\delta, \quad m = 1, 2, 3, \ldots \tag{3.31}$$

Thus, a plot of $\sin^2\theta_m$ (or, approximately the squared maxima positions) versus the order m of the peaks should give a straight line, the slope of which is a sensitive measure of the thickness D [3.60]. The offset of such a plot contains the dispersive correction δ of the refraction index. Its knowledge allows one to calculate the density ρ of the corresponding film according to Eq.(3.24). More generally, δ can be determined independently of the appearance of film interferences from the condition of total reflection for X-rays, yielding a critical angle θ_c below which total reflection occurs

$$\theta_c = \sqrt{2\delta} = \lambda \sqrt{\frac{r_0}{\pi} \rho_e} \qquad\qquad (3.32)$$

Experimentally, θ_c can be determined from the drop in the intensity down to the half value of the plateau in the first part of the XRR curve (Fig. 3.47b). For negligible absorption of the surface layer, the reflected intensity is a step function with a discontinuity at θ_c, whereas absorption yields a decrease in the reflectivity $R \sim \theta^{-4}$ close to θ_c. An additional damping has to be considered for a rough surface by a *Debye–Waller*-like factor [3.61]

$$R^{\text{rough}} = R^{\text{smooth}} \exp(-q^2 \sigma^2) \qquad\qquad (3.33)$$

where R^{smooth} is the reflectivity of a smooth surface, and σ is the r.m.s. roughness parameter.

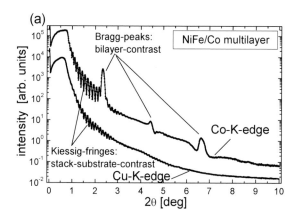

Figure 3.48a: XRR curve of a magnetron-sputtered $[\text{NiFe}_{2\text{nm}}/\text{Co}_{2\text{nm}}] \cdot 15$ multilayer stack (total thickness 60 nm) measured at two different photon energies at the ESRF Grenoble. The curves are shifted vertically for clarity. The resonant contrast enhancement at the Co K absorption edge enables to determine the roughness parameter of the NiFe/Co interfaces from the Bragg peak heights to $\sigma_{\text{rms}} = 0.5$ nm

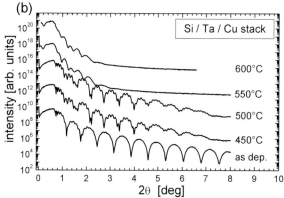

Figure 3.48b: XRR of 10 nm Ta deposited onto Si and capped by 50 nm Cu as prepared and after annealing at different temperatures. The decrease of the Ta oscillations at 450 °C can be formally described by an increased roughness parameter, and the fading of the Ta peaks at 550 °C indicates the complete failure of the Ta diffusion barrier leading to the decay of the layer set-up [3.45]

An example of the more complex case of a multilayer stack consisting of a periodic arrangement of $\text{Ni}_{80}\text{Fe}_{20}/\text{Co}$ bilayers with an individual layer thickness of 2 nm and a repeat number $N = 15$ is represented in Fig. 3.48a. The interference between rays reflected at the top and bottom of the stack results in *Kiessig* fringes, i.e. intensity oscillations from which the total thickness of the stack can be derived. Since the angular distance of such oscillations in the scattering curve behaves reciprocally to the corresponding thickness in real space, the peaks corresponding to the periodicity length (bilayer "Bragg peaks") have a much larger distance and contain N-2 *Kiessig* fringes in between them. However, in the case of the used sample, for most energies of the radiation no such bilayer Bragg peaks appear due to the low scattering contrast between the constituent elements Fe, Co and Ni (Fig. 3.48a, lower curve), according to the low contrast in electron density between these elements. In such cases, the resonant enhancement of scattering factors close to an absorption edge of a constituent element may be employed, resulting in a significant increase in the scattering contrast and enabling one to conclude on the parameters of the individual layers in the multilayer stack (Fig. 3.48a, upper curve).

As shown in Fig. 3.48b, by using XRR the thermal stability of thin layers and interfaces can be sensitively monitored. Even after annealing at 450 °C, where no changes in the phase composition are detectable by wide-angle XRD, the Ta film related oscillations of about 0.7° periodicity length decrease, corresponding to the onset of Cu trace diffusion through the Ta film acting as a diffusion barrier [3.45].

Diffuse Scattering

In contrast to the symmetrical beam geometry represented in Fig. 3.47a, diffuse scans include offsets between the angle of the incident (α_i) and the reflected (α_f) beams to the sample surface. Therefore the diffraction vector is not restricted to a perpendicular alignment to the film surface, and the scattered intensity contains also lateral information about the surface/interfaces. Assuming a description of the surface by a fractal model with the height–height correlation function [3.61]

$$C(x) = \sigma^2 \exp[-(x/\xi)^{2h}] \tag{3.34}$$

the parameters ξ (lateral correlation length) and h (*Hurst* parameter $0 \leq h \leq 1$, related to the fractal dimension $D = 3 - h$) can be derived from diffuse scans. Furthermore, by diffuse scattering the degree of conformal roughness in a multilayer, i.e. the vertical replication of roughness profiles, can be determined, and a distinction between morphological roughness and interdiffusion is possible (for details, see e.g. [3.59–3.61]). A section of a mapping of diffuse scattering from a $Ni_{80}Fe_{20}/Cu$ multilayer is represented in Fig. 3.49. The middle vertical line at $\alpha_i = \theta$ corresponds to the specular reflectometry, and the dark regions on the

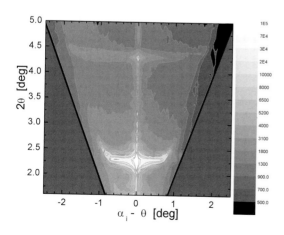

Figure 3.49: Diffuse scattering of a $[NiFe_{2nm}/Cu_{2nm}] \cdot 30$ multilayer stack deposited by PVD. The individual layer thickness was adjusted to the 2nd coupling maximum of the GMR. The photon energy was tuned close to the Cu K absorption edge (8.98 keV)

left and right hand side are not attainable in coplanar geometry due to the shadowing effect of the sample surface. The extension of intensity from the specular line in the transverse direction depends on the r.m.s. roughness, the parameters ξ and h, and also on the vertical replication of the roughness within the multilayer layer stack. For example, significant stripes are visible at the positions of bilayer Bragg peaks (at $2\theta = 2.2°$ and $4.4°$) extending far into the horizontal direction. These far extending sheets indicate concentration of the scattered intensity due to partially coherent scattering at adjacent interfaces, i.e. a correlation between the roughness profiles in the multilayer in a vertical direction.

3.3.3 Soft X-Rays and Magnetic Scattering

Though the interaction of X-rays with the magnetic moment of scattering electrons is considerably weaker than with their charge, magnetic scattering of X-rays has been increasingly exploited in recent years, mainly due to the availability of high-brilliance synchrotron sources enabling also the study of small effects. Furthermore, a strong enhancement of magnetic scattering has been attained by resonant scattering of polarized radiation at those absorption edges that include excitations from magnetic orbitals. In resonance, the magneto-optical effects may reach the magnitude of the charge scattering and are much larger than those for visible light. For $3d$ transition metals huge enhancements were obtained at the L and M absorption edges, which are disposed in the soft X-ray energy range (denoting a photon energy interval between ~ 40 and 2000 eV [3.62]). Whereas the X-ray wavelengths in this range are too long for Bragg diffraction at metal crystal lattice planes, they can be well adapted to the typical nanometer scale of thin magnetic films. The advantages of the resonant magnetic X-ray scattering compared to alternative methods like X-ray absorption techniques, neutron or electron scattering, arise from their combination of element specificity with the simultaneous receipt of structural and magnetic information, the possibility to measure with an applied magnetic field (e.g. hysteresis loops), the tunable depth of information and the simple sample preparation (e.g. no need for thin foil preparation).

An intensively studied X-ray magneto-optical effect is X-ray magnetic circular dichroism (XMCD), which is based on intensity changes corresponding to the different response of a

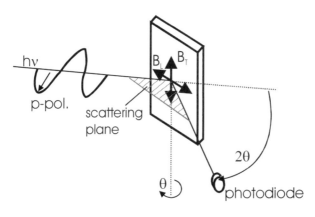

Figure 3.50: Scattering geometry and orientation of magnetic field for a magnetic scattering experiment with longitudinal (B_L) and with transversal (B_T) measurement geometry. For the p-polarized incident radiation and the applied field B_T shown here, the T-MOKE can be measured, whereas consecutive application of magnetic fields B_T and B_L is required to determine the XMLD from a ferromagnetic layer. Utilization of circularly polarized radiation and field component B_L corresponds to the reflection XMCD

magnetized film to left- and right-handed circularly polarized light. When measured in absorption or photoemission, it is related to the imaginary part of the magneto-optical refractive index. From more sophisticated measurements, which include a polarization analysis like those of the *Faraday* or the *Voigt* effect [3.63], additionally the real (dispersive) part of the refraction index can be inferred directly.

The magnetic scattering in the soft X-ray range is mainly performed by exploiting magneto-optical *Kerr* effects (MOKE), for example in the transversal geometry (T-MOKE, [3.64]) or in the longitudinal geometry (L-MOKE, [3.65]), or by utilizing X-ray magnetic dichroism effects in reflection geometry (Fig. 3.50). All these scattering effects are related to both the real and the imaginary parts of the refraction index. MOKE was originally discovered by *J. Kerr* for visible light (1877) and is usually denoted as X-MOKE in the soft X-ray energy range [3.65]. Depending on the scattering geometry, magnetically induced changes of the reflected intensity (as in T-MOKE geometry) or of the polarization of the reflected beam are detected. Comparable to T-MOKE also the reflection XMCD is an intensity measurement depending linearly on the sample magnetization **M** [3.66], whereas the reflection X-ray magnetic linear dichroism (XMLD) or the *Voigt* effect measured in reflection depend quadratically on **M** [3.63]. Therefore, the latter effects also enable the investigation of antiferromagnetic materials showing no macroscopic net magnetic moment.

A microscopic description of the interaction between X-ray photons and magnetic samples was given by *Hannon et al.* [3.67], who calculated the resonant scattering amplitude for dipole transitions in a single atom:

$$f_{res}(E) = f_{charge}(E) + f_{circ}(E) + f_{lin}(E) = \frac{-3}{4\pi q} \{ (\vec{e}_f^* \cdot \vec{e}_i)(F_{-1}+F_{+1})$$

$$+ i(\vec{e}_f^* \cdot \vec{e}_i)\, \vec{m}\, (F_{-1}-F_{+1}) + (\vec{e}_f^* \cdot \vec{m})(\vec{e}_i \cdot \vec{m})(F_0-F_{-1}-F_{+1})\} \tag{3.35}$$

($\vec{e}_f$, $\vec{e}_i$ .. polarization vectors of the scattered and incident beam, respectively, $\vec{m}$ .. local magnetic moment, $F_{\Delta M}$ .. matrix element for dipole transitions with changes of the magnetic quantum number by $\Delta M = -1$, 0 or +1). The resulting terms f_{charge}, f_{circ} and f_{lin} correspond to the nonmagnetic resonant charge scattering, the effects depending linearly on M as the XMCD, and effects quadratic in M as XMLD, measured in reflection.

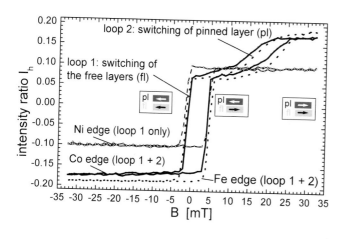

Figure 3.51: Hysteresis loops measured in L-MOKE geometry at different photon energies (Fe edge: 707 eV; Co edge: 778 eV; Ni edge: 853 eV) of a spin-valve system containing 3.5 nm NiFe and 3 nm CoFe films as free and pinned layers, respectively. For normalizing the intensity, the ratio

$$I_h = \frac{2I}{I_{min}+I_{max}} - 1 \text{ was}$$

chosen

Soft X-rays with a wavelength on the nm scale are perfectly suited to investigate both the magnetic and structural properties of layer systems with typical thicknesses in the nanometer range. Fig. 3.51 contains the X-MOKE hysteresis loops measured at an IrMn-based spin-valve structure (cf. Section 5.3) with Co and Fe in the pinned and Ni and Fe in the free layer. Since the intensity of the reflected light depends on the orientation between the magnetization of the layers and the external magnetic field, the switching in the different layers can be observed. For that purpose an appropriate angle of incidence has to be chosen, since the intensity modulation depends on the interference conditions of the photons in the layer stack. The measured loops are partitioned into subloops that are attributed to the switching in the free and pinned layers according to the element specificity of the measurement, controlled by the incident photon energy tuned to the particular absorption edges. Making use of the interference effect within the layer, it is also possible to observe the switching of the free and pinned layers separately at appropriate fixed photon energy, i.e. without switching the photon energy to several absorption edges. Utilizing this feature opens up the possibility for magnetic depth profiling in magnetic layer systems.

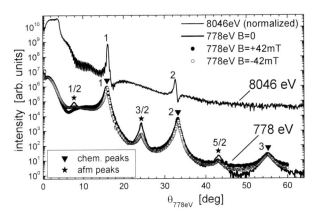

Figure 3.52: XRR of a $[Co_{1nm}/Cu_{2nm}] \cdot 30$ multilayer measured at two different photon energies. The lower curves were measured at the Co L_3 absorption edge (778 eV, circ. polarized light) for demagnetized sample (B=0) and two signs of an applied magnetic field. The applied magnetic field of 42 mT was below the saturation field of about 100 mT for this sample. The upper curve measured with Cu Kα radiation was converted to 778 eV using Eq. (3.21)

Polarized soft X-rays can be also employed to analyze antiferromagnetic (afm) structures. Whereas an observation of the magnetism in natural antiferromagnets like IrMn requires magneto-optical effects quadratic in the magnetization (as, for example, the XMLD or the Voigt effect [3.63]), the nm scaling of artificial antiferromagnets like GMR multilayers yields magnetic scattering effects directly measurable by intensity changes in reflectometry. As an example, Fig. 3.52 shows reflectometry curves of a Co/Cu multilayer optimized for the 2nd maximum of the afm coupling. Apart from a series of structural Bragg peaks arising from the chemical modulation periodicity length of about 3 nm, additional half-integer order peaks appear if the photon energy is tuned to the L_3 absorption edge of Co. Such peaks have been proven to be of purely magnetic origin and to match perfectly with the magnetic periodicity length [3.68], which is about 6 nm for the investigated multilayer. The decrease of the magnetic peaks in an applied magnetic field corresponds to the deterioration of the afm alignment of the Co layers in that field. For example, this effect can be exploited to study the decay of the GMR effect in afm coupled multilayers during annealing. Whereas the macroscopic multilayer structure in the Co/Cu system still remains intact at a moderate temperature of 200 °C as indicated by the persistence of the structural Bragg peaks, the magnetic peaks decrease in intensity already at that temperature in correspondence to a GMR reduction [3.69]. Thus, the dependence of the magnetic substructure on very localized interfacial changes, which are not observed in reflectometry experiments with lateral macroscopical sensitivity, was confirmed.

Furthermore, the study of the multilayer Bragg peaks at varying photon energy opens up the opportunity to determine the near-edge course of the magneto-optical constants (e.g. [3.70]). For example, the peak position of the structural Bragg peaks in Fig. 3.52 shifts far from the edge according to Braggs law with varying photon energy, whereas close to the edge a significant deviation due to resonant changes of the dispersive coefficient δ occurs. This deviation depends on the magnetic field. Thus, also the magnetic part of the refraction index can be determined.

Due to the ongoing increase in the brilliance of modern synchrotron sources and focusing devices, a combination of local imaging methods with magneto-optical scattering techniques will be one of the next steps for the local characterization of magnetic structures.

3.4 Spectroscopic Techniques

3.4.1 Element Distribution Analysis

Spectroscopic techniques for element distribution analysis can be carried out by electrons, X-rays or ions and are frequently used in connection with imaging techniques such as scanning or transmission electron microscopy. In the case of a scanning mode lateral element distributions can be determined by element analysis of individual points, by line scans or by mappings of the whole area. Energy filtered (EF) electrons also allow the production of a simultaneously illustrated TEM image (so called EFTEM, see Section 3.2.1).

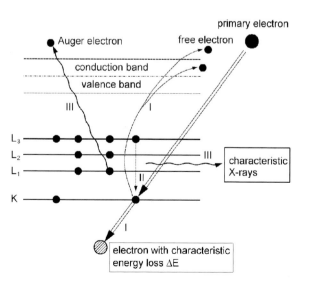

Figure 3.53: Emission of characteristic X-rays and of electrons with characteristic energy loss from an atom after irradiation with primary electrons

Important spectroscopic techniques for the element distribution analysis of thin films and thin layered systems for electronics are electron energy loss spectroscopy (EELS) and spectroscopy with characteristic X-rays, either energy or wavelength dispersive (EDXS, WDXS). The physical associations are shown in Fig. 3.53. If a sample atom is irradiated by primary electrons a lot of interactions take place. The primary electrons may ionize the inner electron shells of the irradiated atom by inelastic pulses. In a primary act, I, an electron from an inner shell is thrown out of the atom, and a gap appears. Within a second act, II, this gap is filled by an electron of an outer shell. In a third act, III, the released energy may lead to the emission of Auger electrons or, alternatively and favored for high atomic numbers, of characteristic X-rays which can be used for element distribution analysis (EDXS, WDXS). Simultaneously, the primary electrons suffer a characteristic energy loss ΔE which is the basis for EELS.

X-ray Spectroscopy

As a prerequisite for the ionization of inner shells, i.e. for functioning X-ray spectroscopy, the primary electron energy must be higher than the critical ionization energy:

$$E_{PE} \overset{!}{>} E_{ion,crit.} \tag{3.36}$$

The jump of an electron from an outer shell into a vacancy in an inner shell may lead to the emission of an X-ray quantum with the energy ΔW, the wavelength of which amounts to

$$\lambda_{char.} = \frac{1.24 \ nm}{\Delta W [keV]} \tag{3.37}$$

Both ΔW and $\lambda_{char.}$ depend on the emitting elements; therefore X-ray signals may be utilized for element distribution analysis. As an example Fig. 3.54 shows the characteristic X-radiation of a Cu anode irradiated with primary electrons.

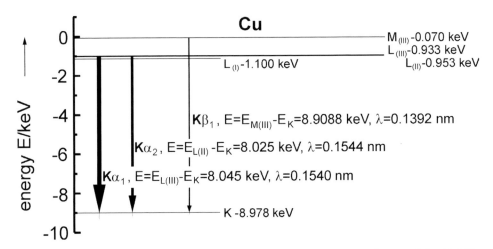

Figure 3.54: Characteristic X-radiation of a Cu anode excited with primary electrons (after [3.71])

For single-electron systems the frequency f of the emitted radiation is determined by the following law:

$$f = R \cdot Z^2 \cdot \left(\frac{1}{n_1^2} - \frac{1}{n_2^2} \right)$$

(3.38)

with n = main quantum number ($n_1 < n_2$),
          R = Rydberg constant factor ($R = 3{,}288 \cdot 10^{15} \, s^{-1}$),
          Z = atomic number.

The partial screening of *Coulomb's* potential by the remaining electrons needs a modification of the function $f(Z)$ by a screening constant whose numerical value is 1 for the K shell.

In this way the emission of the K_α-lines ($n_1 = 1$, $n_2 = 2$) is described by *Moseley's* law that was found as early as 1913 [3.72]:

$$f_{K\alpha} = \frac{3}{4} \cdot R \cdot (Z-1)^2$$

(3.39)

This equation gives the connection between the measured frequency of X-rays and the atomic number of the excited and emitting solid state atoms. This connection is shown graphically in Fig. 3.55.

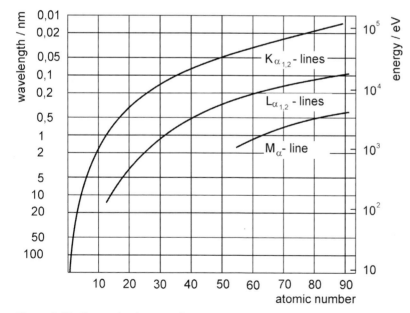

Figure 3.55: Connection between the wavelength or the energy respectively of the emitted X-rays and the corresponding atomic number

In connection with TEM the EDXS mode of X-ray spectroscopy is used in almost every case, whereas in analytical SEM or in electron probe microanalysis (EPMA) respectively either EDXS or WDXS are used. The advantages and disadvantages of both analytical techniques are compared in Table 3.3 [3.73].

Table 3.3: Comparison of EDXS and WDXS for element detection in analytical SEM or EPMA

Analytical technique	EDXS	WDXS
Detection of elements	Parallel	Sequentially
Limit of detection	$\left(\dfrac{m_x}{m_o}\right)_{min} \approx 2 \cdot 10^{-3}$	$\left(\dfrac{m_x}{m_o}\right)_{min} \approx 2 \cdot 10^{-4}$
Lateral resolution	$\delta_{min} = (0.2 \dots 1)\ \mu m$	$\delta_{min} = (1 \dots 3)\ \mu m$
Quantification	$\dfrac{\Delta m}{m} \approx 2 \dots 3\%$ (main components)	$\dfrac{\Delta m}{m} \leq 1\%$ (main components)
Detection of light elements	$Z \geq 5$	$Z \geq 4$

For an EDXS analysis in the TEM the possible lateral resolution is much better because of the small scattering volume in the very thin, electron-transparent films. Distances of $\delta = (2 - 5)$ nm can be resolved. All other parameters possess similar values as in Table 3.3.

The principle of EDXS is schematically illustrated in Fig. 3.56. Characteristic X-rays with an energy quantum $E_x = h \cdot f$ fall into a semiconducting Ge or Si(Li) crystal detector which is coated on both sides with thin conducting gold films. In the semiconductor crystal at first photoelectrons develop, each of which forms N electron hole pairs with an energy E_i, according to

$$E_x = h \cdot f = N \cdot E_i \tag{3.40}$$

The energy E_i for the formation of an electron hole pair amounts to

$$E_{i,Si(Li)} = 3.81\ eV \quad \text{for a Si(Li) crystal and}$$

$$E_{i,Ge} = 2.97\ eV \quad \text{for a Ge crystal .}$$

From Eq. (3.40) results

$$E_x \sim f \sim N$$

By *Moseley's* law (3.39) the atomic number Z of the excited atoms can be determined from the measured frequency f. In this way Eq. (3.40) allows an element calibration of the energy scale that usually serves as the abscissa in the spectrometric diagrams. The ordinate represents the events/time and is therefore a measure of element concentration. Its unit of measurement is counts per second (cps).

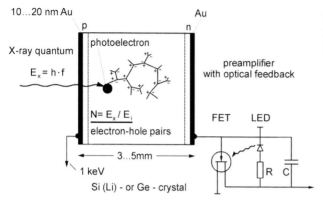

Figure 3.56: Energy dispersive X-ray spectrometer shown schematically

In order to avoid contamination the semiconductor crystal of the EDXS system is situated in a vacuum. The former Be entrance windows had a high absorption for long wave X-rays, thus measurements were limited to elements with atomic numbers not lower than $Z = 11$ (that is for Na). Recently entrance windows from organic materials, e.g. polyimide have been developed, allowing semi-quantitative EDXS analyses down to $Z = 5$ (boron) [3.74].

Figure 3.57 shows an example of an EDX spectrum of a chromium–nickel specimen acquired in an analytical TEM using an accelerating voltage of 200 kV and a Ge EDX detector. Die X-ray lines are broadened to peaks because of the noise of the detector and the limited energetic resolution of about 130 eV.

The peaks are indicated in the following way: the letter K means that the K shell has been ionized by the primary electron and the electron which filled this hole has come from the L

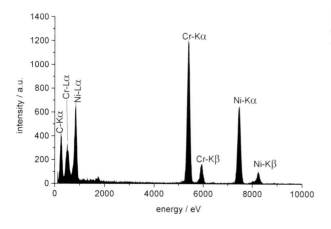

Figure 3.57: EDX spectrum of a Cr–Ni sample, recorded in a TEM with 200 kV accelerating voltage

shell (K-α) or the M shell (K-β). For the L-α peak the L shell has been ionized by the primary electron and the filling electron has come from the M shell (cf. Fig. 3.54). Beside the expected X-rays characteristic for Cr and Ni carbon was also found in the EDX spectrum. This is an effect of carbon hydrogen contamination of the specimen often observed in TEM.

Beside the qualitative element analysis the quantitative element composition of the investigated sample area can also be calculated from the spectra. The ratio of the concentrations c_A and c_B of two elements A and B is given by

$$\frac{c_A}{c_B} = \frac{k_A}{k_B} \cdot \frac{i_A}{i_B} \tag{3.41}$$

with i_A and i_B as the intensities of the characteristic X-ray peaks (i.e. the peak areas) of the elements A and B (e.g. the K-α peaks of Cr and Ni in Fig. 3.57) and k_A and k_B as affiliated factors of sensitivity. The sensitivity factors consider the energy dependent cross sections of all the steps important for the X-ray spectroscopy. These represent the probability of the ionisation of a core shell (cross section for ionisation), the probability of the emission of an X-ray quantum (fluorescence yield), and the probability of the detection of this quantum by the EDX spectrometer (spectrometer efficiency). The sensitivity factors can be calculated using models or measured by standards. Using calculated k-factors ("standardless quantification") an atomic ratio Cr/Ni of 1.7 follows from the spectrum in Fig. 3.57. Because of the assumptions included in the model the relative error can reach up to 30% in the case of EDX spectra acquired in the TEM. It can be reduced to values below 5% by measurements with standards.

Sources for quantification errors are also given by the partial absorption of the X-rays within the specimen and the secondary fluorescence. The absorption has to be considered especially for the low energy X-rays emitted by light elements. The quantification formula with absorption correction for spectra taken in the TEM is given by

$$\frac{c_A}{c_B} = \frac{k_A}{k_B} \cdot \frac{i_A}{i_B} \cdot \frac{(\frac{\mu}{\rho})_A}{(\frac{\mu}{\rho})_B} \cdot \frac{1 - \exp[-(\frac{\mu}{\rho})_B \cdot \frac{\rho \cdot t}{\sin \delta}]}{1 - \exp[-(\frac{\mu}{\rho})_A \cdot \frac{\rho \cdot t}{\sin \delta}]} \tag{3.42}$$

with the mass absorption coefficients (μ/ρ) for the X-ray energies of the elements A and B, the specimen density ρ, and the specimen thickness t. The detector "looks" on the specimen surface keeping the "take off angle" δ (see also [3.75]) .

The problem with using this formula is the determination of the local specimen thickness and the density of the thin film which can be completely different than in the case of bulk samples. Therefore different methods exist for the absorption correction which try to avoid the direct input of thickness and density (cf. Section 4.7).

Another possibility of investigations affects the acquisition of X-ray intensity profiles along a line over the specimen. That allows the investigation of the element distribution along this line.

As an example, the element distribution analysis over a cross-section of a Cu metallization on oxidized Si wafer with intermediate W–Ti diffusion barrier will be discussed [3.76]. The left micrograph of Fig. 3.58 shows a TEM bright field image of magnetron sputtered W–Ti diffusion barrier layer, followed by an 80 nm Cu metallization layer (cross-section by ion thinning preparation). Ti was added to the W barrier layer material for an enhanced adhesive force. To analyze the element distribution an EDXS line was recorded along the marked position arrow, that is across the barrier and the metallization layer (right plot). The element line scans show the typical edge-shaped thickness course, and in the as-deposited state the individual layers are chemically well separated from each other.

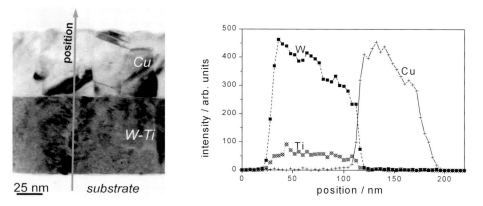

Figure 3.58: TEM cross-section micrograph of a Cu layer with W–Ti barrier on oxidized Si substrate (left) and EDXS element distribution line scans (right, along the marked position line)

In the case of wavelength dispersive X-ray spectrometry the dispersion of the wavelengths λ of the emitted X-rays is realized by a crystal spectrometer with interplanar distance d, which diffracts X-rays corresponding to Bragg's equation (cf. Section 3.3.1):

$$2\, d \cdot sin\,\theta = m\lambda \tag{3.21}$$

Because of that a sequential element detection is possible corresponding to $\lambda = f(\theta)$, by rotation of the crystal lattice around θ. For different angle positions θ of the crystal lattice different wavelengths λ of the characteristic X-rays are in the Bragg position and are therefore maximally diffracted. In order to avoid absorption of long-wave radiation the spectrometer is installed under vacuum. The long-wave radiation of the elements O, N and C is detected by organic crystals with interplanar distances $d = (50 \dots 80) \cdot 10^{-10}\,m$. Sample, spectrometer crystal and detector are arranged on the so-called *Rowland* Circle, and for a spectrometer rotation on θ the detector must be rotated on $2\,\theta$ to attain maximum intensity.

In the last 5 years a new fast generation of EDX detectors has been developed operating close to room temperature, this means no liquid nitrogen is required [3.77, 3.78]. This detector type is called a SDD X-ray detector because it works on the silicon drift diode (SDD) principle. A monolithically integrated on-chip FET acts as a signal amplifier and causes a very high energy resolution, e. g. 127 eV at MnK_α with 1000 cps. The detection of light elements from Boron (Z = 5) upwards is possible, e. g. with the SDD X Flash Detector 3001 of RÖNTEC Company [3.79].

Because of their favorable properties SDD X-ray detectors have opened a wide spectrum of experiments in the field of X-ray spectroscopy: element analysis, diffractometry, materials analysis, but also synchrotron experiments and element mapping in the scanning electron microscope.

The principal mechanism of a silicon drift detector can be understood from the scheme of Fig. 3.59, which shows the central part of a cylindrical SDD with an integrated JFET amplifier for spectroscopic applications. The detector scheme is based on sideward depletion [3.77]. The idea is that a large semiconductor waver of high resistivity, e.g. n-type silicon, can be fully depleted by a small n^+ ohmic contact positively biased with respect of the p^+ contacts covering both surfaces of the silicon waver. Additionally, an electrical field parallel to the surface is added. This is simply achieved by a segmentation of the upper p^+ ring areas to form a strip pattern and superimposing a voltage gradient on the strip system. The direction of the voltage gradient is such that the anode is the point of minimum potential energy for the electrons and therefore collecting all signal electrons generated in the depleted volume. By an integrated voltage divider only the ring # 1 and the last p^+ ring must be contacted and biased externally. The p-n-junction of the back side is used as an entrance window for X-ray radiation.

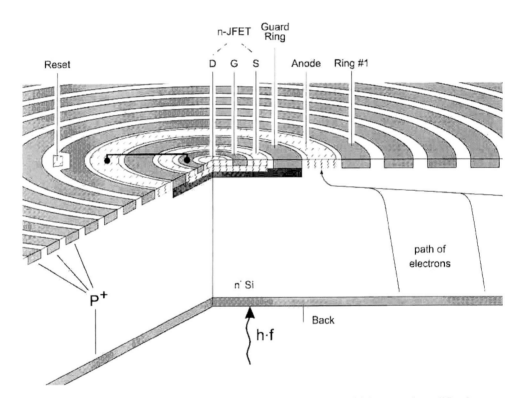

Figure 3.59: Central part of a cylindrical silicon drift detector (SDD) with integrated amplifier for X-ray spectroscopy [3.77]

The sideward depletion of the active detector volume in connection with the integrated drift structure provides an extremely small detector capacitance of SDD X-ray detectors. This results in very short shaping times of the electronics, allowing the operation at extremely high count rates, up to 10^6 cps. Therefore SDD detectors can perform analyses up to 6 times faster than a conventional EDX detector.

Electron Energy Loss Spectroscopy

Because of inelastic scattering processes the primary electrons suffer characteristic energy losses ΔE (see Fig. 3.53). As an example, Fig. 3.60 shows the electron energy loss spectrum of a thin carbon layer taken with an electron spectrometer. According to the size of ΔE, three regions of the energy-loss spectrum can be distinguished . The zero-loss peak includes both unscattered and elastically scattered electrons and those electrons which have excited phonons (ΔE typically smaller than 0.1 eV). The second part of the spectrum is the low loss region with plasmon excitations (ΔE typically some 10 eV). The third spectrum region is characterized by ionisation losses, that means by the edges of inner-shell ionisations ($\Delta E >$ 50 eV).

The energy loss is given by

$$\Delta E = E_{ion} \ (+ \ E_{kin}) \tag{3.43}$$

Its lowest value is the ionization energy E_{ion} of the inner-shell electron, and an additional amount may come from transferred kinetic energy E_{kin}. The energy loss E_{ion} at the ionization edge is characteristic of the corresponding excited element atom and thus EELS allows a qualitative element distribution analysis.

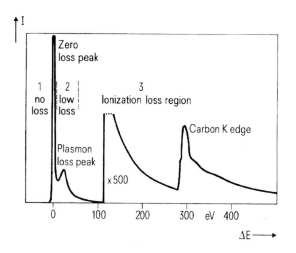

Figure 3.60: Electron energy loss spectrum of a thin carbon layer (after [3.80])

For quantitative analysis the intensities of the ionization losses must be determined. If a specimen is a compound of 2 elements A and B and the corresponding intensities I_A and I_B of the edges of both elements are measured after background subtraction in an energy loss spectrum as shown in Fig. 3.61, the ratio of their mass concentrations c_A/c_B can be calculated after *Egerton* [3.81]:

$$\frac{c_A}{c_B} = \frac{I_A(\Delta E_w, \beta)}{I_B(\Delta E_w, \beta)} \cdot \frac{\sigma_B(\Delta E_w, \beta)}{\sigma_A(\Delta E_w, \beta)} \tag{3.44}$$

with σ_B = ionization cross-section
            ΔE_w = energy window for electron collection
            β = collection angle of the spectrometer

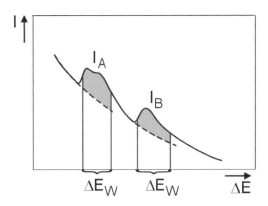

Figure 3.61: Schematic electron energy loss spectrum of a specimen with 2 elements A and B

By a serial registration of I (ΔE) spectra along a position line one can realize a three-dimensional, locally resolved element distribution analysis. As an example Fig. 3.62 demonstrates an investigation of a nanoscale Co/Cu fourfold layer using the linescan method with acquisition of EEL spectra [3.82]. The layer stack was produced by pulse laser deposition with a desired sample layer thickness of 8 nm. The series of 50 spectra shown in Fig. 3.62a was obtained along the layers normal direction. The net intensities of the Co-L and the Cu-L edges marked in the figure were calculated by background approximation and subtraction. These intensities are plotted against the electron probe position in Fig. 3.62b. A partially mixed Co and Cu region can be clearly seen in this example.

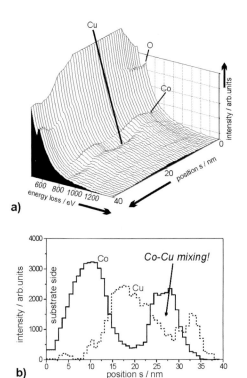

Figure 3.62: Chemical composition of pulse laser deposited Co/Cu fourfold nanolayers;
a) set of 50 EEL spectra along the layers normal direction,
b) corresponding concentration profiles of the calculated Co and Cu intensities

3.4.2   Element Depth Profile Analysis

The structural and chemical depth profile characterization of thin layers and multilayer systems has become more and more important during recent years [3.83]. Also the understanding of interface structures is of growing interest [3.84]. These arrangements are not only determined by the depth profile of element concentrations but also by that of chemical states and of phase concentrations, as evaluated by principal component analysis [3.85]. Because thin films down to monolayers are of growing interest, problems of layer influence during sputtering by mixing, implantation and similar processes become more and more awkward. From these problems follows the demand for a low information depth (few nanometers) and for a good depth profile resolution down to the monolayer range.

The concentration–depth profile $c = f(z)$ must be evaluated from the measured intensity–time profile $I = f(t)$. Two steps are necessary to obtain the $c(z)$ profile:
1. depth calibration $z = f(t)$,
2. element or phase concentration calibration $c = f(I)$.

The sputtering process results in a variation of composition and morphology of the uncovered surface layer which must be taken into account to reveal the true profile $c = f(z)$ [3.86].

The eroded depth z is determined by the sputtering rate $\dot{z}$ according to

$$z(t) = \int_0^t \dot{z}\, dt \tag{3.45}$$

Assuming a constant sputtering rate, Eq. (3.45) can be written as

$$\dot{z} = \frac{z_0}{t_0} = \text{const} \tag{3.46}$$

and the calibration is carried out by a crater depth measurement.

Considering the means of obtaining the depth profile $I(t)$ or $c(z)$, modern methods for depth profiling may be divided into 3 groups:
1. The surface layer is alternately removed by ion sputtering and analyzed by means of Auger or photoelectron spectrometry (AES, XPS).
2. The sputtered part of the surface is immediately used for depth profile analysis. This is the case for secondary ion and sputter neutral mass spectrometry (SIMS, SNMS) and for glow discharge optical emission spectrometry (GDOES).
3. Non-destructive depth profiling methods such as total reflection X-ray fluorescence (TXRF), *Rutherford* backscattering (RBS) or angle resolved photoelectron spectrometry (ARXPS).

Each of these methods has special advantages. AES combines a depth resolution of only a few nm with best lateral resolution values of about 20 nm [3.87]. The main advantage of XPS is the possibility of obtaining chemical bonding information. Trace analyses down to hydrogen ($Z = 1$) with detection limits below the ppm range can be obtained with SIMS and SNMS. GDOES is a quick method with sputtering rates up to 100 nm s^{-1} [3.88], without the requirement of high vacuum. On the other hand, one must renounce microscopic lateral resolution. The following expositions shall be limited to the important methods AES, XPS, SIMS and GDOES, most frequently used for depth profile analysis of thin films for electronics.

Electron Spectroscopy (AES, XPS)

Element depth profile analysis can be carried out by the electron spectroscopic methods AES and XPS. Fig. 3.63 gives the schematic diagrams of the electron emission processes for XPS (a) and AES (b) respectively. The photoelectron process (a) is characterized by the emission of a photoelectron, set off by a photon with energy $h \cdot f$. This process can be split into 3 steps [3.89]:
1. photoionization of an atom by photon absorption,
2. transport of the emitted electron to the surface,
3. transition of this electron through the surface into vacuum.

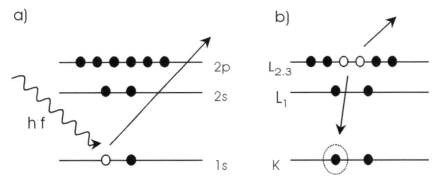

Figure 3.63: Schematic diagrams of electron emission process in solids; a) photoelectron emission, b) Auger process

The final atomic state is single ionized and the kinetic energy E_{kin} of the emitted electron is given by

$$E_{kin} = h \cdot f - E_B - \Phi \qquad (3.47)$$

with E_B =   binding energy of the excited electron,
            Φ =   effective work function.

E_{kin} is measured in an electron energy analyzer, and the element specific binding energy E_B can be calculated from Eq. (3.47). For quantification the atomic concentration c_A of an element A in a sample with n homogeneously distributed elements can be found as

$$c_A = \frac{I_A / S_A}{\sum\limits_{i}^{n} I_i / S_i} \qquad (3.48)$$

with I = measured electron intensity (usually from peak areas),
          S = sensitivity factor.

The sensitivity factors S_i are experimentally determined and include calibrated device specific functions.

As is visible in Fig. 3.63b the hole in the K-shell is filled by an electronic transition from an upper shell, here from the $L_{2,3}$ shell. The energy difference $E_K - E_{L2,3}$ can be given up to an electron in the same or a higher shell, shown in Figure 3.63b also as the $L_{2,3}$. This is called the Auger process. The emitted Auger electron has an element-specific kinetic energy E_{kin} given by

$$E_{kin} = E_K - 2E_{L2,3} + E_R - \Phi \tag{3.49}$$

with E_R = sum of intra-and extraatomic relaxation energies,
         Φ= effective work function.

Auger spectra show the relative intensity $N(E)$, which is received after energy dispersion in an electron energy analyzer, in dependence on the kinetic energy. In order to improve the signal-to-noise ratio and to reveal plasma satellites the energy is usually electronically differentiated (see Fig. 3.64). The absolute peak-to-peak height APPH is a good measure of intensity. Element concentrations can be calculated by multiplication of the measured Auger electron intensities by relative sensitivity factors [3.90].

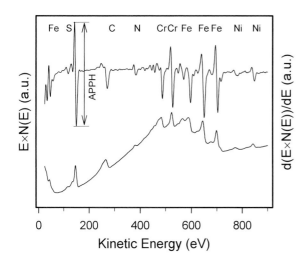

Figure 3.64: AES spectrum of a steel surface after annealing in ultra-high vacuum measured with 10 keV electrons using a cylindrical mirror analyzer. The positions of the Auger peaks are marked. The lower curve is the measured spectrum $E \times N(E)$. The upper curve shows the result of numerical differentiation.

In electron spectroscopy, both XPS and AES, element concentration depth profiles are frequently required. The total depth is then usually greater than the analytical depth resolution. Therefore alternating surface spectroscopy and progressive removal of the surface by ion sputtering is used. As an example Fig. 3.65 shows the AES depth profile of an as-deposited spin-valve sample using 800 eV Ar ions for sputtering [3.91]. At the interface FeMn/CoFe an excess of Fe is calculated as expected for a CoFe alloy layer. This is represented by the bold line in Fig. 3.65. In the other "CoFe" layers a much lower Fe signal was found. Annealing at 350 °C leads mainly to oxidation of the Ta capping layer, but no evidence for a demixing of the Cu and Co layers was found [3.92].

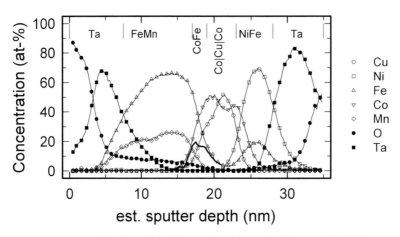

Figure 3.65: AES depth profile of an as-deposited spin-valve sample [3.91]

Chemical states of elements can be determined from both AES and XPS spectra by the so-called chemical shift. This is the observed energy shift of the peak from a particular element when the chemical state of this element changes. The chemical shift is particularly distinct in XPS spectra. It is caused by the alteration in the valence electron density when an atom combines with another atom or atom group. Looking at Eq. (3.47) the binding energy E_B of the core electrons therefore changes, giving rise to shifts in the corresponding photoelectron peaks. For the identification of unknown chemical states tabulations of chemical shifts have been made by many authors (see e.g. [3.93]).

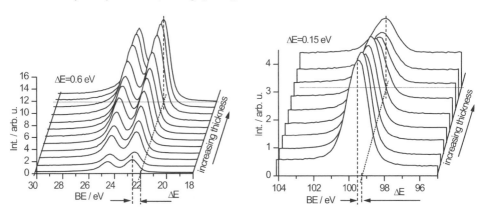

Figure 3.66: Binding energy (BE) shift of XPS peaks for increasing thickness of a Ta film up to 10 nm on sputtered Si, left: photoelectron peaks of Ta 4f, right: photoelectron peaks of Si 2p

In thin films and layer systems the chemical states of elements often change in dependence on the layer depth. This can be shown graphically by the binding energy shift of photoelectron peaks in depth profiles. As an example Fig. 3.66 demonstrates the binding energy shift of the XPS peaks of Ta 4f and Si 2p for increasing thickness of a Ta film on sputtered Si [3.94]. Tantalum was deposited by magnetron sputtering up to a thickness of

10 nm. At the interface, the Ta 4f photoelectron peak shows a significant shift of 0.6 eV to higher binding energy compared to metallic Ta, which is observed at a film thickness of 10 nm (left image). The shifted peak at the interface agrees very well with the $TaSi_2$ reference spectrum. A shift in the binding energy of the Si2p peak was also observed during the deposition series (right image), which is a further hint for silicide formation at the Ta/Si interface.

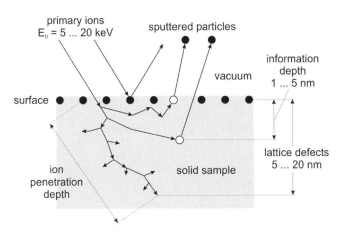

Figure 3.67: Secondary ion mass spectrometry (SIMS); interactions of primary ions with the bombarded surface

In contrast to the electron spectroscopic methods in secondary ion mass spectrometry the sputtered part of the surface is immediately used for depth profile analysis. This means the primary ions have two functions: they act as an analytical source and simultaneously they give rise to sample sputtering. To produce depth profiles with sputter rates of typically 1 nm/s a primary ion current density of about 10^{-4} A/cm^2 is necessary (so-called dynamical SIMS).

Figure 3.67 shows the interaction of primary ions ($E_o = 0.5 - 20$ keV) with the bombarded sample surface. The primary ion beam produces lattice defects in the range of 10 nm. The SIMS information depth is lower, it amounts to only a few nm. The sputtered particles are

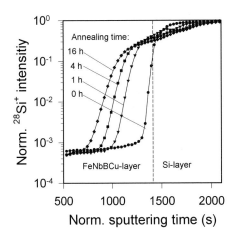

Figure 3.68: SIMS depth profiles of Si diffusion into amorphous FeNbBCu after thermal treatment [3.95]

neutral and charged particles of sample material and resputtered primary ions. The emitted secondary ions are extracted via an electrical potential and analyzed using a mass spectrometer. Though a semiquantitative SIMS analysis is possible only by standardization, the detection limit of SIMS is much better than that of comparable depth profile methods. It lies in the ppm range depending on element and materials composition.

As an application Fig. 3.68 shows SIMS depth profiles of Si diffusion into amorphous FeNbBCu after thermal treatment at 400 °C for different annealing times [3.95].

An important parameter for accuracy in element depth profile analysis is the depth resolution Δz. This resolution is limited not only by the penetration and escape depth of the interacting radiation and particles, but also by surface roughening, atomic mixing and preferential sputtering during surface removal [3.96]. Its value must be minimized. Because

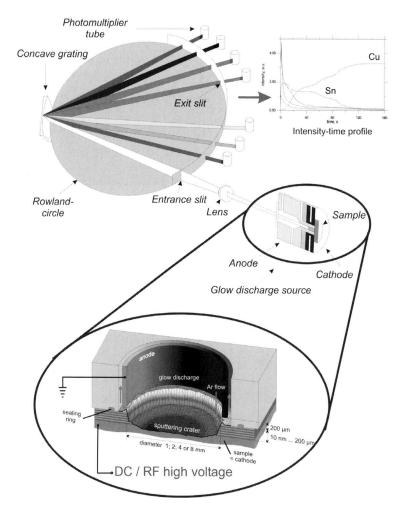

Figure 3.69 Principle of Glow Discharge Optical Emission Spectroscopy (GDOES)

of the non-directed and low energetic sputtering process a narrow depth resolution function down to $\Delta z \approx 5$ nm is available with plasma SNMS and with GDOES [3.83].

Therefore, in recent years GDOES has proved to be a very useful technique for fast depth profile analysis of thin layers and layer systems. In a glow discharge the spectral emission line intensities of all sample constituents are measured simultaneously during the sputtering process. For this purpose a low-pressure electrical discharge in a noble gas is used at pressures from 10^2 Pa to 10^3 Pa. Sputtering the sample surface successively and measuring the emitted light of the sputtered material allows qualitative and quantitative characterization of the element distribution in dependence on the depth.

The principle and the schematic function of GDOES is shown in Fig. 3.69. The glow discharge source, after *Grimm* [3.97], consists of a ring-shaped anode and opposite to it the sample, which serves as the cathode. During the discharge, the cathode is bombarded by positive ions of the noble gas. This bombardment induces a surface erosion of the sample which causes a typical sputtering crater. By ion collision processes and further interactions in the plasma the sputtered material, atoms and ions, is excited for optical transitions in the visible and ultraviolet ranges. The generated photons correspond to the characteristic spectral lines of the elements in the plasma, giving information on the element depth profile of the sample which is removed layer-by-layer by the sputtering process.

Sputtering rates up to 100 nm s^{-1} are possible. The emitted light is analyzed in an optical spectrometer which consists of a concave grating in a *Rowland*-circle and calibrated photomultiplier tubes. The measured intensity–time profiles have to be converted for quantitative analysis into concentration–depth profiles. For that purpose the quantification by *Bengtson* [3.98] is commonly used very successfully.

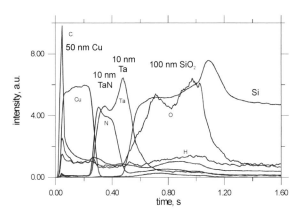

Figure 3.70: GDOES element depth profile of a microelectronic multilayer system with TaN based Cu diffusion barrier

As an example of GDOES analysis Fig. 3.70 shows the depth profile of a multilayer system with 10 nm TaN and 10 nm Ta that was used as a barrier against Cu diffusion from a Cu metallization layer into the SiO$_2$ substrate [3.99]. A radio frequency (rf) source had to be used because of the non-conducting SiO$_2$. The rf-GDOES used system allows depth profile analysis of such layer systems with depth resolution on the nanometer scale due to the fast stability of the electrical parameters. The intensities of the Ta and N lines in the 10 nm TaN layer agree with those of much thicker TaN layers of the same composition. That means that GDOES depth profile analysis allows a reproducible concentration determination, even for layers of about 10 nm thickness.

3.5   Stress Measurement Techniques

In contrast to bulk materials, thin films on substrates are usually under mechanical stress. For an unattached thin film and a substrate, temperature increase, for example, results in different expansion of film and substrate due to the mismatch of their thermal-expansion coefficients. For a thin film attached to a thick substrate, the film is stretched just like the substrate. As a result of this constraint of the substrate, a strain arises in the film and, therefore, a stress is generated. This film stress is revealed in a curvature of the film/substrate package.

A serious technological problem of stressed thin films is the reduction of reliability and lifetime due to degradation, through-plane cracking, delamination, and/or in-plane buckling. Furthermore, stress results in changes in the thin-film properties. The biaxial strain leads to dimensional and shape changes of the unit cell. The changes in interatomic distances, in the main, cause the changes in the properties, as for instance in the magnetic properties via magnetoelastic coupling.

In the following sections, the relation between stress and strain in thin films is considered. The reasons for generating stress are analyzed. The curvature of the substrate under the influence of this stress is considered. The two most important methods for stress measurements, namely laser-optical determination of the substrate curvature and measurement of lattice strain by means of X-ray diffraction (XRD), are demonstrated.

3.5.1   Stress and Strain

Stress–Strain Relation
An extended film of thickness d_f on a very thick substrate is considered. The coordinates x and y are within the film plane, z is perpendicular to this plane. The film material is assumed to be homogeneous and isotropic. As, in general, thin films do not support stresses in the film normal (z) direction, the film stress σ_f is biaxial within the film plane. Thus, the components $(\sigma_x, \sigma_y, \sigma_z)$ are $(\sigma_f, \sigma_f, 0)$. The strain components $(\varepsilon_x, \varepsilon_y, \varepsilon_z)$ can be calculated using *Hooke's* law. For the x direction for example

$$\varepsilon_x = \left(\sigma_x - v_f \sigma_y\right)/ E_f \tag{3.50}$$

where E_f is the *Young's* modulus and v_f is the *Poisson's* ratio of the film. For biaxial stress, the in-plane strain ε_f is given by

$$\varepsilon_f = \frac{1 - v_f}{E_f}\sigma_f \tag{3.51}$$

and for the strain perpendicular to the film plane (z direction), it is

$$\varepsilon_z = -\frac{2v_f}{E_f}\sigma_f \tag{3.52}$$

The magnitude $E_f/(1-v_f)$ is denoted as the biaxial modulus. The in-plane strain and the perpendicular strain have opposite signs. In an arbitrary direction ξ with an angle ψ to the z axis, the strain ε_ξ is

$$\varepsilon_\xi = \left(\frac{1+v_f}{E_f} \sin^2 \psi - \frac{2v_f}{E_f} \right) \sigma_f \tag{3.53}$$

There exists a strain-free direction, i.e. $\varepsilon_\xi = 0$, for the angle ψ_0 according to

$$\sin^2 \psi_0 = \frac{2v_f}{1+v_f} \tag{3.54}$$

independent of the position of the projection of ξ into the film plane. Eqs. (3.53) and (3.54) are the basis of the $\sin^2 \psi$ method for stress measurement (Section 3.5.3). It should be emphasized once more, that the above consideration is valid for isotropic thin-film stress. For anisotropic thin-film stress, see [3.100, 3.101].

Stress Generation

In a thin film on an extremely thick substrate, which is practically not deformed by film stress, each dimensional change ε_r of the film relative to the substrate results in the generation of film stress according to

$$\sigma_f = -\frac{E_f}{1-v_f} \varepsilon_r \tag{3.55}$$

As the film is attached to the substrate, the lateral dimension of the film has to fit the dimension of the substrate. The minus sign in Eq. (3.55) reflects that an in-plane dimensional increase ($\varepsilon_r > 0$) leads to a biaxial compressive stress ($\sigma_f < 0$).

The main reasons for relative dimensional changes are (i) the misfit of the thermal-expansion coefficients of film and substrate (thermal stress) and (ii) changes in the average volume/atom due to structural and compositional changes in the film (structure evolution stress).

Considering thermal stress, the relative strain ε_r is $(\alpha_f - \alpha_s)\Delta T$, where α_f and α_s are the thermal-expansion coefficients of film and substrate, respectively, and ΔT is the temperature increase. Using Eq. (3.55), the thermal stress σ_{th} is

$$\sigma_{th} = -\frac{E_f}{1-v_f}(\alpha_f - \alpha_s)\Delta T \tag{3.56}$$

For metallic thin films on silicon substrates, α_f is usually larger than α_s (at room temperature: $\alpha_{Si} = 2.49 \cdot 10^{-6}$ K^{-1} at 300 K [3.102]). Therefore, compressive stress ($\sigma_{th} < 0$) develops in the film with temperature increase.

If any structural evolution is associated with a uniform volume change ΔV_r within the film, then the relative length change in each space direction is $\Delta V_r/3$. Therefore, the structure evolution stress is in this case

$$\sigma_{str} = -\frac{E_f}{1-\nu_f} \cdot \frac{\Delta V_r}{3} \qquad (3.57)$$

Origins of volume changes in thin films associated with the evolution of the microstructure (atomic rearrangement) with increasing temperature may be

- grain growth (densification due to reduction in the excess volume associated with grain boundaries),
- crystallization of amorphous films, phase transformation, and precipitation,
- annihilation of excess vacancies at grain boundaries perpendicular to the film plane,
- grain boundary relaxation (atomic rearrangement within the grain boundaries, e.g., to produce equilibrium grain boundaries),
- shrinkage of grain boundary voids which may be formed during deposition, and/or
- volume magnetostriction due to magnetic ordering.

These mechanisms and their effect on the stress evolution were considered in detail by *Doerner* and *Nix* [3.103]. As a rule, the mechanisms result in material densification. Therefore, they lead to the development of an irreversible tensile stress ($\sigma_{str} > 0$) during heat treatment, which is superimposed to the (most compressive) thermal stress.

The relatively simple correlation between atomic rearrangement, volume change, and stress generation makes in situ stress measurements during a thermal cycle or isothermal annealing a useful tool to study microstructural processes during heat treatment.

Stress already develops during thin-film deposition (growth stress). This stress depends on the growth mode (e.g., island nucleation and coalescence or layer-by-layer growth), the adatom mobility, the ion bombardment during sputter deposition, and others, as considered, e.g., by *Koch* [3.104]. For the deposition of films on crystalline substrates, epitaxial strain must be taken into consideration. This strain arises if the film grows with an epitaxial relationship to the substrate and a mismatch between the lattice spacing of the two materials exists. It decreases by the incorporation of misfit dislocations with increasing layer thickness. Furthermore, structure evolution processes, as already described for post-deposition annealing, can also occur during deposition and, therefore, contribute to the development of the growth stress. In situ stress measurements during deposition deliver useful results concerning the film growth. For details and models for this very recent area with intensive research, the reader is referred to the literature [3.105–3.109].

Surfaces and interfaces also lead to stress contributions because of the changed atomic binding in these regions. These stress contributions are especially relevant for very thin films or multilayers [3.106, 3.107, 3.110].

The previous paragraphs have only considered that the thin film is elastically deformed by the stress but metallic thin films are often under stress near to or at the plastic yield. It is a characteristic of the mechanical properties of thin films that the flowing strength in thin films is increased in comparison to bulk materials. For mechanical properties of thin films and their measurement, see, e.g., [3.111–3.114]. Upon occurrence of elastic and plastic deformation, the total relative strain $\varepsilon_{r,tot}$ is given by $\varepsilon_{r,tot} = \varepsilon_{r,el} + \varepsilon_{r,pl}$, where $\varepsilon_{r,el}$ and $\varepsilon_{r,pl}$ are the elastic and plastic contributions, respectively.

At low temperatures or high strain rates, plastic deformation usually occurs by the motion of dislocations. At higher temperatures and for stresses lower than the flow strength of the film, it occurs by diffusional flow. For the latter mechanism, the grain boundary diffusion plays an important role in thin films because of the typical small grain sizes (Coble creep). As plasticity is stress-driven, plastic deformation usually acts as a stress-reducing process (stress relaxation). This is opposite to thermal stress and structure evolution stress, which can increase or decrease the total stress, depending on the initial stress conditions and the special process conditions.

The laws for the plastic deformation rate $d\varepsilon(\sigma,T)/dt$ in thin films are expected to be analogous to the bulk laws, as described with deformation–mechanism maps by *Frost* and *Ashby* [3.115] with some thin-film adaptations [3.116]. The considered processes are dislocation glide and climb, power-law creep and breakdown, and diffusional flow.

3.5.2 Substrate Curvature

In the last section, the generation of stress in thin films was considered. This stress develops when the film is attached to the substrate, regardless of whether the substrate is deformed or not due to the film stress. The latter case may occur, if the substrate is extremely thick. In this section, the deformation (curvature) of a substrate with thickness d_s, e.g. a Si wafer, is considered under the influence of the stress of the film. This substrate curvature is the measured quantity of the substrate curvature method for stress measurement.

A simple relation between film stress and substrate curvature can be obtained under the following conditions: A thin, extended film is attached to a thick substrate, so that $d_f << d_s$ is valid (thin film assumption). The film contribution to the deformation behavior of the film/substrate package is negligible. The substrate is bent by the force of the film which is under stress and attached to the substrate surface. The thin film assumption is met under typical conditions: $d_f \approx 0.1$ μm, $d_s \approx 300$ μm.

Now, the stress and strain distribution within the substrate is considered under the influence of the stress σ_f of the attached film. The determination of this distribution is the basis for the determination of the substrate curvature. The total force, as well as the moment of this force in the film/substrate package, are zero in equilibrium. These conditions result in a stress distribution, as is shown in a section through the package in Fig. 3.71a. The stresses on the upper and lower side of the coated substrate, $\sigma_{s,u}$ and $\sigma_{s,l}$, respectively, are

$$\sigma_{s,u} = -4\sigma_f d_f / d_s \tag{3.58}$$

and

$$\sigma_{s,l} = 2\sigma_f d_f / d_s \tag{3.59}$$

The quantity $\sigma_f d_f$ is the force per unit width due to the film stress, F_w. It characterizes the force acting on the film cross-section. The stress in the substrate is much smaller than that in the film, as is expressed by the proportionality to the ratio of the thicknesses of film and substrate. Beneath a thin film with tensile stress, for example, the substrate stress is compressive. The neutral axis, which is characterized by $\sigma_s = 0$, is located at a distance of $d_s/3$ from the uncoated substrate side according to the ratio of the absolute values of $\sigma_{s,l}$ and $\sigma_{s,u}$. The asymmetry of the distribution of the substrate stress in the depth is a result of the one-sided action of the film stress.

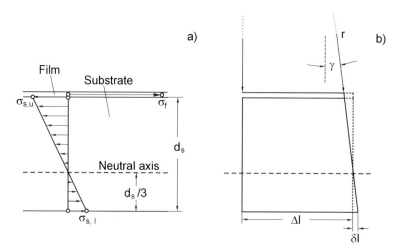

Figure 3.71: Stress (a) and deformation (b) in a film/substrate package under the influence of stress of the film

The different stresses on both sides of the substrate lead to its curvature. The deformation angle γ on a film/substrate element of length Δl is equal to the ratio of the substrate dilation $\delta l = \Delta l\, \sigma_{s,l}\,(1-\nu_s)/E_s$ on the lower side of the substrate to the distance of the neutral axis $d_s/3$ from the uncoated substrate side (Fig. 3.71b). The factor $E_s/(1-\nu_s)$ arises due to the biaxial stress state also in the substrate. This is analogous to Eq. (3.51). Thus, the deformation angle is

$$\gamma = 6 \cdot \frac{1-\nu_s}{E_s} \cdot \frac{\sigma_f d_f}{d_s^2} \cdot \Delta l \tag{3.60}$$

This leads to a curvature with radius r according to

$$r = \frac{\Delta l}{\gamma} = \frac{1}{6} \cdot \frac{E_s}{1-\nu_s} \cdot \frac{d_s^2}{\sigma_f d_f} \tag{3.61}$$

in the considered film/substrate element and, consequently, also in the whole substrate. The relation between film stress and curvature radius of the substrate given in Eq. (3.61) is the well-known *Stoney* equation. It is the base for the determination of film stress by the substrate curvature method. In this equation, the force per unit width $\sigma_f d_f$ of the film appears as reason for the substrate curvature. The substrate deformation is only determined by the properties of the substrate (biaxial modulus and thickness). The elastic properties of the film do not occur in this equation because of the thin film assumption, as mentioned above. Thus, one can determine the film stress without knowledge of the elastic properties of the film. This is an advantage of the substrate curvature method.

The substrate curvature in response to the film stress only results in a very small additional strain ε_a of the film, namely $\varepsilon_a = \varepsilon_{s,u}$. Its ratio to the strain ε_r for an uncurved substrate is

$$\frac{\varepsilon_a}{\varepsilon_r} = -4 \cdot \frac{E_f}{1-v_f} \cdot \frac{1-v_s}{E_s} \cdot \frac{d_f}{d_s} \tag{3.62}$$

This can be neglected in the thin film approximation $d_f \ll d_s$.

Except for effects near to the edges of the film/substrate package, each region of this package is bent with the same curvature radius r according to Eq. (3.62). This leads to a spherical curvature of the whole substrate, independent of its shape. This is an advantage for the measuring technique.

Considering the sign of the curvature, if one looks at the coated side of the substrate, the curvature is concave with a curvature radius $r > 0$ if the film is under tensile stress ($\sigma > 0$), and it is convex with $r < 0$ for compressive stress ($\sigma < 0$).

So far, it has been assumed that a homogeneous thin film with biaxial stress σ_f exists on the substrate. If a stress gradient occurs within the film, or if a thin film configuration like a bilayer or multilayer is deposited, the forces per unit width of the individual layers are summed. For example, for a bilayer with layers 1 and 2 and total thickness d_{tot}, the average stress σ_a (effective for substrate bending) is

$$\sigma_a = \sigma_1 \frac{d_1}{d_{tot}} + \sigma_2 \frac{d_2}{d_{tot}} \tag{3.63}$$

If the thin film approximation $d_f \ll d_s$ is not fullfilled, modifications of the *Stoney* equation have been considered recently by *C.A. Klein* [3.117].

3.5.3 Measurement Techniques

The most important methods for determining film stresses are based either on the measurement of the substrate curvature or on the measurement of strain by XRD. By means of substrate curvature measurement, one determines the stress directly. Only the film thickness, the substrate thickness, and the biaxial modulus of the substrate must be known. By means of XRD measurements of strain, knowledge of the elastic properties of the thin film is necessary in order to convert strain into stress. Separate measurement of stress and strain by utilizing both methods allows one to determine the biaxial modulus of the film.

Substrate Curvature Method

As described in Section 3.5.2, it is possible to measure the substrate curvature due to the film stress on samples of any shape. Stripes or circular wafers are frequently-used sample shapes.

A bending beam which is clamped on one side (cantilever) is considered in the following. Fig. 3.72 shows the cross-section. Each element of the length Δx of the stripe contributes to the bending with a deformation angle γ according to Eq. (3.60). The contribution to the deflection is δz. Integration over the corresponding ranges leads to the deflection z_L and the bending angle Γ_L for a bending beam with length L. It is

$$z_L = L^2 / 2r \qquad (3.64)$$

and

$$\Gamma_L = L/r = 2z_L / L \qquad (3.65)$$

respectively.

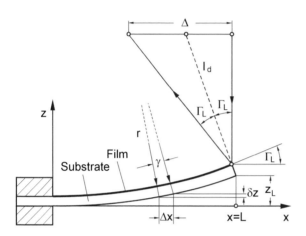

Figure 3.72: Bending beam with a film/substrate package under the influence of film stress. The substrate curvature is exaggerated. Curvature radius and deflection, e.g., for a bending beam of length 40 mm and with a thin film of 100 nm thickness and a stress of 400 MPa, deposited on a 375 μm thick Si (100) substrate, amount to $r = 106$ m and $z_L = 7.6$ μm, respectively

Depending on the measurement technique, either the deflection z_L or the bending angle Γ_L is measured. The quantity z_L is determined, for example, if the free end of the cantilever is used as a moveable plate of a capacitor [3.118]. The quantity Γ_L is measured if the deflection of a laser beam due to the substrate bending is considered [3.119]. The position shift Δ of the laser beam on a detector for the usual case of very small angles is

$$\Delta = 2\Gamma_L l_d \qquad (3.66)$$

where l_d is the distance between sample and detector. With the progress of the laser-optical components, the laser-optical measurement of substrate curvature is the most-frequently-used method for film stress measurement. Other methods, like capacitor measurements or interferometer methods, are increasingly forced onto the sidelines. Therefore, in the following only laser-optical substrate curvature measurement is considered.

The simple method with a clamped bending beam and measurement of the deflection of a laser beam on the unclamped end has two disadvantages:

1. The transverse bending of the bending beam is suppressed at the clamping position which also has influence on the bending along the beam. Leaving a part of the clamped end without deposited film solves the problem.

2. The smallest inclination at the clamping position, e.g., due to changes in the environment temperature, leads to a falsification of the measuring signal. This is why a method with two laser beams is often used. One beam is the reference beam near the clamping position, the other is the measuring beam on the free end of the cantilever. The curvature is measured from the difference signal. Usually, two fixed laser beams are used. Sometimes also one laser beam which scans across the sample is used [3.112]. In many cases, a free (unclamped) bending beam (or wafer), which is supported by two knife-edges or a three-point support is used. Furthermore, by utilizing more than two laser beams, it is, in principle, possible to obtain information about the homogeneity of the curvature across the substrate [3.120]. Deviations from spherical curvature may by caused by inhomogeneities in the film thickness or stress (microstructure).

A sensitive set-up for laser-optical measurement of film stress is demonstrated in Fig. 3.73. The device is subdivided into two vacuum chambers. In one chamber, the laser-optical unit is mounted on a thermostated stage. In the other (sample chamber), the knife-edge sample support, a furnace for high-temperature investigations (up to 500 °C), and a four-point-bending equipment are located. In the laser-optical unit, the laser beam is split into two beams by mean of a system of partially transmitting and normal mirrors. The beams are reflected from the coated substrate onto position-sensitive detectors (photocells). The four-point bending allows one to study stress relaxation after uniaxial loading [3.121]. For more details concerning construction, properties, and sensitivity of the described setup see [3.122].

The substrates, for example silicon wafers, usually have a curvature even without a deposited film. In order to determine the film stress, one has to subtract the curvature of the undeposited substrate from the value for the deposited substrate. Then, the *Stoney* equation is

$$\sigma_f = \frac{1}{6} \cdot \frac{E_s}{1-\nu_s} \cdot \frac{d_s^2}{d_f} \cdot \left(\frac{1}{r_1} - \frac{1}{r_0} \right) \tag{3.67}$$

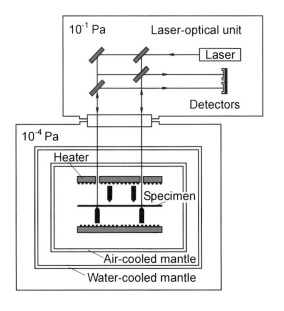

Figure 3.73: Set-up for sensitive measurement of film stress by means of the substrate curvature method. Because of the small curvature of the film/substrate package, the incident beam and reflected beam on the specimen nearly coincide

where r_1 and r_0 are the substrate curvature radii after and before film deposition, respectively. For the laser-optical two-beam measurement, one has to measure the deflection of the two laser beams of the coated substrate compared with that of the uncoated one [3.123].

For the measurement of biaxial stress, one needs substrates having an isotropic biaxial modulus within the substrate plane. Glass substrates meet this requirement. Crystalline silicon has anisotropic elastic parameters, in principle. Nevertheless, for (111)-oriented silicon wafers, E_s and ν_s are isotropic within the wafer plane. For (100)-oriented silicon wafers, the biaxial modulus $E_s/(1-\nu_s)$ is isotropic within the wafer plane, although, separately, E_s and ν_s are anisotropic [3.124]. The biaxial moduli amount to 229.0 GPa and 180.5 GPa for (111)- and (100)-oriented silicon wafers, respectively, at room temperature [3.125]. In good approximation, these values are also valid for oxidized silicon wafers with standard thickness, as long as the oxide thickness remains of the order of some 100 nanometers.

Stress measurements by means of the substrate curvature method are possible on all thin films, independent of their microstructure (amorphous, nanocrystalline, polycrystalline, or single-crystalline). If the reflectance of the film material is insufficient for the laser beams, one can also measure the substrate curvature on a polished back-side of the substrate.

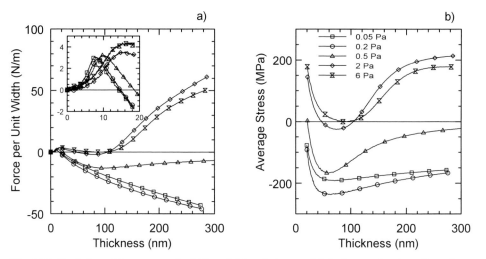

Fig. 3.74: Evolution of force per unit width $\sigma_f d_f$ (a) and stress σ_f (b) during sputter deposition of 300 nm thick Cu thin films on oxidized Si substrates at various sputter pressures as indicated in (b). The deposition rate was 0.1 nm/s.

An advantage of the laser-optical measurement is the registration of the intensities of the reflected laser beams by the photocells. Thus, one may observe changes in the reflectance of the thin film during annealing. In special cases, this allows one to draw conclusions about microstructural changes or film oxidation.

The substrate curvature method is often used for in situ stress measurements during film deposition and, therefore, for obtaining information about the growth mechanism [3.126, 3.127]. Furthermore, curvature measurements on substrates with magnetic thin films under the influence of a magnetic field give information about the magnetostriction effect of the thin-film material [3.128]. Many stress measurements on thin films for electronics have been reported in the literature (e.g. [3.129], [3.130]).

The stress evolution of Cu thin films during magnetron sputtering under various sputter pressures is shown in Fig. 3.74 [3.131]. The features of the stress evolution during the early stage of deposition [insert in Fig 3.74a] were ascribed to the *Vollmer–Weber* mechanism (i.e. nucleation of discrete islands, island growth, and coalescence, see, e.g., [3.132]). The transition from tensile to compressive stress with decreasing sputter pressure for greater film thicknesses is caused by the peening effect due to particle bombardment onto the growing film as considered by *Windischmann* [3.133]. Figure 3.74b demonstrates that a stress that is inhomogeneous in depth is developed during the deposition that is not considered, e.g., in grain growth and plasticity processes at higher temperatures up to now.

Figure 3.75 shows two typical examples of the stress development during thermal cycling of a 400 nm thick Cu metallization film and of a 320 nm thick alloy NiCr(60 wt.%) resistive film. Stress measurement during thermal cycling with a constant heating and cooling rate (e.g. 5 K min^{-1}), sometimes combined with isothermal annealing, is often used to study stress generation and relaxation processes.

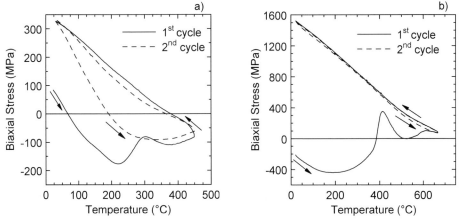

Figure 3.75: Stress evolution in a 400 nm thick Cu thin film (a) and a 320 nm thick NiCr(60 wt.%) thin film during two thermal cycles at heating and cooling rate of 4 K min^{-1}

For the Cu thin film (Fig. 3.75a) the stress evolution starts with a thermal-stress generation according to Eq. (3.56). Above 100 °C, the distinct irreversible stress change reflects plastic flow as well as material densification especially associated with grain growth. The former mechanism, in combination with the thermal-stress development, determines the stress evolution during the cooling branch as well as the second thermal cycle. The hysteresis behavior is typical for plastic deformation in thin films. Grain growth and plasticity of Cu thin films have been extensively studied in the last years (see, e.g., [3.134, 3.135]).

For the NiCr thin film (Fig. 3.75b), the stress evolution during the first thermal cycle is associated with the crystallization of the as-deposited, amorphous thin film [3.136]. Two metastable phases, namely a bcc supersaturated solid solution and a tetragonal σ phase, and the equilibrium α(Cr)+γ(Ni) phase mixture are formed with increasing temperature. The absence of a pronounced hysteresis in the second thermal cycle shows that plasticity in the two-phase alloy does not play an important role up to 650 °C.

$\sin^2 \psi$ Method

A thin film, e.g. under biaxial tensile stress, is expanded within the film plane. Perpendicular to this plane, it is compressed. Measuring the lattice parameter by means of XRD with the symmetrical *Bragg–Brentano* geometry, one determines the lattice spacing d from the reflection angle according to the Bragg equation (3.21). This lattice spacing in the film under stress is changed in comparison with the spacing in a stress-free thin film, d_0, according to Eq. (3.52). Therefore, one measures a lattice spacing according to

$$d = d_0 \left(1 - \frac{2v_f}{E_f} \sigma_f \right) \qquad\qquad (3.68)$$

So, the lattice spacing and, therefore, the lattice parameters for thin films are changed due to the biaxial stress. Lattice parameters in thin films can be compared with bulk values only after correction according to Eq. (3.68).

The characteristic of the $\sin^2 \psi$ technique is the determination of lattice spacings for various angles ψ between the surface normal and the diffraction vector (cf. Section 3.3.1 and Fig. 3.41), e.g., by rotating the surface normal out of the scattering plane. In this case, the lattice spacings d_ψ result from Eq. (3.53):

$$d_\psi = d_0 \left(1 - \frac{2v_f}{E_f} \sigma_f + \frac{1+v_f}{E_f} \sigma_f \sin^2 \psi \right) \qquad\qquad (3.69)$$

Plotting d_ψ versus $\sin^2 \psi$, one obtains a straight line which allows one to determine the biaxial stress σ_f and the unstrained lattice spacing d_0 according to Fig. 3.76. A detailed review article series about X-ray stress analysis has been published by *Eigenmann* and *Macherauch* [3.137].

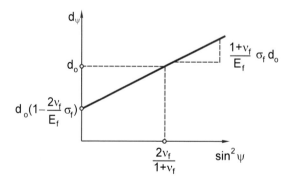

Figure 3.76: Schematic diagram of the $\sin^2 \psi$ technique for stress measurement

Stress measurements by means of the $\sin^2 \psi$ technique are possible regardless of the optical reflectivity of the thin film. Reference measurements are not necessary. Local stresses can be determined. However, the X-ray diffraction requires a crystalline microstructure of the film. As the $\sin^2 \psi$ technique is basically a strain-measurement technique, knowledge of the biaxial modulus is necessary for conversion of the strain into stress. One should consider that the elastic parameters of the film are influenced by the texture due to the anisotropy of these parameters and that they vary for different reflections and measurement directions. Stress determination in textured thin films has been reported, e.g., by *Clemens* and *Bain* [3.138].

In comparison to the substrate curvature method, it is favorable that the $\sin^2 \psi$ technique allows one to determine stresses in individual layers of a thin-film configuration, for example, of a bilayer or trilayer. Results on a NiFe/Cu/NiFe trilayer are reported in [3.139].

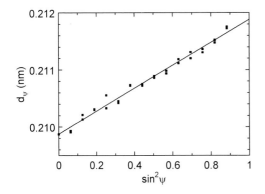

Figure 3.77: Stress measurement in NiMn with the $\sin^2 \psi$ method on a 4 nm Ta/ 200 nm NiMn/ 4 nm Ta/ 1 nm Cu thin film configuration after annealing at 400 °C

As an example, Fig. 3.77 shows the stress measurement on a 200 nm thick NiMn layer in a 4 nm Ta/200 nm NiMn/4 nm Ta/1 nm Cu thin film configuration after annealing at 400 °C. From the slope of the plot d_ψ versus $\sin^2 \psi$ and with the bulk biaxial modulus $E_f/(1-\nu_f) = 165$ GPa, the stress was determined to be 1200 MPa. The obtained strain-free cubic lattice constant amounts to 0.365 nm, which is near to the bulk value (0.364 nm).

3.6   References

[3.1]   J.L. Vossen, W. Kern, *Thin Film Processes II*, Academic Press Boston, 1991.
[3.2]   R. F. Bunshah, *Deposition Technologies for Films and Coatings*, Noyes Publications, Park Rich, New Jersey, USA, 1982.
[3.3]   R. A. Haefer, Oberflächen- und Dünnschichttechnologie, Teil1 Beschichtungen von Oberflächen, Werkstoff-Forschung und -Technik, Springer-Verlag Berlin, 1987.
[3.4]   M. Madou, *Fundamentals of Microfabrication*, CRC Press, Boca Raton, London, 1997.
[3.5]   K. Seshan, *Handbook of Thin Film Deposition*, Noyes Publications, Norwich, 2002.
[3.6]   A. Gorbunoff, *Layer-Assisted Synthesis of Nanostructured Materials*, Fortschritt-Berichte VDI, Reihe 9 Elektronik/Mikro- und Nanotechnik, Nr. 357, VDI Verlag, Düsseldorf, 2002.
[3.7]   G. Carter, J.S. Colligon, *Ion Bombardment of Solids,* Elsevier, Amsterdam, 1968.
[3.8]   Klaus K. Schuegraf, *Handbook of Thin Film Deposition Processes and Techniques*, Noyes Publications, Park Rich, New Jersey, 1988.
[3.9]   M. Konuma Film, *Deposition by Plasma Techniques*, Springer Verlag Berlin, Springer Serie on Atoms + Plasmas, vol. 10, 1992.
[3.10]   B. Chapman, *Glow Discharge Processes*, John Wiley and Sons, New York, 1980.
[3.11]   J.J. Cuomo, S.M. Rossnagel, H.R. Kaufmann, *Handbook of Ion Beam Processing Technology*, Noyes Publications Park Ridge, New Jersey, 1989.

[3.12]   A. Sherman, *Chemical Vapor Deposition for Microelectronics*, Noyes Publications Park
          Ridge, New Jersey, USA, 1987.
[3.13]   R. A. Haefer, Oberflächen- und Dünnschichttechnologie Teil 2, Oberflächen-modifikation
          durch Teilchen und Quanten, Springer-Verlag Berlin, 1991.
[3.14]   Inder P. Batra, *Metallization and Metal-Semiconductor Interfaces*, Plenum Press, New
          York, 1988.
[3.15]   Wolf-Joachim Fischer, *Mikrosystemtechnik*, Vogel Buchverlag, Würzburg, 2001.
[3.16]   B.A. Movchan, A.V. Demchishan, *Fizika Metallov*, **28**, 653 (1969).
[3.17]   J.A. Thornton, *J. Vac. Sci. Technol.*, **11**, 666 (1974).
[3.18]   B.I. Chin, G. Yao, P. Ding, J. Fu, L. Chen, *Semicond. Int.*, May 2001, 107.
[3.19]   O.A. Novodvorski, C. Wenzel, J.W. Bartha, O.D. Chramova, E.O. Filipova, *Opt.Lasers
          Eng.*, **36**(3), 303 (2001).
[3.20]   A.E. Braun, *Semicond. Int.*, Oct. 2001, 52.
[3.21]   M. Haider, S. Uhlemann, E. Schwan, H. rose, B. Kabius, K. Urban, *Nature* **392**, 768
          (1998).
[3.22]   B. Freitag, S. Kujawa, P. M. Mul, J. Ringnalda, P. C. Tiemeijer, *Ultramicr.* **102**, 209
          (2005).
[3.23]   H. Lichte, in *Advances in Optical and Electron Microscopy*, ed. R. Barer, V. E. Cosslett,
          Academic Press, London, 1991, vol. 12, p. 25.
[3.24]   A. Thust, M. Lentzen, K. Urban, *Ultramicroscopy*, **53**, 101 (1994).
[3.25]   P. Stadelmann, *Ultramicoscopy*, **21**, 131 (1987).
[3.26]   International Tables for Crystallography: Vol. A, Space-Group Symmetry, ed. T. Hahn,
          Academic Press, London, 1996.
[3.27]   J. C. H. Spence, J. M. Zuo, Electron Microdiffraction, Plenum Press, New York, 1992.
[3.28]   M. Knoll, *Z. Techn. Phys.*, **16**, 767 (1935).
[3.29]   M. v. Ardenne, *Z.Techn. Phys.*, **19**, 407 (1937).
[3.30]   A. V. Crewe, *Science*, **168**, 1338 (1970).
[3.31]   K. Wetzig, D. Schulze, In situ Scanning Electron Microscopy in Materials Research
          Akademie Verlag, Berlin, 1995.
[3.32]   E. Arzt, O. Kraft, R. Spolenak, *Z. Metallk.*, **87**, 934 (1996).
[3.33]   K. Wetzig, H. Wendrock, A. Buerke, Th. Kötter, in *Stress Induced Phenomena in
          Metallization,* AIP Conference Proceedings, ed. O. Kraft et al., American Institute of
          Physics, , Melville, 1999, vol. 491, p. 89.
[3.34]   S. Lipp, Ph. D. thesis, Universität Erlangen-Nürnberg, (1997).
[3.35]   R. Spolenak, B. Heiland, Ch. Witt, R. M. Keller, P. Mullner, E. Arzt, *Practical
          Metallography,* **37**, 90 (2000).
[3.36]   S. Kikuchi, *Jpn. J. Phys.*, **5**, 83 (1928).
[3.37]   D. J. Dingley, *Scanning Electron Microsc.*, **11**, 569 (1984).
[3.38]   A. J. Schwartz, M. Kunar, B. L. Adams, *Electron Backscatter Diffraction in Materials
          Science,* Kluwer/Plenum Press, New York, 2000).
[3.39]   D. Rauser, IFW Dresden (unpublished).
[3.40]   C.G. Darwin, *Phil. Mag.*, **27**, 315 (1913).
[3.41]   P. Ewald, *Z. Phys.*, **2**, 332 (1920).
[3.42]   A. Schubert, B. Kämpfe, E. Auerswald, *Mater. Sci. Forum*, **133-136**, 117 (1993).
[3.43]   C. Cabral, K. Barmak, J. Gupta, L. Clevenger, B. Arcot, D. Smith, J. Harper, *J. Vac. Sci.
          Technol. A*, **11**, 1435 (1993).
[3.44]   M. Hecker, W. Pitschke, D. Tietjen, C.M. Schneider, *Thin Solid Films*, **411**, 234 (2002).
[3.45]   M. Hecker, D. Fischer, V. Hoffmann, H.-J. Engelmann, A. Voss, N. Mattern, C. Wenzel,
          C. Vogt, E. Zschech, *Thin Solid Films*, **414**, 184 (2002).
[3.46]   G.K. Williamson, W.H. Hall, *Acta Mater.*, **1**, 22 (1953).
[3.47]   B.E. Warren, B.L. Averbach, *J. Appl. Phys.*, **21**, 595 (1950).
[3.48]   H. Wendrock, W. Brückner, M. Hecker, T. Kötter, H. Schloerb, *Microelectron. Reliability*,
          40, 1301 (2000).

[3.49]   E. E. Fullerton, I.K. Schuller, H. Vanderstraeten, Y. Bruynseraede, *Phys. Rev. B*, **45**, 9292
          (1992).
[3.50]   G. Schneider, G. Denbeaux, E.H. Anderson, B. Bates, A. Pearson, M.A. Meyer, E.
          Zschech, D. Hambach, E.A. Stach, *Appl. Phys. Lett.*, **81**, 2535 (2002).
[3.51]   X. Zhang, H. Solak, F. Cerrina, B- Lai, Z. Lai, P. Ilinski, D. Legnini, W. Rodrigues, *Appl.
          Phys. Lett.*, **76**, 315 (2000).
[3.52]   P.C. Wang, G.S. Cargill, I.C. Noyan, C.-K. Hu, *Appl. Phys. Lett.*, **72**, 1296 (1998).
[3.53]   H.-K. Kao, G.S. Cargill, C.-K. Hu, *J. Appl. Phys.*, **89**, 2588 (2001).
[3.54]   P.-C. Wang, I.C. Noyan, S.K. Kaldor, J.L. Jordan-Sweet, E.G. Liniger, C.-K. Hu, *Appl.
          Phys. Lett.*, **76**, 3726-8 (2000).
[3.55]   R. Spolenak, N. Tamura, B.C. Valek, D.L. Barr, M.D. Morris, J.F. Miner, W.L. Brown,
          A.A. MacDowell, R.S. Celestre, H.A. Padmore, J.C. Bravman, B.W. Batterman, J.R. Patel,
          in *Sixth International Workshop on Stress Related Phenomena in Metallization*, Cornell
          University, Ithaca NY, AIP, **612**, 217 (2002).
[3.56]   N. Tamura, A. A. MacDowell, R. S. Celestre, H. A. Padmore, B. Valek, J. C. Bravman,
          R. Spolenak, W. L. Brown, T. Marieb, H. Fujimoto, B. W. Batterman, J. R. Patel, *Appl.
          Phys. Lett.*, **80**, 3724 (2002).
[3.57]   J. S. Chung, G. E. Ice, *J. Appl. Phys.*, **86**, 5249 (1999).
[3.58]   B.C. Larson, W. Yang, G. E. Ice, J. D. Budai, J. Z. Tischler, *Nature*, **451**, 887 (2002).
[3.59]   V. Holy, U. Pietsch, T. Baumbach, *High-Resolution X-Ray Scattering from Thin Films and
          Multilayers,* Springer Verlag, Berlin, 1999.
[3.60]   B. Lengeler, *Microchim. Acta*, **I**, 455 (1987).
[3.61]   S.K. Sinha, E.B. Sirota, S. Garoff, H.B. Stanley, *Phys. Rev. B*, **38**, 2297 (1988).
[3.62]   M. Sacchi, *Surf. Rev. Lett.*, **7**, 175 (2000).
[3.63]   P. M. Oppeneer, H.-Ch. Mertins, D. Abramsohn, A. Gaupp, W. Gudat, J. Kunes, and
          C. M. Schneider, *Phys. Rev. B* **67**, 052401 (2003).
[3.64]   D. Knabben, N. Weber, B. Raab. Th. Koop, F. U. Hillebrecht, E. Kisker, and G. Y. Guo,
          J. Magn. Magn. Mat. **190**, 349 (1998).
[3.65]   O. Hellwig, J.B. Kortright, K. Takano, E.E. Fullerton, *Phys. Rev. B*, **62**, 11694 (2000).
[3.66]   C. C. Kao, C. T. Chen, E. D. Johnson, J. B. Hastings, H. J. Lin, G. H. Ho, G. Meigs,
          J.-M. Brot, S. L. Hulbert, Y. U. Idzerda, and C. Vettier, *Phys. Rev. B* **50**, 9599 (1994).
[3.67]   J. P. Hannon, G. T. Trammel, M. Blume, D. Gibbs, *Phys. Lett.* **61**, 1245 (1988).
[3.68]   T.P.A. Hase, I. Pape, D.E. Read, B.K. Tanner, H. Dürr, E. Dudzik, G. van der Laan,
          C.H. Marrows, B.J. Hickey, *Phys. Rev. B*, **61** (2000) 15331.
[3.69]   M. Hecker, U. Muschiol, C.M. Schneider, H.-Ch. Mertins, D. Abramsohn, F. Schäfers,
          J. Magn. Magn. Mater., **240**, 202 (2002).
[3.70]   M. Sacchi, C.F. Hague, L. Pasquali, A. Mirone, J.-M. Mariot, P. Isberg, E. M. Gullikson,
          J.H. Underwood, *Phys. Rev. Lett.*, **81**, 1521 (1998).
[3.71]   K. Wetzig, H.-J. Ullrich, in *Industrielle Vakuumtechnik,* ed. W. Teubner Deutscher Verlag
          für Grundstoffind., Leipzig, 1980, p. 343.
[3.72]   G. J. Moseley, *Phil. Mag.*, **26**, 1024 (1913).
[3.73]   K. Wetzig, in *Analytiker Taschenbuch,* Springer-Verlag, Berlin, 2000, vol. 21, p. 65.
[3.74]   D. Brandon, W. D. Kaplan, *Microstructural Characterization of Materials,* J. Wiley,
          Chichester, 1999.
[3.75]   N. J. Zaluzec, Quantitative X-ray Microanalysis: Instrumental Considerations and
          Applications to Materials Science, in: *Introduction to Analytical Electron Microscopy,*
          ed. J. J. Hren, J. I. Goldstein, D. C. Joy, Plenum Press, New York, 1979.
[3.76]   J. Burschik B. Adolphi, *Fresenius' J. Anal. Chem.*, **365**, 269 (1999).
[3.77]   L. Strüder, N. Meidinger, D. Stotter, J. Kemmer, P. Lechner, P. Leutenegger, H. Soltau,
          F. Eggert, M. Rhode, T. Schülein, *Microsc. Microanal.* **4**, 622 (1999).
[3.78]   T. Terborg, M. Rhode, *Microsc. Microanal.* **9** (Suppl. 2), 120 (2003).
[3.79]   RÖNTEC GmbH, Product Information: X-ray Detectors and Spectrometers (2004).

[3.80]   E. Fuchs, H. Oppolzer, H. Rehme, *Partielle Beam Microanalysis,* VCH
         Verlagsgesellschaft, Weinheim, 1990.
[3.81]   R. F. Egerton, *Ultramicroscopy*, **4**, 169 (1979).
[3.82]   K. Wetzig, J. Thomas, H.-D. Bauer, *Appl. Surf. Sci.*, **179**, 143 (2001).
[3.83]   K. Wetzig, S. Baunack, V. Hoffmann, S. Oswald, F. Präßler, *Fresenius' J. Anal. Chem.*,
         358, 25 (1997).
[3.84]   W. J. Lorenz, W. Plieth, *Electrochemical Nanotechnology*, Wiley-VCH, Weinheim, 1998.
[3.85]   S. Oswald, S. Baunack, *Surf. Interface Anal.*, **25**, 942 (1997).
[3.86]   S. Hofmann, *Pract. Surf. Anal.*, **1**, 143 (1990).
[3.87]   Ch. Linsmeier, *Vacuum*, **45**, 673 (1994).
[3.88]   P. W. J. M. Boumans, *Anal. Chem.*, **14**, 1219 (1972).
[3.89]   S. Hüfner, Photoelectron Spectrocopy: principles and applications, Springer-Verlag, Berlin,
         1995.
[3.90]   H. Bubert, H. Jenett, *Surface and Thin Film Analysis,* Wiley-VCH, Weinheim, 2002.
[3.91]   S. Baunack, S. Menzel, W. Brückner, D. Elefant, *Appl. Surf. Sci.*, **179**, 25 (2001).
[3.92]   M. Menyhard, A. Sulyok, K. Pentek, A. M. Zeltser, *Thin Solid Films*, **36**, 129 (2000).
[3.93]   M. P. Seah, and D. Briggs, in *Practical Surface Analysis*, ed. D. Briggs, M. P. Seah, J.
         Wiley, Chichester, 1990, p. 1.
[3.94]   M. Zier, S. Oswald, R. Reiche, K. Wetzig, *Anal. Bioanal. Chem.*, **375**, 902 (2003).
[3.95]   S. Oswald, S. Baunack, G. Henninger, D. Hofmann, *Anal. Bioanal. Chem.* **374**, 736 (2002).
[3.96]   C. W. Magee, R. E. Honig, *Surf. Interface Anal.*, **4**, 35 (1982).
[3.97]   W. Grimm, *Spectrochim. Acta, Part B*, **23**, 443 (1968).
[3.98]   A. Bengtson, *Spectrochim. Acta Part B*, **40**, 631 (1985).
[3.99]   V. Hoffmann, R. Dorka, L. Wilken, K. Wetzig, *Proceedings 8th Symposium on Laser
         Spectroscopy*, Taejon/Korea, 2000, p. 78.
[3.100] E. van de Riet, *J. Appl. Phys.*, **76**, 584 (1994).
[3.101] R. L. Engelstad, Z. Feng., E. G. Lovell, A. R. Mikkelson, J. Sohn, *Microelectronic
         Engineering*, **78-79**, 404 (2005).
[3.102] D. R. Lide (ed.), *CRC Handbook of Chemistry and Physics 2004-2005*, CRC Press Boca
         Raton, 2004, p. 12-97.
[3.103] M.F. Doerner, W.D. Nix, *CRC Crit. Rev. Solid State Mater. Sci.*, **14**, 225 (1988).
[3.104] R. Koch, *J. Phys.: Condens. Matter*, **6**, 9519 (1994).
[3.105] W.D. Nix, B.M. Clemens, *J. Mater. Res.*, **14**, 3467 (1999).
[3.106] F. Spaepen, *Acta Mater.*, **48**, 31 (2000).
[3.107] R.C. Cammarata, T.M. Trimble, D.J. Srolovitz, *J. Mater. Res.*, **15**, 2468 (2000).
[3.108] L.B. Freund, E. Chason, *J. Appl. Phys.*, **89**, 4866 (2001).
[3.109] J.A. Floro, S.J. Hearne, J.A. Hunter, P. Kotula, E. Chason, S.C. Seel, C.V. Thompson,
         J. Appl. Phys., **89**, 4886 (2001).
[3.110] J.A. Ruud, A. Witvrouw, F. Spaepen, *J. Appl. Phys.*, **74**, 2517 (1993).
[3.111] W.D. Nix, *Metall. Trans. A*, **20**, 2217 (1989).
[3.112] S.P. Baker, W.D. Nix, *SPIE Proc. Ser.*, **1323**, 263 (1990).
[3.113] F.R. Brotzen, *Int. Mater. Rev.*, **39**, 24 (1994).
[3.114] R.-M. Keller, S.P. Baker, E. Arzt, *J. Mater. Res.*, **13**, 1307 (1998).
[3.115] H. J. Frost, M.F. Ashby, *Deformation-mechanism maps*, Pergamon Press, Oxford, 1982.
[3.116] M.D. Thouless, J. Gupta, J.M.E. Harper, *J. Mater. Res.*, **8**, 1845 (1993).
[3.117] C. A. Klein, *J. Appl. Phys.*, **88**, 5487 (2000).
[3.118] R. Koch, H. Leonhard, G. Thurner, R. Abermann, *Rev. Sci. Instrum.*, **61**, 3859 (1990).
[3.119] P.A. Flinn, D.S. Gardner, and W.D. Nix, *IEEE Trans. Electron Devices*, **ED-34**, 689
         (1987).
[3.120] J.A. Floro, E. Chason, S.R. Lee, *Mater. Res. Soc. Symp. Proc.*, **406**, 491 (1996).
[3.121] V. Weihnacht, W. Brückner, *Mater. Res. Soc. Symp. Proc.*, **673**, P1.10 (2001).
[3.122] V. Weihnacht, W. Brückner, C.M. Schneider, *Rev. Sci. Instrum.*, **71**, 4479 (2000).
[3.123] W. Brückner, H. Grießmann, *Rev. Sci. Instrum.*, **69**, 3662 (1998).

[8.124] A. Heuberger, *Mikromechanik,* Springer-Verlag, Berlin, 1991 pp. 39 ff.

[8.125] W.A. Brantley, *J. Appl. Phys.*, **44,** 534 (1973).

[8.126] R. Abermann, *Mater. Res. Soc. Symp. Proc.*, **239**, 25 (1992).

[8.127] V. Ramaswamy, W.D. Nix, B.M. Clemens, *Mater. Res. Soc. Symp. Proc.*, **505**, 589 (1998).

[8.128] W. Brückner, C. Lang, C.M. Schneider, *Rev. Sci. Instrum.*, **72**, 2496 (2001).

[8.129] *Thin Films: Stresses and Mechanical Properties, Mater. Res. Soc. Symp. Proc.*, **130** (1989), **188** (1990), **239** (1992), **308** (1993), **356** (1995), **436** (1997), **505** (1998), **594** (2000), **695** (2001).

[8.130] EMSR Spring meeting 2001, Symposium Strains and Stresses in Materials – Origins, Analysis, Effects, *Adv. Eng. Mater.,* **4**, No. 8 (2002).

[8.131] M. Pletea, W. Brückner, H. Wendrock, R. Kaltofen, *J. Appl. Phys.,* **97**, 054908 (2005).

[8.132] M. A. Phillips, V. Ramaswamy, B. M. Clemens, W. D. Nix, *J. Mater. Res.*, **15**, 2540 (2000).

[8.133] H. Windischmann, *J. Vac. Sci. Technol.* A **9**, 2431 (1991).

[8.134] R.-M. Keller, S.P. Baker, E. Arzt, *Acta Mater.*, **47**, 415 (1999).

[8.135] V. Weihnacht, W. Brückner, *Acta Mater.*, **49**, 2365 (2001).

[8.136] W. Brückner, W. Pitschke, J. Thomas, G. Leitner, *J. Appl. Phys.*, **87**, 2219 (2000).

[8.137] B. Eigenmann, E. Macherauch, *Mat.-wiss. und Werkstofftechnik*, **26**, 148 and 199 (1995), **27**, 426 and 491 (1996).

[8.138] B.M. Clemens, J.A. Bain, *MRS Bull.*, July 1992, 46.

[8.139] W. Brückner, S. Baunack, M. Hecker, J.-I. Mönch, L. van Loyen, C.M. Schneider, *Appl. Phys. Lett.*, **77**, 358 (2000).

4 Challenges for Thin Film Systems Characterization and Optimization

4.1 Electromigration in Metallization Layers

4.1.1 Fundamentals

The trend in the architecture of integrated circuits is characterized by a continuing downscaling process [4.1]. Typical dimensions of interconnects are now less than 100 nm. This miniaturization is accompanied by very high electrical current densities of up to some 10^6 A cm^{-2} in the interconnects. A further effect is high temperatures, which are also caused by high clock frequencies of integrated circuits (today in the GHz range). Therefore, the miniaturization process results in high mechanical and electrical loading of interconnect lines. On the one hand, high mechanical stresses develop as a result of mismatch with the SiO$_2$/Si substrate material in thermal expansion. On the other hand, the high electrical current density of about 10^6 A cm^{-2} gives rise to high impulse forces of the conduction electrons onto the interconnect metal ions, additionally to the Coulomb force of the electric field (see Fig. 4.1a).

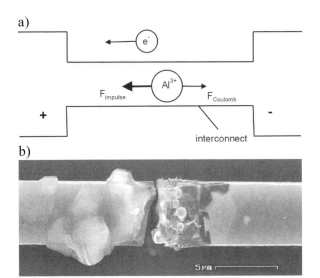

Figure 4.1: Electromigration in an Al interconnect line: a) resulting forces on metal ions, b) SEM micrograph of a totally damaged interconnect

Metal Based Thin Films for Electronics, Second Edition. Klaus Wetzig and Claus M. Schneider (Eds.)
Copyright © 2006 WILEY-VCH Verlag GmbH & Co. KGaA, Weinheim
ISBN: 3-527-40650-6

The resulting momentum transfer from conduction electrons to atoms of the interconnect line generates a so-called "electron wind", which can give rise to a net material transport. This effect is known as electromigration [4.2] and can create open circuits in the form of voids and short circuits through so-called hillocks. As an example Fig. 4.1b shows a totally damaged Al interconnect with distinctly marked hillocks in the left part of the line.

The phenomenon electromigration can be studied either by experiments as described in the following sections or by model experiments where an electric current passes through short metal stripes the displacement of which is observed. A model structure for measuring material displacement by electromigration was first suggested by *Blech* [4.3].

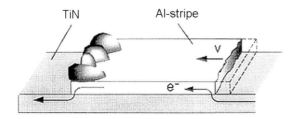

TiN   Al-stripe

v

e⁻

Figure 4.2: Model structure for material displacement by electro-migration, suggested by *Blech* [4.3]

The principle is shown in Fig. 4.2. An Al stripe is deposited on a substrate with lower electric conductivity (here TiN). An applied electric field then causes a displacement of the right edge of the Al stripe and an accumulation of atoms at the left side, where extrusions in the form of hillocks develop. For long stripes the mass flow J, the product of atomic density N and drift velocity v is given by:

$$\vec{J} = N \cdot \vec{v} = \frac{N \cdot D}{k \cdot T} \cdot e \cdot Z^* \cdot \rho \cdot \vec{j} \qquad (4.1)$$

with D = effective diffusivity,
        k = Boltzmann's constant,
        T = absolute temperature,
        e = elementary charge,
        Z^* = effective charge number,
        ρ = electrical resistivity of stripe material,
        j = current density.

From experiments it was found that no drift occurs at short stripes beneath a critical length l_c. This so-called "*Blech* effect" results from the development of mechanical stress gradients in the interconnect lines which cause a mechanical driving force for back-diffusion of atoms [4.4]. This additional force leads to a correction term in Eq. (4.1):

$$\vec{J} = N \cdot \vec{v} = \frac{N \cdot D}{k \cdot T} \left(e \cdot Z^* \cdot \rho \cdot \vec{j} - \Omega \cdot \frac{\overrightarrow{\Delta\sigma}}{l} \right) \qquad (4.2)$$

with Ω = atomic volume,
         $\Delta\sigma$ = stress difference near the ends of the stripe,
         l = stripe length.

From Eq. (4.2) the critical product of current density and stripe length for a disappearing mass velocity $\vec{J}(l_c) = 0$ can be determined:

$$\left(j \cdot l\right)_c = \frac{\Omega \cdot \Delta\sigma}{e \cdot Z^* \cdot \rho} \tag{4.3}$$

Inserting Eq. (4.3) into Eq. (4.2) the drift velocity is given as a function of the stripe length:

$$v(l) = \frac{D}{k \cdot T} \cdot e \cdot Z^* \cdot \rho \cdot j \left(1 - \frac{l_c}{l}\right) \tag{4.4}$$

As an example, Fig. 4.3 shows drift velocities, measured at Al stripes of different lengths and fitted with the $v(l)$ curve of Eq. (4.4) [4.5]. As a result electromigration damage, accompanied by formation of voids and hillocks, occurs only for stripe lengths above the critical value l_c. By extrapolating the curve $v(l)$ to $l \rightarrow \infty$ the product of diffusivity and electric charge can be obtained.

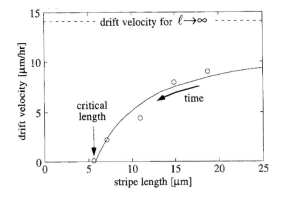

Figure 4.3: Drift velocity in Al stripes as a function of stripe length [4.5]

As a presumption for the development of interconnect damage with voids and hillocks formation local gradients of the mass flow must exist, caused by gradients of the temperature, the mechanical stress, the activation energy and/or the diffusivity. In modern polycrystalline metallizations 4 different diffusion paths can have a share in the mass flow. As shown schematically in Fig. 4.4 these are grain boundary diffusion D_{GB}, volume diffusion D_V, interface diffusion D_{IF} and surface diffusion D_S. Surface diffusion takes place at unpassivated interconnect lines, at poor adhesive layers and at inner surfaces around pores. It has the lowest activation energy [4.6]. For passivated polycrystalline interconnects under operational conditions grain boundary diffusion is the determining diffusion mechanism. As will be discussed in the following sections this is still valid for enhanced test loading up to 350 °C, whereas at higher temperatures volume diffusion plays the decisive part. Interface diffusion becomes important for missing grain boundaries in the mass flow direction, as for the so-called bamboo microstructure.

a)   b)

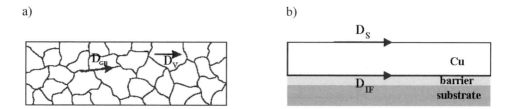

Figure 4.4: Diffusion paths for electromigration in an interconnect line; a) top view with grain boundary and volume diffusion, b) cross section with surface and interface diffusion

Divergences in the current induced mass flow lead to regions of atom accumulation and depletion [4.7]. At places with positive divergence removal of material takes place which can cause the formation of voids with subsequent interconnect failure. Conversely, a negative divergence results in material accumulation connected with hillock or whisker formation. Mass flow divergences can be caused by temperature gradients, materials inhomogeneities or changes in the interconnect geometry [4.8, 4.9].

Inhomogeneities in the microstructure also give rise to mass flow divergences and therefore to void and hillock formation. The smallest unit of microstructure that can cause a mass flow divergence is a grain boundary triple point. Under the presumption of equal diffusivities for all 3 grain boundaries the mass flux reaching a triple point, is twice the size of the mass flux leaving it. The resulting negative divergence may result in hillock formation at the triple point. An opposite direction of the electron current or of the mass flux may cause voids at the triple point. The interconnect damage is highly affected by abrupt changes in grain sizes [4.10]. This is particularly valid for the existence of so-called blocking grains, which extend across the whole line width. In such "bamboo" structures the mass flux along grain boundaries is interrupted. As discussed in the next section pure bamboo structures show therefore a low material damage, whereas at the transition from bamboo-like to polycrystalline areas the damage is very intensive if the polycrystalline sections exceed the *Blech* length.

4.1.2   Methods for Quantitative Damage Analysis

The most frequently used method for quantification of electromigration is lifetime measurement. The results give a valuable statistical measure for metallization reliability, and furthermore the activation energy can be determined. The lifetime of interconnect lines depends on various parameters. A mechanism-based lifetime model that was discussed by Arzt et al. incorporates the influences of grain size and line width [4.11]. The effects are illustrated in Fig. 4.5. In the case of a polycrystalline microstructure, that is line width/grain size > 1, the lifetime scales with the line width because numerous diffusion paths are available. If grain sizes and line width are comparable, a so-called "near-bamboo" structure exists with strong flux divergences at the ends of polycrystalline areas and with a lifetime minimum. In the case of real bamboo structures no grain boundaries exist in the direction of the mass flux. Electromigration is now restricted to slower diffusion paths, such as metal/barrier interfaces or the lattice, and therefore lifetimes increase accordingly.

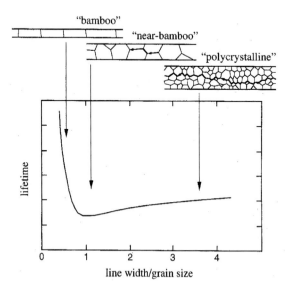

Figure 4.5: Lifetime of metallizations under electromigration versus line width/grain size ratio [4.11]

In lifetime measurements the electromigration resistance of test interconnect lines is determined by subjecting the lines to enhanced current densities (up to some 10^6 A cm^{-2}) at elevated temperatures (e.g. 500 K for Al and somewhat higher for Cu). The lifetime, that is time-to-failure, of each line is then determined under these accelerated conditions by measuring the resistance of the test structure up to the failure of each individual test line. Usually the lifetime distribution is approximately log-normal. The median time-to-failure (MTF) as a characteristic lifetime parameter can be empirically related to the current density j and the absolute temperature T by *Black's* equation [4.12]:

$$MTF = A \cdot j^{-n} \cdot e^{E_a / kT} \tag{4.5}$$

with E_a = activation energy,
  k = Boltzmann's constant.

The exponent n of the current density amounts to approximately 2. The activation energy E_a is not only material specific but also depends on the dominating diffusion mechanism and therefore on the test conditions such as current density and temperature. The activation energy follows from the slope of the logarithmic *Black's* equation. As an example Fig. 4.6 shows MTF curves determined on samples of Cu–W–Al chains [4.13]. Failure distributions at 5 different stress temperatures showed, for 200 °C and 225 °C, single log-normal (mode 1)

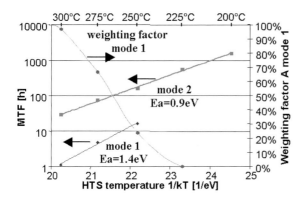

Figure 4.6: Determination of activation energies E_a from MTF curves of Cu–W–Al samples [4.13]

and for higher T, bimodal distributions (mode 2). Bimodal distributions were fitted by a weighted superposition of two log-normal distributions. In this way both the activation energy of the two modes and also the weighting factor could be determined. As Fig. 4.6 shows, the Arrhenius plot of the MTF gives straight lines. From their slopes the activation energies could be determined as $E_a = 1.4$ eV (mode 1) and $E_a = 0.9$ eV (mode 2).

In order to guarantee a certain reliability level of microelectronic products, numerous techniques and models for statistical evaluation of electromigration behavior have been developed. In general, all techniques use *Black's* equation and have to determine the activation energy E_a and the current density exponent n by series of measurements. The most popular methods will be described briefly in the following.

Lifetime Measurement (LTM)

This is the method most frequently applied for electromigration measurement. Test lines are loaded with constant heat and current, the current densities used are slightly higher than in service conditions. Tests are carried out at package level, i.e. the test lines are housed and bonded like the real chips. The resistance of test lines is measured periodically and time to failure is determined for a significant number of interconnects (usual failure criterion is a resistance change of 10 %). From two series of such experiments with variation of temperature and of current density, the two parameters E_a and n of *Black's* equation are estimated [4.14]. The time of these tests is of the order of 10–1000 hours, because test conditions should not deviate too much from conditions under service. Typical loading values are 2 MA cm^{-2} and 250 °C for Al lines, or 350 °C for Cu lines, respectively.

Standardized Wafer-Level Electromigration Accelerated Test (SWEAT)

These kind of measurements are done at wafer level (using probe tips for contacting without bonding and housing). A very high current density without external heating is applied to the test lines, reaching high test temperatures only by Joule's heating of the lines. The time to failure measured in these tests is in the range of some minutes. The results of these tests can be interpreted only when assuming certain values of E_a and n which must be known from other tests, therefore its main usage is monitoring the reliability during the fabrication process [4.15].

Breakdown Energy Measurement (BEM)

In this technique current density is increased continuously or stepwise until the tested line fails. By measurement of resistivity over time, the total energy per length E_{Fail} transferred to the line can be evaluated using following equation:

$$E_{Fail} = \int_0^{t_{fail}} \frac{R(t)}{L} \cdot I^2(t) \cdot \exp\left[-\frac{E_a}{kT}\right] dt \qquad (4.6)$$

where $R(t)$ is resistance, $I(t)$ is currrent, L is length of the line, E_a is activation energy and t_{Fail} is time to failure. These failure energies are then interpreted in the same statistical manner as time-to-failure-values in the SWEAT technique.

For both of these methods it must be considered that additional gradient effects (e.g. thermal gradients) cannot be neglected and results can only be used for comparison under similar conditions of test and specimen design [4.16].

Temperature Ramp Resistance Analysis to Characterize Electromigration (TRACE)

This method consists of very precise measurement of resistance changes with very high current densities during a time-constant increase of temperature during some hours. This technique makes use of *Matthies'* law and assumes that vacancy concentration in a metal line increases due to electromigration, from which a resistance increase ΔR follows. This ΔR includes a term depending linearly on the temperature and a term showing the early effects of electromigration [4.17].

Early Resistance Change (ERC)

Hereby the resistance increase in the early stage of electromigration is determined by precise measurement of resistance over time. After some hours a linear increase is observed, and the slope depends on the temperature of the tested line. From such results, with the assumption of Arrhenius behavior the activation energy of the early electromigration process can be derived rather quickly. The accuracy neccesary for this technique is about some 100 ppm of resistance change, this is achievable with modern digital voltmeters.

The results of the ERC method are discussed controversially, it is questioned if the early phase of electromigration has the same activation energy as the final stage. However, the value of the effective charge number Z^* and the effective diffusion coefficient D can be determined with low effort by this technique, and a number of comparative measurements with the ERC and LTM techniques showed similar values for the activation energy, too [4.18].

4.1.3 Al Interconnects

Microstructure

As already explained in Chapter 2, Al interconnects are fabricated by a subtractive process. Thus their texture and grain size correspond to those of a homogeneous thin film. Therefore we find columnar grains with a grain size of the order of the interconnect thickness. The texture can be influenced by intermediate layers such as Ti and the deposition parameters. It has been found experimentally [4.19] that a narrow (111) texture significantly improves the interconnect reliability. This observation can be explained by the reduction of flux divergences in the diffusion path along the interconnect. The kind of flux divergence that emerges depends strongly on the interplay between geometry (line width) and grain size.

As the application requires alloys rather than pure materials the distribution and form of the alloying elements plays an essential role. In the case of Cu as an alloying element different microstructures, depending on the heat treatment, can be observed. A relatively homogeneous distribution can be observed when the alloy is annealed at high temperature (e.g. 400 °C) and subsequent rapidly cooled. This is beneficial for processing as the line edges can be defined precisely. For electromigration issues, however, a subsequent long anneal at about 250 °C (aging) results in a coarse distribution of precipitates, which act as a reservoir for Cu, which is beneficial for electromigration [4.20].

Diffusion Paths

In general there are different classes of diffusion paths in interconnects: the grain boundaries, the interfaces and the volume. One could have surface diffusion as well which would be fastest with the lowest activation energy. However, Al forms a native oxide when exposed to air and in practice the interconnects are encapsulated. Thus the grain boundary diffusion path is the predominating one at service and test temperatures. Now the role of geometry comes into play. In order for this diffusion mechanism to be active a sufficient number of grain boundaries have to be oriented parallel to the current. When now the line width is decreased to an extent that one grain spans the width of the line, flux divergences are created. At this blocking grain the diffusion path changes to interfacial diffusion.

These quasi-bamboo structures show a minimum in lifetime [4.21] as a function of line width (at constant grain size). As has been explained before, the regions of enhanced diffusivity (polycrystalline) have to be longer than the Blech length in order to lead to failure. When the line width is decreased to below the grain size, the reduced diffusivities increase the lifetime again. Naturally, single crystalline lines would be most resistant to damage, which has been shown by *Joo et al.* [4.22]. However, it is not a technologically feasible route. When locally damaged mechanically these lines show a pronounced Blech effect [4.23]. The longer damaged segments that show higher diffusivities by dislocation pipe diffusion nucleate voids whereas the shorter ones do not.

The electromigration process in Al interconnects was studied *in situ* in a scanning electron microscope by *Wetzig et al.* [4.24]. As an example Fig. 4.7 shows a SEM micrograph series of an 8 μm wide polygranular Al interconnect, that was loaded with a current density $j = 2 \cdot 10^6 \, A \, cm^{-2}$ at $T = 230$ °C. Polygranular areas are interrupted by blocking grains, extending across the whole line width. The generated interruption of continuous diffusion paths along grain boundaries in the current direction leads to mass flow divergences and therefore to void formation at the end of the blocking grains. As Fig. 4.7 shows, the electromigration process is characterized by the generation of both voids and hillocks, by healing effects and finally by total failure (here after 28 hours loading).

The diffusion path between void and hillock is marked as a bright ellipse in the orientation map of Fig. 4.7. The neighboring dark grain deviates distinctly from the preferred <111> direction, and therefore has a considerable misorientation to the surrounding grains. The marked grain boundaries run almost exactly parallel to the direction of current flow. Because of their large misorientation, these boundaries possess a high diffusivity, which enables increased material transport.

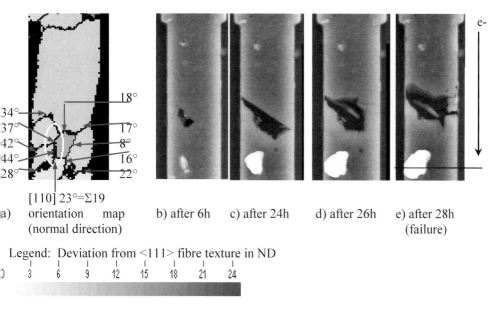

18°

34°—
37°—   17°
42°—   8°
44°—   16°
28°   22°

[110] 23°=Σ19

a)   orientation   map   b) after 6h   c) after 24h   d) after 26h   e) after 28h
      (normal direction)   (failure)

Legend: Deviation from <111> fibre texture in ND
         | | | | | | | |
0   3   6   9   12   15   18   21   24

Figure 4.7: SEM micrograph series of electromigration at an 8 μm wide polygranular Al interconnect $j = 2 \cdot 10^6$ A cm^{-2} at $T = 230$ °C)

Hillock formation is observed at the end of the preferred diffusion path at a triple point with two added grain boundaries crosswise to the current direction. Therefore the triple point acts as a material depression, which causes material accumulation in a hillock.

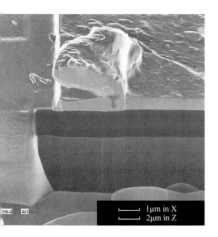

Figure 4.8: FIB cut through the hillock of Fig. 4.7 (along the line of 4.7e)

The hillock in Fig. 4.7 is <100> oriented, which is different to the grain beneath, as verified by EBSD [4.24]. This indicates that the hillock must be a grain that is newly formed, contrary to an existing model of *Gladkikh* [4.25] where hillocks grow from initially existing grains. This assumption was supported by a focused ion beam (FIB) cut along the dark line of Fig. 4.7e (see also Fig. 4.8).

Obviously, the hillock grows like an extrusion from the interconnect surface and is connected with it only by a thin neck. Therefore it is reasonable that the hillock is not correlated in orientation with the interconnect grains beneath.

Alloying Effects

As mentioned in Section 4.1.3, Cu was found to be the alloying element to retard electromigration in Al. Al_2Cu precipitates act as Cu reservoirs. As long as a sufficient Cu concentration is present [4.20, 4.26, 4.27] Cu diffuses preferentially rather than Al (see Fig. 4.9). This effect has been postulated empirically by *Rosenberg et al.* [4.28] and modelled by Monte Carlo simulation by *Dekker et al.* [4.29]. This leads to an incubation time for Al electromigration. First, all Cu reservoirs have to be emptied over a distance of at least a Blech length. Then Al diffusion starts and macroscopic damage (hillocks and voids) may form [4.30].

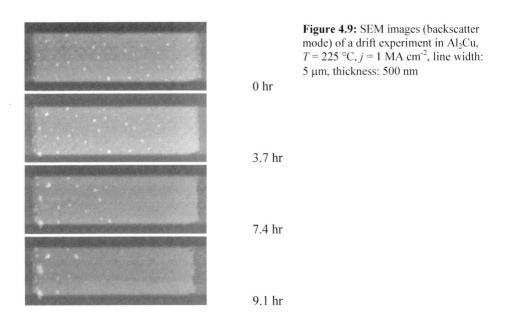

Figure 4.9: SEM images (backscatter mode) of a drift experiment in Al_2Cu, $T = 225$ °C, $j = 1$ MA cm^{-2}, line width: 5 μm, thickness: 500 nm

0 hr

3.7 hr

7.4 hr

9.1 hr

However, even after Cu depletion Al does not show the same macroscopic mass transport as pure aluminum. This can be explained by sink and source action [4.31]. Al atoms do not only need to diffuse from one end of the line to the other, but also need to be incorporated into the lattice. This incorporation can be envisioned for instance as dislocation climb. If dislocation climb is inhibited by the segregation of alloying elements such a process may become rate limiting. Cu and, for instance, Mg show the necessary size difference to the lattice to account for such an effect. Mg has been reported [4.32] to be at least as effective as Cu in retarding electromigration. However, as it stays in solid solution the resistivity increase on Mg addition is technologically not acceptable. The concepts found for Al can be generalized and also applied to Cu as the primary interconnect material. In general, alloying always comes at the cost of increased resisitivity.

Early Plasticity

One of the most recent advances in electromigration research is the observation of plasticity even before macroscopic damage like voids and hillocks are observed. In the standard models (e.g. [4.33]) all processes causing stress build up are reversible as long as the interconnect length is shorter than the Blech length. *Valek* [4.34] recently observed effects of early electromigration induced plasticity in Al lines. The (220) reflection of a blocking grain in the center of the line showed a pronounced streaking, which is due to dislocation nucleation and their possible organization into small angle grain boundaries. Streaking usually occurs across the line width, i.e. the boundaries formed are oriented parallel to the length of the line (Fig. 4.10).

Al (220)

Figure 4.10: Evolution of Al (220) reflection: from left to right under EM conditions at 205 °C: 0 MA cm^{-2}, 1 MA cm^{-2} for 14 h, 1 MA cm^{-2} for additional 6 h, –1 MA cm^{-2} for additional 5 h, –1 MA cm^{-2} for additional 16 h

This process could only be partially recovered upon current reversal and a permanent effect remained. This observation is especially interesting as it provides supporting experimental evidence for sink/source models [4.31, 4.32] as mentioned above. The effect of alloying elements can then be seen in a clearer light.

The experimental advance needed to make these experimental observations was the focusing of white X-rays (6–20 keV) down to smaller than one micron. Laue diffraction then allowed determination of the 3D deviatoric strain tensor (see Section 3.3) on the one hand and the characterization of peak shapes on the other. The observation of the latter as a function of time demonstrates the effect of early plasticity (Fig. 4.10).

4.1.4   Cu Interconnects

Microstructure

Cu interconnects are commonly produced by electroplating, also called electrochemical deposition (ECD), into previously etched trenches, and by subsequent removal of excess material by chemical–mechanical polishing (CMP). This technology together with differences in some material constants (low stack fault energy, yield strength, elastic anisotropy) leads to another microstructure and electromigration behavior than in the case of Al. Cu interconnects made by ECD show a strong twinning and only a weak (111) texture, and often an unwanted texture component from sidewall grain growth is observed [4.35].

Immediately after deposition the ECD Cu films have small grains of 50–100 nm with a weak (111) fibre texture, but after some 10 h the microstructure changes at room temperature, and grains grow up to some µm including strong twinning, accompanied by a resistance decrease of about 20 % [4.37]. This phenomenon, called self-annealing, is

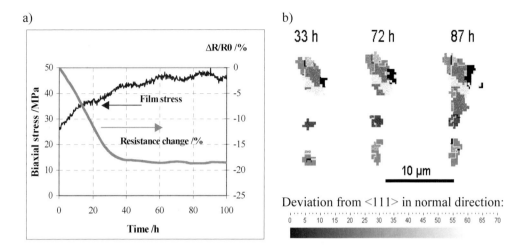

Figure 4.11: Self annealing effect of a 1 μm thick ECD Cu layer on a TaN barrier,
a) development of biaxial stress and electrical resistivity until 100 h after deposition;
b) repeated EBSD maps of the same site show grain growth with strong twinning
(white lines - Σ3–grain boundaries) [4.36]

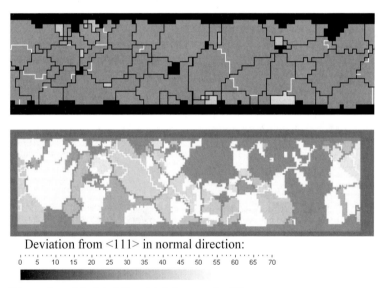

Line colors: black/white = high/low angle GB

Figure 4.12: EBSD mappings for microstructure and texture of a 2 μm wide PVD copper line (top) and
a 2.8 μm wide ECD copper line (bottom)

explained by desorption of impurities from the grain boundaries, especially of the organic additives which are necessary for void-free trench filling [4.38]. In Fig. 4.11 typical results of such behavior of a 1 μm thick ECD film on TaN are shown. By EBSD measurements and also by FIB imaging a very strong twinning was found [4.36].

In contrast, Cu films sputtered by PVD show a strong (111) fibre texture similar to Al. After an annealing step (usually 450 °C) the grain size is comparable to film thickness, and a large amount of low-angle and high-angle grain boundaries is found by EBSD measurement (Fig. 4.12).

Diffusion Paths

In contrast to Al, Cu does not form a self-passivating oxide layer, thus the surface and interface areas can act as diffusion paths in addition to the grain boundaries and the volume. The activation energies for diffusion along the questionable paths in Al and Cu, respectively are compared in Table 4.1 [4.6]. It is evident that Cu has a steady diffusion path along the interfaces, and free surfaces e.g. stress induced inner nanopores are also active.

Table 4.1: Activation energies (in eV) for different diffusion pathways (after [4.6])

Metal	Lattice	Grain boundary	Interface	Surface
Al	1.4	0.6	- -	- -
AlCu	1.2	0.7	0.9 ... 1.0	- -
Cu	2.1	1.2	0.9 ... 1.2	0.7

In the first years of Cu interconnect fabrication, very low values of the activation energy for electromigration failure were also reported [4.39], which from today's point of view was caused by non-perfect or non-homogeneous interfaces. Indeed, the interfaces between diffusion barrier and Cu line are decisive for good reliability, but nevertheless the grain boundaries also have an influence on the damage formation.

A number of in situ experiments with unpassivated PVD and ECD Cu interconnects and a Ta/TaN barrier showed rather similar behavior in spite of the differences in microstructure mentioned above [4.40]. As can be seen in Figs. 4.13 and 4.14, defects are formed not only at interfaces but also at the whole interconnect area. More than 60% of the voids and hillocks observed could be correlated with interruptions of the high angle grain boundary network by "blocking regions". In the case of PVD Cu such blocking regions contained only low angle grain boundaries, whereas in ECD Cu they were formed by Σ3-twin boundaries. For the PVD Cu lines, an activation energy of 0.78 eV was found, and for the ECD lines values of 0.85–0.95 eV are reported. The lines in Figures 4.13 and 4.14 are 500 nm thick and have a mean grain size of 0.6–0.9 μm, so they are clearly polycrystalline in the context of Fig. 4.5.

Because of the active role of interfaces, the mass flux divergences can be maximal at the intersections of sidewall and high angle grain boundaries. This is consistent with the fact, that the formation of voids was observed mostly at the interconnect edges.

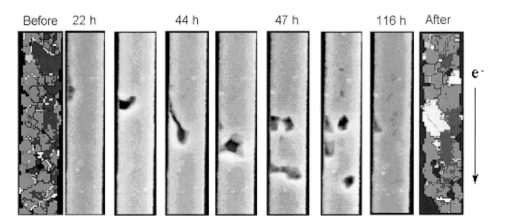

Figure 4.13: EBSD mapping of a 2 μm wide PVD Cu line and corresponding SEM micrographs of void formation during EM loading with 7 MA cm⁻² at 315 °C [4.41], colors of grain areas and lines are coded in the same manner as in Fig. 4.12

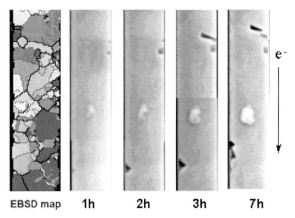

Figure 4.14: EBSD mapping of a 2.8 μm wide ECD Cu line and corresponding SEM micrographs of hillock formation during EM loading with 3.5 MA cm⁻² at 295 °C [4.40]

On the other hand, hillocks were often found to form in the inside of the line at GB triple junctions. At unpassivated lines they can reach a rather large height compared with the underlying line segment. Micro-cuts through such hillocks with a FIB device reveal that they are in general polycrystalline and only rarely have epitaxial connection to the base microstructure (Fig. 4.15) [4.40].

These different forms of material agglomeration give rise to the supposition that local variations in stress can also affect the damage formation in Cu. Indeed, the anisotropy of elastic and plastic properties of Cu can lead to stress concentration near certain grain boundary configurations [4.42]. For elastic behavior the ratio of *Young's* modulus for {111} related to {100} grains is about 2.3, and it was shown for 0.3–1 μm thick PVD Cu films by dedicated XRD measurements that {111} grains have nearly the double stress of {100} grains.

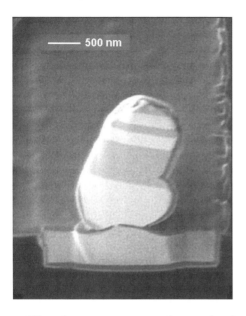

500 nm

Figure 4.15: FIB cut through a hillock grown on a 2 µm Cu PVD line after electromigration test at 3 MA cm^{-2} and 295 °C

Hints for stress concentrations at incoherent {322} twin boundaries were also found in ECD and PVD Cu thin films after thermal cycling [4.43]. By means of TEM analysis stress voids were observed at these metastable incoherent twins, and FEM calculations showed that at such grain boundaries a stress gradient follows from the anisotropic elastic constants of Cu.

These incoherent twins were present more frequently in films with a weak (111) texture, but the reasons for that are still not clear. *Ramanath et al.* [4.44] studied the interface influence without grain boundaries by epitaxial growth of Cu on appropriate single crystalline substrates (MgO), but with the commonly used diffusion barriers (Ta, TaN, TiN). With these single domain (001) Cu films, 2 µm wide interconnects were fabricated and tested for electromigration lifetime. Compared to polycrystalline interconnects, the single domain lines with nitride underlayers showed much higher lifetime, and activation energies of 0.8 – 1.2 eV were found, whereas the lines with Ta underlayer were much worse in lifetime and activation energy.

Thus, one can conclude that both the interfaces and the grain boundaries are active diffusion paths during electromigration, and for a higher EM resistivity both parts of the lines have to be optimized. For interfaces, the TaN barriers are a good choice, whereas ways for optimizing microstructure have not yet been found. The aim should be to avoid incoherent twins and high angle grain boundaries.

For optimization of reliability and mechanical strength, alloying effects are also of interest. Recently some results on Cu alloying with Ag [4.45] and Sn [4.46] have been published, showing that electric resistance does not increase too much, but grains become smaller, and the strength increases. However, EM results are not yet published for alloyed Cu.

X-ray Microdiffraction

At the beginning of this chapter the importance of stress build up in hindering electromigration on the one hand (Blech effect) and as the cause of void nucleation on the other hand was described. In the standard models (e.g. [4.33]) focus is set on the hydrostatic component of stress. In some more recent models this is slightly modified to normal stresses on atomic sinks or sources [4.31]. However, at high temperatures when diffusive processes are active all deviatoric components of stress are expected to be equilibrated and thus to disappear as described in the models. However, when measurements are performed in an intermediate temperature range e.g. electromigration test temperatures, the diffusive processes leading to stress build-up (current induced) and those responsible for relaxation (stress gradient driven) may happen at a different time scale. Deviatoric stress components may and do arise (compare Fig. 4.17).

As mentioned in Section 3.3 several groups have successfully measured the stress evolution due to electromigration on a local basis. However, they all were averaging over a set of grains or merely investigating one reflection. When one needs to investigate bamboo-type lines the stress measurement of single grains becomes essential. This is possible by Laue microdiffraction (see Section 3.3). At the Advanced Light Source in Berkeley experiments were performed to elucidate the character of flux divergences in damascene Cu lines. As described in Chapter 2 the top interface of a damascene line (Cu/SiN$_x$) is different from the other three interfaces (e.g. Cu/Ta). Thus the diffusivities are also different. There is experimental evidence that the top interface is the fastest diffusion path at test temperatures. Electroplated Cu possesses a significant twin component (115) in addition to the (111) texture [4.37]. Figure 4.16 shows that the (115) surface has a strong in-plane anisotropy ,unlike the (111) surface. Thus the grains that are oriented (115) out of plane can act as blocking grains or not depending on their in-plane orientation; i.e. the grain blocks the diffusion path when the <1-1 0> direction is oriented perpendicular to the line direction. As a consequence stress concentrations are to be expected at blocking sites.

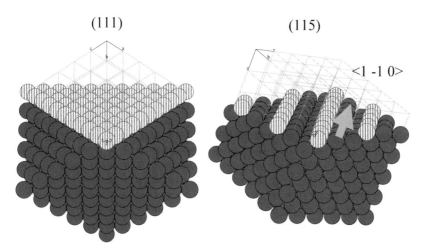

(111)   (115)

<1 -1 0>

Figure 4.16: Schematic of (111) and (115) surfaces (without reconstruction)

Laue microdiffraction allowed for the experimental observation of such flux divergences. Figure 4.17 shows that the sites that show stress build-up as a result of electromigration (see black lines) can be correlated to a change in orientation, which is exactly along the lines of the argument discussed. As the kind of 3D stress state that evolves is strongly dependent on the kind of atomic sink or source (e.g. grain boundary, interface, single dislocations, etc.) and varies as a function of the location on the line, its interpretation is highly non-trivial and will be the subject of further research.

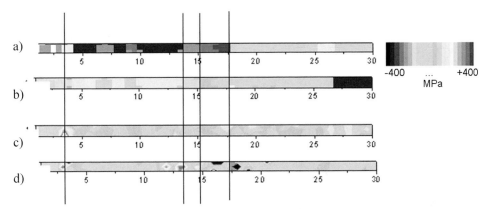

Figure 4.17: 0.8 μm wide and 30 μm long Cu damascene lines from top to bottom: a) out-of-plane orientation: dark (111) light (115); b) in-plane orientation: the center (115) grains show a blocking boundary; deviatoric stress across the line after EM at 223 °C; c) 0.5 MA cm^{-2} for 7 h; d) 1 MA cm^{-2} for 12 h. Dark colors indicate high stresses (compressive as well as tensile)

AC Induced Damage

Another current related phenomenon [4.47] in Al as well as Cu conductor lines is the, in some cases misleadingly called, 'AC electromigration'. In theory an alternating current carrying line should not be subject to electromigration as the frequency of the current is much higher than the electromigration time scale. This is true in general as long as the amplitude of the current is small and there is no DC offset to the current. Nevertheless, when the amplitude of the current is increased damage is observed (Fig. 4.18).

The morphology of the damage resembles that of fatigue induced failures. As matter of fact the plastic deformation observed (strongly correlated to (111) slip planes) is caused by cyclic mechanical loading (thermo-mechanical fatigue). The current causes a local temperature increase within the interconnect and thus a thermal stress due the mismatch in the coefficients of thermal expansion between the line and the substrate. The amplitude in strain can be controlled by the current amplitude and frequency. In general a frequency range from the Hz to the kHz regime is possible. This process also bears resemblance to damage mechanisms in SAW devices that operate in the GHz regime (compare Section 4.3).

The main future challenges in metallization for ULSI chips are the progress of downscaling and the application of low-k dielectrics together with Cu interconnects. Recently, for very small ECD Cu interconnects, it was shown that resistivity increases dramatically due to inelastic electron scattering at surfaces and grain boundaries [4.48]. Thus, for line widths below 50 nm, Al becomes interesting again, because in Al the resistivity increase due to grain boundary scattering of electrons is much lower. Further serious problems for interconnect reliability arise with the use of intermetal dielectrics with low dielectric constant (low-k material), such as porous SiO$_x$. Here the stresses caused in the lines by electromigration can lead to dielectric cracking and failure [4.49], another serious

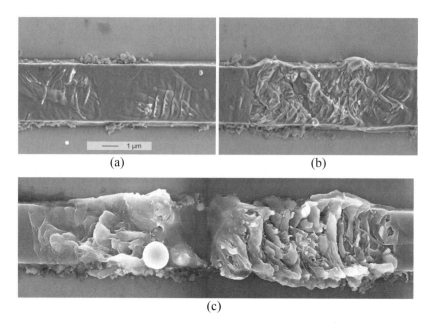

(a)   (b)

(c)

Figure 4.18: 3.3 μm wide unpassivated Al-1at.%Si line after 9×10^6 AC cycles at 100 Hz and an rms current density of 11 MA cm^{-2}. Region (a) is only lightly damaged, while region (b) is more severely damaged. Region (c) shows the site where open failure of the line occurred. All images are shown at the same magnification [4.47]

problem is the adhesion of the barrier layer. However, the majority of problems with downscaling the circuits to critical dimensions of 45 nm can be solved with the techniques known today [4.50]. Some of the requirements cannot be satisfied by better materials, but only with new and more complex design concepts such as hierarchic structure of wiring, minimizing of wire length and others.

4.2   Barrier and Nucleation Layers for Interconnects

4.2.1   Introduction

Advanced integrated circuits (ICs) contain tens to hundreds of millions of logic devices to achieve complex functions. Metal interconnects which are embedded in isolating dielectric material are the wire connections to supply electrical signals to these devices (see Section 2.1). With the introduction of copper as interconnect material and with the substitution of silicon dioxide by low-*k* dielectric materials, the manufacture of advanced ICs faces huge challenges because reliability-limiting degradation mechanisms and new failure types have to be understood.

The barrier integrity is one of the challenges. The liners have to fulfill several tasks:

- protect dielectrics and active devices from copper diffusion,
- protect dielectrics from process gas and chemical penetration,

- provide sufficient adhesion to the dielectric layers,
- provide thermo-mechanical stability, particularly for Cu/low-k dielectrics systems.

For the implementation of porous ultra low-k (ULK) materials as isolating dielectrics between the copper interconnects, pore sealing is a critical issue since defective sealing will lead to holes in the barrier.

It was demonstrated in Section 2.1 that barriers have to provide a very complex functionality. With the ongoing scaling-down of device feature sizes and consequently of interconnect dimensions (particularly of local interconnects, see Section 5.1), the push for thinner barriers is enormous, to free up space for lower-resistivity copper. For thinner barriers, the choice of the barrier material with optimized microstructure, the development and introduction of advanced deposition processes like atomic layer deposition (ALD) and the barrier/seed step coverage analysis using transmission electron microscopy (TEM) are challenges. The purpose of this chapter is to provide information about ongoing research and development work as well as implementation of these challenging tasks in the semiconductor industry.

4.2.2 PVD Barrier Layers for Copper Interconnects

The introduction of copper as conducting material for interconnects requires effective diffusion barriers, since copper readily diffuses into silicon oxide and most other dielectrics as well as into silicon, and since copper is known to form a deep level trap in silicon (acceptor levels at 0.24, 0.37 and 0.52 eV above the valence band [4.51]). This means that copper interconnects must be encapsulated with metallic (e.g. tantalum) or dielectric (e.g. silicon nitride, silicon carbide) layers to prevent electrical leakage between metal interconnects and degradation of transistor performance. Another function of the liner is to provide a good adhesion, particularly to the dielectric material since copper has poor adhesion to most dielectric materials. If low-k dielectrics are used, barriers should be selected in such a way that they contribute to the mechanical stability of the copper/low-k system.

Figure 4.19 shows an *Arrhenius* plot for the Cu diffusion coefficient D_{Cu} in Si, in SiO_2 and in some potential, Ta-based diffusion barrier materials [4.52]. Copper is characterized by fast diffusion in silicon and silicon oxide: $D_{Cu/Si} = 4.7 \cdot 10^{-3}$ cm^2 s^{-1}, $E_a = 0.43$ eV (for $T = 300 - 700$ °C), $D_{Cu/Si} = 4.0 \cdot 10^{-2}$ cm^2 s^{-1}, $E_a = 1.0$ eV (for $T = 800–1100$ °C) [4.53], $D_{Cu/SiO2} = 2.5 \cdot 10^{-8}$ cm^2 s^{-1}, $E_a = 0.93$ eV (for $T = 250–300$ °C) [4.54], $D_{Cu/SiO2} = 2.0 \cdot 10^{-11}$ cm^2 s^{-1}, $E_a = 1.2$ eV (for $T = 423$ °C) [4.55]. In addition to the high mobility of copper atoms in silicon, Cu forms silicides at about 200 °C [4.56, 4.57].

Tantalum and Ta-based compounds have been widely used as high-performance liners for inlaid copper interconnects. For both Ta-based barriers and nucleation layers, PVD has been mainly applied for ultrathin film deposition. This technique is applicable at least up to the 65 nm technology node by using ionized metal plasma (IMP) [4.58] and self-ionized plasma (SIP) [4.59] techniques. Using IMP, the metal ions are generated by RF power. The SIP approach takes advantage of the magnetron source design, which provides an efficient transfer of energy from the secondary electrons to the plasma waves to increase the ionization of the sputtered metal atoms. In both cases, the flux of the metal ions and the amount of resputtering can be controlled by the wafer bias. Overhang at the top corner of a trench or via structure, step coverage and bottom resputtering are determined by the directionality of ionized metal atoms, which again is controlled by the wafer bias.

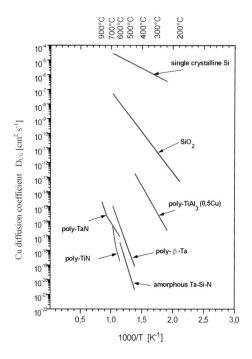

Figure 4.19: *Arrhenius* plot of Cu diffusion in several materials [4.52]

Tantalum provides high thermal and chemical stability. The stability of the Cu-Ta system at high temperatures [4.60, 4.61] is based on the fact that the refractory metal Ta ($T_{melting} = 3020\ °C$) does not react with Cu and that the solubility of Ta in Cu and vice versa is very low in the solid state [4.62]. Silicide formation occurs at relatively high temperatures only [4.63]. In thin Ta films, three modifications can be observed: the thermodynamically stable body centered cubic (bcc) α-Ta (space group: $Im\bar{3}m$ (229) [4.64], resistivity: $\rho_{\alpha\text{-Ta}} \approx 15$–$30\ \mu\Omega$ cm [4.65]), the metastable tetragonal β-Ta (space group: $P42_1m$ (113) [4.66], resistivity: $\rho_{\beta\text{-Ta}} \approx 150$–$220\ \mu\Omega$ cm [4.65]), and sometimes a face centered cubic (fcc) phase [4.65]. Depending on the deposition parameters, a deposited Ta layer consists of either one or a mixture of the above-mentioned three phases [4.65]. At room temperature, the metastable β-Ta usually grows on Si or SiO_2 [4.61, 4.63]. During thermal treatments, the metastable β-Ta phase can be transformed into the equilibrium α-Ta phase. Parameters determining this phase transformation are annealing temperature and time [4.60, 4.63], annealing ambience, substrate material [4.63], and film thickness [4.67].

The Cu/Ta interface and the possible formation of an amorphous "intermixing layer" with about 2 nm thickness has not yet been completely understood. The interface between Ta and Si should keep its integrity up to at least 550 °C since Ta silicidation does not occur until this temperature [4.61, 4.68].

It has been demonstrated that Tantalum and Ta-based barriers are very stable against copper interdiffusion [4.61, 4.69]. The β–Ta barrier was first shown to be an excellent barrier in 1986 by *Hu et al.* [4.70]. However, grain boundary diffusion of Cu atoms becomes relevant at elevated temperatures due to the polycrystalline structure of Ta [4.71]. One way to improve the thermal stability of Ta barrier layers is the addition of oxygen or nitrogen N during the deposition process [4.72]. Oxygen impurities in Ta films are believed to decorate grain boundaries and, consequently, block the diffusion pathes along grain boundaries [4.72]. In addition to being good copper diffusion barriers, it is essential to improve adhesion at all interfaces to resist delamination during processing or thermal stressing, and for electromigration resistance of small copper structures. Furthermore, the grain size is reduced if tantalum oxides or nitrides are formed, leading to a nanocrystalline or amorphous thin film [4.73].

TaN/SiO$_2$ adhesion is excellent, i.e., much better than Ta/SiO$_2$ adhesion [4.74], but Cu/TaN adhesion is poor, which is consistent with better electromigration resistance for Cu/Ta interfaces. Since the liner/interlayer dielectrics (ILD) adhesion and the Cu/liner adhesion have conflicting dependences on the nitrogen content [4.74], Ta/TaN$_x$ layer stacks or graded TaN$_x$ layers have been proposed to meet the mentioned requirements [4.75, 4.76]. It has been shown that Ta/TaN$_x$ bilayers can be deposited that have a low in-plane resistivity. This layer stack can provide the needed thermal stability, a good adhesion to the interlayer dielectrics and a strong interface bonding to copper that reduces the atomic transport along the Cu/Ta interface [4.77]. The SIP technique allows one to produce a stacked TaN/Ta layer. Films of varying nitrogen content can be deposited in the same chamber.

For reduced copper interconnect dimensions, thinner liners with low in-plane resistivity are required [4.1, 4.78]. To fulfill their very complex functions (diffusion barriers for copper, adhesion layers for interlayer dielectrics, thermo-mechanical stability), the microstructure of these layers or layer stacks has to be optimized. The Ta/TaN bilayer meets this requirement since the low-resistivity α-Ta phase (bcc-Ta) with a resistivity of $\rho_{\alpha-Ta}$ = 15 - 30 $\mu\Omega$cm is spontaneously formed when Ta is deposited on a TaN surface. This phenomenon occurs on TaN films at least as thin as 2 nm [4.77]. It was shown for PVD-TaN/Ta bilayers that the reduction of the TaN thickness up to 1 nm is possible for the α-Ta formation on top without any loss in adhesion [4.79]. The resistivity of the stable α-Ta phase is about 10 times lower than that of the metastable β-Ta phase with $\rho_{\beta-Ta}$ = 150 – 220 $\mu\Omega$cm, which is normally formed when Ta is deposited directly on the dielectrics.

Chen et al. [4.80] investigated the impact of the PVD Ta/TaN (10 nm TaN, 15 nm Ta) liner microstructure on several process-related properties of Cu seed and ECD copper films including via resistance and interconnect reliability. They found that Cu seed and ECD Cu films can be manipulated by the liner microstructure. Based on transmission electron microscopy (TEM) and atomic force microscopy (AFM) studies *Zschech et al.* [4.81] have shown that the barrier type has a significant influence on the grain size distribution of Cu seed films. But there was no significant change observed for the grain size distribution of ECD Cu, determined from ion images using focused ion beam (FIB) techniques. The Cu seed grains show a broader distribution for TaN barriers than Ta barriers. Phase and texture formation of thin layers were studied using XRD. It was shown in [4.81], that the use of Ta and TaN single layer barriers results in different texture of Cu seed and EP Cu layers. In particular, the Cu seed <111> texture is enhanced on an α-Ta film with strong Ta <110> orientation, resulting in a seed layer with low in-plane resistivity and better wetting property [4.80]. According to [4.81], thin Ta films that crystallize in a metastable tetragonal β-phase reveal a weak <001> texture, whereas the TaN film reveals a weak <110> texture. The strength of the Ta texture is influenced by the layer thickness, i.e., the <001> texture becomes weaker with decreasing barrier layer thickness. This dependence has not been observed for TaN. The Cu <111> texture of both PVD Cu films (seed layer) and ECD Cu films (electroplated and subsequently annealed) is influenced by the type of barrier material and the barrier thickness. It is much stronger using a Ta barrier than using a TaN barrier (see Fig. 4.20). The texture of the ECD copper is strongly influenced by the crystal structure and the thickness of the Ta barrier. In addition, the copper texture can also be adjusted by the barrier deposition parameters. With decreasing Ta layer thickness, the Cu <111> texture gets stronger. That means that the texture of the ECD copper can be adjusted by the barrier deposition process. Combining the results for barrier and Cu textures, it was shown in [4.81] that the Cu texture gets sharper if the Ta texture gets weaker. Again, this effect has not been observed for the TaN barrier layer.

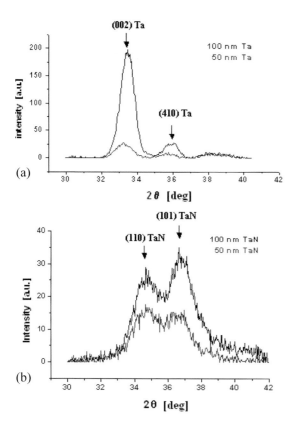

Figure 4.20: X-ray texture study: I-2θ scans of a Ta barrier layer (a) and a TaN barrier layer (b) with different thickness [4.81]

Binary and ternary Ta alloys increase the total interconnect resistance more than pure α-Ta, but they usually provide a higher thermal stability as well as sometimes improved electromigration behavior and increased mechanical stability, as experimentally proven for barriers deposited onto Si [4.75]. *Bian et al.* [4.82] studied TaSiN diffusion barriers (with a thickness of 16 nm) for Cu/SiO$_2$ and Cu/SiLK structures. The thermal stability of the system against Cu interdiffusion and the interface stability is given up to 450 °C.

Microstructure, crystal structure and degradation of TaN$_x$ thin films (10 nm thickness) and Ta/TaN layer stacks were studied for different annealing temperatures between T_{an} = 300 °C and 700 °C [4.68, 4.83]. A small addition of N$_2$ to the sputtering gas results in the transition from metastable tetragonal β-Ta to nanocrystalline bcc TaN$_x$ [4.84]. Further increase in the

N_2 flow leads to the formation of nanocrystalline fcc TaN. The annealing experiments showed stable TaN_x films until $T_{an} = 500$ °C. For the pure Ta barrier, Ta diffusion through the Cu capping layer to the sample surface is observed at $T_{an} = 500$ °C, and the transformation of initially grown metastable β-Ta into the equilibrium α-Ta phase occurs at $T_{an} = 600$ °C. The barrier degradation starts at $T_{an} = 550$ °C for the pure Ta and for the $Ta_{80}N_{20}$ films. For stoichiometric fcc TaN ($Ta_{50}N_{50}$) barriers, the degradation starts at 700 °C only. The cross-sectional TEM micrograph shown in Fig. 4.21 displays an intact TaN barrier. An amorphous Si layer appears beneath the TaN layer. This layer has also been observed in the as-deposited state, and is caused by the TaN deposition process.

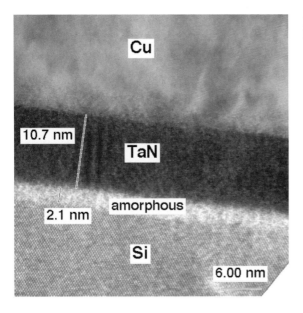

Figure 4.21: TEM image of the TaN barrier after annealing at 600 °C [4.68]

Depth profiling techniques were used for barrier stability evaluation. In particular, glow discharge optical emission spectroscopy (GDOES) was used for copper depth profiling within the barrier and within the substrate. The depth profiles for the copper distribution in Ta, $Ta_{80}N_{20}$ and $Ta_{50}N_{50}$ and in the Si substrate are shown in Figs. 4.22 a–c [4.68].

Generally, with increasing N content, both the barrier stability against Cu interdiffusion and the critical temperature for significant barrier failure increase. In the $Ta_{80}N_{20}$ film, the initially amorphous Ta(N) transforms to Ta_2N at 500 °C. In contrast to the β-Ta film, no Cu interdiffusion through the Ta(N) barrier was detected below 550 °C. The highest thermal stability has been determined for the 10 nm - $Ta_{50}N_{50}$ films due to the high melting temperature ($T_{melting} = 3090$ °C [4.85]). The nanocrystalline fcc TaN structure is preserved at least until annealing at 800 °C. Compared to Ta, its adhesion to SiO_2 is increased [4.74, 4.68]. However, TaN has a high electrical resistivity. The grain growth in TaN leads to higher Cu diffusion through the barrier at $T_{an} = 600$ °C and 700 °C, which however remains lower than through the Ta and Ta(N) barriers annealed at 550 °C [4.68]. Fig. 4.23 shows the results of the quantitative analysis of copper trace concentration in the substrate and in the barrier using electrothermal atomic absorption spectrometry (ETAAS) [4.68].

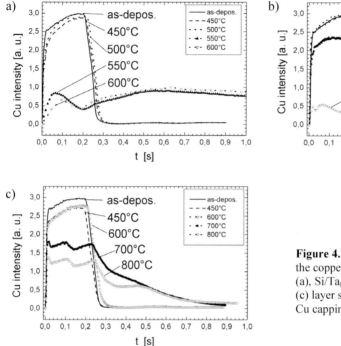

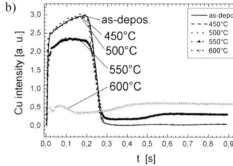

Figure 4.22: GDOES depth profiles of the copper distribution of the Si/Ta/Cu (a), Si/Ta$_{80}$N$_{20}$/Cu (b) and Si/Ta$_{50}$N$_{50}$/Cu (c) layer stacks (barrier thickness 10 nm, Cu capping layer thickness 50 nm) [4.68]

The thermo-mechanical stability of TaSiN diffusion barriers with varying nitrogen content is discussed in [4.87]. Local microstructure characterization by TEM revealed two main degradation processes that cause TaSiN barriers with lower N content to lose their stability against Cu diffusion at elevated temperatures: On the one hand, Ta atoms with a weak chemical bonding diffuse through the Cu layer to the sample surface forming a continuous Ta/Ta oxide layer. On the other hand, Ta silicide and Ta nitride crystallites are formed depending on the chemical composition of the barrier layer. Both processes result in a thinning of the original TaSiN barrier (Fig. 4.24a). Local sites were found where Cu grains contact the underlying SiO$_2$ layer. A Cu diffusion into the SiO$_2$ layer could be proved there. With increasing N content, more and more Ta atoms are chemically bonded in Ta$_2$N and TaN crystallites. As a consequence, Ta diffusion to the surface of the Cu layer can no longer be observed. Additionally, the tendency to form irregular Ta nitride crystallites decreases with increasing N content so that more homogeneous and conformal barriers are obtained. Although Ta$_{56}$Si$_{19}$N$_{25}$ and Ta$_{41}$Si$_{20}$N$_{39}$ barriers are completely crystallized after an annealing at 600 °C for 100 h, no Cu diffusion into the SiO$_2$ layer could be detected using analytical TEM techniques. The highest barrier stability is expected for the Ta$_{30}$Si$_{18}$N$_{52}$ barrier because this barrier remains in an amorphous state for all annealing conditions (Fig. 4.24b).

Shamiryan et al. [4.88] studied the pore sealing behavior of PVD-Ta(N) on ultra-low-*k* (ULK) dielectrics. Since it is easier to seal pores for ULK materials with high carbon concentration, the pore sealing was less difficult for SiCO:H material with about 20 % C compared to HSQ (no carbon).

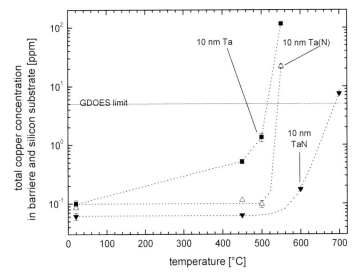

Figure 4.23: Quantitative analysis of copper trace concentration in the substrate and in the barrier using electrothermal atomic absorption spectrometry (ETAAS) [4.68]

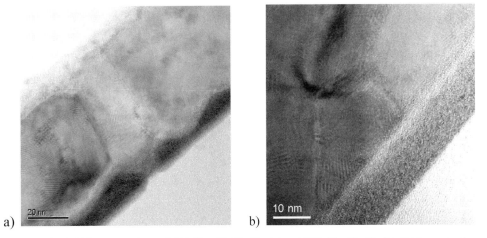

Figure 4.24: TEM cross-sections of $SiO_2/Ta_{62}Si_{20}N_{18}/Cu$ (a) and $SiO_2/Ta_{30}Si_{18}N_{52}/Cu$ (b) layer stacks annealed at 600 °C for 100 h

4.2.3 Barrier/Seed Microstructure and Step Coverage

A minimum overhang of barrier and seed material at a top corner of a trench or via, which limits the amount of material that can be deposited within the structure, and a conformal step coverage of barrier and seed layers in deep narrow trenches and vias are challenges for process technology and, therefore, highly sophisticated process control is needed.

This means that barrier and seed layers have to be deposited in such a way that the resulting layer stack has a high conformality. Consequently high stability of barrier and seed deposition processes are essential to ensure that the thickness variations of these ultrathin layers are within the defined tolerances. The measurement of thickness and uniformity of the barrier and sometimes of the seed layer at the most critical structures, the sidewalls of vias, commonly known as barrier/seed step coverage analysis, is one of the most challenging analytical tasks in today's semiconductor manufacturing. Only transmission electron microscopy (TEM) has sufficient spatial resolution to evaluate the thickness distribution of non-planar barrier and seed layers [4.89]. Figure 4.25 shows an overview TEM bright field image of a vertical focused ion beam (FIB) cut through a via chain with deposited Ta barrier and Cu seed layers.

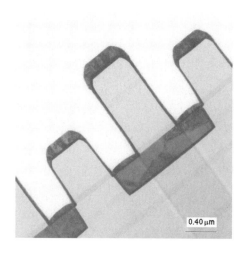

Figure 4.25: TEM cross-section of part of a via-chain test structure (brightfield image, overview) [4.90]

Several TEM sample preparation techniques have been developed for barrier/seed step coverage analysis, however, only the FIB technique makes it possible to prepare via chain cross-sections with high accuracy in a reasonable timeframe (see Section 3.2). It has been demonstrated that a complete specific site preparation of a TEM lamella can be carried out in less than one hour with a success rate of nearly 100 percent using the procedure described above [4.90]. Since the TEM technique is not an in-line/insitu technique, it will probably never become an in-fab metrology method but sample extraction can be done non-destructively in-line using FIB tools and the lift-out technique.

More highly magnified TEM brightfield images of a vertical cross-section through a via chain after several process steps, as shown in Fig. 4.26, are commonly used for step coverage analyses. The film thickness is determined from the gray value changes caused by the barrier/seed layers. Ta and its compounds, are characterized by a high mass–thickness contrast in the TEM bright field image. If only the barrier layer has been deposited, then step coverage analysis is possible with acceptable results but it is much more difficult to distinguish between Ta barrier and Cu seed layers at via sidewalls by TEM bright field imaging, because Cu also shows a high mass–thickness contrast. An additional difficulty results from the contrast variations in the Cu seed layer due to the orientation distribution of the Cu grains.

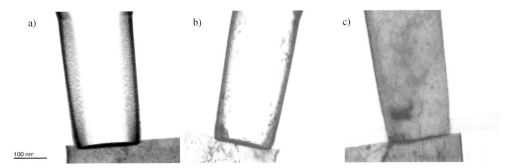

Figure 4.26: TEM brightfield images of a vertical cross-section through a via: a) after Ta barrier deposition, b) after Cu seed deposition, c) after ECD-Cu fill and CMP [4.90]

A special imaging method with a weak orientation-dependent diffraction contrast of crystals is the so-called Z-contrast imaging technique. Z-contrast images are taken in the high resolution scanning TEM (STEM) mode. The electron beam is scanned in a raster across the specimen and the intensities of electrons that are scattered incoherently to very high angles are integrated over a large angular range for each illuminated volume fraction using a high angle annular dark field (HAADF) detector. Furthermore, the high angle electron scattering probability depends strongly on the atomic number Z of the investigated material. As shown in the STEM-HAADF image in Fig. 4.27, the Ta barrier can now be clearly distinguished from the Cu grains. The total measuring error for barrier/seed thickness, considering both the instrumental and sample thickness errors as well as the subjective error which is mainly caused by the operator-to-operator deviation, is in the $2 - 3$ nm range [4.90]. To increase the reproducibility of barrier/seed step coverage analyses and to make them independent of the operator, automated measuring routines can be used that are based on simulations and image processing algorithms.

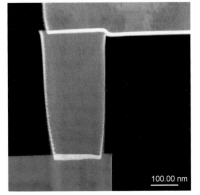

Figure 4.27: STEM-HAADF image of a vertical cross-section through a via [4.90]

Measurement of the barrier/seed step coverage at vertically cut vias using TEM brightfield as well as Z-contrast imaging is complicated because the thin film stack is curved. For typical via diameters (200 nm) and TEM lamella thickness (80 nm), the projection of the via curvature leads to image intensities which cannot be analyzed clearly due to the superposed contributions from different layers. Since the lamella thickness achieved using the FIB technique that is widely applied to prepare TEM samples of via structures is limited to a thickness of 30 – 40 nm, the lamella thickness/via diameter ratio increases with further shrinking of the interconnect dimensions. If the projection effects of the via curvature on the image gray value changes are not taken into account adequately, the error of film thickness measurements is increased dramatically. Additionally, the deviation of the via from the ideal cylindrical form and the via sidewall roughness can lead to further complication (Fig. 4.28).

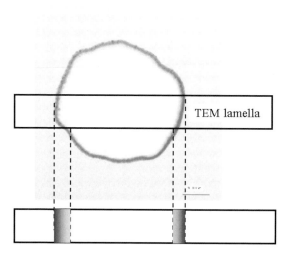

Figure 4.28: TEM bright field image of a horizontal cross-section through a via after barrier deposition: effect of the via curvature and geometrical explanation. [4.89, 4.90]

Horizontal or tilted plan view cuts through via chains are not influenced by the via curvature. However, conicity and side-wall roughness of the vias make barrier/seed step coverage analysis difficult. Nevertheless, the error in barrier/seed thickness measurements is lower than for step coverage analyses at vertical cuts. The big disadvantage of horizontal cuts through via chains is that the step coverage at the via/trench transition and in the corners at the via bottom, which is most critical for process control, cannot be monitored sufficiently.

For smaller interconnect dimensions and increasing aspect ratios (> 4:1), the thickness of the ultrathin layers decreases at the side-walls and, in particular, the Cu seed layer becomes more sensitive to temperature-induced agglomeration. Therefore, TEM-based techniques for evaluation of the roughness of the Ta and/or Cu surfaces within the via become more important. Electron tomography and HAADF-imaging of "inner surfaces" are potential approaches.

A recently introduced technique for the characterization of interconnect structures and for the measurement of physical dimensions is to create a three-dimensional (3D) reconstruction of the object using electron tomography [4.91]. In contrast to the methods described above, a FIB-cut specimen is prepared with a thickness which contains a large part of the via or even the entire via. The approach of this technique is to reconstruct the 3D structure of the object from a series of its 2D projections recorded at different projection angles. TEM brightfield images or, for increased chemical contrast, HAADF images of the vias can be used. In order to achieve sufficient resolution in the reconstruction, a large number of projections has to be collected by tilting the specimen over a large angle range up to ± 70°. Figure 4.29 shows a 3D surface rendering of a FIB-cut copper contact with Ta barrier/Cu seed layers. This 3D reconstruction is based on a series of 134 images recorded over a tilt angle range of -67 to

+67° at two perpendicular tilt axes orientations. Layer thickness and uniformity can be evaluated at any position from such a reconstruction. As device feature sizes continue to shrink, electron tomography has a high potential to become the technique of choice for barrier thickness and uniformity characterization at contacts. The automatization of tilt series acquisition is a prerequisite for this purpose and has been developed to a high level. Further development of these extensive data processing procedures is necessary until it can become a standard TEM technique, especially regarding its usefulness for automated routine application of high quality 3D reconstructions.

Figure 4.29: 3D-TEM surface representation of a via (diameter 200 nm) after Ta barrier/Cu seed deposition [4.90]

A fast and simple method to image the homogeneity of the via side-wall coverage is provided by the *Z*-contrast imaging technique. HAADF images yield a high contrast for small mass thickness changes. The HAADF image of the via inner surface shows clearly the Ta barrier roughness at the via side-wall. The reason for this is the strong *Z*-dependence of the HAADF signal. Using the theoretical electron scattering cross sections of the transmitted materials, layer thickness can be calculated from intensity changes in the HAADF images.

4.2.4 New Barrier/Seed Concepts Using CVD and ALD

The current inlaid copper processes utilize PVD Ta-based barriers and Cu nucleation layers. The shrinking of device feature sizes and consequently of the interconnect dimensions, and the introduction of new low dielectric constant materials goes along with the need to reduce thickness and to achieve higher conformality for barriers and nucleation (seed) layers. Ultrathin barriers are needed to maintain the effective conductor resistivity in extremely small, high A/R features. Nucleation layer conformality requirements become more stringent to enable Cu ECD filling of inlaid structures. This task will require the development of other or optimized barrier materials as well as the introduction of new barrier and nucleation layer deposition solutions.

According to the 2003 ITRS [4.1], the most difficult near-term challenges for barrier and nucleation layers include the rapid introduction and integration of new materials and processes, dimensional control and reliability. Surface segregation, CVD, ALD, and dielectric barriers represent intermediate potential solutions. Zero thickness barriers are desirable but not required.

The future barrier materials for copper interconnects will be selected from Ta, Ti and W as well as their nitrides and silicon nitrides using ionized PVD as well as CVD deposition techniques. For the sub-100 nm technology nodes, and particularly for deposition on porous ULK materials, CVD techniques will probably have to be introduced to meet the requirements for step coverage in geometrically critical structures. However, the introduction of CVD barrier and nucleation layers, and possibly also a CVD trench and via fill process, provides significant process and process integration challenges. Interfaces, contamination, adhesion, mechanical stability, thermomechanical stress, electrical parameters, and thermal budget, create a difficult-to-manage complexity.

TiSiN films are a proven barrier solution using a TDMAT (tetrakis dimethyl amino titanium) based CVD process [4.92]. A conformal step coverage and Cu diffusion barrier properties equivalent to PVD-Ta(N) were demonstrated. CVD-TiN films would be an ideal candidate material because of the experience from tungsten plug/aluminum wiring applications [4.93, 4.94]. However, CVD-TiN and CVD-W films are less effective as a copper diffusion barrier compared with PVD-Ta and PVD-TaN films, and they provide poorer adhesion [4.95]. However, it seems possible to integrate the unique film properties of PVD-Ta and PVD-TaN with CVD barrier films, too.

Barriers will undoubtedly be a key focus area for research for the long term, since engineering the smoothness and other properties of the copper barrier interface will be key to ameliorating the expected copper resistivity increase due to electron scattering effects. The nucleation layer for ECD copper is usually deposited using PVD techniques. Since either long throw or various ionized PVD techniques yield marginal coverage for dual inlaid structures, electroless and ECD copper seed layer repair are being developed as nucleation solutions over the next few years.

Further extension of copper interconnect technology includes alloying of the copper seed layer, i.e., depositing copper alloy like CuSn. Alloying elements are introduced to improve electromigration resistance of copper interconnects (see Section 2.1).

4.2.5 Atomic Layer Deposition (ALD)

For sub-100 nm technology node barrier and nucleation layers deposition, ALD is evaluated [4.96, 4.97] (see Section 3.1). It is supposed to be the dominant solution for the deposition of ultrathin layers in extremely narrow inlaid structures with high A/R because of their superior conformality and controlled thickness. In addition, the barrier deposition process could be replaced by in situ treatments of the etched low-*k* dielectric side-walls to form effective barriers against copper diffusion.

ALD is based upon a self-controlled, sequential surface reaction process using different precursors, with the potential for controlling the film growth and the film composition on a monolayer scale [4.98]. The challenge of the ALD technique is to select the optimized precursor chemistry to meet all required properties of barrier and seed layers [4.99].

Typical thicknesses of barriers deposited using ALD are in the range of 1–2 nm. Ti, W and Ta-based barriers are formed using TDEAT (tetrakis diethyl amino titanium), TEB (thiethyl borane) and TBT DET (tertbutylimidotis diethyl amino tantalum) precursors with NH_3 and other reactants for the deposition process. The application of ALD-WCN barriers on porous CVD-SiCO:H dielectrics has been reported [4.100]. It was shown for TaN_x and $Ta(C)N_x$ barriers that the resistivity decreases with increasing C concentration. [4.101]. Ta based barriers were deposited in a highly conformal manner up to A/R = 6 using PE-ALD processes [4.101, 4.102]. Etch-pit tests demonstrated better barrier behavior for ALD-TaN barriers compared to amorphous PVD-TaN barriers [4.103].

4.3   Acoustomigration in Surface Acoustic Waves Structures

4.3.1   General Remarks

Surface acoustic wave (SAW) structures are increasingly offered for high-power applications, especially for IF, RF and GPS filters, duplexers, wireless LAN, Bluetooth IF or RF filters, resonators, basestation filters and optical communications networks in the frequency range from some 10 MHz to 10 GHz [4.104, 4.105]. Trends to higher frequencies above several 100 MHz result in smaller surface acoustic wave (SAW) structure dimensions to below 1 µm. Higher electrical input power values above 1 W are also currently observed in applications of SAW devices. Additionally, reducing the packaging sizes of SAW products below 2 mm x 2 mm is of interest for future product generations. Such trends result in higher power density levels and, consequently, in higher SAW stress loading of metallization [4.106]. However, increasing power density load causes both higher operating temperatures above 100 °C and mechanical stress loads of the SAW structures, damaging the finger electrodes by grooving, void and hillock formation due to the so-called acoustomigration effect. Delamination effects can also be observed, preferably at the edges of the finger electrodes. An extreme power density level can even cause damage of the piezoelectric substrate material by cracking and spalling effects. Such damage gives rise to undesirable effects in SAW propagation, limited lifetime or power durability of SAW devices.

Therefore, important challenges due to increasing SAW power durability as well as technical advances in microelectronics [4.107, 4.108] are

- new materials developments for
  - finger electrodes (Cu, Al-based or Cu-based alloys and multilayers)
  - barrier layers necessary for copper metallizations preventing diffusion of oxygen, copper and substrate elements (see Section 4.2)
  - hard and insulating layers protecting the surface or suppressing material transport at the "free" surface
  - piezoelectric substrate materials e.g. single crystalline bulk materials like quartz, lithium niobate ($LiNbO_3$) or lithium tantalate ($LiTaO_3$), or piezoelectric thin films like ZnO, AlN, or lead zirconate titanate (PZT)
- new technology developments as
  - lift-off or damascene technique for patterning the copper/barrier thin film systems
  - techniques for long-term stable bondings.

The following sections will focus on materials damage of the finger-shaped SAW metal electrodes due to the acoustomigration effect.

4.3.2 Acoustomigration Mechanism

The mechanism of acoustomigration is not yet widely understood. However, it can be assumed that the damaging effects in SAW structures mainly originate from the excited SAW rather than from thermal effects [4.109]. Therefore, the propagation of the SAW including its temporal and spatial variations in magnitude is important for the damage of SAW structures. In terms of SAW propagation one can classify these structures into those with standing or with traveling waves. However, the principle of acoustomigration damage of SAW finger electrodes by a high cyclic mechanical stress field is quite similar for all SAW structures (Fig. 4.30): A RF voltage that is applied to every other finger electrode forms a strong varying electric field between these electrodes. Concerning the piezoelectric, or more precisely the inverse piezoelectric, effect the substrate will be constrained, resulting in a cyclic particle displacement at the substrate surface between these electrodes.

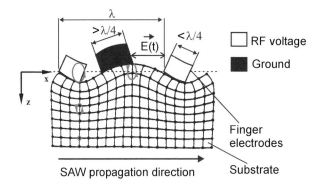

Typical dimensions:
- thickness of electrodes 0.05–0.3 μm
- width of electrodes 0.5–3 μm
- SAW amplitudes < 1 nm
- wave length 2–12 μm

Figure 4.30: Schematic diagram of stress generation in SAW structures for λ/4 electrodes

For the simple case of Rayleigh-type waves such a particle motion can be described by a set of wavefunctions

$$U_n = A_n \cdot \exp(kbx_3) \cdot \exp(j(2pft - kx_1)) \qquad (4.7)$$

giving evidence of an elliptical motion with longitudinal and transverse components [4.110]. In Eq. (4.7) U_n is the displacement, A_n is the amplitude, x_1 and x_3 are coordinates where x_1 is the propagation coordinate, k is the wave number (2p/L), f is the frequency, and b is the decay constant. Because the SAW is a surface propagating wave the SAW energy is confined at the substrate surface with a penetration depth of the order of the wavelength, i.e. typically in the μm range. The substrate particle displacement generates a high dynamic stress field in the electrode material, superimposing the internal residual stresses. As an example the calculated alternating SAW stress components σ_{ij} for a standing wave on STX quartz are shown in Fig. 4.31 [4.118]. As a result, the total stress in the finger electrodes varies rapidly with time over a distance of λ/4. This means that the mechanism of the acoustomigration could be interpreted as an ultra high cyclic bending load combined with a high dynamic tensile/compression load at elevated temperature (due to the self-heating effect) running at

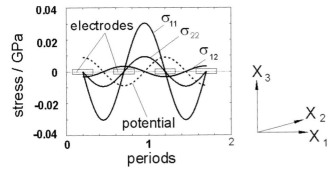

Figure 4.31: Alternating stress components σ_{ij} for standing acoustic waves on STX quartz, calculated for 100 MHz, 1 W/mm [4.118]

SAW frequency. Additionally to plastic effects in the metal as described by *Mughrabi* [4.111] or *Kraft et al.* [4.112], such a stress situation can release the phenomenon of drift-induced diffusion where atomic transport is initiated by a stress gradient force. Some other potential sources that can influence the mass transport in SAW finger electrodes are

- temperature gradients (thermal transport)
- concentration gradients
- charged carrier gradients (electrotransport).

One can assume that drift-induced migration occurs preferentially along diffusion paths with low activation energy, such as grain boundaries and interfaces or "free" surfaces. From this point of view the acoustomigration mechanism shows similarities to that of the electromigration effect in metallization layers (see Section 4.1).

4.3.3 Metallization Concepts for Power SAW Structures

Due to increasing demands on power-durable SAW structures many efforts have been made to improve the power durability of the commonly used Al metallizations by different means, e.g. by alloying or by application of multilayer structures and dielectric coatings. Different materials have been reported in the literature to possess a higher power durability and reliability than pure aluminum. It is well known from microelectronics that adding one or more other elements to Al in order to obtain Al-based alloys (such as AlCu or AlCuSi) can result in strengthening and hardening effects [4.113]. Especially, AlCuSi alloy thin films have a higher resistance to electromigration in interconnects and also to stress migration [4.114]. By analogy with electromigration several Al alloy thin films and Al-based multilayer structures were tested in SAW devices technology for high power or/and high frequency applications. In order to study the effect of alloying elements on acoustomigration damage, finger electrodes of different sputtered or e-beam evaporated polycrystalline Al alloys on usual piezoelectric substrates (quartz, LiNbO$_3$, LiTaO$_3$, etc.) have been compared in the literature. The materials were usually characterized by their time-to-failure (TTF) behavior. The TTF indicates the lifetime of a SAW structure under accelerated load conditions due to its temperature and SAW stress. As a failure criterion an irreversible destruction of important electrical signal characteristics (e.g. a defined value of peak frequency shift of filter frequency curves) is usually applied. Some authors also determine an activation energy using the *Arrhenius* equation for the relation of the TTF and the sample temperature [4.115, 4,116], or the *Eyring* model [4.117], which is a very powerful model for acceleration tests. Relevant articles are related to binary systems such as AlCu [4.117–4.120], AlSi [4.119], AlCuSi [4.119], AlMg [4.118], AlTi [4.116, 4.120, 4.121], AlGe [4.121], and AlZn [4.121]. Beside bilayered Al/Ti [4.118, 4.122, 4.123] or Al/Mg metallizations [4.124] one or more other materials like Ti, Cu or Mg are used as interlayers in such aluminum-based multilayered thin film systems [4.125]. A titanium interlayer is also commonly used as an adhesive layer due to the piezoelectric substrate material. Binary or ternary alloy thin films used in multilayered structures have also been investigated. Metallizations such as AlCu/Cu/AlCu [4.124], AlScCu/Ti/AlScCu/Ti, as well as AlMgCu/Ti/AlMgCu/Ti and AlZrCu/Ti/AlZrCu/Ti [4.126] have a high resistance against the acoustomigration effect. Furthermore, amorphous AlY layers were deposited and characterized in terms of high-ower SAW applications [4.127]. Such AlY thin films deposited at room temperature (RT) using the e-beam evaporation technique show high strength and ductility, low density, and a relatively low electrical resistivity compared to other amorphous metal alloys. The results of *Yuhara et al.* [4.120], *Matsukura et al.* [4.128], and *Ebata et al.* [4.129] showed clearly the relevance of the thin film microstructure on SAW power durability. In this regard sputtered or highly textured films showed lower acoustomigration damage than e-beam evaporated metallization systems.

However, acoustomigration effects in different SAW thin film materials, including evaluation by their activation energy, have not been systematically investigated, but it seems that alloying additions to Al or Al-based multilayers may improve SAW power durability and reliability. The effect of solid solution formation and segregation of alloying elements into the grain boundaries results in both improved thermomechanical properties (increase in hardness and yield stress, suppression of stress relaxation effects) of the finger electrodes material and blocking of grain boundary and volume diffusion paths for the stress induced atomic transport. Concerning the relevance of grain boundaries for drift-diffusion processes, epitaxial film deposition forming single crystalline thin metal films would be also of interest for SAW device technology, but this concept is sensitive to substrate materials and thus is not of practical relevance for a low-cost process.

One reason for the improved power durability of Al alloys is the higher diffusion coefficient of the material added to Al compared with the self-diffusion coefficient of Al. Furthermore, in annealing processes after the formation of the layered film, favorable diffusion and alloying reactions such as the formation of intermetallic phases may also progress between Al and the metal used for the interlayer [4.115]. This can lead to a hardening of the multilayer system. Metallization systems based on Al–Ti have been frequently investigated for interfacial reactions in layered Al/Ti interconnects [4.123, 4.130].

As an example, Fig. 4.32 demonstrates the lateral element distribution over a crosssection of alternating Al/Ti metallization layers on a LiTaO$_3$ substrate [4.123]. The left micrograph shows a TEM bright field image of electron beam evaporated Ti and Al layers in the as-deposited state (crosssection by FIB preparation, at the top of the image the structurally amorphous Pt protection layer is visible). Columnar grains are formed in the layers. To analyze the elemental distribution an EDXS line scan was recorded along the bright arrow, that is across the metallization layers (right plot). The element line scans show that the individual layers are chemically well separated from each other. The Ta signal is caused by the LiTaO$_3$ substrate. The discussed Al/Ti multilayer structure is stable up to 400 °C.

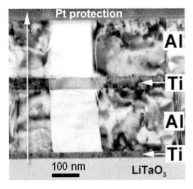
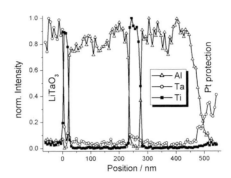

Figure 4.32: TEM bright field micrograph (left) and EDXS element distribution line scans (right, along the bright arrow) of alternating Al/Ti metallization structures for SAW applications [4.123]

During a heat treatment of 5 h at 450 °C the initial layered structure is destroyed and globular Al$_3$Ti grains are formed that can be regarded as embedded in an Al matrix. The formation of the thermodynamic stable intermetallic phase Al$_3$Ti was also observed in [4.131], consistent with the phase diagram of the Al–Ti system. However, the effect of Al$_3$Ti phase formation in terms of the acoustomigration behavior of such Al/Ti multilayers is not yet understood in detail. The effect of thermal treatment on Al/Ti multilayers can be different if deposited on other piezoelectric substrates than LiTaO$_3$, for example deposited on quartz or LiNbO$_3$ (Fig. 4.33). After annealing of an Al/Ti/Al/Ti layer stack in air at 450°C for 5 h, both Ti layers are completely dissolved in the case of quartz whereas the bottom Ti layer remains stable for LiNbO$_3$ substrates. As a result, an almost pure Al layer has been observed instead of the intermediate and, only in the case of quartz, also the adhesion of the Ti layer [4.132]. Therefore, the microstructure of the multilayer system on quartz substrate observed after annealing is similar to that observed for the LiTaO$_3$ substrate but is completely different from that for LiNbO$_3$.

The present tendency in microelectronics to replace Al by Cu as metallization because of its better physical properties seems to be also relevant for the SAW technology. That was why Cu-based thin film systems were considered as metallizations for high power durable SAW structures [4.118]. The significantly higher resistance of Cu metallizations to

a)

b)

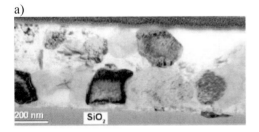

Figure 4.33: TEM bright-field images of the microstructure of Al/Ti/Al/Ti-layer stacks on (a) quartz, and (b) LiNbO$_3$ after annealing at 450°C for 5 h, original layer thicknesses: 22 nm for bottom adhesive Ti layer, 213 nm for the first Al layer, 41 nm for intermediate Ti layer, and 195 nm for the top Al layer [4.132]

acoustomigration in comparison with the Al/Ti metallization system will be demonstrated in detail in Section 4.3.5. However, due to the diffusivity and oxidation behavior of Cu thin films, the Cu technology for SAW metallization requires an effective thin diffusion barrier layer against diffusion of both copper or oxygen and elements from the piezoelectric substrate material (e.g. Si, Li, Nb, O, Ta). Furthermore, the diffusion barrier should guarantee a stable film adhesion to the metallization layer and to the substrate, it should possess high thermal stability and should remain adequately conductive. Ta-based barrier layers, particularly amorphous TaSiN thin films, have been considered for use as Cu diffusion barriers in the copper interconnect technology due to their lack of fast diffusion paths and high crystallization temperature [4.133].

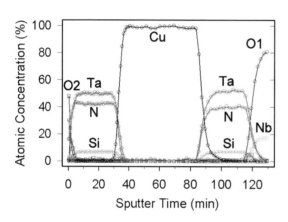

Figure 4.34: AES element concentration depth profiles of a TaSiN/Cu/TaSiN/LiNbO$_3$ crosssection after annealing (400 °C for 5 h) [4.135]

The properties of TaSiN barriers depend strongly on their chemical composition. *Cabral et al.* found the optimum TaSiN composition for low oxygen diffusion in the range Ta (20–25 at.%), Si (20–45 at.%), N (35–60 at.%) [4.134]. *Baunack et al.* investigated the interdiffusion behavior of the system TaSiN(50 nm)/Cu(150 nm)/TaSiN(50 nm) on LiNbO$_3$ by means of AES depth profiles [4.135]. The TaSiN layers were sputtered reactively from a Ta$_5$Si$_3$ target and possess the designated composition Ta$_{30}$Si$_{18}$N$_{52}$. Samples annealed for 5 h at 400 °C did not show any significant interdiffusion effect, as can be concluded from Fig. 4.34. The microstructure of the layered system remains unchanged even at 500 °C as the TEM crosssections in Fig. 4.35 demonstrate. Therefore, TaSiN is a suitable barrier to avoid Cu diffusion at high temperatures.

Another technological disadvantage for the use of copper films is their higher density, which is about 3.3 times greater than that of Al. Because of the mass-loading effect, variation in film thickness over the wafer diameter has to be kept very small, but as shown in Section 4.3.5 the significantly higher SAW power durability and reliability seem to suggest further developments in deposition techniques.

As in the case of aluminum, copper-based alloying is also supposed to lead to an enhancement of migration resistance. For CuSn alloy this was demonstrated concerning electromigration and stress migration [4.46]. The addition of Sn did not increase the intrinsic resistivity of copper significantly, while improving the microstructural stability to stress-thermal treatment. Preliminary examinations were also done with Ag as an alloying element [4.45, 4.136]. Such alloy materials have generally the smallest increase in resistivity of all copper alloys. Both the electrical and the mechanical properties of Cu–Ag systems were

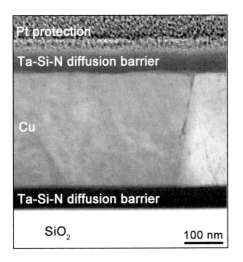

Figure 4.35: TEM crosssections of a TaSiN/Cu/TaSiN system on SiO$_2$ substrate after 5 h annealing at 500 °C

investigated. As first results, improved strength values of the alloyed layers could be reached at low resistivity (< 2.5 μΩ cm at Ag content < 5 wt.%).

A certain amount of the input power is commuted into device heating that increases its temperature and, consequently, influences the acoustomigration damage and the SAW velocity. Due to their higher thermal conductivity the Cu-based metallizations are advantageous to SAW applications, but temperature stability is also a critical issue for stable operation of SAW devices, especially in communication and sensor applications. *Wu et al.* used the negative temperature coefficient of delay (TCD) of SiO$_x$ as a thermal compensation layer in a multilayered ZnO/SiO$_2$/Si structure where the piezoelectric ZnO layer possesses a positive TCD [4.108]. Results of simulation using the transfer-matrix method showed that temperature compensation is achieved in the GHz range for a proper design of the multilayer system.

4.3.4 Experimental Setup and SAW Technology

Usually, acoustomigration tests are carried out using commercial filter devices where an irreversible change in acoustical specifications (e.g. a shift of the peak frequency of the resonance curve) is used as a device failure criterion [4.137, 4.138]. However, the device itself degrades during such a test and, consequently, electrical properties and stress loading of the metallization change. Special SAW structures are desirable that give the possibility to test metallizations independently of the influence of the exciting IDT on measurements [4.117, 4.118]. Metallizations that must be compared have to be held under constant or defined varied stress. Furthermore, power durability or lifetime experiments due to acoustomigration effects under direct microscopic observation are also advantageous. Therefore, in situ experiments are done in the optical or in the scanning electron microscope [4.117, 4.122, 4.139]. *Feuerbaum et al.* even described an experimental setup using a stroboscopic-controlled SEM for visualization of the SAW [4.140].

Power SAW Test Structure

In order to separate the exciting IDT from structures under investigation a test structure for power SAW loading was designed by *Schmidt et al.* [4.139]. This power SAW (PSAW) test structure allows a direct comparison of any metallization systems with the following advantages:

- separation of the SAW excitation and IDT measurement structures
- simultaneous measurements of resistance and signal behavior of two IDTs
- immediate comparability of the damage behavior of any metallization systems
- excitation of standing or, preferentially, traveling waves that results in homogeneous damage over the metallizations including test lines
- symmetrical input power dissipation that means in the case of traveling waves exactly the same acoustic power loading is observed for the two metallizations
- both metallizations have the same operating temperature
- accelerated measurements at elevated temperature or cooling are possible .

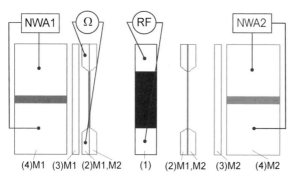

(4)M1   (3)M1 (2)M1,M2   (1)   (2)M1,M2   (3)M2   (4)M2

Figure 4.36: Scheme of the PSAW test structure layout, from center to outside: (1) the SAW transmitter (any metallization), (2) a pair of resistor lines (M1 and M2), (3) a fully metallized region (M1 or M2), and (4) a test IDT realized for M1 (left) and M2 (right) metallization, M1: metal 1, M2: metal 2, RF: input power, NWA: network analyzer

For the design concept shown in Fig. 4.36 a $\Lambda/4$-electrode source IDT for 130 MHz on a piezoelectric substrate placed in the center of the device was used as the acoustic source, while the test transducers at each side of the device are designed for a wavelength half of the first one. Because of its large wavelength the source IDT is rather robust in its electric signal behavior. Concerning the bidirectionality of the device both test transducers will be loaded with the same acoustic level of 50 % of the total acoustic power, allowing the simultaneous investigation of two different metallizations under comparable SAW stress loading. Both test transducers are realized with a large number of finger pairs for sensitive detection of even small changes in their acoustic properties using admittance measurements. The test transducers are almost nonreflecting at the excitation frequency, but more sensitive to changes in the metallization properties than the source IDT due to their smaller wavelength and higher number of electrodes.

When the acoustic power is dissipated at both sides of the device using a damping mass the SAW propagation is almost without reflections forming a traveling wave (Fig. 4.37a). Otherwise, the propagating SAW will be reflected, e.g. at sample edges forming a standing wave (Fig. 4.37b), which shows acoustomigration damage as a striped pattern. The advantage of traveling SAW is the homogeneous damage across the aperture of 1500 μm that is determined by the overlapping range of the finger electrodes of the source IDT.

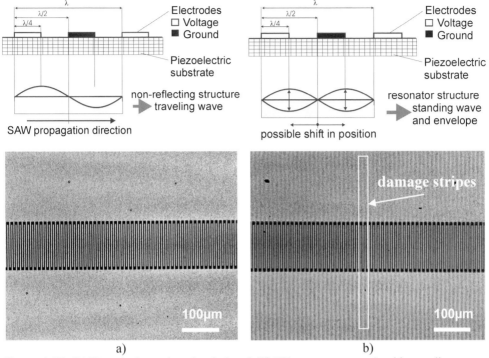

Figure 4.37: SAW excitation using the designed PSAW test structure; a) with traveling wave, b) with standing wave

Traveling wave excitation also allows one to measure the resistivity variation of small-resistance lines (see Fig. 4.36) due to thermal effects and the acoustomigration damage, respectively. However standing waves can easily generate large mechanical stress in a low power range and at relatively low sample temperature. However, stress fields using standing waves are strong and inhomogeneous, resulting in a spatial damage profile.

SAW Thin Film Technology

In addition to high SAW power durability and reliability the required metallization thin films for SAW structures have to satisfy certain requirements:
- adaptability to the most important deposition methods for microelectronics
- low electrical resistivity
- good adhesion on polished piezoelectric substrates (quartz, LiNbO$_3$, LiTaO$_3$, etc.)
- high flatness at the metal surface and high finger edge quality after patterning
- compatibility with usual bonding or flip-chip techniques
- easy patterning in chemicals not reactive with photoresists
- dry etching or lift-off structuring
- high quality (e.g. small variation in film thickness, long-time stability of its physical properties of ohmic contacts, e.g. by bonding, high reproducibility)
- no diffusion effects or chemical reactions between the metallization layer and the ambience or the substrate material

Metallizations are vacuum deposited with a resistivity below 3 $\mu\Omega$ cm, this is comparable to that of the bulk material (2.4 $\mu\Omega$ cm). In SAW technology the Al thin film systems are deposited preferably by the e-beam evaporation technique. In comparison to sputtering the evaporation process is more efficient. Furthermore, e-beam films show a better homogeneity in film thickness (e.g. better than 1% for a 4 in wafer). Concerning the relatively high energy of evaporated particles such films also show lower side-wall deposition that is necessary for the usual high-quality lift-off patterning.

In developing the SAW copper technology both the TaSiN barrier and the Cu thin film were deposited by magnetron sputtering in a cluster tool without a vacuum break. Such an in situ technique prevents oxide formation at the interfaces between the copper and the barrier layers. For structuring of the Cu layer systems an optimized lift-off patterning technique as for Al was used, achieving high-quality patterns in the μm range up to a total film thickness of about 350 nm. As seen from cross-sectional geometry the barrier layers also cover the side walls of Cu-finger electrodes preventing their oxidation completely.

Another advanced SAW technology was firstly proposed by *Menzel et al.* who fabricated the Cu finger electrodes into trenches of the piezoelectric substrate material (Fig. 4.38) [4.141].

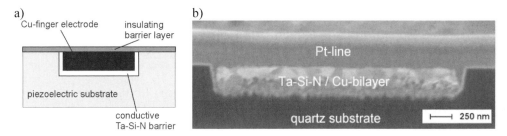

Figure 4.38: a) Schematic cross section and b) FIB cut (tilt 45°) of a copper damascene SAW structure, from [4.141]

Using such a so-called damascene structure some new features are expected in comparison with conventional finger electrodes regarding power durability, technological compatibility to microelectronics, pattern quality especially in the sub-μm range, as well as SAW propagation properties. Consequently, the SAW load cannot effect fatal failures by short circuits between adjacent fingers, which is important for high power density load of SAW devices operating in the frequency range above 1 GHz. The trench structuring for the damascene technology can be carried out by dry etching (ion beam or reactive ion etching) or wet etching techniques. Thereafter, these trenches are filled with a metallization, e.g. with a magnetron sputtered copper system, which may consist of a 20 nm thick conductive Ta-Si-N layer at the bottom, a 100 – 250 nm Cu film, and a 20 nm thick Ta-Si-N,O bilayer as an insulating covering layer. After deposition, the fingers are structured by a chemical-mechanical polishing (CMP) process. Such a metallization system enables sufficient bonding properties using Al wires that were bonded to the embedded copper pads through the covering bilayer.

Accelerated Power Durability and Life Tests

Failure of SAW structures due to acoustomigration can be determined by power durability or life tests under accelerated conditions showing how time-to-failure (TTF) at a particular operating stress level can be used to predict the equivalent TTF at different operating stress levels. Such measurements are carried out using complex experimental equipment as shown in Fig. 4.39.

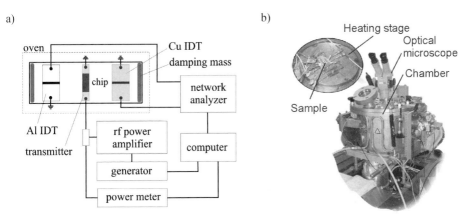

Figure 4.39: Typical experimental setup for lifetime measurement with the appropriate electrical block diagram [4.142]

Acceleration means that there must be the same causes of failure as under normal stress, and only the time scale is different. At given rf power the TTF or, in the case of more statistical testing, the median time-to-failure (MTF) can be determined from an equation similar to Eq. (4.5) in Section 4.1.2

$$TTF = A\exp\{E_a / kT\} \tag{4.8}$$

where parameter A represents the influence of the thin film properties (microstructure, finger geometry, etc.) and the substrate material, k is Boltzmann's constant, T is the temperature, and E_a is an activation energy for initiating the dominant failure mechanism given under test conditions. Note that TTF instead of MTF enunciates a smaller number of samples for durability or life tests, which means that TTF does not represent a statistical investigation exactly. A more general and powerful model for acceleration tests is known as the *Eyring* model based on quantum mechanics and chemistry [4.117]. This model describes how the TTF varies not only with temperature but also with relevant stresses. Therefore the *Eyring* model is also used in a modified manner to describe the electromigration as a significant interconnect failure mechanism (see Section 4.1). However, for lifetime experiments due to acoustomigration a simple formalism based on the *Eyring* model is not yet derived in the literature.

4.3.5   Acoustomigration Experiments

Results on Al-Based Metallizations

Since damage of the SAW finger electrodes becomes important for the lifetime of power SAW structures investigations have been made to determine the amount of damage in different metallization systems. The basic objectives of such experiments on SAW structures are:

1. Loading tests determining limits for operation (temperature, input power, frequency) including lifetime measurements.
2. Study of the mechanism of the mass transport due to acoustomigration.
3. Development and characterization of new high performance metallization systems.
4. Development and optimization of the deposition and structuring technology.

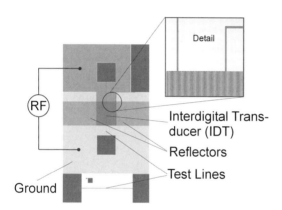

Figure 4.40: Scheme of the SAW layout of a one-port resonator structure for 430 MHz, width of electrodes is 1.81 μm (= $\lambda/4$)

A literature overview is given in Section 4.3.3. Further investigations on Al-based metallizations on quartz, LiNbO$_3$, or LiTaO$_3$ were described by *Menzel et al.* [4.118, 4.122] and *Schmidt et al.* [4.139, 4.143] using optical or electron microscopy. During the SEM study of a power loaded one-port resonator structure (schematically shown in Fig. 4.40) the admittance was simultaneously measured during imaging of the Al-IDT. It could be observed that the damaging intensity varies over the IDT depending on the SAW envelope properties.

However, the local damage rate is a function of the metallization material and also the adhesive titanium interlayer (Fig. 4.41). After loading with 520 mW input power for only 5 min AlMg finger electrodes (without an adhesive layer) showed both void and hillock formation and also delamination damage. Voids and hillocks could almost be avoided by Cu alloying of Al, whereas delamination remained, although to a lesser extent (loading time 20 min). Therefore delamination must be avoided by improved adhesion between the finger electrodes and the quartz substrate using interlayers. For a 9 nm Ti interlayer between quartz and Al the result is shown in the lower micrograph of Fig. 4.41. No delamination can be observed even after 20 min loading with 520 mW. However, in the Al/Ti finger electrodes void and hillock formation still takes place. In addition to a reversible thermal shift in frequency of the measured admittance curve of the resonator due to the self-heating effect, a significant signal destruction caused by a nonthermal frequency shift, variation of the curve shape, and a decrease in the amplitude of the curve were observed (Fig. 4.42). These effects were almost irreversible when the power was switched off.

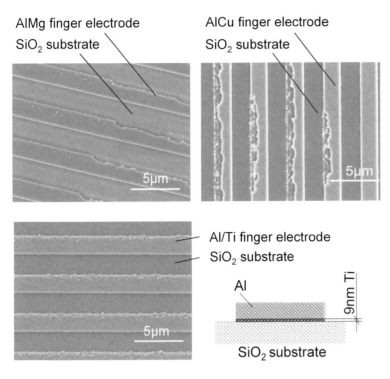

Figure 4.41: SEM micrographs of damaged finger electrode metallizations on SiO_2 substrate of a SAW resonator structure after loading with 520 mW

The measured electrical damage could be correlated with the microstructural damage of the finger electrodes. All kinds of investigated Al-based metallizations showed more or less strong damage.

Microstructural effects of acoustomigration damage could be seen in more detail using transmission electron microscopy. As an example, an AlMg-alloy finger electrode on quartz after power loading at 1.3 W for 71 min is shown in Fig. 4.43. The TEM lamella was prepared at the side of the IDT by the FIB technique. Because of the local confined high stress gradient the left side of the finger electrode is almost unaffected by acoustomigration whereas its right side is strongly damaged. In the marked circle the formation of materials accumulation, a so-called hillock, is clearly visible. Microstructural changes destroy the finger geometry and its physical properties and, as a result, the electrical characteristics of the resonator (see signal behavior in Fig. 4.42).

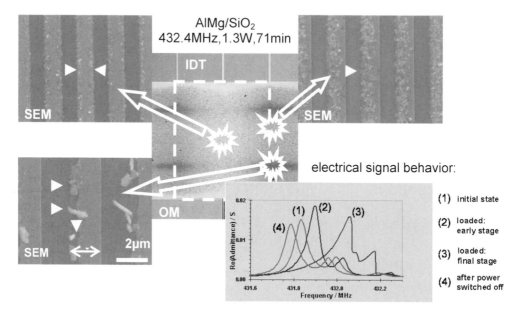

Figure 4.42: In situ SEM investigation of damages of AlMg alloy electrodes at different positions of the IDT, one port resonator structure on quartz, after loading with 1.3 W for 71 minutes. Measured admittance curves in the lower right diagram, damages are indicated by marks

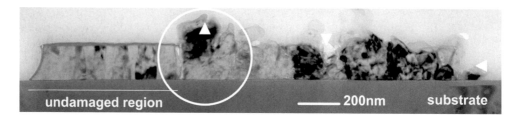

Figure 4.43: TEM micrograph of a damaged AlMg alloy resonator finger electrode on quartz after loading, marked details of microstructure are strongly damaged (TECNAI 30, 300 kV)

Comparison of Al- and Cu-Based Metallizations

High performance copper-based metallizations for SAW applications were first investigated by *Menzel et al.* [4.118]. In that work a test structure with unidirectional IDTs ensuring high SAW amplitudes at given input power was used.

Beside transmitting IDTs, receivers have been incorporated in the test structure for monitoring the electric behavior during microscopic observation. The first results from a comparative study of Al- and Cu-based metallizations did clearly indicate the high acoustomigration resistance of copper films.

A considerably revised PSAW test structure as shown in Fig. 4.36 allowed much better test conditions of two different metallizations with traveling or standing waves. Some experiments were carried out under microscopic observation using either optical or scanning electron microscopy (see Section 3.2.3).

At first, Ta/Cu/Ti-finger electrodes were tested in comparison with Al/Ti bilayer electrodes characterizing the dependence of electrode damage on applied SAW power or on time. These studies were supplemented by simultaneous electrical measurements such as the frequency shift and variation in film resistivity during the migration experiments as well as numerical analysis of the SAW stress field. Firstly, the test devices were altered by thermal treatment (130 °C, air, 1 h) without application of any acoustic power. This procedure was similar to the heat treatment that is necessary, e.g., for conditioning of the damping mass. Measurements of the electrical impedance as a function of temperature allowed the determination of the linear temperature coefficient of resistance (TCR) for the Al line ($TCR_{Al} = 3.7 \times 10^{-3}\ K^{-1}$) and the Cu line ($TCR_{Cu} = 3.2 \times 10^{-3}\ K^{-1}$).

a)

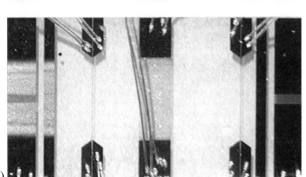

b)

Figure 4.44: Sequence of optical micrographs of a PSAW test structure on 128° XY LiNbO$_3$ during high power SAW loading, a) initial state, b) after 1 h loading time; damaged Al/Ti structures (turn from dark to bright) on the left, unaffected Ta/Cu/Ti structures on the right side; transport of small particles towards the transmitter center due to large SAW amplitudes, aperture width 1.5 mm [4.139]

During the first temperature cycle a variation in resistivity of both the aluminum and the copper thin film was observed caused by microstructural effects [4.144, 4.145]. Unlike Al samples that were not markedly influenced by heating to 130 °C the resistivity of the Cu lines decreased by $\approx 8\%$. This can be caused by different relaxation effects in sputtered copper films [4.145].

High power loading experiments with traveling waves were performed at the PSAW test structure. Preannealed samples (130 °C, 1 h, air) were loaded at RT for 1 h, applying an acoustic power of 7 W that corresponds to a loading of 3.5 W for each test IDT. The sample temperature was kept constant below 30 °C by water cooling. Significant changes in the optical properties of the Al/Ti metallization were observed in the aperture region because of increasing surface roughness. By contrast the Cu metallization system remained optically unaffected (Fig. 4.44).

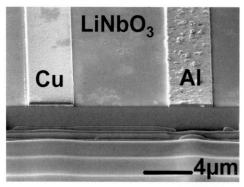

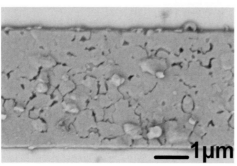

Figure 4.45: SEM micrographs of SAW test structures on LiNbO$_3$ after high power loading. Top image: FIB overview, below: Al (left) and Cu (right, unaffected)

For that experiment a layered Ta(50 nm)/Cu(250 nm)/Ti(50 nm) multilayer structure was used. This result is confirmed by FIB and SEM investigations on the metallizations (Fig. 4.45): The microstructure of the copper thin film system looks undamaged, whereas the Al/Ti line shows significant damage by void and hillock formation as well as a strong grooving effect. Corresponding to such visible damage the measured real part of the admittance of the Al/Ti-IDT shifts irreversibly towards lower frequency (Fig. 4.46).

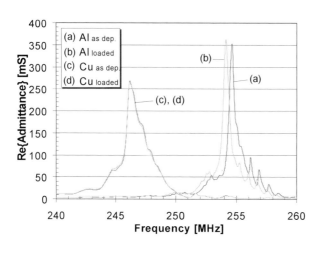

Figure 4.46: Real part of measured electrical admittance of the Al/Ti- and the Ta/Cu/Ti-IDT of the PSAW test structure before (as deposited) and after loading experiment at input power of 7 W for 1 h (loaded)

This effect is similar to the corresponding measurements on Al alloy resonator structures in Fig. 4.42. On the other hand, the frequency curve of the Cu-IDT remains constant. Al-structures are homogeneously damaged as also shown in Fig. 4.37a because the traveling wave was excited. More details in terms of microstructural damage are visible at higher magnifications using a long distance objective lens. Figure 4.47 shows the high power loaded SAW test structure using the described Al(350 nm)/Ti(9 nm) layer system on the left side (AL-IDT) and a TaSiN(50 nm)/Cu(250 nm)/TaSiN(50 nm) on the right side (Cu-IDT). The Cu-IDT was unaffected up to more than 4 W (input power 8 W) while the AL-IDT was strongly damaged by void and hillock formation at much lower SAW power. Loading of preannealed samples at elevated temperature results in a maximum sample temperature increase of 70 K measured at steady state by a thermocouple at the sample holder. According to this the acoustomigration effect was strongly enlarged at the Al lines and their electrical resistance increased more strongly than by a thermal treatment without power loading. After cooling an enhanced resistance of about 117% of the initial value was reached. Contrary to Al, resistivity of the Cu lines showed only an increase according to the TCR effect, and it decreased to the starting value after cooling to RT.

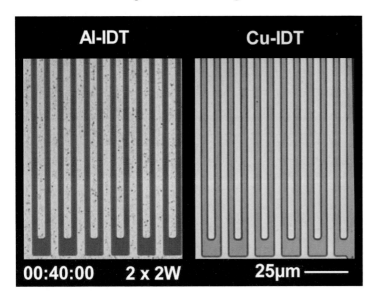

Figure 4.47: Power SAW test structure after power loading with traveling waves, input power 4 W, 40 min; left: damaged Al/Ti-IDT, right: TaSiN/Cu/TaSiN-IDT, unaffected (optical micrographs)

Similar results were observed for standing wave excitation. In Fig. 4.48 two damage stages of a systematic power durability study were selected, demonstrating the Al/Ti layer damage in comparison with the TaSiN/Cu/TaSiN layer system. The first stage of acoustomigration damage is observable at the relatively low power of 0.5 W after about 4 h. The damage increases with power and time, resulting in a larger area of damage that, in the final stage, is similar to that observed for traveling waves (see Fig. 4.47). However, for the standing wave the damage intensity is much stronger because of higher amplitudes in mechanical stresses. Void and hillock formation spreading over a large area is observed. But even for a loading of 3 W for more than 11 h the Cu-IDT will remain unaffected in terms of acoustomigration effects.

Some results of irreversible peak frequency shift as well as change in resistance estimated from frequency curves as shown in Fig. 4.46 are summarized in Fig. 4.49.

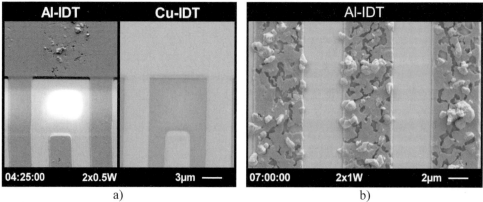

a)   b)

Figure 4.48: PSAW test structure after power loading with standing waves, a) input power 1 W, after 4 h 25 min, b) input power 2 W, after 7 h; Al-IDT: Al(350 nm)/Ti(9 nm), Cu-IDT: TaSiN (50 nm)/Cu (250 nm)/TaSiN (50 nm), SEM/SE micrographs

a)

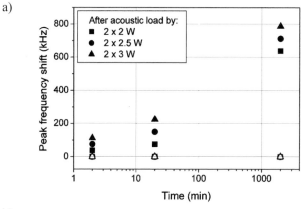

b)

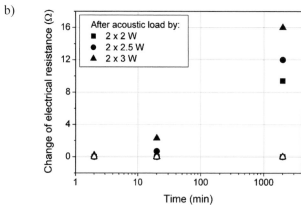

Figure 4.49: Peak frequency shift (a) and change of resistance (b) of Al/Ti (solid symbols) and Ta-Si-N/Cu IDT (empty symbols) during acoustic loading for different times, after *Pekarcikova et al.* [4.142]

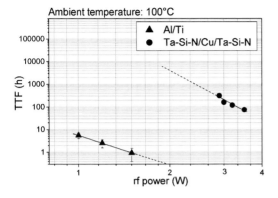

Figure 4.50: Time-to-failure (TTF) of loaded PSAW test structure at elevated temperature of 100°C, failure criterion Δf=300 kHz [4.142]

For the Al-test IDT a maximum shift of 800 kHz (at an rf input of 3W) was observed, whereas no change took place for the Cu IDT. Results from time-dependent measurement of electrical resistances during loading indicate a strong increase by about 28 % that correlates with the damage of the microstructure and, consequently, the change of elastic properties of the Al/Ti metallization. In contrast, at this load the Cu resistor lines are affected only by heating (TCR effect) at the beginning of loading. Figure 4.50 shows the results of estimated TTF applying a failure criterion of $\Delta f = 300$ kHz. Note that the TTF of the Ta-Si-N/Cu system is more than three orders of magnitude higher in comparison with the Al/Ti system.

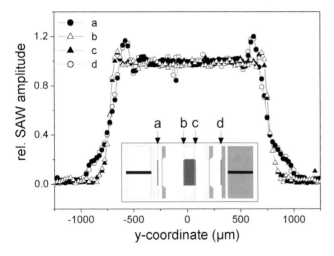

Figure 4.51: Measured SAW field on different positions (a–d) of the power SAW test structure as shown in Fig. 4.36, measured perpenticular to the SAW propagation direction [4.146]

Generally, the Δf_0 and ΔR rise with increasing loading time or input power, but for Cu-based metallization all samples remained unaffected up to 3.5 W. During these experiments the calculated metallization temperature was below 150 °C determined from TCR measurements of the Cu metallization (TCR= 2.8×10^{-3} K^{-1}). This temperature was much higher than the sample holder, temperature due to the heating effect by SAW power. From TTF measurements an activation energy of 0.4 eV to 0.6 eV was estimated for Al/Ti confirming the relevance of transport processes via interphases (surface, Al–Ti interphase, grain boundaries). By optical and scanning electron microscopy it was observed during these experiments that the damage was not homogeneously distributed across the sample surface, especially at extremely high power loads. Imaging the SAW field by a laser optical method it could be shown in [4.146], that such an effect is caused by a more inhomogeneous wave field at high rf power exemplarily shown in Fig. 4.51.

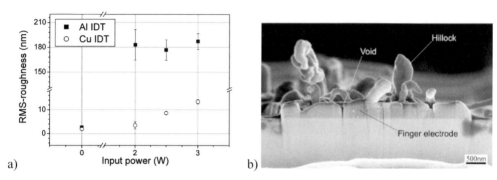

a)   b)

Figure 4.52: RMS-roughness vs. input power of Al/Ti and Ta-Si-N / Cu metallizations after acoustic loading for 2000 min (note: scale enlarged for Al), right picture: SEM/SE image after FIB crosssection, 90° tilted sample of an Al finger electrode after load at 3 W for 2000 min [4.147]

The surface roughness of surface areas of uniform size (10 × 10 μm) of loaded Al- and Cu-IDTs was studied systematically using atomic force microscopy [4.147]. It increases dramatically with time and rf power for Al/Ti in comparison with the as deposited state during load experiments as demonstrated in Fig. 4.52. The root mean square (RMS) roughness for the initial state of both metallizations is comparable (2 nm to 2.5 nm). This lies only about 1 nm above the surface roughness of a polished LiNbO₃ wafer. After loading for 2000 min the Cu metallization shows only a small increase of the roughness with increasing input power, but the Al metallization was significantly changed by about 180 nm even for a small load. Here, the hillocks rise almost perpendicular from some 10 nm up to more than 1000 nm. As a consequence, a quantitative discussion of material transport as a function of the hillock density as done by [4.148, 4.149] may result in a discrepancy at high hillock extrusion length. However, the microstructural damage of the Cu metallization system was so tiny that the electric behavior of the Cu-IDTs did not change.

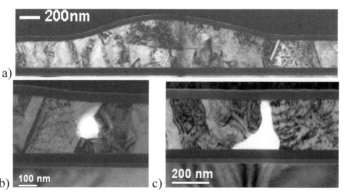

a)

b)   c)

Figure 4.53: TEM cross-sectional micrographs of damaged Ta-Si-N/Cu metallization, a) hillock, b) and c) voids at grain boundaries generated in the middle (b) of the copper film, and at lower interface to the barrier (c) as a slit void formation [4.132]

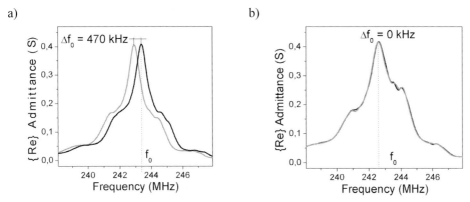

Figure 4.54: Comparison of the frequency curves of Cu metallization on 128°LiNbO₃ a) without and b) with covering Ta-Si-N (50 nm) layer after SAW load of 3 W for 2000 min [4.142]

The formation and growth of hillocks and voids with growing time or rf power can be observed preferentially at grain boundaries or grain boundary triple junctions. One of the reasons for observing only flat bumps and suppressed voiding is the complete covering of the copper film with an amorphous Ta-Si-N diffusion barrier. In Fig. 4.53 some typical TEM images of damaged Cu metallization demonstrate that the upper barrier layer remained undamaged. Hillocks in the Cu metallization have a more rounded form without sharp edges in comparison to Al films (see Fig. 4.43). In addition, such a Cu-barrier interface has a lower diffusivity (or a higher activation energy) for mass transport than the "free" copper film surface, and it prevents chemical reactions with ambient. This means that covering layers with a sufficient thickness as well as mechanical strength can decrease the frequency shift (see Fig. 4.54) and, therefore, enhance the power durability significantly by the fact that they withstand the complete stress state (intrinsic stress and SAW load stress). One can also deduce from SEM/EBSD investigation on Al/Ti that microstructural damage and material transport are correlated to grain boundaries of high misorientation angles (Fig. 4.55). In the grey step classification these are black as shown in Fig. 4.55.

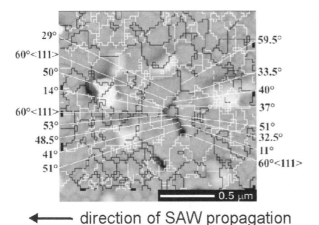

Figure 4.55: SEM micrograph and high resolution EBSD mapping of an SAW loaded Al/Ti metallization on a 128°LN-substrate [4.142]

← direction of SAW propagation

Results on Embedded Cu Fingers

Dependence of the TTF on temperature and stress means the acoustomigration mechanism is influenced not only by the SAW stress but also by geometry influencing the temperature of the finger material. In Fig. 4.56 an example of first results of radiation impedance measurement on an embedded Cu-test IDT using a quartz substrate is shown where a sharp peak at about 207 MHz can be detected [4.141]. The relatively strong peak resistance is a result of internal reflections inside the transducer. A similar study was carried out for a 128°XY LiNbO$_3$ substrate showing a much higher peak value [4.148]. In LiNbO$_3$ substrates the copper damascene structures were realized in good quality, too.

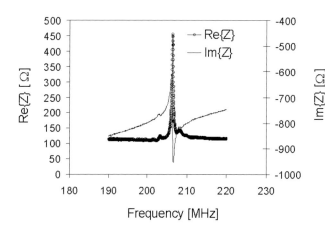

Figure 4.56: Radiation impedance of an embedded IDT on quartz substrate using copper electrodes as a function of frequency (real part of impedance, Re{Z}, denotes radiation resistance; imaginary part, Im{Z} denotes radiation reactance), from [4.141]

Modeling of SAW Fields

The diversity of SAW features (see Section 5.2) implies detailed modeling of ac stress fields for better understanding of the acoustomigration process. This can be done in the framework of numerical studies based on the equations of motion and the constitutive equations with regard to elastic, dielectric, and piezoelectric material parameters as well as mechanical and electrical boundary conditions at interfaces and surface. The state-of-the-art is characterized by the simulation of SAW excitation and propagation in metallization gratings on piezoelectric substrates using FEM/BEM methods. However, for many purposes an approach considering the wave propagation in a continuously metallized substrate is fully informative using a partial wave analysis [4.118].

Such an analysis was applied in order to understand both the damage patterns of Fig. 4.42 and SAW stress distribution across the metallization layer stacks. As already mentioned, the calculated distribution of ac stress components and their local position with respect to the electrodes shows a high intensity of the compressive/tensile stress component σ_{11} in the wave propagation direction and of the component σ_{22} in the finger direction. The shear component σ_{12} is 90° shifted in phase, but has only a small amplitude. Considering σ_{11} and σ_{22} as responsible for the acoustomigration effect the damage of finger edges seems reasonable. The asymmetric hillock formation at only one finger edge outside the center of the resonator is caused by the fact that the resonance frequency is slightly different from the synchronous frequency of the grating structure, as discussed above. The stress amplitudes correspond to an input power of 1 W per 1 mm SAW beam width at 100 MHz. It would be hazardous to estimate a damage threshold for the ac stress in the face of the number of loading cycles of the order of 10^{12} that is beyond the firm knowledge of high cycle fatigue [4.111, 4.112]. However, dislocation effects could influence the material transport due to acoustomigration as discussed by *Kubat* [4.149] or *Eberl* [4.150].

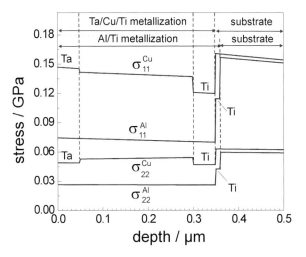

Figure 4.57: Depth profiles of dominant stress components in the Ta(50 nm)/Cu(250 nm)/Ti(50 nm) system and in the Al(350 nm)/Ti(9 nm) layer system on 128° XY LiNbO$_3$ substrate for 127 MHz, 3.5 W / 1.5 mm at both sides of the device

The SAW-induced stress components in the Al/Ti and the Ta/Cu/Ti metallizations applied to the PSAW test structure were calculated by means of a partial wave analysis. The stress magnitudes in both metallizations are of the same order (Fig. 4.57) but in the copper system the value is about twice that of the aluminum system. In both systems the character of the stress figure is also comparable. This means that a dominant σ_{11} (in the SAW propagation direction) component and a notable σ_{22} (in the surface plane, perpendicular to the SAW propagation direction) component exist, whereas all other stress components are negligible. Nevertheless, the SAW-induced stress values are very small compared to those of the classical fatigue theory.

4.4 Thermal Stability of Magnetoresistive Layer Stacks

The temperature plays a crucial role in both the manufacturing and operation of microelectronic devices. During the manufacturing processes in silicon-based technology, oxidation or passivation steps may require a temperature as high as 350 °C. In addition, applications in automation or automotive engineering do often require long-term operation at more than 100 °C. The development of devices with a high tolerance against elevated temperatures is therefore both a technological challenge and an economic opportunity.

A similar situation is encountered in spintronics. The complete magnetic layer stacks are usually deposited at room temperature. Elevated temperatures are applied mainly in the post-growth processing in several circumstances. Firstly, depending on the specific application, a reference magnetization must be defined using the exchange biasing effect. For this purpose, the layer stack is heated above the blocking temperature T_B of a pinning layer and subsequently field-cooled. Secondly, some materials employed as pinning layers (NiMn or PtMn) are deposited in a nonmagnetic structural modification, which must be transformed into the proper antiferromagnetic phase by extended annealing at $T \geq 250\,°C$. Thirdly, following the present scenario of introducing spintronics functionality onto the market, magnetic components will likely be integrated into silicon-based devices. The magnetic layer stacks must thus survive the temperatures involved in standard silicon technology back-end processing.

The temperature may affect a magnetic thin film system in several respects. First, the magnetic ordering has an intrinsic temperature dependence characterized by the Curie and Néel temperatures, T_C and T_N, for ferro- and antiferromagnets, respectively. In addition, magnetic coupling phenomena, such as magnetic anisotropies or interlayer coupling may change strength and even sign with temperature. In all of these cases, the influence of the temperature can be considered reversible, i.e., at a given temperature we find the same magnetic behavior. However, an increase in temperature may also lead to irreversible changes in the magnetic behavior which are caused by changes in the structure, morphology and chemistry of the films during the heat treatment (annealing). This situation is much more difficult to analyze and will be the topic of the present section.

4.4.1 Metallic Multilayers as GMR Model Systems

Considering a typical spin valve design (cf. Section 2.4), the individual layer thicknesses are mostly in the nanometer range and most of the functionality is related to the interfaces. This is particularly true for the metallic layers carrying the giant magnetoresistance (GMR) effect, for example, in Co/Cu-based film stacks, or the insulating barrier in tunneling magneto-resistance (TMR) systems. Therefore, a considerable amount of work has been devoted to understanding thermally induced deterioration effects in GMR multilayers and complex magnetoelectronic layers.

Principal materials used in GMR layer stacks are 3d transition metal compounds and elements, such as cobalt, copper, or permalloy – a nickel–iron ($Ni_{80}Fe_{20}$) alloy. The minimum system showing giant magnetoresistance is formed by a trilayer stack: two ferromagnetic layers of the same type (Co or $Ni_{80}Fe_{20}$) separated by a nonmagnetic spacer layer (Cu). In fact, both material combinations are of high technological interest. Co/Cu multilayers provide GMR values in excess of $\Delta\rho/\rho = 50\%$ [4.151, 4.152], whereas the $Ni_{80}Fe_{20}$/Cu system shows small saturation fields even in the first antiferromagnetic (afm) coupling maximum [4.153–4.156].

More complex spin valve layer stacks, of course, contain a variety of other materials, such as NiO or Mn-based antiferromagnetic layers for exchange biasing purposes, oxide layers to increase the specular reflectivity, or Co/Ru/Co trilayers as artificial antiferromagnets. A similar situation is found in magnetic tunneling contacts where Al_2O_3 or MgO are employed as tunneling barrier materials. A concise treatment of thermally induced degradation effects in these systems requires the consideration of all possible material combinations and the respective interfaces. This will by far exceed the framework of this contribution. In the following, we will therefore concentrate on the most prominent materials in GMR systems, such as cobalt, copper and permalloy.

With respect to device applications, the choice of the GMR system will also be dictated by thermal stability aspects. It is thus very instructive to compare the material combinations Co/Cu and $Ni_{80}Fe_{20}$/Cu with respect to thermally induced degradation of the magnetic and magnetotransport properties, as one of them is immiscible (Co/Cu), whereas the other one contains miscible constituents (Cu–Ni) [4.157]. In order to adress the role of the interfaces in the degradation process it is very useful to investigate systems with multiple equivalent interfaces. This situation is best realized in multilayer structures. The multilayer samples discussed in the following were prepared by a turntable DC sputtering technique in order to obtain high-quality structures with reproducible layer thicknesses throughout the multilayer. They usually consisted of 40 periods Co/Cu and $Ni_{80}Fe_{20}$/Cu, respectively. These films were annealed at different temperatures in vacuum in order to avoid oxidation. All transport and structural characterization was carried out at room temperature.

4.4.2   Co/Cu Multilayers

The Co/Cu multilayer system in the as-deposited state yields magnetoresistance values of $\Delta\rho/\rho \approx 50\%$ in the first, and $\Delta\rho/\rho \approx 25\%$ in the second afm coupling maximum. These GMR values refer to a current-in-plane (CIP) geometry. It should also be noted that the maximum GMR values are obtained for a well-defined combination of Co layer and Cu interlayer thicknesses (for further details see Section 4.5). This is due to the quantum well origin of the interlayer exchange coupling.

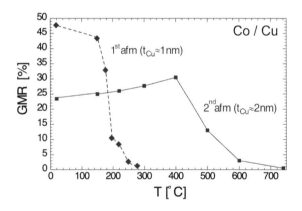

Figure 4.58: Variation of the GMR in Co/Cu multilayers with the annealing temperature. The thermal stability is strongly determined by the Cu inter-layer thickness

In the annealing experiments the samples were kept at each chosen temperature for 1h then cooled and measured at room temperature. The results show a strong influence of the Cu interlayer thickness (Fig. 4.58). In the case of samples with 1 nm thick Cu interlayers the GMR signal already degrades significantly for annealing temperatures below $T_{an} = 200$ °C. For samples with 2 nm thick interlayers, however, the GMR even increases with the annealing temperature up to 400 °C. Only above this critical value does the GMR degrade slowly and finally disappear around $T_{an} \approx 750$ °C.

The value of the giant magnetoresistance is determined by two factors. Firstly, the interlayer coupling stabilizes the antiparallel alignment of the magnetization in the magnetic layers in the field-free ground state. If the coupling is reduced, the antiparallel ground state may be destroyed, resulting in a much reduced GMR signal. Secondly, the spin dependent transport behavior is strongly governed by the quality of the Co/Cu interfaces. A thermally induced degradation of the interfaces will therefore also directly affect the spin dependent transport and thus the GMR signal. In order to clarify the microscopic mechanisms underlying the GMR degradation extensive magnetic and structural characterizations must be carried out.

Electrical Resistivity

Some insight can be gained already from the variation of the overall resistivity of the annealed samples. Typical resistances measured in zero magnetic field (ρ_{max}) and at magnetic saturation (ρ_{sat}) continuously decrease with increasing annealing temperature (Fig. 4.59). The decrease indicates that the Co and Cu layers do not intermix – a finding consistent with the immiscibility of these two materials predicted by the bulk phase diagram. On the contrary, the layer system should be stable up to high temperatures, which was later confirmed by the XRD and TEM investigations [4.158, 4.159]. These studies also revealed that the resistance decrease is rather connected to a grain growth, i.e., a reduction in the grain boundary density.

The GMR is conventionally given as a quantity normalized to the resistivity at saturation field, ρ_{sat}. Therefore, any reduction of ρ_{sat} will be directly reflected in an increase in the GMR value. In fact, this explains most of the GMR increase up to 400 °C annealing temperature in the Co/Cu system (cf. Fig. 4.58).

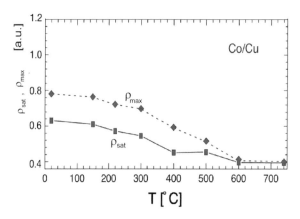

Figure 4.59: Change in the room temperature resistivity in Co/Cu multilayers with 2 nm Cu interlayers (2. afm coupling maximum) upon annealing. Both the values at zero magnetic field (ρ_{max}) and for magnetic saturation (ρ_{sat}) are compared, the difference being directly related to the spin-dependent scattering contribution

In order to see the influence of the spin-dependent contribution to the charge transport, the magnetically induced resistance change $\Delta\rho = \rho_{max} - \rho_{sat}$ must be considered separately. In the above case $\Delta\rho$ shows only a minute increase at low annealing temperatures and remains almost constant up to 400 °C, indicating only marginal changes in the spin-dependent scattering behavior. These may be associated with a change in the interfacial roughness caused by interfacial inter- or demixing. A demixing process has indeed be observed in careful depth profiling investigations on multilayers with larger individual layer thicknesses [4.160]. The major breakdown of the GMR at high annealing temperatures, however, must be due to a different type of structural rearrangement.

Magnetic Measurements

The magnetic properties of the multilayers, as seen by magnetooptical *Kerr* effect (MOKE) and ferromagnetic resonance (FMR) techniques, also exhibit characteristic changes with the annealing temperature. The hysteresis loops of the as-grown Co/Cu multilayers show a typical hard axis shape for both the first and second afm maximum situations and thus clearly reflect the antiferromagnetically coupled ground state (Fig. 4.60). Upon annealing at increasing temperatures the hysteresis loops gradually change, and reveal a quite different behaviour for the two coupling maxima.

For the multilayer samples with 1 nm thick Cu interlayers (1st afm maximum) the as-grown films exhibit a small hysteresis at low fields which is also visible in the GMR curves. Upon annealing at 200 °C the hysteresis becomes more pronounced and develops sharp transitions. These loop shapes are consistent with the formation of a sizable ferro-magnetically coupled contribution in the multilayer. Presumably, this contribution is due to a local breakdown of the antiferromagnetic coupling in the vicinity of pinholes and defects in the layer structure. These ferromagnetically coupled regions no longer contribute to the giant magnetoresistance, thereby causing the drastic reduction observed in the GMR signal. With increasing annealing temperature this ferromagnetic coupling contribution dominates and the antiferromagnetic coupling is completely destroyed.

In contrast, the antiferromagnetic coupling in the Co/Cu multilayers with 2 nm interlayer thickness increases upon annealing at moderate temperatures (Fig. 4.60b). The hysteresis loops after the 200 °C anneal become more slanted and saturation is achieved at higher fields than in the as-grown state.

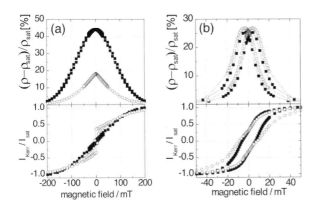

Figure 4.60: Change in the magnetic properties of Co/Cu multilayers upon annealing at 200 °C for the 1st (a) and 2nd (b) afm coupling maximum. The GMR curves (top panels) are compared to the respective Kerr effect hysteresis loops (bottom panels). Note the reduced magnetic field range in (b)

■—■—■ RT
○—○—○ 200 °C

The increase in coercive fields also shows up clearly in the GMR curves. In addition, the GMR increases, as already discussed in the context of Fig. 4.58.

The considerably smaller saturation field of the Co/Cu system in the 2nd afm coupling maximum makes this type of film very interesting for technological applications in magnetic sensors. We will therefore discuss the microstructural development in this type of system in more detail in the following. Nevertheless, Co/Cu trilayer stacks in the 1st afm coupling maximum are employed as synthetic antiferromagnets to improve the stability of a reference magnetization (Infineon Sensor).

Microstructural Investigations

In order to shed some light on the microscopic mechanisms of the GMR degradation, extensive microstructural studies have been carried out. It is useful to distinguish two aspects which require dedicated experimental approaches: (i) the layered structure of the multilayer stack; (ii) the crystalline microstructure (grains and grain boundaries) of the individual layers. The first aspect may be investigated by X-ray diffraction techniques, whereas the second is addressed best by means of transmission electron microscopy.

The global evolution of the layered structure can be deduced from X-ray reflectivity (XRR) or small angle diffraction experiments. The angular variation of the diffracted intensity of the as-grown Co/Cu multilayers (Fig. 4.61) reveals pronounced 1st and 2nd order Bragg diffraction peaks which stem from the periodicities of the multilayer (2 nm Co and 2 nm Cu). In addition, a rapid oscillation pattern ("*Kiessig* fringes") is superimposed. This is due to multiple scattering processes between the layer stack and the substrate/multilayer boundary. The appearance of these features indicates a good multilayer quality with sharp interfaces and a flat boundary to the substrate (cf. Section 3.3).

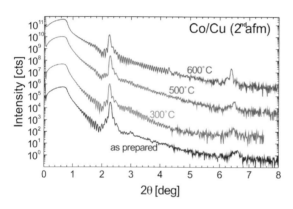

Figure 4.61: X-ray reflectometry from Co/Cu multilayers in the as-grown state and after subsequent annealing up to 600 °C (spectra referring to higher annealing temperatures are shifted vertically)

Both the Bragg peaks and the *Kiessig* fringes remain almost unchanged upon annealing up to $T \leq 600$ °C (Fig. 4.61). This result suggests that the gross properties of the multilayered structure still persist up to this temperature. Only for annealing temperatures 600 °C $< T < 700$ °C do the Bragg peaks disappear, indicating the transformation of the layered structure into a granular-like state. The GMR breakdown above 400 °C is clearly not related to the loss of the layered structure and must therefore have a more subtle origin.

Inspection of the crystalline structure by means of large angle X-ray diffraction (XRD) reveals a clear signature of a face-centered cubic lattice. This means that the Co layers do not crystallize in the hcp lattice of bulk cobalt, but grow in a quasi-epitaxial registry with the Cu layers within a crystalline grain. The average vertical grain size as inferred from the peak width is of the order of 10–20 nm and thus significantly larger than the individual layer thicknesses. Furthermore, the grains exhibit a pronounced {111} fiber texture with a single diffraction peak (Fig. 4.62). The latter finding indicates that the individual Co and Cu layers must have a common lattice constant leading to a coherently strained state of the multilayer structure.

The {111} textured state is preserved up to annealing temperatures of 300 °C. Above 300 °C, however, the XRD studies reveal dramatic changes in the texture (Fig. 4.62). Up to 500 °C, the {111} orientation of the grains is completely transformed into a {100} texture.

This transformation proceeds via an abnormal growth of {100} oriented grains at the expense of grains with {111} orientation [4.159]. The persistent multilayer stacking and the large grains give rise to pronounced satellite diffraction peaks. The average grain size reaches values in the micrometer regime laterally and may encompass the entire layer stack in the vertical direction (≈ 120 nm). The presence of a unique (Co, Cu) diffraction peak suggests that the multilayer is still coherently strained. Only at annealing temperatures well above 600 °C do individual Co and Cu diffraction peaks appear, marking the break-up of the multilayer structure into a granular-like state.

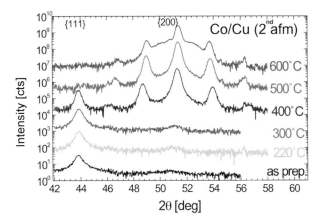

Figure 4.62: X-ray diffraction from Co/Cu multilayers in the as-grown state and after subsequent annealing up to 600 °C (spectra referring to higher annealing temperatures are shifted vertically)

It is tempting to correlate the change in texture with the breakdown of the GMR signal above $T_{an} \approx 400$ °C. In this context it is useful to recall that the maximum of the GMR appears only close to the antiferromagnetic coupling maximum, which requires a well-defined Cu interlayer thickness. From studies on single crystalline systems it is known that the optimum interlayer thickness varies strongly with the crystalline orientation due to the quantum mechanical nature of the interlayer coupling phenomenon [4.161, 4.162]. In polycrystalline films the averaging over the various grain orientations washes out the crystalline dependence of the interlayer coupling. In highly textured systems, however, some of this behavior may still be preserved. One should therefore expect a more or less pronounced change of the GMR during the texture transformation. On the other hand, according to the X-ray analysis, the texture transformation is already completed below 500 °C [4.159]. It can thus be only partially responsible for the GMR degradation.

Since the X-ray data cannot fully explain the GMR breakdown, further degradation mechanisms must be sought on a local nanoscopic scale. This task is best addressed by transmission electron microscopy (TEM). This approach involves a considerable challenge in the case of Co/Cu or FeNi/Cu multilayers for several reasons. Firstly, the image contrast in TEM depends on the atomic number, which differs only slightly from Fe to Cu. Secondly, the individual layer thickness is in the nanometer range, thus demanding high lateral resolution. Thirdly, the procedures employed in the preparation of the TEM samples may themselves affect the structural integrity of the multilayer stack (for further discussion, see Section 3.2).

From conventional TEM images it is not possible to distinguish between Co and Cu layers, because the image contrast lacks chemical information. This problem can only be solved by means of energy-filtered TEM (EFTEM). In this case the electron beam is filtered

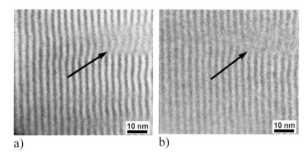

Figure 4.63: EFTEM images of a Co/Cu multilayer with 2 nm single layer thickness in cross section: as-deposited state a) Co, b) Cu). The arrows mark a grain boundary with small layer shifts. (Images recorded with friendly assistance by B. Freitag, University of Bonn)

a)   b)

with respect to energy losses after it has passed the sample. These energy losses are caused by electronic core-level excitations in the sample and are thus characteristic for a certain element or even chemical state. In Co/Cu multilayers it is convenient to exploit the L-edge losses at $\Delta E = 779$ eV (Co) and at $\Delta E = 931$ eV (Cu). An example for the type of information obtained by the EFTEM technique is given in Fig. 4.63, showing the very regular 2 nm layer stacking in a Co/Cu multilayer in the as-grown state. The image is composed of two individual images taken at the respective energy losses of Co and Cu. The arrow marks a grain boundary. At the grain boundary the layer stacking is locally interrupted and the layers are partially shifted with respect to each other. It is thus reasonable to assume that grain boundaries may act as nucleation points for the layer degradation upon annealing. This is particularly true if the mass transport can be provided by grain boundary diffusion.

At first sight, the EFTEM images after annealing at $T_{an} \leq 400$ °C do not exhibit major changes in the layered structure. This is consistent with the preservation and even slight increase of the GMR up to this temperature. A closer inspection of the images, however, reveals the development of local defects, so-called "pinholes". It is interesting to note that these pinholes are forming in the Co layers (Fig. 4.64) [4.163]. This finding is in contrast to previous suggestions that the GMR degradation may be related to pinholes in the Cu interlayers (the direct contact of adjacent Co layers should lead to an effective ferromagnetic coupling, thus destroying the antiferromagnetically ordered ground state) [4.164, 4.165].

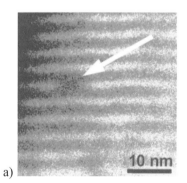
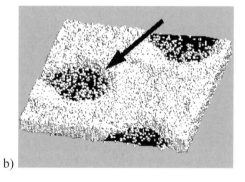

a)   b)

Figure 4.64: Co/Cu multilayer with 2 nm single layer thickness after annealing at 400 °C. a) Co EFTEM image of a cross section, b) Monte-Carlo simulation, in agreement with the measurement (white: Co atoms) [4.160]. The arrows mark a pinhole within the Co layer

A possible driving force for the pinhole formation may be the lower surface free energy of copper compared to cobalt, as will be discussed in more detail below. Apparently the density and spatial extension of these pinholes is still too small to have a significant negative effect on the magnetoresistance at $T_{an} \leq 400$ °C.

The mechanism of pinhole formation in the Co/Cu system with 2 nm thick individual Co and Cu layers has also been theoretically investigated by means of Monte Carlo simulations [4.160]. The results show that the initial defects are indeed formed in the Co layer, in the vicinity of or directly at grain boundaries, due to thermally induced fluctuations of the interfacial roughness (Fig. 4.64b). Obviously, some local structural defects in the as-grown state are necessary to start and promote the multilayer degradation during annealing. It should be noted in this context that the interfacial intermixing created by the sputter deposition process may also take the role of such local structural defects in the case of very thin (≤ 1 nm) single layer thicknesses. The spontaneous formation of pinholes may supply a similar mechanism to explain the early degradation of the Co/Cu multilayers in the first afm coupling maximum.

In order to understand the physical mechanism driving the pinhole formation it is useful to recall the situation at surfaces. It is well-known that Co overlayers on a Cu surface spontaneously form pinholes upon annealing. Through these pinholes Cu segregates onto the surface because of its lower surface free energy. A similar mechanism is proposed to be effective at internal surfaces, i.e., grain boundaries, on the assumption that the grain boundary energy in Co is higher than that in Cu [4.160, 4.163]. As a consequence, the system tries to minimize the contribution of internal Co surfaces in the grain boundary. This is achieved by a diffusion of Cu atoms into the grain boundary and may act as the first step in the formation of a pinhole in the Co layer.

With increasing temperature the pinhole is predicted to grow by a recession of the Co layer, i.e., a larger Cu interconnect is formed. The Co atoms which are repelled from the interconnect area remain at the pinhole boundary, thereby causing a local increase in the Co layer thickness in the vicinity of the pinhole. At the same time, the respective Cu interlayer thickness is reduced as the Cu atoms diffuse into the pinhole region. This process of an "interfacial grooving" is similar to the grain boundary grooving observed at surfaces. As a result the interlayer coupling shifts locally away from the antiferromagnetic coupling maximum, thereby increasing the ferromagnetic coupling contribution. Consequently, the giant magnetoresistance will be reduced.

If the temperature is further increased the thickening of the Co layers at the pinhole boundary becomes more and more pronounced. In this course, adjacent Co layers start to coalesce around the pinhole region thereby leading locally to a strong ferromagnetic coupling of adjacent layers. With the growth of these ferromagnetically coupled regions the average interlayer coupling is shifted more and more away from the antiferromagnetic coupling condition. The higher the ferromagnetic contribution to the interlayer coupling, the smaller will be the GMR signal observed from these films.

These predictions from the numerical simulations are mainly confirmed by the transmission electron microscopy studies. An example for the development of the layer stacking in the vicinity of a grain boundary after annealing at 600 °C is given in Fig. 4.65. In particular in the EFTEM image (Fig. 4.65b), one can clearly discern several regions. The originally 2 nm layer stacking is only preserved in the bottom center part of the structure. In the top part of the structure the coarsening of the layer stacking is very pronounced. The individual Co and Cu layer thicknesses appear to have grown by a factor of 2–4, and the Co layers in particular exhibit a larger amount of discontinuities. The latter finding may justify the assumption of a higher Co grain boundary energy. It is obvious that the system shown in Fig. 4.65 can no longer provide the well-defined layer morphology required for an optimum interlayer coupling, and thus it will yield only a small residual GMR signal $(\Delta\rho/\rho < 5\%)$. However, the total amount of regions with the original layer stacking must still be quite large in order to explain the persistence of the multilayer Bragg peaks in the XRD data of these annealed films.

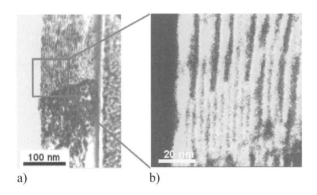

a)   b)

Figure 4.65: Defocused TEM cross section micrograph of the 2 nm Co/Cu multilayer (a) and EFTEM image of Co (b) after annealing at 600 °C (see [4.160])

Upon annealing at $T_{an} = 750\ °C$ the layer structure is completely lost. The morphology consists of separated Co and Cu regions and plate-like Co precipitates in a Cu matrix (Fig. 4.66). This state corresponds to the complete loss of the GMR signal.

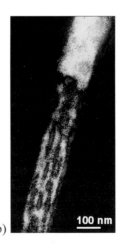

a)   b)

Figure 4.66: EFTEM images of the microstructural state of the original 2 nm Co/ 2 nm Cu multilayer after annealing at 750 °C; a) Co, b) Cu

The above results show clearly that the GMR decay in Co/Cu multilayers during a high temperature treatment is dominated by local rather than global degradation mechanisms. The driving force is presumably the higher free energy of the internal Co surfaces which promotes grain boundary diffusion of Cu atoms. This mechanism causes the initial formation of pinholes in the Co layer and a subsequent local change in the Co and Cu layer thicknesses. As a consequence, the antiferromagnetic coupling condition is gradually destroyed thus explaining the relatively high temperature stability of the 2 nm Co/Cu multilayers.

4.4.3 Ni$_{80}$Fe$_{20}$/Cu Multilayers

The GMR ratios in the Ni$_{80}$Fe$_{20}$/Cu system are about a factor of 2–3 smaller than in Co/Cu, i.e., $\Delta\rho/\rho \approx 20\%$ in the 1st and $\Delta\rho/\rho \approx 10\%$ in the 2nd afm coupling maximum. At the same time, relatively small fields are needed to obtain magnetic saturation. This property makes the Ni$_{80}$Fe$_{20}$/Cu system attractive for sensitive magnetic field sensors. The thermal stability, however, is clearly less than in the Co/Cu combination. In particular, for interlayer thicknesses of 2 nm the critical temperature for the breakdown of the GMR is only slightly above 200 °C (Fig. 4.67). In addition, the degradation takes place much faster than in Co/Cu multilayers resulting in an almost complete loss of the GMR at $T_{an} \approx 300$ °C.

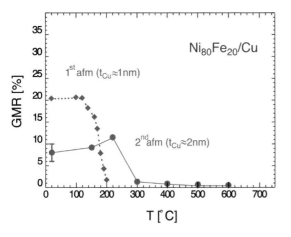

Figure 4.67: Variation of the GMR in Ni$_{80}$Fe$_{20}$/Cu multilayers with the annealing temperature. The thermal stability is strongly determined by the Cu interlayer thickness

Electrical Transport

The absolute resistivities ρ_{max} and ρ_{sat} of Ni$_{80}$Fe$_{20}$/Cu multilayers in the as-deposited state are higher than those of the Co/Cu system. In Fig. 4.68 the results are shown for the 2nd afm coupling maximum. This higher resistivity can be related to the intrinsically higher charge scattering in a binary alloy (Ni$_{80}$Fe$_{20}$) as compared to a single element. Upon annealing the resistances show an initial marginal reduction. With further increasing T_{an} subsequently up to 500 °C the resistivity of the samples increases monotonically by more than 50%. The onset of this behavior is located around 150 °C and saturates at $T_{an} > 500$ °C. This behavior is

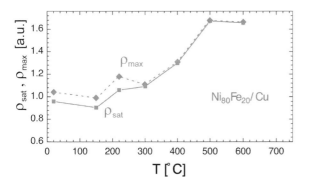

Figure 4.68: Change in the resistance in Ni$_{80}$Fe$_{20}$/Cu multilayers with 2 nm Cu interlayers (2nd afm coupling maximum) upon annealing. The values at both zero magnetic field (ρ_{max}) and for magnetic saturation (ρ_{sat}) are compared

opposite to that found in Co/Cu multilayers. We also note that in the temperature interval of the GMR breakdown (180 °C < T_{an} < 300 °C) the resistivity increases only slightly.

In order to understand this behavior one has to take into account a possible intermixing between the $Ni_{80}Fe_{20}$ and Cu layers. According to the bulk phase diagram, Ni and Cu are easily miscible, whereas Fe and Cu have a large miscibility gap [4.157]. Incorporation of Ni into the Cu layer will increase the defect scattering contribution. This results in a higher average resistivity of the multilayer stack.

Magnetic Measurements

The annealing of the 1st afm $Ni_{80}Fe_{20}$/Cu multilayers at $T_{an} \approx 200$ °C leads to an almost complete loss of the GMR (Fig. 4.67). The magnetic measurements reveal that the antiferromagnetic coupling of the neighboring permalloy layers has been destroyed by the temperature treatment (Fig. 4.69a). The characteristic hard axis type hysteresis loop of the as-grown state is replaced by a jump-like magnetization reversal in very low applied fields. This indicates that the adjacent $Ni_{80}Fe_{20}$ layers are ferromagnetically coupled and the multilayer reverses its magnetization as a whole. However, a minute amount of afm coupling survives and is responsible for the rounding of the hysteresis loop at higher fields and the residual GMR of 2–3%.

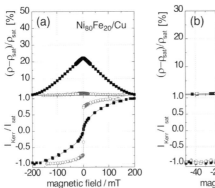
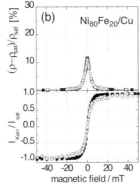

Figure 4.69: Change in the magnetic properties of $Ni_{80}Fe_{20}$/Cu multilayers upon annealing at 200 °C for the 1st (a) and 2nd (b) afm coupling maximum. The GMR curves (top panels) are compared to the respective Kerr effect hysteresis loops (bottom panels). Note the reduced magnetic field range in (b)

The same temperature treatment has only marginal effects on the GMR in the 2nd afm $Ni_{80}Fe_{20}$/Cu multilayers (Fig. 4.69b). Both the maximum GMR signal and the saturation fields remain basically unchanged. Note the absence of hysteretic behavior in the GMR curve, in contrast to the Co/Cu system (Fig. 4.60b). The Kerr effect data indicate a slight increase in the saturation fields and a more pronounced rounding of the magnetization loops after annealing. This trend is compatible with an increase in the antiferromagnetic coupling and may explain a small improvement in the GMR signal found in some samples. Anneals in excess of $T_{an} \approx 250$ °C finally lead to significant loop shapes indicating the break-down of the afm coupled ground state.

Microstructural Investigations

Further in-depth studies of the microscopic degradation mechanisms show that the layer stacking is indeed only stable to $T_{an} \approx 300$ °C. Above this temperature the multilayer Bragg peaks start to disappear indicating the progressive dissolution of the layers due to intermixing

[4.166]. In the $Ni_{80}Fe_{20}/Cu$ multilayer system this intermixing process is governed by two competing chemical aspects and proceeds in a peculiar way [4.167]. Whereas Ni and Cu have a strong tendency to form solid solutions, Fe and Cu are immiscible. Depth profiling studies on thicker trilayer systems reveal that the intermixing of Ni and Cu occurs predominantly via diffusion of Ni atoms from the $Ni_{80}Fe_{20}$ layer into the adjacent Cu layer. The diffusion of Cu atoms into the $Ni_{80}Fe_{20}$ layer is significantly smaller. This process changes the multilayer system in several respects. Firstly, the chemical composition of the magnetic layer is shifted to higher Fe concentration, i.e., both the magnetic properties and electronic structure are changing. Secondly, the Ni atoms may form isolated magnetic moments ("loose spins" [4.168]) in the Cu layer. Thirdly, the effective thickness of the magnetic layer is reduced. Finally, the effective nonmagnetic layer thickness increases. All of these changes influence the optimum ratio of the magnetic/nonmagnetic layer thicknesses and thus push the system out of the antiferromagnetic coupling maximum. In addition, the changes in the electronic structure will also influence the spin-dependent scattering processes in the layers and at the interfaces.

The thermally induced GMR decay in the $Ni_{80}Fe_{20}/Cu$ system obviously differs substantially in origin from that observed in Co/Cu multilayers. The primary microscopic process is the diffusion of Ni into the Cu interlayer. We are thus dealing with a *global* degradation mechanism, as the Cu–Ni interdiffusion is taking place throughout the entire layer and is not localized at structural defects. It is also important to note that the GMR signal of the 2nd afm multilayers disappears already at annealing temperatures ($T_{an} > 250\ °C$) at which the multilayer stacking is not yet measurably affected. This suggests that the subtle changes in the electronic structure and magnetic properties of the individual layers are responsible for the GMR decay. An alteration of the electronic structure of the layers directly affects the formation of quantum well states which mediate the interlayer exchange coupling (see Section 2.4). As a consequence, the system moves away from the antiferromagnetically ordered ground state and the field-induced changes in the resistivity become smaller.

4.5 Functional Magnetic Layers for Sensors and MRAMs

Spintronic applications, such as magnetic sensors or MRAMs, require layered systems with precisely defined functionality. At present, these systems may be grouped into three categories: multilayers, spin valves and tunnel junctions. The main interest in multilayers and spin valves comes from the field of magnetic sensors, whereas the tunnel junctions are favored mainly for MRAM applications, but also as read heads in high-density magnetic data storage. The fast progress in these research areas continuously brings about new challenges and questions. A concise review is therefore beyond the scope of this contribution. Instead, we will concentrate on some selected subjects of general importance for the functionality of magnetic layer stacks.

4.5.1 Magnetic Multilayers: Layer Thickness Dependence of the GMR Parameters

Conventionally the giant magnetoresistance (GMR) is defined as

$$\text{GMR} = \frac{\Delta\rho}{\rho_{\text{P}}} = \frac{\rho_{\text{AP}} - \rho_{\text{P}}}{\rho_{\text{p}}} \tag{4.9}$$

where ρ_{AP} denotes the resistivity in the state with antiparallel alignment and ρ_{P} the resistivity in the magnetically saturated state, i.e., for parallel alignment (cf. Section 2.4). In order to obtain a large GMR signal both a high ρ_{AP} and a low ρ_{P} are required. The most critical parameter found to affect the GMR ratio is the individual layer thickness of the constituents. Tailoring a multilayer system for magnetic sensor applications is a challenging task when a single layer thickness precision in the Å range is considered. In the following example of Co/Cu multilayers, we discuss the influence of the Co and Cu layer thickness on $\Delta\rho/\rho_{\text{P}}$, ρ_{AP}, and ρ_{P}, respectively.

GMR Amplitude and Layer Thickness of the Components

Results published on sputtered [Co/Cu]·n multilayers (n denotes the number of bilayer repetitions) show a dominant oscillatory GMR variation with the spacer layer thickness, whereby for optimum antiferromagnetic (afm) exchange coupling the maximum GMR amplitudes are located at about 10, 20, and 30 Å, respectively (see also Fig. 2.13 in Section 2.4) [4.169, 4.170]. A less pronounced GMR variation is connected to the magnetic layer thickness, which mainly modulates the GMR amplitude. For identically prepared multilayers the $\Delta\rho/\rho_{\text{P}}$ values change from $\approx 30\%$ ($t_{\text{Co}} \cong 4$–5 Å) over 65% ($t_{\text{Co}} \cong 8$–9 Å) to $\approx 40\%$ ($t_{\text{Co}} \cong 38$–40 Å), if the optimum Cu layer thickness of about 9 Å in the 1st afm maximum is kept in each case [4.171]. The GMR amplitude reduction towards very thin Co layer thicknesses is caused by at least two mechanisms. Firstly, the resistivity ρ_{P} in the aligned state (Eq. (4.9)) increases with smaller multilayer periods (compare Fig. 4.72). Secondly, the fraction of misaligned magnetic moments (e.g., canted moments at crystal defects) increases with decreasing t_{Co}, thus causing a high field magnetoresistance with superparamagnetic-like behavior [4.171]. On the other hand, the GMR amplitude also drops for larger t_{Co} values, because of parasitic nonmagnetic scattering inside the magnetic material. The optimum magnetic layer thickness yielding a maximum GMR amplitude therefore depends also on the crystalline perfection and chemical composition (see also Fig. 4.70).

The spacer layer and magnetic layer thickness variations of the GMR are superimposed and obey different finite size characteristics. Thus, a representative picture of the system is obtained by comparing film stacks with equal thicknesses of the two components and by changing their thicknesses simultaneously (see also below). This approach led to the data in Section 2.4 (after [4.172]). These data stress that the afm coupling (e.g., the 1st afm maximum) changes into a ferromagnetic coupling upon spacer thickness variations of only a few Å. Therefore, for sensor fabrication the multilayer deposition has to maintain an accuracy of ± 1 Å of the spacer layer thickness in order to reproducibly obtain a high GMR amplitude.

GMR Field Dependence and Layer Thickness of the Components

For practical application not only the maximum GMR value, but also the magnetic field dependence of $\Delta\rho/\rho_P$ is important. This aspect is illustrated in Fig. 4.70 compiling $\Delta\rho(H)/\rho_P$ curves from a series of Co/Cu multilayers around the 1st afm coupling condition. These curves again reflect the characteristic variation of the GMR peak value with the magnetic and nonmagnetic layer thicknesses. In addition, strong changes in the shapes of the $\Delta\rho(H)/\rho_P$ curves were observed. The magnetic Co layer thickness mainly affects the saturation field and the slope of the $\Delta\rho(H)/\rho_P$ curves (the sensitivity in %/Oe), respectively (note the different x-axis scaling in Fig. 4.70).

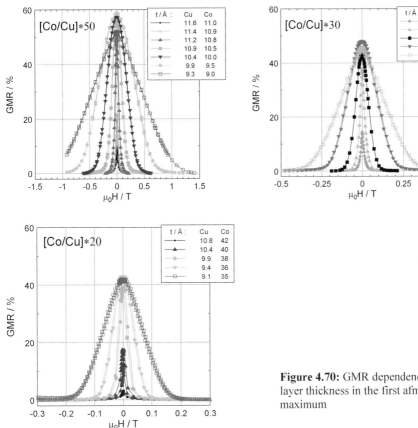

Figure 4.70: GMR dependence on the Co layer thickness in the first afm coupling maximum

This influence follows the relation:

$$J_{bil} = M \cdot H_{Sat} \cdot t / 4 \tag{4.10}$$

where J_{bil} is the bilinear afm coupling strength, H_{Sat} the saturation magnetic field, M the magnetization, and t the thickness of the magnetic component in a multilayer with a large number of periods. For a constant J_{bil} in Eq. (4.10), H_{Sat} can be tuned via the magnetic layer thickness t_{Co}. This is confirmed when comparing the three panels of Fig. 4.70. J_{bil} itself, however, depends strongly on the spacer layer thickness t_{Cu}. Fig. 4.70 shows a monotonic decrease of J_{bil} by an order of magnitude for a t_{Cu} increase in the 1 Å range. ($J_{bil} \cong -0.03 - -0.3$ mJ m^{-2}, in [4.171] for the identical multilayer family up to $J_{bil} \cong -0.5 - -0.7$ mJ m^{-2}). On epitaxial MBE grown Co/Cu/Co sandwiches *Johnson et al.* [4.173, 4.174] found oscillatory J_{bil} values up to -1.1 mJ m^{-2} from magnetization measurements. In this case, the rather thick single crystalline Cu substrate made a GMR measurement impossible.

The layer thicknesses t_{Cu} and t_{Co} in the sputtered and textured multilayers must always be considered as values averaged over large lateral distances and over the n multilayer periods. In the case of very stable and calibrated deposition rates, the individual layer thickness may be computed from the total multilayer thickness in the 100 nm range. This procedure yields nominal thicknesses in fractions of Å. These values include the influence of roughness on different length scales. This aspect has to be kept in mind when comparing the results from sputtered multilayers to measurements and theoretical predictions (RKKY coupling, see Section 2.4) based on a single crystalline situation, especially for dimensions in the range of lattice plane distances.

It can be stated from Fig. 4.70, that the GMR amplitude is nearly constant within a spacer thickness range $\Delta t_{Cu} \cong \pm 1$ Å - but the coupling strength J_{bil} (Eq. (4.10)) can change by more than an order of magnitude and with it the saturation field and the sensitivity. Thus for engineering multilayers as magnetic sensor materials it is necessary to control the magnetic layer thickness in the Å range and the spacer layer in the range of tenths of Å. The oscillatory magnetic layer thickness dependence of J_{bil} found in epitaxial Co/Cu/Co layers [4.175, 4.176] seems to be washed out in sputtered multilayers (see the t_{Cu} and t_{Co} discussion above).

For some applications it is more favorable to use multilayers in the 2nd afm coupling maximum. The loss in the GMR amplitude by a factor of 2 –3 is opposed by a strongly enhanced sensitivity, because of a small J_{bil} in Eq. (4.10). By employing NiFe instead of Co as magnetic component one can avoid strong hysteresis effects and realize sensitivities up to the range of %/Oersted with a reduced GMR amplitude $\cong 12\%$ (see Fig. 4.71). These

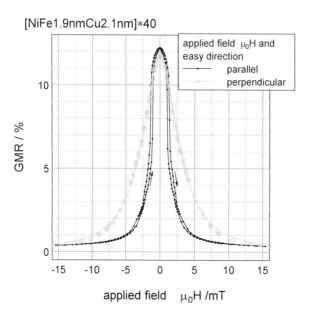

[NiFe1.9nmCu2.1nm]*40

applied field $\mu_0 H$ and easy direction
— parallel
○ perpendicular

Figure 4.71: GMR dependence on applied fields for a permalloy–copper multilayer in the 2nd afm coupling maximum. The measuring current is perpendicular to the easy direction. The hysteresis is shown only for the applied field parallel to the easy direction

multilayers are also stable against moderate thermal treatments in air [4.154] (see also Section 4.4), an important factor for the application as sensor material.

Resistivity Variation of the Layer Thickness in the Aligned State

As mentioned above, the resistivity $\rho_P = \rho_{sat}$ in the magnetically aligned state has a great impact on the maximum GMR ratio (see Eq. (4.9)). Nevertheless, it is scarcely discussed in the literature, especially not with respect to the layer thicknesses of the multilayer components. It can be shown [4.172] that in Co/Cu multilayers with equal constituent thicknesses $t_{Co} \cong t_{Cu}$ the resistivity ρ_P is proportional to the reciprocal bilayer thickness $(t_{Co} + t_{Cu})^{-1}$ (see Fig. 4.72). This behavior is well known from size effect theories in thin films [4.177] and emphasizes the role of the interfaces and the layer thicknesses. The temperature dependence of ρ_{sat} (Fig. 4.72), however, suggests that this size effect is multiplicative rather than additive, as is known from theories (e.g. [4.178]) for angle selective electron scattering at surfaces (here interfaces). From the data in Fig. 4.72 one can deduce:

$$\rho(t, T) / \rho(t, T_0) = f(T) \tag{4.11}$$

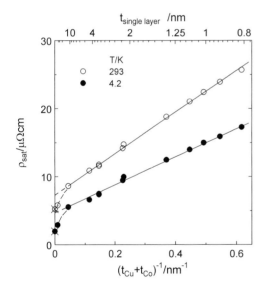

Figure 4.72: Resistivity dependence of equidistant Co/Cu multilayers on the bilayer or single layer thickness, respectively, for room temperature and $T = 4.2$ K [4.172]

Figure 4.73 shows, for very different thickness of the components, the normalized (to $T = 293$ K) resistivities in the temperature range $T = 4.2–300$ K in comparison to single layers of Cu and Co, deposited under similar conditions.

For multilayers with single layer thicknesses $t_{sl} \leq 10$nm the T-dependences closely agree despite changing the Cu/Co ratio as well as the absolute Cu layer thickness by an order of magnitude or more. This means, in particular, that no shunting effect by the higher Cu contribution appears, contrary to what should be expected from the much lower resistivity and stronger T-dependence of Cu in comparison to Co (RRR: residual resistance ratio):

$\rho_{Cu}(293\ K) = 3.1\ \mu\Omega\ cm,$ $\rho_{Cu}(4.2\ K) = 1.1\ \mu\Omega\ cm,$ RRR = 2.8
$\rho_{Co}(293\ K) = 15.9\mu\Omega\ cm,$ $\rho_{Co}(4.2\ K) = 9.0\ \mu\Omega\ cm$ RRR = 1.8

The results on two NiFe(Py)/Cu multilayers have been included in Fig. 4.73, and it is remarkable that they also coincide with the Co/Cu data.

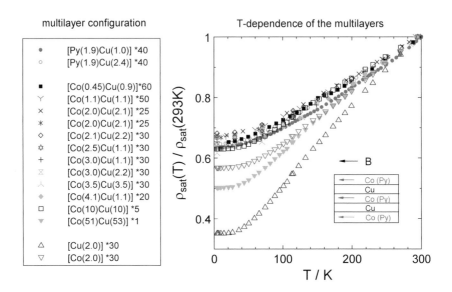

Figure 4.73: Temperature dependence of the resistivity (normalized to $T = 293\ K$) in the saturated state of Co–Cu multilayers with strong variations in the Co and Cu thickness. The data are compared to the resistivities of thick films of the individual components. The values of two NiFe(Py) –Cu multilayers are also included

In conclusion, to reproducibly obtain a large GMR effect in magnetic multilayers, the following aspects must be taken into account:

- The multilayer deposition technique and parameters must permit control of the spacer and magnetic layer thickness in the 0.1 Å and 1 Å range, respectively.
- A low resistivity ρ_P in the aligned state is a prerequisite for a large GMR amplitude. This is favored by structural (lattice) matching of the multilayer components. It was shown that in Co/Cu multilayers the Co layers crystallize like Cu in the fcc lattice structure, and form coherently grown columnar grains (see also Section 4.4). The Fermi surfaces for Cu and for the majority ↑ electrons in fcc Cobalt are very similar and in this way an "electronic" matching of the components is achieved (Section 2.4). A large part of the conduction electrons can pass the interfaces, whereas in the afm state (because of the completely different Fermi surfaces for majority and minority electrons) most electrons are reflected at the interfaces of the oppositely oriented magnetic layers, resulting in a high GMR. In the case of Fe/Cr multilayers both components crystallize in the bcc structure and the lattice mismatch is even smaller than in the Co/Cu system. The Fermi surfaces of Cr and of the minority electrons of Fe are very similar, therefore the above discussed mechanism to create the GMR is also effective. The density of states and the

mean free path of minority electrons in Fe are high enough. The systems Fe/Cr and Co/Cu show the largest GMR values realized up to now (for an overview see [4.179]). Therefore the search for future GMR material combinations has to take into consideration the aspect of electronic matching.

- The deposition method (MBE, e-beam deposition, DC/RF magnetron sputtering, IBS, and so on) should guarantee an optimal structural matching, whereby surface properties may play an important role.

4.5.2 Spin Valves

Spin valves (SV) are complex magnetic layer stacks which are already employed as sensor materials in hard disk read heads (see Sections 2.4 and 5.3). Challenges for further applications are, among others, the enhancement of the GMR amplitude, the increase of exchange biasing of the hard layer, and an optimum combination of coercivity and orange peel coupling of the free layer to get a sharp minor loop centered at zero applied field. Resolving these issues requires a dedicated and detailed magnetic engineering of the individual layers and the entire layer stack. This often leads to the introduction of additional magnetic or nonmagnetic layers which enhance the performance of the adjacent layers or introduce an original functionality.

The maximum signal of a spin valve sensor is mostly determined by the spin-dependent scattering in the GMR active trilayer. Therefore, the GMR amplitude enhancement is pursued in several approaches, for example, the following (cf. Section 2.4):

- Combination of a top and bottom SV to a dual SV sharing the same free layer.
- Increase in the spin dependent scattering contribution via dusting the interface spacer - free layer (in most cases NiFe) with 0.7–1 nm of Co or CoFe (in the case of a Cu spacer).
- Improved control of shunting effects by reduction of the thicknesses of the protective layers, seed layers, afm layers, spacer layers and so on.
- Introduction of nanooxide layers (NOL) on one or both sides of the GMR active trilayer in order to enhance the specular scattering and decrease the resistivity ρ_P in the aligned state.

Using these optimization strategies the GMR amplitudes in spin valves can be driven into the 20% range for selected cases.

Another important issue in the optimization of a spin valve concerns the properties of the magnetic reference layer. Generally, its magnetic stability is enhanced by exchange coupling ("biasing") to an antiferromagnet. This coupling is strongly temperature dependent and depends on the magnetic properties of the antiferromagnet and on many details of the antiferro-/ferromagnet interface structure and morphology. Exchange biasing has mainly been optimized by two approaches:

- An appropriate deposition and growth control to obtain and to characterize the optimum interface structure. In this way exchange energy densities of $J_{EX} = 0.36$ mJ m^{-2} at the interface IrMn/CoFe could be realized [4.180]. For a $t = 2$ nm CoFe layer (with $M \cong 1.8$ T) this corresponds to an effective exchange field H_{EX}

$$H_{EX} = J_{EX} / M \cdot t \tag{4.12}$$

with $\mu_0 H_{EX} = 120$ mT at room temperature. Similar methods have been used to realize blocking temperatures above 300 °C [4.181].

- The other well-known way to improve the exchange coupling parameters is to incorporate one or two SAFs (synthetic antiferromagnet, see Section 2.4) into the spin valve. This gives the possibility of combining the large exchange coupling from the SAF with a high blocking temperature in the natural antiferromagnet.

The above-mentioned improvements of the spin valve parameters are partly diametral and one has to consider the most important property relating the applications.

In the following we will concentrate on one of the most serious problems, the operation of spin valves at elevated temperatures, which is of great importance for sensor applications.

Spin Valve Characteristics above Room Temperature

The first generation of spin valves was based on the antiferromagnets FeMn and NiO. Impressive fundamental results have been obtained on these systems which were also employed in hard disk read heads. However because of the relatively low Néel and blocking temperatures (see Fig. 4.74 and Section 2.4), respectively, the magnetic reference goes soft at elevated temperature and at $T = 100\ °C$ most characteristics are no longer suitable in sensors. Obviously, one has to employ afm layers with higher blocking temperatures. Figure 4.74 gives an overview of the exchange bias as a function of temperature for several afm materials considered as pinning layers in spin valves (after *Nozières et al.* [4.182]). It should be kept in mind, as outlined above, that the exchange bias values strongly depend on the preparation conditions, but the general material trend towards higher blocking temperatures is clear.

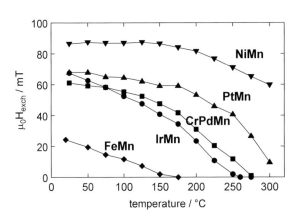

Figure 4.74: Temperature dependence of the exchange field for several afm layer materials, after *Nozières et al.* [4.182]

A more detailed example is discussed in Fig. 4.75. The characteristics of three types of IrMn-based spin valves are compiled for measuring temperatures from room temperature up to 200 °C. Figure 4.75a relates to a top SV, Fig. 4.75b to a bottom SV, and Fig. 4.75c to a bottom SV with SAF. After deposition of the film stacks the SVs are annealed at $T = 280\ °C$ in an applied field $\mu_0 H = 1\ T$ for a time of $t = 1\text{min}$ and field-cooled to define an exchange bias.

From the experimental results one can draw the following conclusions:

- The conventional SVs exhibit suitable magnetic characteristics up to $T = 100\ °C$. At $T = 200\ °C$ the blocking temperature for the SV with a small exchange bias (Fig. 4.75a) is not exceeded, whereas for the SV with a three times larger exchange bias at room temperature (Fig. 4.75b) the SV characteristics have already disappeared, the blocking temperature is exceeded.

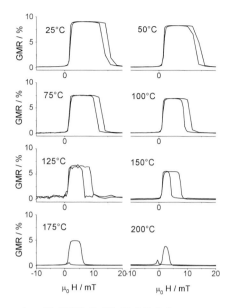

a)   Ta5NiFe3.5CoFe5Cu2.5
      CoFe3IrMn8Ta10

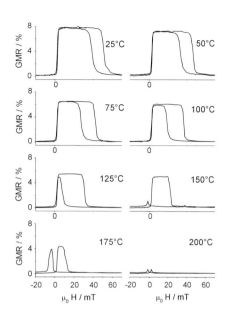

b)   Ta5Cu1CoFe1IrMn7CoFe3Cu2.5
      CoFe1NiFe5Ta10

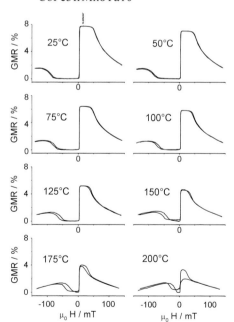

c)   Ta3.5NiFe2IrMn10CoFe4.5Ru0.8
      CoFe4Cu3CoFe0.8NiFe5Ta10

Figure 4.75: Temperature dependence of the
characteristics of some IrMn based spin valves
a) top spin valve
b) bottom spin valve
c) bottom spin valve with SAF
Layer thickness of the components of the layer
stacks is given in nm

- In the SV with SAF the temperature regime with suitable characteristics is extended to at least $T = 150°C$ and the exchange bias is still present at $T = 200\ °C$ (Fig. 4.75c).

These data illustrate that in conventional spin valves the parameters blocking temperature T_B and exchange bias H_{EB} for an afm/fm layer system may depend on each other: the growth of one parameter can decrease the other one (cf. [4.181]). On the contrary, in spin valves with SAF the blocking temperature and the exchange field can be optimized as follows: the afm/fm interface region should provide a high blocking temperature, an eventual loss in the coupling energy J_{EX} is compensated for the desired effective coupling strength after Eq.

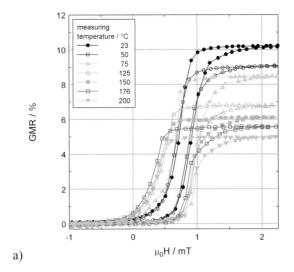

a)

Figure 4.76: Temperature dependence of the minor loops of
a) a conventional PtMn-based bottom spin valve
b) a PtMn-based top spin valve with SAF.
For details of the layer stacking see text

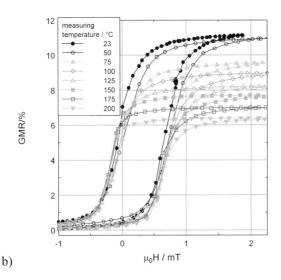

b)

(4.12), because of the reduced effective parameter t in the SAF. This is a challenging optimization problem.

The comparison in Fig. 4.74 also suggests that Pt/Mn based spin valves should respond to the high temperature demands even better than Ir/Mn based SVs. This is supported by the results in Fig. 4.76 where the temperature dependence of the minor loops for a conventional PtMn-based SV and a SV with SAF are compared.

The conventional bottom spin valve has the layer structure Ta3NiFe4.5PtMn20CoFe4.5Cu3CoFe2NiFe6Ta3, and the top spin valve with SAF: AlOx4.5PtMn20CoFe2Ru0.6CoFe4.5Cu3CoFe2NiFe6Ta3 (thicknesses in nm). From the results in Fig. 4.76 one can clearly extract the following information:

- The minor loops do not collapse up to $T = 200\ °C$ for both spin valve types: the blocking temperature is not exceeded.
- The GMR signal (loop height) reduces with increasing temperature, but its amplitude is suitable for sensor applications also at $T = 200\ °C$.
- The loops of the SAF spin valves are less influenced by the temperature increase.

About half of the GMR reduction with temperature is caused by the regular resistivity ρ_P increase with temperature in the aligned state according to Eq. (4.9), the reduction of the magnetically induced resistivity change $\rho_{AP} - \rho_P$ is only about 25%. One can summarize, that PtMn-based spin valves are promising candidates for sensor applications at elevated temperatures.

According to Fig. 4.74 a further improvement in the high temperature parameters should be expected from other materials, such as NiMn. The formation of the proper afm fct phase of NiMn, however, requires complicated post-deposition annealing treatments which are currently under investigation ([4.183], see also below).

4.5.3 Magnetic Tunnel Junctions

Magnetic tunnel junctions (MTJs) represent a new key technology for magnetic recording heads, magnetic random access memories (MRAMs) and picotesla field sensors. The optimum tunneling magnetoresistance (TMR) ratio in the range 20–50 % and the switching characteristics of magnetic tunnel junctions depend crucially on the interface roughness and the quality of the thin insulating barrier layer and on the spin polarization and microstructure of the two ferromagnetic electrodes (see Section 2.4, [4.184]). For applications the main challenges are not only the height of the TMR effect, but also a low enough resistance×area product (RA) requiring very thin tunneling barriers. Thermal stability of the layer stack up to about 400 °C is required for compatibility with the temperature regimes of CMOS back-end processes [4.185, 4.186]. Two topics of general interest are the dipolar magnetic coupling which increases with decreasing barrier thickness, and the annealing behavior of exchange-biased magnetic tunnel junctions.

Néel Coupling in Magnetic Tunnel Junction

An independent magnetization reversal of the layers in a TMR stack is necessary to obtain a high magnetoresistance. It may be strongly affected by various magnetic coupling phenomena, in particular, by those related to the film morphology. The surface roughness of the two ferromagnetic interfaces next to the nonmagnetic barrier leads to a dipolar magnetic coupling, also known as *Néel* coupling or "orange-peel" coupling (see Section 2.4 [4.187, 4.188]). The modified *Néel`s* model [4.189] of orange peel coupling in thin ferromagnetic layers predicts that the dipolar coupling field H_d increases with the height h of the film roughness, with the magnetization of the hard magnetic film M_H, with the decreasing correlation length λ of the film roughness, with the decreasing thickness of the hard and soft magnetic films t_S and t_H, and with the decreasing barrier thickness d:

$$H_d = \frac{\pi^2 h^2 M_H}{\sqrt{2}\lambda t_S} \exp(-2\pi\sqrt{2}d/\lambda) \cdot [1 - \exp(-2\pi\sqrt{2}t_S/\lambda)] \cdot [1 - \exp(-2\pi\sqrt{2}t_H/\lambda)] \qquad (4.13)$$

Extended STM and MOKE investigations confirmed the correspondence between the measured values and those calculated within this model [4.190]. The results clearly indicate that the interfacial morphology of the TMR stacks is of crucial importance for their magnetic behavior. The effect of the dipolar coupling can be clearly seen in Fig. 4.77. The TMR stacks of type [Ta5/Al50/Cox/Al(oxid.)1.8/NiFe50/Al50/...] (thickness in nm) with different Co film thickness were deposited by e-beam evaporation onto cooled substrates (80 K). The magnetization and TMR curves show that the orange peel coupling markedly affects the magnetization reversal field for Co and NiFe in these samples: the reversal field for Co decreases and for NiFe increases with decreasing Co film thickness. In the case of strong coupling between the films, the magnetization of the layers cannot align in a perfectly antiparallel manner and the normal TMR curve is "cut off" at the magnetization reversal field of the hard magnetic layer. This behavior is clearly improved in the sample with 10 nm Co, where the TMR curve is more symmetric and the TMR effect is higher.

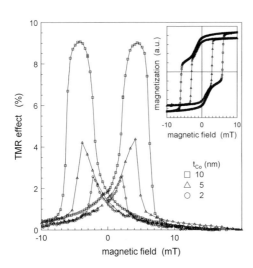

Figure 4.77: TMR curves for a stack (Ta5/Al50/Cox/Al(ox.)1.8/NiFe50/Al50) with different Co film thickness (t_{Co}=2, 5, 10 nm). The 2 nm Co layer is coupled most strongly to the NiFe film, thus the TMR effect is small. Only for the 10 nm Co layer is there a stable antiparallel state of the two magnetic layers and the effect is 9%. The magnetization curves (see inset) also show, that the thinner Co layer is coupled most strongly to the NiFe film

For better device performance it is desirable to reduce the dipolar coupling (see Section 5.3).

A further challenge is the reduction of the RA values in MTJs. A lower RA parameter can be achieved by means of very thin insulating barriers ($d < 1$ nm). In order to avoid the negative influence of the dipolar coupling in this case, the interface must be very smooth (h low and λ high). In sputtered MTJ stacks (see also the next subsection) it is possible to prepare thin and smooth barriers without losing the TMR effect [4.191].

Exchange-Biased Magnetic Tunnel Junctions

In most applications it is desirable to switch between two stable magnetic configurations. The switching characteristics of magnetic tunnel junctions can be influenced and optimized by an exchange bias layer, similar to that in spin valves (Section 2.4, [4.192]). In fact, most of the improvements and developments in spin valve engineering are applicable also for the case of TMR junctions. For MTJs the manganese-based antiferromagnetic (afm) materials PtMn and NiMn are interesting pinning materials, because they have the highest blocking temperatures (see Fig. 4.74, [4.193]). The as-deposited films of these materials, however, are not antiferromagnetic. An annealing treatment in a magnetic field is necessary for a phase transformation from the paramagnetic face-centered cubic (fcc) to the afm face-centered tetragonal (fct) crystal structure [4.183, 4.194].

The annealing behavior of a bottom NiMn tunnel junction of the type (Ta4/NiFe4/NiMn50/CoFe4/Al(ox.)1.6/CoFe1/NiFe50/Cu40/Au20) grown by magnetron sputtering will be discussed in the following in more detail. The tunnel barrier is formed by plasma oxidizing the 1.6 nm Al layer. For TMR measurements the layer stacks were patterned by photolithography and reactive ion etching into square structures. It is a non-trivial task to pattern the small junctions without causing damage to the tunnel barrier. Most of the milling-based dry etching processes leave metallic bridges behind (redeposition during the etching), which may short circuit the tunnel barrier. The HRTEM images of the edge of an etched MTJ isolated with Si_3N_4 (Fig. 4.78) show examples of a perfect barrier down to the Si_3N_4 isolation and a barrier having short circuits due to a metallic coating at the edges.

Additional plasma and/or chemical treatments after the etching process are necessary to eliminate these conducting bridges.

For the as-deposited film stack the TMR curve is symmetric and the coercive field of the hard magnetic CoFe electrode is small (Fig. 4.79). After annealing at 350 °C for 1 min and cooling in a magnetic field of $\mu_0 H = 1$ T the TMR effect is more than two times higher and the TMR curve shows the characteristic asymmetric behavior.

a)   b)

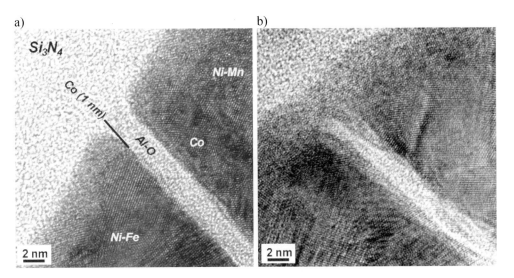

Figure 4.78: HRTEM images of the edge of TMR structures a) without and b) with metallic bridges over the barrier, isolated with Si_3N_4 (by courtesy of J. Thomas)

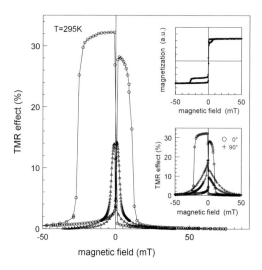

Figure 4.79: TMR vs. magnetic field at room temperature for a 4Ta/4NiFe/50NiMn/4CoFe/1.6Al(ox.)/1CoFe/50NiFe/40Cu/20Au stack before (triangles) and after (circles) annealing at 350 °C for 1 min in a magnetic field $\mu_0 H = 1$ T. The upper inset shows the magnetization hysteresis loops and the lower one the unidirectional magnetic anisotropy for an annealed MTJ

The shift (exchange bias) and broadening of the TMR curves and hysteresis loops are characteristic for pinned ferromagnetic films (cf. Section 2.4). The exchange bias is caused by a unidirectional magnetic anisotropy. Only for the magnetic field applied parallel to the pinned magnetization direction is an antiparallel alignment of the magnetization in the junctions necessary for a high TMR effect to be obtained (lower inset in Fig. 4.79).

The magnetization measurements (Fig. 4.79, upper inset) also demonstrate that in this sputter-deposited MTJ the interlayer coupling is much smaller than in the electron-beam deposited junction in Fig. 4.77. This difference and the additional exchange bias allow one to achieve a symmetric minor loop centered at zero field. This symmetric minor loop constitutes the two stable switching states necessary for applications (see upper inset in Fig. 4.80).

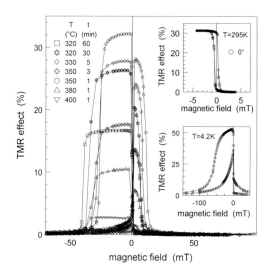

Figure 4.80: The effect of annealing at different temperatures and for different duration times on the TMR effect and on the switching characteristics of bottom NiMn junctions. The upper inset shows the symmetric minor loop centered at zero field and the lower inset the TMR curve at 4.2 K for the annealed tunnel junction of Fig. 4.79.
Junction area: 100×100 μm²

For the formation of the desired afm (fct)-phase in PtMn or NiMn two strategies are possible: extended annealing at lower temperature or a rapid annealing at higher temperatures [4.183]. The annealing results for different temperature–time combinations (Fig. 4.80 and Table 4.2) show that with both methods high exchange bias fields (H_{ex}) and high pinned layer coercivities (H_c) can be achieved. However, these are not necessarily connected with the highest TMR effect. Rapid annealing at intermediate temperatures, for the example in hand 350 °C, leads to the highest TMR effect of 32%.

The TMR effect of many other MTJ material combinations starts to degrade at temperatures above 300 °C and drops sharply at 400 °C [4.195]. The reason is the diffusion of the metal atoms (e.g. Mn) into the ferromagnetic layers or into the barrier material. The increase of the TMR effect to 52% at 4.2 K (lower inset in Fig. 4.80) demonstrates that the barrier and the interfaces of this junction are still not ideal or are already damaged, because the TMR effect of a MTJ is temperature independent, at least according to the Jullière model (cf. Section 2.4).

Further improvements of the thermal stability in MTJs are achieved with interfacial layers which help to suppress segregation processes. Large TMR effects (40 %) are achieved after annealing at 400 °C [4.196].

Table 4.2: Characteristic values for the annealed MTJs of Figure 4.80

T (°C)	t (min)	R (kΩ)	TMR (%)	$\mu_0 H_c$ (mT)	$\mu_0 H^{ex}$ (mT)
as-deposited	-	17	15	2	-
320	60	5.5	17	23	17
320	30	5.7	26	21	15
330	5	6.4	28	17	12
350	3	6.4	17	18	13
350	1	8.0	32	18	7
380	1	6.8	10	19	14
400	1	4.2	3	20	17

4.6 Multilayers for X-Ray Optical Purposes

4.6.1 Multilayers as Reflectors for X-Rays

Artificial multilayers with periods of several nanometers are the one-dimensional analogue of natural crystals with a three-dimensional periodicity. As already explained in Sections 2.5 and 3.3, X-rays can be reflected on multilayer structures according to the generalized Bragg equation similarly to the diffraction on crystal lattice planes. The advantage of using nm-multilayers for the reflection of X-rays instead of crystals is that the structural parameters of the multilayers can be freely chosen. Depending on the intended application which requires the reflection of X-rays of a certain wavelength λ_X and at a special angle of incidence α, the material combination of absorber and spacer layers, the layer thickness and the minimum number of periods N_{min} can be selected [4.197] (cf. Section 2.5).

Compared to visible light, X-rays span a much larger wavelength range, from 0.01 nm to 30 nm. The most commonly used wavelengths in laboratory applications are the Kα emission lines of the elements Cu, Mo, Cr, Co. For these wavelengths commercial reflectometers and diffractometers are offered by several companies. Further large fields of interest are the fluorescence analysis for material science with wavelengths in the range 0.8–2.0 nm and investigations of biological structures in the so-called water window region between the absorption edges of carbon and oxygen in the range 2.4–4.4 nm. However, the most important driving force for the development of nm-multilayers within the latest decade has been extreme ultraviolet lithography (EUVL), which works around $\lambda = 13$ nm [4.198]. EUVL will be used in the semiconductor industry to print patterns of integrated circuits with typical feature sizes of <50 nm. The currently used photolithography, working with laser light (248 nm, 193 nm) and transparent lenses and masks, cannot be applied when the wavelength of the radiation, which is one of the parameters that determines the minimum size of the printed patterns, is reduced further. Since no transparent material can be found for $\lambda = 13$ nm, the wavelength change to such short values requires the application of reflection optics, which consist of substrates with precise aspheric surfaces and extremely low micro-roughness and of reflection coatings consisting of Mo/Si multilayers. Today, hundreds of publications deal with this system, the first results on Mo/Si multilayers were presented already in 1985 by *Barbee et al.* [4.199].

Assuming ideal layers (bulk densities of the layers, no interface roughness and diffusion, no surface reaction and contamination) two aspects have to be balanced in order to get optimum multilayers as reflectors (Section 2.5):

- highest possible X-ray optical contrast $\Delta\delta = \delta_{absorber} - \delta_{spacer}$ between absorber and spacer layers,
- low absorptions $\beta_{absorber}$ and β_{spacer} of the X-rays in both layers.

A high contrast $\Delta\delta$ increases the reflectance of the X-rays on every interface. However, the best optical contrast is useless if the absorption of X-rays in the individual layers is high. In this case only a few periods would contribute to the entire reflection since the X-rays cannot penetrate into deeper layers. We will illustrate this effect with the example of nm-multilayers for the reflection of radiation with $\lambda = 13.5$ nm at an angle of incidence $\alpha = 1.5°$. Calculations of the reflectance of model systems consisting of two layers per period show that molybdenum and silicon is the combination with the highest reflectance. We can find combinations with higher contrast between the individual layers (e.g. Ru/Si) and with lower absorption (e.g. Nb/Si), but in all cases the resulting reflectance is lower than for the Mo/Si multilayer. The higher contrast of Ru/Si is overcompensated by the higher absorption of Ru compared to Mo. The gain that is reached due to the improved contrast disappears due to the higher absorption, which decreases the number of reflecting periods N_{min} to 54 and results in a reflectance of only 70.6 %. The situation is opposite to that with the Nb/Si combination. Due to the lower absorption of Nb compared to Mo, the number of reflecting periods $N_{min} = 66$ is higher than for the Mo/Si multilayer. However, this time the decreased contrast between Nb and Si results in a lower reflectance of Nb/Si ($R_{max} = 74.1$ %) compared to Mo/Si ($N_{min} = 57$, $R_{max} = 75.4$ %). This example shows that, always, a compromise between the highest possible contrast and the lowest absorption of the layers must be found.

For the total wavelength region between $\lambda_x = 0.01$ nm and 30 nm, due to the appropriate optical constants of the layer materials, preferred material combinations can be assigned to the particular λ_x regions, as explained by Table 4.3.

Table 4.3: Wavelength regions of optimum operation for various multilayer types

Regime of operation	λ_x / nm	Materials	Reference
GIR	<0.01	Pt/C, Cu/Si, W/Si, Ni/Si	[4.200–4.204]
	0.059	Ru/B$_4$C	[4.205]
	0.01-0.071	Mo/B$_4$C, Mo/C	[4.206]
	0.01-1	Ni/C, Ni/B$_4$C, Mo/C, Mo/B$_4$C	[4.207, 4.208]
	0.01-1	W/B$_4$C, W/C, W/Si	[4.209, 4.210]
NIR	1.4-2.4	W/B$_4$C	[4.211]
	2.2-4.4	Cr/Sc	[4.212, 4.213]
	6.76	La/B$_4$C	[4.214]
	6.8	B$_4$C/Si	[4.215]
	7.8-11.4	Mo/Y	[4.216]
	11.4-12.5	Mo/Be, Rh/Be, Ru/Be	[4.217, 4.218]
	12.5-25	Mo/Si	[4.219–4.222]
	18-25	B$_4$C/Si	[4.215]
	25-30	Si/C	[4.223]
	36-42	Sc/Si	[4.224]

4.6.2 Real Structure of nm-Multilayers

Model calculations give a first indication of which material combinations are promising candidates for the reflection of a certain wavelength. The final decision which material combination is best suited for a special wavelength has also to take into account aspects of the real multilayer structure. The preparation of a real nm-multilayer can be carried out using different deposition techniques, which have been considered in Section 2.5. However, none of the techniques gives completely perfect multilayer structures. Depending on the deposition method used, the following deviations from the mathematically ideal structure are more or less pronounced:

- deviations from strict periodicity within the multilayer stack,
- interface roughness,
- interface interdiffusion,
- differences between thin film and bulk density,
- surface oxidation and contamination.

Consequently, it can be seen that, besides the correct choice of the materials of the multilayer for the reflection at a given wavelength, several real structure effects have to be taken into account for the right choice of a nm-multilayer system. In the following sections the different kinds of multilayer imperfections will be discussed.

Deviations from Strict Periodicity within the Multilayer Stack

Any deviation from the periodicity of the layer stack causes distortion of the constructive interference of all the reflected X-rays from the individual interfaces and results in reflectance losses of the whole multilayer. Concerning the different kinds of periodicity distortion, two phenomena have to be distinguished:

- drifts of the period thickness along the depth of the multilayer,
- statistical deviation of the period thickness without correlation to the depth.

After careful investigation of systematic drifts, e.g. due to drifts of the deposition rate, they can be compensated by proper control of the deposition regime. Statistical deviations can be reduced by the stabilization of all deposition parameters. Not only the particle source but also all other parameters like substrate movement, vacuum conditions and thermal conditions have to be kept constant over the deposition interval.

The detection of stack irregularities can be performed by X-ray reflectometry with a high sensitivity (cf. Section 3.3). Any deviation from the periodic stack results in line broadening of the observed Bragg reflection peaks and in a decrease of the regularity of the Kiessig fringes between the Bragg reflection peaks. To illustrate this effect, Fig. 4.81 shows a typical Cu-Kα reflectograph of a Mo/Si multilayer having 40 periods (a) and a corresponding model calculation with a perfect periodical stack (b), a stack with a period thickness drift from 6.80 nm to 6.68 nm (corresponds to a drift of 0.003 nm per period) (c) and a stack with statistical period thickness deviations of 0.01 nm (d). It can be seen that systematic drifts of the multilayer period can clearly be identified by investigation of the half widths of the Bragg reflection peaks. A clear peak broadening occurs if drifts exist, whereas the shape of the Kiessig fringes between the Bragg peaks is only weakly influenced by such a period drift. In contrast to this, statistical fluctuations of the period within the multilayer results in distortions of the Kiessig fringes, whereas the half widths of the Bragg peaks are not altered. Hence, both effects, systematic drifts and statistical fluctuations, can be separated.

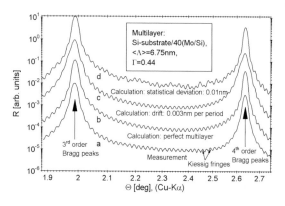

Figure 4.81: Section between 3rd and 4th Bragg reflection peaks of measured (a) and calculated (b–d) Cu-Kα reflectographs showing the influence of layer thickness irregularities

By comparison of the measured curve and the model calculations it can be determined that the real multilayer must have thickness deviations of < 0.003 nm, otherwise, the Kiessig fringes between the Bragg reflection peaks would not show this regularity. Hence, it is legitimate to say that nm-multilayers have to be and can be deposited with picometer accuracy!

Interface Roughness

In real multilayers, the transition from one layer to another is not an ideal step function. In contrast to the mathematically ideal situation, where the interface is described as the abrupt transition from one material to the second across a plane parallel to the substrate surface, real layers exhibit a certain waviness. This is called interface roughness (Fig. 4.82, left-hand side). Interface roughness has generally two causes. Firstly, each substrate has a finite roughness, which propagates through the multilayer and which, in some cases, can be amplified. Substrates having the currently highest surface quality show rms-roughnesses of typically < 0.1 nm. Secondly, the deposition process itself results in a certain layer roughness even if the substrateis perfectly smooth. With lower energies of the incoming particles on the substrate, i.e. with lower normalized temperatures in the structure zone model (Section 2.5), this effect increases. How can interface roughness be reduced?

The initial substrate roughness must be improved further. This is particularly true for large substrates with curved surfaces. The challenge is to meet the specifications for the figure and the micro-roughness of the substrates at the same time. This problem will be solved by the substrate manufacturer.

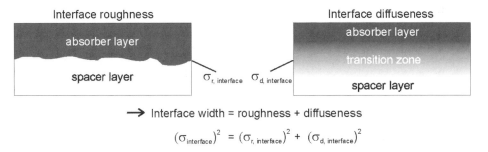

$$\left(\sigma_{interface}\right)^2 = \left(\sigma_{r,\,interface}\right)^2 + \left(\sigma_{d,\,interface}\right)^2$$

Figure 4.82: Schematic view of interface imperfections in nm multilayers. Left-hand side: interface roughness, right-hand side: interface diffuseness. The parameter $\sigma_{interface}$ describing the interface width, consists of both contributions

With a given substrate surface, only a slight improvement in high-frequency roughnesses is possible by the deposition of so-called buffer layers, which are deposited using electron evaporation in combination with ion polishing [4.225]. Another way to smooth substrate high-frequency roughnesses is to deposit buffer layers by PLD, where the particle energies can also be high (compare Section 2.5).

Additional roughness originating from the deposition process can be reduced by the proper choice of the deposition parameters. Generally, the increase in the mobility of the particles condensing on the substrate results in a decrease in the layer roughness. Considering the most common deposition methods, it turns out that typical energies of the particles are very different (Table 4.4). Consequently, methods to smooth the growing layers are in some cases necessary and have to be adapted to the specific deposition technique. Layers deposited by e-beam evaporation can be polished by high-energy ions, which are directed onto the layer surface under an optimized angle of incidence.

Using this method the typical columnar growth of such layers can be compensated and multilayers having high reflectances can be prepared [4.221, 4.226, 4.227].

Table 4.4: Common deposition methods with typical kinetic energies of the particles condensing on the substrate surface and methods to reduce the roughness of the growing layers

Deposition method	Typical kinetic energy of condensing particles / eV	Methods to decrease the layer roughness
e-beam evaporation	~0.1	Ion beam polishing
		Increase of substrate temperature
Magnetron sputtering	~10	Decrease in the sputter gas pressure
		Application of BIAS
Ion beam sputter deposition	~50	Adjustment of the ion energy
Pulsed laser deposition	~100	Not necessary

Using magnetron sputter deposition, much higher particle energies can be reached than with e-beam evaporation. Nevertheless, the columnar growth of the layers can still be observed. The degree of the columnar growth can be strongly influenced by the sputter gas pressure and the target–substrate distance. Particles coming directly from the target, which do not lose energy due to collisions with sputter gas atoms or ions on their way to the substrate, have a mobility high enough to form smooth layers for most of the materials. Hence, such collisions must be reduced as far as possible. This can be realized by the reduction of the sputter gas pressure and the target–substrate distance. Both lead to an increase in the mean free path of the particles. However, there are materials (Ag, Au, …) that form rough layers even for the lowest possible sputter gas pressure and target–substrate distance. For such materials the application of a bias voltage can accelerate ions from the plasma to the substrate surface and smooth the growing layer.

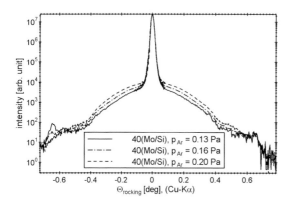

Figure 4.83: Rocking scans of 1st Bragg reflection peaks of Mo/Si multilayers deposited using different Ar sputter gas pressures

Quantitative determination of the roughnesses of all layers within the multilayer stack is very difficult. The description of the layer roughness using only one parameter $\sigma_{interface}$ is just an approximation. A more realistic description has to take into account the distribution of the roughness frequencies, which is characterized by the so-called power spectral density (PSD) function, the Fourier transform of the surface heights $z(x,y)$ [4.197]. Since the determination of the PSD function requires a high effort (interferometry, AFM), in most cases only the simplified description with only one parameter $\sigma_{interface}$ is used and the problem remains that no distinction between interface roughness and interdiffusion can be made. Both effects influence $\sigma_{interface}$ in the same way. Nevertheless it is possible to compare multilayers with respect to their interface roughness by diffuse scattering, which is only sensitive to layer roughness and not to diffuseness. Performing non-specular scans, the intensity of the diffuse scattered light is a measure of the interface roughness (cf. Section 3.3).

Comparing two multilayers, it can be concluded which stack has the higher roughness. An experimental example is shown in Fig. 4.83, where Mo/Si multilayers prepared using different Ar sputter gas pressures have been measured in the rocking scan mode around the first Bragg peak. It can clearly be seen, that the multilayer deposited with $p_{Ar} = 0.15$ Pa shows a higher diffuse background than that prepared at $p_{Ar} = 0.10$ Pa.

Interface Interdiffusion

One of the main reasons for the discrepancy between the reflectances of real multilayer structures and their theoretical limits is the formation of transition layers between the original layers due to interdiffusion and reaction of the individual layer materials (Fig. 4.82, right-hand side). These transition layers reduce the optical contrast between spacer and absorber layers and cause a reflectance loss. Consequently, a lot of attention must be paid to reducing the interdiffusion to its smallest possible extent. The first obvious way to do this is to reduce the mean energy of the particles coming from the source and forming the layers on the substrates. This is particularly true for the PLD process where the highest mean particle energies are observed. However, even with deposition methods, like e-beam evaporation or sputter deposition, which emit particles with much lower energies the interdiffusion of the individual layers cannot be completely avoided for every material combination. In this case the introduction of tiny barrier layers is a further option to reduce interdiffusion and interface reactions. In connection with applying such barrier layers the question arises as to how the calculated reflectance is influenced. The spacer and absorber material are always chosen as the optimum combination for a given photon energy. The introduction of a third material for the barrier layers normally reduces the calculated reflectance. Hence, the best compromise between the reduction of interface interdiffusion and the calculated reflectance loss must be found. In the following it will be shown how this can be done for the example of Mo/Si multilayers with period thicknesses of 6.9 nm, which are used as reflectors for EUV lithography.

When the Bragg condition is fulfilled, a standing electromagnetic wave field with wavelength equal to the period thickness is formed within the multilayer (Fig. 4.84, [4.219]). Considering the amplitude of this standing wave it can be stated that the nodes and antinodes are at fixed positions of the multilayer stack. This results in the fact that at the locations of the nodes of the electrical field, i.e. at the Mo-on-Si interface, the addition of EUV-absorbing barrier layers will only weakly affect the electrical field and the EUV reflectance. Hence any material that is suitable as a diffusion and reaction barrier can be used at the Mo-on-Si interface provided that the barrier layer prepared is thin enough. Calculations show that, regardless of the material that is used as barrier, the EUV reflectance is >75% for barrier layers with thicknesses $d \leq 0.6$ nm [4.219]. Nevertheless, a final decision has to consider the suitability of the particular material as an inhibitor of diffusion and reaction between the individual layers of Mo and Si.

Looking at the Si-on-Mo interface the situation is completely different: the standing wave field has an antinode at this interface and therefore absorption of barrier layers at this stack position will strongly affect the EUV reflectance. From calculations we can conclude that even materials with a lower EUV absorption like boron carbide or carbon cause a stronger loss of reflectance than does $MoSi_2$ if the thickness of these barrier layers approaches the size of the $MoSi_2$ intermixing zones. Knowing that the thickness of the $MoSi_2$ transition layer is about 0.7 nm at the Si-on-Mo interface, an increase in the EUV reflectance can only be realized if barrier layers thinner than 0.5 nm for B_4C or 0.3 nm for C will prevent the interdiffusion [4.219].

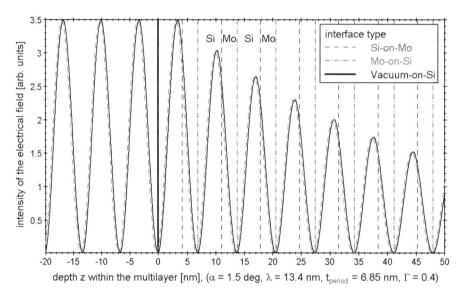

Figure 4.84: Calculation of the standing wave field within the multilayer for the example of Mo/Si multilayers for the reflection at $\lambda_X = 13.4$ nm and $\alpha = 1.5°$ [4.219]

The procedure described can analogously be applied to every other material combination of spacer and absorber layers. It is always valid that the nodes of the electrical field are on the interfaces absorber-on-spacer when the Bragg condition is fulfilled. Consequently, on this interface the choice of the barrier layer material has not to take into account itsabsorption. It can be freely optimized with respect to the inhibition of interdiffusion on these interfaces.

Since the thickness of the diffusion barrier layers has to be at least >0.2 nm to form closed layers, this method can be applied preferably in multilayers with a period thickness >4 nm. If the period thickness is smaller, the relative contribution of interface imperfections increases. The period of multilayers with moderate period thickness (>4 nm) typically consists of an absorber layer A, a spacer layer S and two transition layers T1 and T2. The thickness of the transition layers T1 and T2 can depend on the material combination, the deposition method used and the kind of the interfaces (spacer-on-absorber or absorber-on-spacer). In multilayers with period thickness <4 nm the thickness of the pure absorber and spacer layers tends to zero and the whole multilayer only consists of transition layers. Consequently, this state is far from the theoretically ideal situation and the reflectance can strongly differ from the calculated values.

For Mo/Si multilayers the optimum barrier layer thickness can be determined by investigating the multilayer period contraction that is caused by the $MoSi_x$ formation

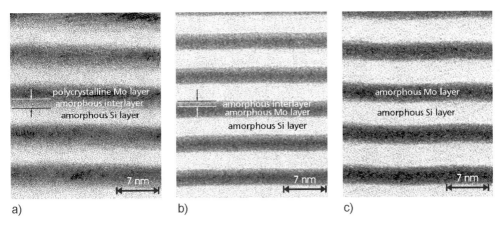

a)   b)   c)

Figure 4.85: TEM cross section of multilayers with barrier layers on the interfaces: a) B$_4$C on the Si-on-Mo interface, b) B$_4$C on the Mo-on-Si interface, c) B$_4$C on both interfaces (t_{B4C} = 0.5 nm in all cases) (HRTEM investigations made by courtecy of R. Scholz, MPI Halle)

depending on the barrier layer thickness and results in values of d_{B4C} = 0.2 nm on the Si-on-Mo interface and d_C = 0.4 nm on the Mo-on-Si interface [4.228].

For Mo/Si multilayers with period thickness of typically 7 nm, the introduction of barrier layers effectively reduces the interdiffusion. This has been found by comparing TEM cross sections of Mo/Si multilayers with barrier layers on different interfaces (Fig. 4.85). This improvement also results in an increase in the maximum EUV reflectance of typically > 1 % [4.219, 4.220, 4.228] from < 69 % to ≈70 %. Having in mind that at least 6 reflectors have to be used in an EUVL illumination system, the overall throughput would increase by > 10 %.

Deviations of Thin Film and Bulk Density

The assumption of bulk densities for nm layers can be another source of discrepancies between the calculated and measured reflectances of multilayer mirrors. In general, the layer densities are decreased compared to the bulk values. The differences can be up to 10 %. This is particularly true for PLD prepared multilayers, where vitreous amorphous layers are formed, which are far from the thermodynamic equilibrium.

The agreement of layer and bulk densities can be assumed if the layers show a polycrystalline structure. This is the case if a critical thickness t_c of the layer at which crystallization starts is reached. Often a texture of the crystallites is observed in which the most densely packed lattice planes are parallel to the substrate surface.

Since the reflectance of a multilayer depends on the optical contrast between spacer and absorber layers, a decrease in the spacer layer density or an increase in the absorber layer density will result in higher reflectances (Fig. 4.86). The densities of absorber layers can only be increased up to the bulk values of the pure element or compound. As mentioned above, this is roughly the case when polycrystalline layers are formed. However, a great potential exists in the reduction of the spacer layer density. One approach to reduce the density of nm layers is the implantation of H$^+$ ions by low energy ion bombardment. In Mo/Si multilayers a reduction of the Si layer density to (64 ± 5)% of the Si bulk value can be reached [4.229]. In addition, a minor smoothing effect of e-beam evaporated multilayers has been observed after H$^+$ ion bombardment.

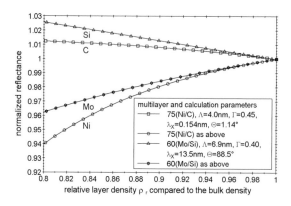

Figure 4.86: Calculation of normalized reflectances of Ni/C and Mo/Si multilayers depending on the spacer and absorber densities

Surface Oxidation and Contamination

Ultra-high vacuum (UHV) deposition techniques are used to prepare nm-multilayers. Except for the MSD process, the pressure during the multilayer deposition is in the range 10^{-7}– 10^{-6} Pa. For the MSD, the base pressure is also in this range and the use of sputtering gas of high purity (typical purities of Ar are 4N8 or 5N) ensures that mainly the partial pressure of the sputtering gas contributes to the total pressure. This pure environment during the coating process is necessary to obtain the required chemical purity of the individual layers, which is another parameter that influences the optical performance of the multilayer. However, after the deposition process the multilayers usually leave the UHV environment and are exposed to atmospheric and radiation attacks. Consequently, at least the top layer of the multilayer is subject to surface oxidation and contamination. Therefore additional capping layers (oxides, carbides, …), which cause absorption of the X-rays and phase mismatch between the partial waves reflected on the multilayer interfaces and on the capping layer, must be taken into account. This has to be borne in mind during the optimization of a certain multilayer design.

The decrease in reflectance due to undesired capping layers is stronger the higher the absorption of the radiation within the capping layer. The calculation of the reflectance depending on the capping layer thickness shows that the influence of capping layers is much more important at lower photon energies, where the absorption is high. Ni oxide capping layers with a thickness of up to 10 nm reduce the reflectance of Ni/C multilayers for Cu-Kα radiation ($\lambda_X = 0.154$ nm) only by < 5 %, whereas capping layers on Mo/Si multilayers for $\lambda_X = 13.4$ nm cause reflectance decreases of > 30 % and > 15 % for Mo oxide and Si oxide layers of similar thickness (Fig. 4.87). Therefore the choice of proper top layers is essential for multilayer reflectors, e.g. for EUV radiation.

Model calculations of the reflectance of Mo/Si multilayers show that higher values can be reached if Mo is used as top layer. However, taking the unavoidable oxidation and the results from Fig. 4.87 into account, we can conclude that Si should be more suitable as a top layer. This has been experimentally confirmed [4.217].

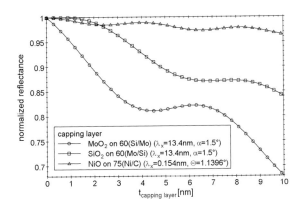

Figure 4.87: Calculation of normalized reflectances of Ni/C and Mo/Si multilayers depending on the type and the thickness of the oxide layer on top of the multilayer

If a certain oxide layer of thickness t_{oxid} is formed, this can be allowed for in the design of the top layer underneath the oxide layer. From TEM investigations it can be concluded that the Mo/Si multilayer with Si top layer exhibits a Si oxide layer of thickness ≈ 2.6 nm. Knowing this, the top Si layer thickness can be aligned in such a way that the reflection of X-rays on the top multilayer period is also in phase with all the other partial waves coming from the inner interfaces. The gain in reflectance can be of the order of 1 %.

4.6.3 High-Resolution Multilayers

Typical multilayers have resolving powers $E/\Delta E$ in the range 10 to 100, whereas perfect crystals usually show values of the order of 10000 [4.230]. Therefore multilayer X-ray optics are preferably used in applications where a high photon flux is required instead of a high resolution. However, there are cases where medium resolution optics with high integral reflectances are needed [4.231]. This was the driving force for the development of multilayer mirrors that can fill the resolution gap between conventional multilayers and single crystals.

Model Considerations

The comparison of the resolution of natural crystals and commonly used multilayers shows that crystal resolution is more than two orders higher. This originates from the fact, that X-rays penetrating the multilayer are absorbed at a certain depth. Therefore only a limited number of periods N_{min} contribute to the reflection of the X-rays. Since $E/\Delta E$ is proportional to N_{min}, it is also limited. To increase the resolution, multilayers consisting of layers with low absorption coefficients have to be used [4.231, 4.232]. Typical solid state materials with low absorption coefficients are light elements with low electron densities or compounds of them: B_4C, BN, C, Al, Al_2O_3, Si, SiO_2, Sc. Exceptionally low absorption coefficients can be used in the vicinity of absorption edges of the particular elements, e.g. for Si below the L absorption edge at $E = 99.4$ eV ($\lambda_X = 12.5$ nm). Far from any absorption edges the actual absorption coefficients have to be considered. A comparison of the absorption coefficients of the individual materials is shown in Table 4.5. The combination of the materials of column 1 and 2 results in multilayers with the lowest absorption and the highest resolution.

Table 4.5: List of solid state materials with the lowest absorption coefficients in different wavelength/ energy ranges (column 1 = lowest absorption). Combinations of materials in column 1 and 2 result in multilayers with the highest possible resolution

Wavelength (nm)	Photon energy (eV)	1	2	3	4	5	6	7
0–0.67	1839-$\propto$	B_4C	C	BN	Sc	SiO_2	Al	Si
-0.80	1560-	B_4C	Si	C	BN	SiO_2	Sc	Al
-2.28	543-	Al	Si	B_4C	C	BN	SiO_2	Sc
-3.02	410-	SiO_2	Al	Si	B_4C	C	BN	Sc
-3.11	399-	BN	SiO_2	Al	Si	B_4C	C	Sc
-4.37	284-	Sc	BN	SiO_2	Al	Si	B_4C	C
-6.59	188-	C	Sc	BN	SiO_2	B_4C	Si	Al
-12.5	99.2-	B_4C	C	BN	Sc	SiO_2	Si	Al
-17.1	72.5-	Si	B_4C	C	Sc	BN	SiO_2	Al
-29.8	41.6-	Al	Si	B_4C	C	BN	SiO_2	Sc

Besides the materials listed in Table 4.5 there exists a further possibility to realize high-resolution multilayers. It is well-known that carbon layers can exist in different modifications: graphite- or diamond-like. The ratio of sp^2 and sp^3 bonds is the main difference and this is also connected with a density change. This can be used to synthesize carbon/carbon multilayers [4.233, 4.234]. Particularly for energies >1.8 keV carbon/carbon multilayers can be of great benefit. According to Table 4.5 the combination B_4C/C should result in the highest resolution. However, it has been proved that B_4C/C multilayers are temporally-unstable. Their half-life at room temperature is only one month [4.235]. Because of the absence of chemical driving forces for interdiffusion between two different neighboring layers, carbon/carbon multilayers should be very long-term stable. If we assume densities of $\rho_{C1} = 2.2$ g cm^{-3} and $\rho_{C2} = 2.5$ g cm^{-3} for the two carbon layers, reflectance of up to 85 % ($t_{period} = 2$ nm, $N = 3000$, $E_{photon} = 8$ keV) can be predicted (Fig. 4.88).

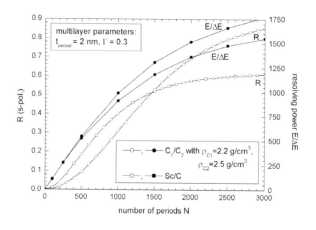

Figure 4.88: Calculated reflectance R and resolving power $E/\Delta E$ at $E_{photon} = 8$ keV of Sc/C and C_1/C_2 multilayers with $t_{period} = 2$ nm. Up to a period number of 1500 the Sc/C system shows higher reflectance. Up to a period number of 500 the resolution of both multilayer systems does not differ

From the theoretical point of view no alternative to carbon/carbon multilayers with high period numbers exists. However, practically it is difficult and takes a lot of time to deposit multilayers with up to 3000 periods. Additionally, it is not yet known how the internal stress of the layers affects multilayers with such a high number of periods. Therefore the question arose as to whether a combination with a higher density contrast and only a slightly higher absorption could be found for photon energies >1.8 keV. The analysis of the δ-values of all the materials listed in Table 4.5 shows that the highest contrast to carbon can be realized when scandium is used as the second material. Since scandium is the fourth material after B_4C, C and BN the resolution should also be high. The detailed calculation shows that, up to period numbers of 500, no resolution difference is observable comparing C_1/C_2 and Sc/C multilayers (Fig. 4.88). Looking at the calculated reflectance for $N = 500$, which shows $R \approx 10\%$ for C_1/C_2 and $R \approx 20\%$ for Sc/C, it can clearly be stated that Sc/C should be favored for period numbers $N \leq 500$. The comparison of multilayers with period numbers >1500 shows, just as clearly as before, that the C_1/C_2 system has higher reflectance as well as higher resolution. For period numbers between $N = 500$ and $N = 1500$ the choice of the material system depends on the application and the question of which property is more important: reflectance or resolution.

Experimental Results

Due to the high number of periods that is necessary to get maximum resolution and reflectance, the stability of the deposition process has to be extremely high. Any drifts or statistical fluctuations of the deposition rate would immediately result in an unacceptable peak broadening. These stringent stability requirements are particularly challenging for the PLD process and electron-beam evaporation, where the stability of the particle source is lower than with sputter deposition methods. The permanent altering of the target surface by the irradiation with the laser beam in the PLD results in an unavoidable drift of the deposition rate. However, the drift is reproducible and can be precisely measured and compensated. Using e-beam evaporation, small deviations of the vapor pressure cause considerable changes in the deposition rate. Therefore in-situ control of the deposition process is mandatory. Moreover, the stable e-beam evaporation of materials like C or B_4C is very demanding.

Up to now no systematic investigations have been carried out for all wavelength ranges listed in Table 4.5. Most of the interest has been focused on the spectral range with photon energies > 1560 eV. For Al_2O_3/B_4C multilayers with 680 periods a resolving power of $E/\Delta E = 300$ has been observed at $E = 12$ keV [4.230]. The peak reflectance has been determined with $R = 18\%$, but it was explained that the real reflectance and resolution could be better than the measured values because of the fact that the mirror surface is not perfectly flat. A certain sample curvature can cause angular broadening of the reflection peak. This also shows that not only the preparation of high-resolution multilayers is challenging, but also the accurate determination of resolution and reflectance of the prepared multilayers needs a lot of effort and requires X-ray beams with low divergences (e.g. at synchrotron radiation sources), high-quality monochromators and goniometers and perfectly flat samples, i.e. deformations of the substrate surface due to multilayer stress must be avoided.

Further practical examples of high-resolution multilayers are B_4C/Si and Mo/Si combinations which can be favorably used in the energy range below the Si-K absorption edge at $E = 1823$ eV. Depending on the multilayer period (composition and thickness) and the number of periods, various resolving powers and reflectance have been measured (Fig. 4.89). The highest resolution is reached for B_4C/Si with the lowest multilayer period $\Lambda = 1.98$ nm and the highest number of periods $N = 500$: $E/\Delta E = 492$ ($2\Theta = 20.37°$, $E = 1.8$ keV). However this is connected with a comparatively low reflectance of $R \approx 3.7\,\%$. The reflectance increases with increasing multilayer period thickness, but this happens at the expense of resolving power: multilayer combination B_4C/Si, $N = 300$, $t = 4.0$ nm, $R \approx 24\,\%$, $E/\Delta E = 203$ ($2\Theta = 10.29°$, $E = 1.8$ keV). A further increase in reflectance is possible by changing the absorber material from B_4C to Mo. Then, reflectance of up to 65 % has been measured (Fig. 4.89).

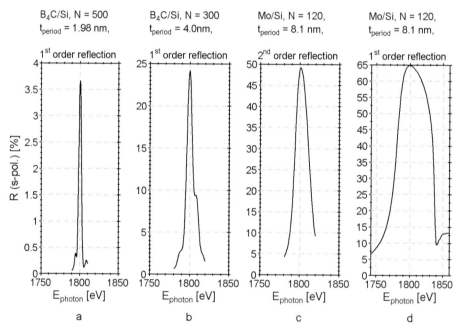

Figure 4.89: Experimental results of high-resolution multilayers at 1.8 keV: a) $E/\Delta E$=492, R=3.7%, b) $E/\Delta E = 203$, $R = 24\%$, c) $E/\Delta E = 89$, $R = 49\%$, d) $E/\Delta E = 31$, $R = 65\%$ (courtesy of F. Scholze, B. Beckhoff, J. Tümmler, PTB/BESSY2 Berlin)

4.6.4 Multilayers with Uniform and Graded Period Thickness

The application of nm multilayers as X-ray optical elements not only requires high reflectance of the multilayers, but also the accurate lateral distribution of the multilayer period thickness across the substrate surface. The optimum X-ray optical performance of a mirror can only be reached if the Bragg condition is exactly fulfilled at every mirror position. Consequently, different angles of incidence require adapted multilayer period thickness.

The most challenging specifications concerning precision of the lateral thickness and size of the substrates are given by the EUVL. Substrates with diameters of up to 400 mm have to be coated with multilayers, the layer thickness of which differs only by 0.01 nm from the theoretical value on every surface position. Since aspherical substrate surfaces have to be coated, demanding gradients are necessary on almost every mirror of the whole system.

Other applications only require uniform thickness profiles (e.g. monochromators, mask blanks for the EUVL), linear gradients (e.g. *Goebel* mirrors) or gradients with rotational symmetry (e.g. mirrors for *Schwarzschild* objectives) of the multilayer period.

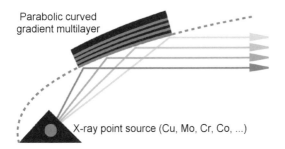

Parabolic curved
gradient multilayer

X-ray point source (Cu, Mo, Cr, Co, ...)

Figure 4.90: Schematic view of the collection and collimation of divergent X-rays emitted by a point source by a so-called *Goebel* mirror. On the parabolic-shaped substrate surface a gradient multilayer is required to fulfill Bragg´s equation at every mirror position

Goebel mirrors are used to collect and collimate divergent X-ray beams emitted by commercial point X-ray sources (e.g. Cu, Mo, Cr, Co, …) [4.236]. A schematic view of the arrangement is shown in Figure 4.90. The shape of the mirror surface is parabolic in the first direction and flat in the second. According to this, the multilayer period must have a gradient in the first direction and a uniform thickness in the second one. Typical dimensions of the mirrors are 60 x 20 mm². For Ni/C multilayers prepared by PLD, the difference between actual and nominal values of the multilayer period thickness is typically < 0.02 nm at every mirror position.

For layers with uniform or rotational symmetric thickness distributions, an arrangement with rotating substrates is the most suitable one, whereas a linear movement is well-adapted to one-dimensional gradients. However, the realization of a special thickness profile requires the optimization of substrate motion and/or baffles or masks.

This can be shown for the example of multilayers prepared by MSD (for the principle see Fig. 2.79). The deposition of uniform multilayers on flat substrates is a rather simple task, but even this needs optimization. If a constant frequency f_R over the sputter sources is chosen, a decrease in the thickness of typically 2 % over the substrate radius of 75 mm will be obtained. Uniform layers can only be deposited if this behavior is taken into account and if the outer parts of the substrates are kept slightly longer over the source. Hence, the substrate speed ω_R depending on the actual rotation angle α has to be controlled. Doing this, typical uniformities with relative standard deviations of the multilayer period in the range 0.04–0.05 % can be achieved on substrates with 150 mm diameter [4.219]. The potential of this rather simple target–substrate arrangement has also been shown on large curved substrates [4.237].

4.7   Functional Electric Layers

This section is devoted to frontiers in thin film research in the field of realization of films with well-defined electronic and electrical transport properties. Generally, high-precision performance of thin films for electronics means the combination of a whole set of transport properties required for the technical design of electronic devices. Thus, for *resistance layers* applied in a microelectronic or hybrid-electronic environment a defined resistivity should be combined with an as low as possible temperature coefficient of resistivity (TCR) including a high long-term stability of resistivity. For *thermoelectric sensor and generator devices* thin film materials are needed which are characterized by a high *Seebeck* coefficient, low resistivity and low thermal conductivity which are contradictory demands from the point of view of the transport parameter physics. Therefore, in this section the optimization of films with defined electrical and thermoelectric performance will be addressed.

4.7.1 Resistance Layers

Resistance layers are needed for resistor applications, both in discrete devices, e.g. for the surface mount, and as a component of the integrated passives (resistors, capacitors, inductors) in integrated thin film resistor layouts. The most intensive impulse is coming from the integrated passive technology due to the general motivation to replace the surface mount devices. The aim of the research and development for passive devices is, analogous to the situation for active devices some decades ago, characterized by the realization of near-zero fabrication costs due to simultaneous production, by use of the scaling-down advantages, by the elimination of individual mechanical interconnect technology, and by achievement of a high yield. The advantage of integration of passive devices lies in the following effects:
(i)   reduction of the system mass thanks to the embedding of the devices in the substrate,
(ii)   improvement of electrical performance arising from the lowering of parasitic effects,
(iii) enhancement of the design flexibility due to the high parameter variability in the
       integrated-circuit technology,
(iv)   improvement of reliability thanks to the elimination of soldering, and
(v)   reduction of costs provided by mass production [4.238].

Table 4.6:. Thin film resistor materials (parameters demanded)

Sheet resistance, kΩ/square	Materials	Material tolerances $\Delta R/R_{nom}$, %	Thickness range t, nm	Temperature coefficient of resistivity |TCR|, ppm K^{-1}	Stability $\frac{\Delta R}{R}(1000\ h), \%$
$5\cdot10^{-4}$–0,1	CuNi [4.239], NiCrAl [4.240]	5–10	500–4000	10–50	< 0.1
0.1–1	TaN$_x$ [4.241], NiCr/NiCrO [4.241], CrSi [4.243],	5–10	10–200	< 15	< 0.1
1–5	CrSi(O,N) [4.244], CrSiAl(O,N) [4.245]	5 –10	10–100	5–15	< 0.1
5–10	Cermets [4.246]	< 10	10–50	25	< 0.1

The performance of thin film resistors, which can be realized as discrete-like, integral or on-chip components, is determined by a complex correlation of composition, structural constitution, mechanical stress and further factors. The fabrication of high-precision and high stability resistors requires a sophisticated technology characterized by a well-optimized deposition technology including a well-defined heat treatment for the adjustment of the resistivity and TCR. To meet the current requirement of the circuit technique resistor values in a very broad range should be realized: from several $m\Omega$ to some $M\Omega$. Due to the limitations of the thickness and the length of resistor layout this broad resistivity spectrum can be realized only by a whole set of different materials (see Table 4.6). The most challenging tasks are the realization of the lower and the upper resistance limit of thin film resistor devices. Therefore in the following these two areas will be considered in more detail.

High-Resistivity Layers

The rapid technological progress in many branches of electronics, especially in the information and telecommunication technology, requires continuous improvement of the electronic components to achieve higher precision, reliability and integration. So, for thin film resistors the need for a higher component performance has become absolutely urgent in recent years. Novel integral resistors need research and development for higher resistivity materials that can deliver a broader range of square resistance values. A higher parameter precision (TCR, tolerance, stability) will be required to reduce size. The materials requirements are fixed in the Technology Roadmap for Integral Passive Components [4.247]. The increasing importance of the qualification of resistor materials arises from the gradual replacement of discrete resistors by integrated devices. At present the resistor fabrication is increasingly realized by a multi-step lithography process using several separate masks [4.248]. The most important requirement for practical use of resistors is the sustaining of the nominal resistance over the whole time of use within the specified operating temperature range $-50\ °C \le T \le 150\ °C$. Therefore, low TCR and high stability $\Delta R/R$ (1000 h) are the main parameters for thin film resistors. For characterization of the temperature dependence of TCR the parameter ΔTCR, the so-called parabolicity according to equation (2.4), is commonly used. The demanded precision properties depend strongly on the resistance range. Basically, it is observed that there is a tendency for degradation of precision properties with increasing resistance [4.249]. Therefore, the material qualification and parameter optimization have to be done for all materials listed in Table 4.6. Because the resistor performance depends in a complex way on the film and substrate properties as well as on the construction and layout of the resistor device the preparation and formation process of the films should be understood in correlation with the electrical transport properties as deeply as possible. In this context, only by comprehensive investigations of the electrical film properties together with analytical studies allowing the characterization of composition, structure and crystallization state can a systematic optimization of the precision properties be achieved.

   In the following, the system Cr–Si–Al is considered as an example for the demonstration of the state of the art in thin film resistor materials. In compositional studies a Si/Cr atomic ratio of about 2 has been chosen in order to get a parabolicity drift within the permissible limit of ΔTCR $\approx \pm 25$ ppm K^{-1} over a wide resistivity range of 1 mΩ cm $\le \rho \le 10$ mΩ cm [4.250]. For fabrication of high-resistivity films the sputtering process is commonly carried out in the presence of oxygen, causing the formation of an insulating matrix. However, not only oxide components but also nitrides or oxide–nitride mixtures are able to contribute effectively to the lowering of the conductivity. Additionally, nitride-containing films are generally characterized by a significantly higher thermal and chemical stability in comparison with oxide containing films.

   Thus, besides Ta–N films (see Table 4.6) the system Cr–Si–N has also attracted interest as a stable material for resistive films [4.244, 4.250]. In Fig. 4.91 Cr–Si- based oxide and

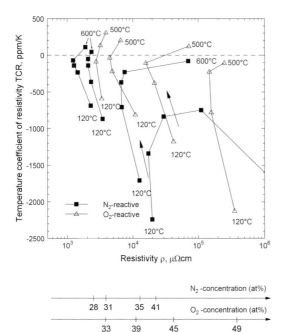

Figure 4.91: Annealing behavior of resistivity and TCR of $Cr_{26}Si_{74}(O,N)$ films. The annealing temperature is increased along the curves starting from 120 °C stepwise:
300 °C → 400 °C → 500 °C and
400 °C → 500 °C → 550 °C →
600 °C for O-doped and N-doped films, respectively

nitride films are compared. The analysis of the annealing behavior data indicates that there are some serious disadvantages of nitride-containing films. As shown in Fig. 4.91 the resistivity range with a small TCR is much smaller in Cr–Si–N than in Cr–Si–O. Analogous behavior is found in the system Cr–Si–Al and at deposition in a partially reactive

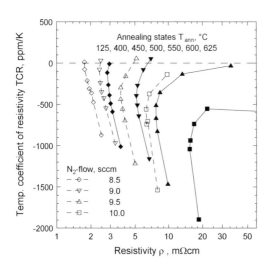

Figure 4.92: Annealing behavior of resistivity and TCR of N_2 –reactively deposited films on oxidized Si wafers (open symbols) and Al_2O_3 ceramic (filled symbols) sputtered from the target SiCrAl (30 at.%)

N_2-atmosphere a high stability can be expected, arising from the mixed dielectric consisting of the insulator SiN_x and the wide gap semiconductor AlN. Therefore, summarizing this discussion, one can state that high-ohmic resistive films with high stability and precision can be realized only in a very restricted resistivity range.

In Fig. 4.92 the annealing behavior of Cr–Si–Al–N films sputtered from the target CrSiAl (30 at.%) is shown in dependence on the N_2 flow determining the concentration of the dielectric phases in the films. We observe a decrease in the TCR in the as-deposited state with increasing resistivity. The resistivity itself increases with increasing reactivity, i.e. with increasing nitrogen content. The comparison of different substrates shows that on Al_2O_3 ceramic TCR values close to zero can be achieved up to larger resistivities than on oxidized Si wafers. Figure 4.92 illustrates that precision properties can be realized in films with as-deposited TCR values in the range -1200 to -1400 ppm K^{-1}.

Figure 4.93 presents the precision properties in more detail. In the left diagram the reduced resistance change is shown in dependence on the temperature for annealed films of variable resistivity ρ (25 °C) on ceramic. The weak temperature dependence indicates small TCR values over the whole temperature range considered. The right diagram presents the parabolicity ∆TCR as a function of resistivity. Films on Al_2O_3 ceramic show ∆TCR in the vicinity of zero up to about 20–25 mΩ cm. Thus, high precision with /TCR/ < 15 ppm K^{-1} can be realized up to a sheet resistance of about 4 –5 kΩ/square. These values are a verification of the physical limit of the high-ohmic precision achievable with nitride films of the composition CrSiAl (30 at.%) (see Section 2.3.6).

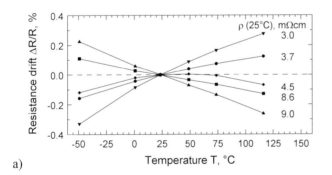

a)

b)

Figure 4.93: Precision properties of N_2-reactively deposited films, sputtered from the target SiCrAl(30 at%) with Si/Cr ratio c_{Si}/c_{Cr}=2. a) normalized resistance drift ∆R/R as function of temperature; b) parabolictiy ∆TCR as function of resistivity

The long-term stability of uncovered films, i.e. the resistance drift of the films after exposure at 155 °C in air for 1000 h is shown in Fig. 4.94. This resistance drift is found to be well below 0.05% and only weakly dependent on the resistivity. Thus, the stability of the nitrogen-containing films is about one order of magnitude higher than in O_2 sputtered films from the same target, demonstrating the usability of these films for extreme conditions.

The electrical behavior of resistive material is characterized by a strong correlation with both the structural and the compositional status of the films. Only from knowledge of the structural film properties can a strategy for parameter optimization be achieved. Therefore, special morphological studies are required for interpretation of the transport properties.

The goal of the TEM characterization is the observation of the morphological development and the phase formation in the films, especially on heat treatment. Normally, the thin films are deposited on special substrates, e.g. on silicon or silicon oxide. For the TEM investigation the films should be deposited on a NaCl single crystal with a very smooth surface. The advantage is an easy possibility of electron microscopic specimen preparation: The thin film can be dissolved from the surface in distilled water and transferred onto a Cu support grid or onto a Mo grid if the specimen is to be annealed in situ. Cu tends to a high surface mobility at higher temperatures and thus the formation of Cu compounds within the thin film is possible if a Cu grid is used for heat treatment experiments.

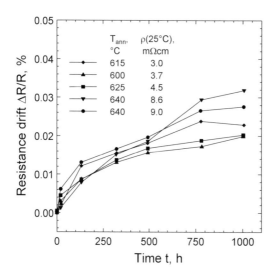

Figure 4.94: Resistance drift of N_2-reactively sputtered films, sputtered from the target CrSiAl(30at%) with $c_{Si}/c_{Cr}=2$ on Al_2O_3 ceramic during long-term exposure to 155 °C

Figure 4.95 shows an example of TEM investigation of mixed Cr–Si thin films with additives of Al and O, exemplary for films with high resistivity.

In the as-deposited state the film is amorphous, as demonstrated in Fig. 4.95a. The bright field TEM micrograph shows only a small phase contrast by a suitable defocusing. The diffraction pattern indicates two diffuse rings. By quantitative analysis of this pattern it is possible to determine the probability of distances between the next neighboring atoms in the amorphous film. Using this method it could be shown that these distances are influenced by the oxygen content (see Fig. 4.96).

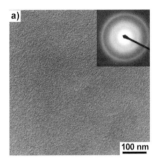
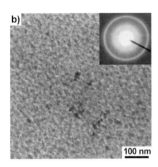
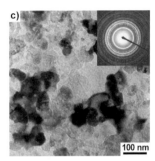

Figure 4.95: TEM micrographs and diffraction patterns of Cr–Si–Al thin films sputtered in O atmosphere with different compositions. a) $Cr_9Si_{51}Al_{40}$, as-deposited state, b) $Cr_{15}Si_{45}Al_{40}$ at 600 °C, c) $Cr_9Si_{51}Al_{40}$ at 600 °C

The Fourier transformation of the scattering curves (Fig. 4.96a) leads to reduced radial density functions (RDF – Fig. 4.96b, for method cf. e.g. [4.253]). It can be seen that the RDF allows one to determine the probability of distances between the next neighboring atoms in the distances. This distance can be assigned to the Si–O spacing within the SiO_2 tetrahedron (Fig. 4.96c). Hence, it can be assumed that the oxygen is bonded on the silicon in this case. Upon annealing, dependent on the oxygen content, the equilibrium phases $CrSi_2$ (also CrSi), Cr_5Si_3 and Cr_3Si are formed [4.251]. In the case of an aluminum content in the film mixed phases $Cr(Al,Si)_2$ could be observed [4.252].

The annealing behavior of sputtered Cr–Si thin films depends on fluctuations of the initial composition. To prove the homogeneity it is necessary to analyze this composition with high lateral resolution, typically by energy dispersive X-ray spectroscopy (EDXS) in an analytical TEM. Using the common standardless routines without any corrections for the quantification of EDX spectra acquired in the TEM both accuracy and reproducibility are low and the spread of the results can reach some atom-%.

This can be improved by consideration of the partial absorption of the X-rays within the specimen. For an absorption correction the mass absorption coefficients, the density and the

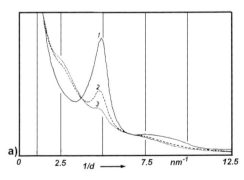
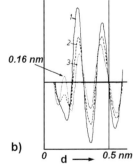
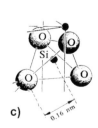

Figure 4.96: Analysis of diffraction patterns of amorphous Cr–Si–O films. a) electron scattering curves of 3 samples with oxygen contents of (1) $c_O \approx 0$ at.%, (2) $c_O \approx 43$ at.%, and (3) $c_O \approx 56$ at.%, b) corresponding RDF curves [4.251], c) structure model of the SiO_2 tetrahedron [4.252]

thickness of the specimen foil must be known. Commonly the density of thin films is not comparable with that of bulk materials and the lateral foil thickness is not known with sufficient accuracy. Therefore experimental methods try to avoid the direct input of density and thickness and use measurements at different take-off angles [4.254], on wedge-shaped sample areas [4.255] or they are based on the electro-neutral ionic compounds [4.256].

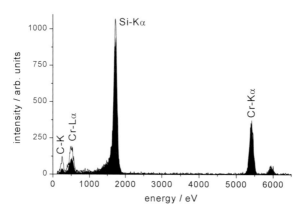

Figure 4.97: EDX spectra of two Cr–Si thin films with the same composition, normalized on the same height of the Cr-Kα peak. Hollow: film thickness 50 nm, filled: film thickness 150 nm

Another method bases on the comparison of the quantification results obtained by use of different peak pairs as demonstrated on Cr–Si thin films in Figs. 4.97 and 4.98. Figure 4.97 shows two different EDX spectra of films with the same composition but at different film thickness. The ratio of the peak heights of Cr-Kα and Cr-Lα is clearly shifted to a higher value for the thicker film. This can be explained by a stronger absorption of the Cr-Lα radiation within the specimen because of its lower energy.

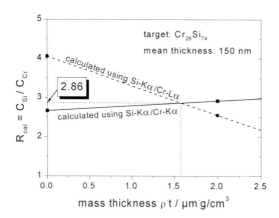

Figure 4.98: Absorption correction by fitted mass thickness using different pairs of characteristic X-ray peaks in EDX [4.257]

This fact can be used for the absorption correction. The condition for the method is an exact calibration of the Cliff–Lorimer k-factors for the K-peaks as well as for the L-peaks. Using the quantification formula with consideration of absorption in thin films (cf. Section3.4) and setting different values for the (unknown) mass thickness for the pairs Si-Kα/Cr-Lα and Si-Kα/Cr-Kα the diagram in Fig. 4.98 can be calculated. Assuming a value lower than the real mass thickness the absorption effect is considered inadequate and the Si content is overvalued if the pair Si-Kα/Cr-Lα is used and undervalued for the pair Si-Kα/Cr-Kα and vice versa for an assumed mass thickness higher than the real value.

The intersection of both curves gives the correct mass thickness and the corrected result for the Si/Cr ratio on the ordinate. The value 2.86 is in excellent agreement with the composition of the sputter target (2.85). Using this method the accuracy could be improved and the spread of the results could be decreased from about 4 at.% to 1.2 at.%.

Low-Resistivity layers

In contrast to high-ohmic films the realization of low-ohmic devices needs layers with large thickness. To obtain a sheet resistance of the order of 1 Ω/square, films with a thickness of > 500 nm should be prepared. As a rule, films in this thickness range need an adhesion layer for reliable mechanical contact with the substrate. Additionally, the materials applicable preferably for low-ohmic resistors, e.g. Ni, Cu–Ni, Cu–Cr, Ni–Cr–Al, tend, as thin films, to the formation of high internal stress which increases the risk of delamination. Further, to ensure a low contact resistance in low-ohmic resistor devices the surface oxidation should be suppressed completely. Therefore low-resistivity films need both adhesion and protection layers in order to fulfill the requirements for electronic applications.

Typical examples for low-ohmic systems are the three-layer arrangement NiCr/Cu-Ni/NiCr and two-component films Cu–Cr characterized by a wide miscibility gap [4.258].

The triple layer NiCr/Cu–Ni/NiCr is based on the well-known bulk resistive material Constantan ($\rho \approx 50$ μΩ cm) [4.259] and the thin film material NiCr ($\rho \approx 140$ μΩ cm) [4.260]. Such a three-layer stack fulfills the requirements of the performance parameters *sheet resistance* ($R_F < 1$ Ω/square), *temperature coefficient of resistivity* (TCR < 50 ppm K⁻¹) and *stability* ($\Delta R/R[1000h] < 1\%$). However, the exposure to elevated temperatures during the technological process up to 300 °C can cause interdiffusion processes within the stack, thus influencing the device parameters. Therefore the study of the correlation of electrical properties and the interface quality of the stacks is very important for the understanding of the resistor behavior.

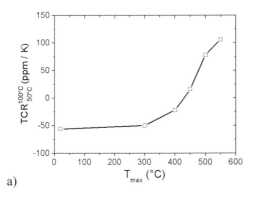
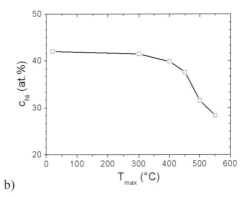

a)   b)

Figure 4.99: Annealing behaviour of NiCr/CuNi(Mn)/NiCr trilayers [4.261]; a) TCR between 50 and 100 °C versus T_{max}, b) Ni concentration in the CuNi(Mn) sublayer versus T_{max}

Auger electron spectroscopic (AES) depth profiles can give decisive hints for changes in the TCR in the NiCr/CuNi(Mn)/NiCr trilayers [4.261]. For quantitative investigation the TCR has been determined between 50 °C and 100 °C by the formula

$$TCR_{50°C}^{100°C} = \frac{R(100\ °C) - R(50\ °C)}{R(75\ °C) \cdot 50\ K} \qquad (4.14)$$

The result is shown in Fig. 4.99a. The zero crossing occurs at about 420 °C. The behavior "TCR = 0" is extremely important for applications of such films as resistors. By AES investigations it could be shown that the increase in the TCR is connected with a decrease in the Ni content within the CuNi(Mn) sublayer (see Fig. 4.99b).

In this way the increase in the TCR can be explained by structural changes in the film configuration. Its shift from negative to positive values is caused by the reduction of the Ni content in the CuNi(Mn) sublayer and the formation of a $Cu_{0.70}Ni_{0.28}Cr_{0.02}$ phase.

Another example concerns the investigation of sputtered Cu–Cr thin films representative of a low resistivity with application potential as material for interconnects in semiconductor devices. In the as-deposited state the morphology is fine-crystalline, associated with a metastable bcc structure (see Fig. 4.100).

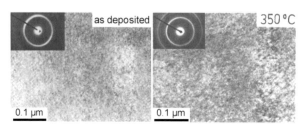

Figure 4.100: Results of in situ annealing of mixed Cu–Cr films: TEM bright field images and diffraction patterns

At temperatures of 350 °C a slight grain growth is observable with maintained bcc structure. This is clearly visible from additional rings in the diffraction pattern. These changes are combined with a decrease in the electrical resistance and an increase in the temperature coefficient. In that way it is possible to adjust the resistance and the temperature coefficient by a special heat treatment procedure [4.262]. In dependence on heat treatment two different performance levels could be adjusted in the films:

(i) After annealing at $T_{ann} < 300$ °C the films are characterized by a TCR close to zero, a good stability ($\Delta R/R \approx 0.1\%$) and a low internal stress. Thus, on the basis of the resulting resistivities $\rho \approx 60–80\ \mu\Omega$ cm, Cu–Cr films can be used as low-ohmic resistors.

(ii) The heat treatment in the temperature range 300–500 °C produces films with $\rho \approx 30\ \mu\Omega$ cm, a low positive TCR in comparison with pure metals and good stability, which make the film attractive as a metallization system in electronic devices [4.258].

4.7.2   Thermoelectric Thin Films

The performance of thermoelectric materials is described commonly by the dimensionless figure of merit $Z \cdot T$ expressed as

$$Z \cdot T = \frac{S^2 \cdot T}{\rho \cdot \lambda} \qquad (4.15)$$

where S is the *Seebeck* coefficient, ρ is the resistivity, and λ is the total thermal conductivity characterized by an electronic and lattice contribution. The free-carrier concentration in thermoelectric semiconductors has to be adjusted within an optimum interval (10^{19}–10^{20} cm^{-3}) in order to minimize the opposite influence of the thermoelectric power and electrical conductivity with temperature. The use of thin films as active thermoelectric components offers some advantages over bulk materials: (i) the realization of the multi-quantum well concept and multi-layered structures for lowering the thermal conductivity needs the availability of well-defined films, (ii) surface and micro-sensors as well as miniaturized low-power generators and on-chip coolers can be prepared only on the base of thin films, and (iii) the combination of the thermoelectric device technology with the IC technology allows a very convenient signal processing in sensor arrays, allows on-chip generation and cooling and offers a possible way for the mass production of thermoelectric modules. The optimization of the thermoelectric device performance needs a careful materials selection because the band gap should be large enough to prevent the onset of intrinsic conductivity in the intended operating temperature range. For the low-temperature range $100 \text{ K} \leq T \leq 300 \text{ K}$ the narrow gap semiconductors Bi–Sb and Bi$_2$Te$_3$ have been established as classic materials. In the following, materials for the temperature range $T \geq 300 \text{ K}$ are considered, characterized by high stability and compatibility with the IC technology. As described in Section 2.3 the compound class of silicides offers attractive features for the realization of this aim. As an example, the two wide-gap materials β-FeSi$_2$ and Ir$_3$Si$_5$ and the narrow-gap silicide ReSi$_{1.75}$ are considered.

For the development of thermoelectric thin films with high thermoelectric efficiency knowledge of the correlations between the thermoelectric properties (thermoelectric power, electrical and thermal conductivity) and the micro- and nanostructure is extremely important. In the following it will be demonstrated that a complex understanding can be achieved by the combination of transport property studies with structure investigations by TEM.

Iron–Silicon Thin Films

For medium and high temperatures, thin films of β-FeSi$_2$ have been established as thermosensitive material due to their high thermoelectric power and good thermal stability. The films were prepared by magnetron compound target sputtering with an Si/Fe ratio of 2.0 and 2.2 on oxidized Si wafers and aluminum oxide ceramic. For n-type films Co was added as doping element, and for p-type films Cr, Al, or Mn were added in order to reach a doping level of about 0.2–2.4 at.% in the films. The amorphous films in the as-deposited state were transformed into the polycrystalline semiconducting β-FeSi$_2$ phase during an annealing treatment at about 750 K, as it is found in X-ray and TEM electron diffraction studies (see Fig. 4.101). There are indications for the formation of an additional phase due to composition shifts arising from the doping treatment. In-situ measurements of resistivity ρ and thermoelectric power S during the heat treatment in the temperature range up to 1100 K reflect the transformation of the initial amorphous state into the polycrystalline state characterized by semiconducting behavior with pronounced thermoelectric properties (see Fig. 4.102).

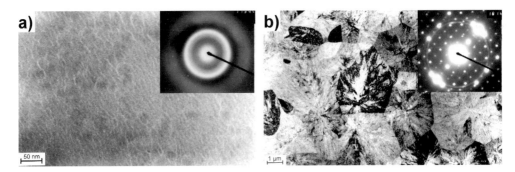

Figure 4.101: TEM bright field images and electron diffraction patterns for magnetron sputtered β-FeSi$_{2.15}$ films doped by Co; a) amorphous as-grown state, b) completely crystallized film

From the structural and transport studies a relatively complete picture of the thermoelectric performance of the films is achieved. Passing through an optimized thermal treatment the main component of the films is the orthorhombic β-FeSi$_2$ phase crystallized in polycrystalline 3 – 5 μm grains.

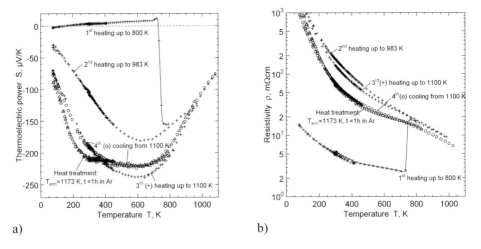

Figure 4.102: Thermoelectric power (a) and resistivity (b) of sputtered Co doped β-FeSi$_{2.15}$ thin films (thickness t = 370 nm and dopant concentration 1.2 at. % Co) on Al$_2$O$_3$ ceramic substrates

In dependence on the doping elements added in the deposition process, samples with both types of conductivity can be prepared, whereas the type of conductivity is identified by the sign of the *Seebeck* coefficient. At high doping concentrations (> 2 at.%) additional minority phases, e.g. CoSi$_2$ or Co$_2$FeSi, can be identified from the electron scattering ring pattern. Analyzing the resistivity as a function of temperature (ln $\rho \sim T^{-1}$) we find for temperatures $T \geq 900$ K, i.e. in the range of intrinsic conductivity, the band gap of β-FeSi$_2$ with $E_G \approx 0.8$ eV, whereas in the extrinsic conductivity range the activation energies of the dopants can be determined, e.g. $E_a(Co) \approx 50$ meV.

As shown in Fig. 4.102 the thermoelectric transport parameters can be optimized by a defined repeated high-temperature heat treatment. In this context it is important to refer to the $\beta \rightarrow \alpha$ phase transition within the Fe–Si system which had to be avoided in order to sustain the performance level of the films, because the high-temperature α phase has metallic conductivity. Calculations based on a simple parabolic band model within the relaxation time approximation can be used for interpretation of experimental resistivity data. Under the assumption that charged point defects are the dominant scattering centers a sufficient correspondence between measured and calculated values can be achieved [4.263].

Power factor values $S^2/\rho = 5$–7 μW K^{-2} cm^{-1} can be achieved for working temperatures above 600 K, making these films suitable for high-temperature sensor devices.

Iridium–Silicon Thin Films

In the Ir–Si system the semiconducting Ir$_3$Si$_5$ is a promising material for thermoelectric applications at higher temperatures [4.264]. Additionally, this wide-gap material has attracted large interest for the realization of Schottky barrier infrared sensors for long-wave applications between 8 and 14 μm [4.265, 4.266]. Due to the existence of a whole set of binary phases within the Ir–Si system a broad interval of Schottky barrier heights is well-known from the literature.

The thin film growth of this material and its correlation with the film properties is not yet completely understood because the current phase diagram does not allow the undoubted description of crystallization studies and needs further clarification [4.267].

In accordance with the phase diagram, beside Ir$_3$Si$_5$ some other phases exist in this system, e.g. IrSi, Ir$_4$Si$_5$, Ir$_3$Si$_4$, orthorhombic and monoclinic IrSi$_3$. The phase formation process can be studied by X-ray diffraction [4.267] and transmission electron microscopy [4.268]. By TEM different phases can be determined in the Ir–Si samples after annealing, as illustrated in Fig. 4.103. It shows an Ir–Si thin film on a molybdenum support grid after annealing to 900 °C and demonstrates a good correlation between the diffraction patterns and the quantification results of the EDX spectra. The discrepancies of only a few atomic percents between the stoichiometry and the EDX quantification results lie within the error limits expected for the standardless EDX quantification. From TEM images it can be seen that the different phases have also a different morphology.

Figure 4.104 shows the resistivity ρ and thermoelectric power S of iridium silicide films with variable compositions in the vicinity of the semiconducting phase Ir$_3$Si$_5$. The diagrams show ρ and S as a function of the temperature for the heating and cooling treatment of as-deposited sputtered films. Thus, in both diagrams we find abrupt property changes in the vicinity of the crystallization temperature 970 K, varying with composition.

Furthermore, the diagrams demonstrate that all samples with Si contents above the composition of Ir$_3$Si$_5$ show semiconducting behavior whereas for the sample with 60.1 at.% Si a qualitatively changed behavior is found.

At high temperatures the power factor of these films is relatively low; it is found to be about 4 μW K^{-2} cm^{-1} due to the low carrier concentration. By doping these films with Fe and Ni the power factor can be enhanced up to 8.5 μW K^{-2} cm^{-1} at temperatures above 1100K [4.269]. Further improvement of the performance seems to be possible because despite the doping the resistivity level reached implies more lowering potential. At the temperature of 1100 K the maximum of the thermoelectric performance seems to be not yet reached, emphasizing the role of this material for working temperatures above 1100 K.

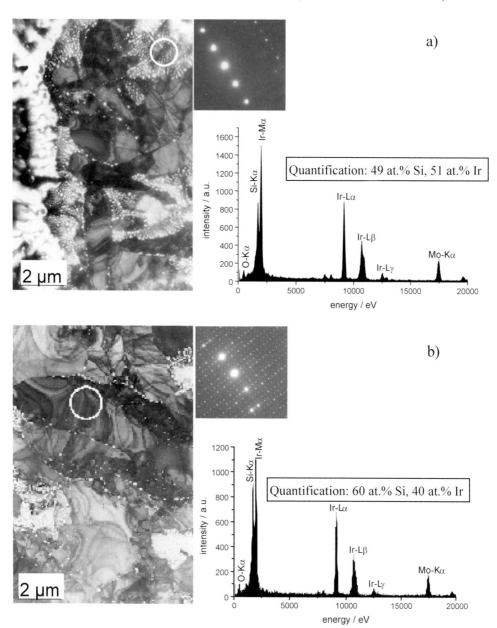

Figure 4.103: TEM micrographs, diffraction patterns and EDX spectra from the areas marked as circles in the TEM micrographs. Diffraction patterns and EDX spectra confirm the phases a) IrSi, zone axis [051], b) Ir₃Si₅, zone axis [100]

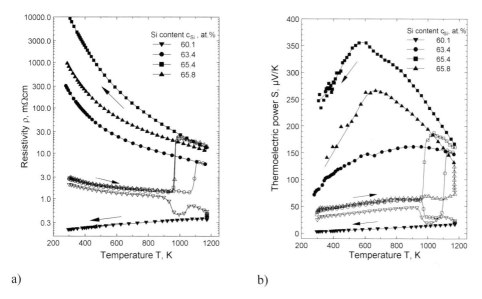

a)   b)

Figure 4.104: Resistivity (a) and thermoelectric power (b) as function of annealing treatment for co-sputtered Ir-Si films with different composition. The open symbols represent the heating-up, the filled symbols the cooling-down process

Rhenium–Silicon Thin Films

In contrast to β-FeSi$_2$ and Ir$_3$Si$_5$ the semiconducting phase ReSi$_{1.75}$ is a narrow-gap semiconductor. This fact opens good possibilities for the preparation of low-resistivity material and, together with the high mobility values known from single-crystal studies, rhenium silicide can possibly be established as a prospective thermoelectric material [4.270].

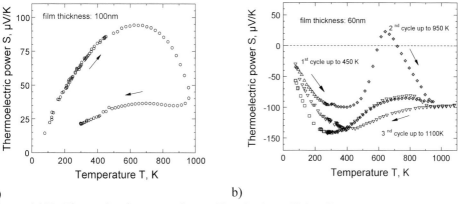

a)   b)

Figure 4.105: Thermoelectric power of crystalline rhenium silicide films prepared by facing target sputtering; a) polycrystalline ReSi$_{1.95}$, b) epitaxial ReSi$_{1.75}$ on silicon on sapphire

The investigation of poly- and monocrystalline films has confirmed that by means of variation of composition, heat treatment and kind of substrates it is possible to prepare films with both types of conductivity and relatively high values of the Seebeck coefficient. However, as is seen in Fig. 4.105 the films show a complicated crystallization behavior requiring a well-defined thermal treatment for parameter stabilizing. Despite the small energy gap the films were found to be stable up to relatively high temperatures, but the process of structure formation and doping needs further intensive investigation.

In contrast to the comparably large grain phases β-FeSi$_2$ and Ir$_3$Si$_5$ the Re–Si films can also be deposited in the nanocrystalline state if deposition and thermal handling takes place in a defined regime (see Section 2.6.2). Assuming that the reason for this special behavior is given already in the as-deposited amorphous state the characterization of the whole crystallization process including the amorphous state is very important.

The calculation of the reduced radial density function (RDF) from electron scattering curves is a well-known method for the structural characterization of the amorphous state [4.253, 4.271]. Figure 4.106 shows the RDF results for three different Re–Si thin film compositions.

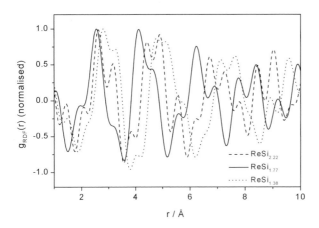

Figure 4.106: Reduced radial density function for three different Re–Si compositions: 42 at.% Re (ReSi$_{1.38}$), 36 at.% Re (ReSi$_{1.77}$) and 31 at.% Re (ReSi$_{2.22}$) [4.269]

The next neighboring atomic distances in the amorphous state, identifiable as the abscissa value of the first maximum of the RDF, are in good correlation with those of the monoclinic ReSi$_{1.75}$ lattice in the case of ReSi$_{1.77}$ and ReSi$_{2.22}$ (0.255 nm). For ReSi$_{1.38}$ the agreement is better for the hexagonal Re lattice (0.273 nm [4.272]). These structural properties are reflected in the electrical behavior of the amorphous thin films. In the same way as the next neighboring distances are shifted from values typical for metallic Re to those typical for Re–Si relations in semiconducting ReSi$_{1.75}$, if the silicon content increases, the temperature coefficient of the electrical resistance changes from about 0 to $-5 \cdot 10^{-3}$ K^{-1}, i.e. from a smaller to a larger bandgap semiconductor (see Fig. 4.107). In fact, with the changing of the next neighboring distances in the amorphous state the electronic structure is also changed.

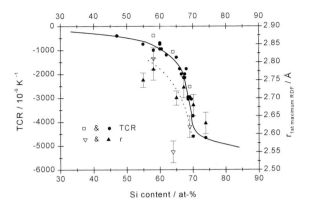

Figure 4.107: Temperature coefficient of the electrical resistance (TCR) versus Si-content of amorphous Re–Si thin films and correlation with distances r of next neighboring atoms (1st maximum of RDF) [4.272, 4.273]

Upon annealing the grain growth shows unexpected behavior: At the beginning of the crystallization process large grains up to 100 nm grow preferentially whereas in the later stages of the crystallization only smaller grains are formed. This means that the mean grain size decreases with increasing annealing time and amounts to 7–18 nm. Using the electron diffraction diagram as shown in Fig. 4.108c the phase which is formed can be identified as $ReSi_{1.75}$, in agreement with the results of X-ray diffraction experiments [4.274]. This behavior can be explained as follows: The crystallization is determined by a high nucleation rate based on the $ReSi_{1.75}$ interatomic distances in the amorphous state. If the composition is non-stoichiometric the diffusion of Re or Si to the nuclei, as well a high activating energy for changes of atomic locations in the case of the stoichiometric composition, limit the growth rate.

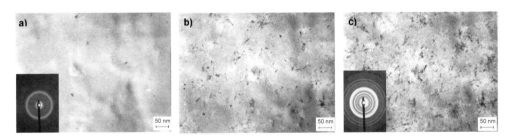

Figure 4.108: Formation of nanocrystals in a $ReSi_{2.22}$ thin film. TEM micrographs and diffraction patterns of in situ observations during annealing at 630 °C; a) after 5 min, b) after 30 min, c) after 50 min [4.273]

The first nucleation will take place at energetic and compositional "advantageous places" with short diffusion paths, and therefore in the beginning comparatively large grains are growing [4.271]. In this way the nanocrystalline fraction in the amorphous matrix increases with increasing time. This means that Re–Si can be used as a self-assembling nanocrystalline system. In this nanocrystalline state anomalous behavior of thermoelectric transport properties is observed, allowing the optimization of the thermoelectric efficiency as it is described in Section 2.3.6 [4.274].

4.8 References

[4.1]   *The InternationalTechnology Roadmap for Semiconductors (ITRS)*, 2003 edition.

[4.2]   H. Wever, *Elektro- und Thermotransport in Metallen*, J. A. Barth, Leipzig, 1973.

[4.3]   I.A. Blech, *J. Appl. Phys.*, **47**, 1203 (1976).

[4.4]   I.A. Blech, C. Herring, *Appl. Phys. Lett.*, **29**, 131 (1976).

[4.5]   R. Spolenak, O. Kraft, E. Arzt, *Microelectron. Reliability*, **38**, 1015 (1998).

[4.6]   J.R. Lloyd, J. Clemens, R. Snede, *Microelectron. Reliablitity*, **39**, 1595 (1999).

[4.7]   D.G. Pierce, P.G. Brusius, *Microelectron. Reliability*, **37**, 1053 (1997).

[4.8]   K.N. Tu, C.C. Yeh, C.Y. Liu, C. Chen, *Appl. Phys. Lett.*, **76**, 988 (2000).

[4.9]   R. Frankovic, G.H. Bernstein, *IEEE Trans. Electron Devices*, **43**, 2233 (1996).

[4.10] G.L. Baldini, I. de Munari, A. Scorzoni, F. Fantini, *Microelectron. Reliability*, **33**, 1779 (1993).

[4.11] E. Arzt, O. Kraft, R. Spolenak, Y.C. Joo, *Z. Metallk.*, **87**, 934 (1996).

[4.12] J.R. Black, *IEEE Trans. Electron Devices*, **16**, 338 (1969).

[4.13] A. von Glasow A.H. Fischer, in *Proceedings of the International Interconnect Technology Conference*, (IEEE, 2002), p. 274.

[4.14] *Isothermal Electromigration Test Procedure*, EIA/JEDEC Standard, EIA/JESD61 (1997).

[4.15] B.J. Root, T. Turner, in *Proceedings 23th Int. Rel. Phys. Symp.*, (IEEE, 1985) , p. 100.

[4.16] C.C. Hong, D.L. Crook, in *Proceedings 23th Int. Rel. Phys. Symp.*, (IEEE, 1985), p.108.

[4.17] R.W. Pasco, J.A. Schwartz, *Solid State Electron.*, **26**, 445 (1983).

[4.18] A.H. Verbruggen, M.J.C. van de Homberg, L.C. Jacobs, A.J. Kalkman, J.R. Kraayeveld, S. Radelaar, in *AIP Conf. Proc., New York*, Woodbury, 1998, p. 135

[4.19] D.B. Knorr, K.P. Rodbell. *J. Appl. Phys.*, **79**, 2409 (1996).

[4.20] C.-U. Kim, J. K. Morris, H.-M. Lee, *J. Appl. Phys.*, **82**, 1592 (1997).

[4.21] O. Kraft, E. Arzt, *Acta Mater.*, **46**, 3733 (1998).

[4.22] Y.-C. Joo, C.V. Thompson, in *Materials Reliability in Microelectronics*, IV. Symp. Mater. Res. Soc., 1994, p.319.

[4.23] Y.-C. Joo, C.V. Thompson, S.P. Baker, E. Arzt, *J. Appl. Phys.*, **85**, 2108 (1999).

[4.24] K. Wetzig, H. Wendrock, A. Buerke, and Th. Kötter, in *Proc. 5th Int.. Workshop Stress Induced Phenomena, Stuttgart*, AIP 491, Melville, 1999, p. 89.

[4.25] A. Gladkikh, E. Glickman, M. Karpovski, Y. Lereah, A. Palevski, J. Shubert, *Mater. Res. Soc. Symp. Proc.*, **391**, 283 (1995).

[4.26] M.C. Shine, F.M. d'Heurle, *IBM J. Res. Dev.*, **15**, 378 (1971).

[4.27] C.-K. Hu, P.S. Ho, M.B. Small, *J. Appl. Phys.*, **72**, 291 (1992).

[4.28] R. Rosenberg, *J. Vac. Sci. Technol.*, **9**, 263 (1972).

[4.29] J.P. Dekker, C.A. Volkert, E. Arzt, P. Gumbsch, *Phys. Rev. Lett.*, **87**, 035901/1-4 (2001).

[4.30] R. Spolenak, PhD thesis, University of Stuttgart, 1999.

[4.31] R. Spolenak, O. Kraft, W.D. Nix, E. Arzt, *AIP* **418**, 147 (1998).

[4.32] R. Spolenak, O. Kraft, E. Arzt. in *Proc. 5th Int.. Workshop Stress Induced Phenomena, Stuttgart*, AIP 491, Melville, 1999, p. 126.

[4.33] R. Kirchheim, *Acta Metall. Mater.*, **40**, 309 (1992).

[4.34] B.C. Valek, J.C. Braman, N. Tamura, A.A. Macdowell, R.S. Celestre, H.A. Padmore, R. Spolenak, W.L. Brown, B.W. Batterman, J.R. Patel, *Appl. Phys. Lett.*, **81**, 4168 (2002).

[4.35] D.P. Field, D. Dornisch, H.H. Tong, *Scr. Mater.*, **45**, 1069 (2001).

[4.36] H. Wendrock, S. Menzel, T.G. Koetter, D.Rauser, K. Wetzig, in *Proc. 6th Int. Workshop Stress Induced Phen. Ithaca*, AIP CP 612, 2002, p 86.

[4.37] C. Lingk, M.E. Gross, W.L. Brown, *J. Appl. Phys.*, **87**, 2232 (2000).

[4.38] S.H. Brongersma, E. Richards, I. Verhoort, H. Bender, W. Vanderhorst, S. Lagrange, G. Beyer, K. Maex, *J. Appl. Phys.*, **86**, 3642 (1999).

[4.39] C-K. Hu, R. Rosenberg, K.Y. Lee, *Appl. Phys. Lett.*, **74**, 2945 (1999).

[4.40] T.G. Koetter, Ph.D. thesis, TU Dresden, 2002.

[4.41]   T.G. Koetter, H. Wendrock, H. Schuehrer, C. Wenzel, K. Wetzig, *Microelectron. Reliability*, **40**, 1305 (2000).
[4.42]   S.P. Baker. A. Kretschmann, E. Arzt, *Acta Mater.*, **49**, 2145 (2001).
[4.43]   J. Koike, A. Sekiguchi, M. Wada, K. Maruyama, in *Proc. 6th Int. Workshop Stress Induced Phen. Ithaca, AIP CP 612*, 2002, p 177.
[4.44]   G. Ramanath, H. Kim, H.S. Goindi, M.J. Frederick, C.S. Shin, R.Goswami, I. Petrov, J.E. Greene, in *Proc. 6th Int. Workshop Stress Induced Phen. Ithaca, AIP CP 612*, 2002, p 10.
[4.45]   S. Menzel, S. Strehle, M. Herrmann, H. Schloerb, H. Wendrock, K. Wetzig, in Proc. AMC 2002, San Diego, MRS 2003, p. 379
[4.46]   D. Padhi, S. Gandikota, C. McGuirk, H.B. Ngyuen, S. Ramanathan, S. Gandikota, K. Musaka, S. Parikh, G. Dixit, in *Proc. AMC 2002*, San Diego, MRS 2003, p. 337
[4.47]   R.R: Keller, R. Monig, C.A. Volkert, E. Arzt, R. Schwaiger, *AIP 612*, 119 (2002).
[4.48]   G. Schindler, W. Steinhoegl, G. Steinlesberger, M. Traving, M. Engelhardt, in Proc. AMC 2002, San Diego, MRS 2003, p. 13
[4.49]   R.F. Cook , J. Thurn, *Acta Mater.*, **50**, 2627 (2002).
[4.50]   M. Engelhardt, G. Schindler, W. Steinhögl, G. Steinlesberger, *Microelectron. Eng.*, **64**, 3 (2002).
[4.51]   S. Sze, J.C. Irvin, *Solid State Electron.*, **11**, 599 (1968).
[4.52]   C. Wenzel, H.J. Engelmann, *Vakuum in Forschung und Praxis*, **13**, 20 (2001).
[4.53]   S. Sze, *VLSI Technology*, McGraw Hill, New York, 1988, p. 309.
[4.54]   Y. Shacham-Diamand, A. Dedhia, D. Hoffstetter, W. G. Oldham, *J. Electrochem. Soc.*, 140, 2427 (1993).
[4.55]   J.D. Mc Brayer, R.M. Swanson, and T.W. Sigmon, *J. Electrochem. Soc.* **133**, 1424 (1986).
[4.56]   L. Stolt, F. M. d'Heurle, *Thin Solid Films*, **189**, 269 (1990).
[4.57]   A. Cros, M.O. Aboelfotoh, K.N. Tu, *J. Appl. Phys.*, **67**, 3328 (1990).
[4.58]   J. Forster, in *Ionized Physical Vapor Deposition*, ed. J. A. Hopwood, Academic Press, New York, 2000, p. 141.
[4.59]   P. Gopalraja, J. Forster, *Appl. Phys. Lett.*, **77**, 3526 (2000).
[4.60]   L.A. Clevenger, A. Mutscheller, J.M.E. Harper, C. Cabral, K. Barmak, *J. Appl. Phys.*, **72**, 4918 (1992).
[4.61]   K. Holloway, P.M. Fryer, C. Cabral, J.M.E. Harper, P.J. Bailey, K.H. Kelleher, *J. Appl. Phys.,* **71**, 5433 (1992).
[4.62]   P.R. Subramanian, D.E. Laughlin, *Bull. Alloy Phase Diagrams*, **10**, 652 (1989).
[4.63]   L. Liu, Y. Wang, H. Gong, *J. Appl. Phys.*, **90**, 416 (2001).
[4.64]   M.H. Mueller, *Scr. Metall.*, **11**, 693 (1977).
[4.65]   P.N. Baker, *Thin Solid Films*, **14**, 3 (1972).
[4.66]   A. Arakcheeva, G. Chapuis, V. Grinevitch, *Acta Crystallogr. Sect. B*, **58**, 1 (2002).
[4.67]   R. Hoogeveen, M. Moske, H. Geisler, K. Samwer, *Thin Solid Films*, **275**, 203 (1996).
[4.68]   M. Hecker, D. Fischer, V. Hoffmann, H.-J. Engelmann, A. Voss, N. Mattern, C. Wenzel, C. Vogt, E. Zschech, *Thin Solid Films*, **414**, 184 (2002) .
[4.69]   T. Laurila, K. Zeng, J. K. Kivilahti, J. Molarius, I. Suni, *J. Appl. Phys.*, **88**, 3377 (2000).
[4.70]   C. K. Hu, S. Chang, M. B. Small, J. E. Levis, in *Proceedings VMOC*, 1986, p. 181.
[4.71]   D. Gupta, *Mater. Chem. Phys.*, **41**, 119 (1995).
[4.72]   L. A. Clevenger, N.A. Bojarczuk, K. Holloway, J.M.E. Harper, C. Cabral, Jr., R.G. Schad, F. Cardone, L. Stolt, *J. Appl. Phys.*, **73**, 300 (1993).
[4.73]   X. Sun, E. Kolowa, J.S. Chen, J.S. Reid, M.A. Nicolet, *Thin Solid Films*, **236**, 347 (1993).
[4.74]   M. Lane, R.H. Dauskardt, N. Krishna, I. Hashim, *J. Mater. Res.*, **15**, 203 (2000).
[4.75]   D. Edelstein, C. Uzoh, C. Cabral, P. de Haven, P. Buchwalter, A. Simon, E. Cooney, S. Malhotra, D. Klaus, H. Rathore, B. Agarwala, D. Nguyen, in *Proc. Int. Interconnect Technology Conf., San Francisco*, IITC, 2001, p. 9.
[4.76]   M. Tagami, A. Furuya, T. Onodera, Y. Hayashi, in *Proc. IEEE Int. Electron. Devices, New York*, IEDM, Piscataway, 1999, p. 26.7.1.

[4.77]   D. Edelstein J. Heidenreich, R. Goldblatt, W. Cote, C. Uzoh, N. Lustig, P. Roper,
          T. McDevitt, W. Motsiff, A. Simon, J. Dukovic, R. Wachnik, R. Rathore, R. Schulz, L. Su,
          S. Luce, J. Slattery, in *Proc. IEEE Int. Electron. Devices, New York,* IEDM, Piscataway,
          1997, p. 773.
[4.78]   A.C. Diebold, in *Proceedings Int. Conf. Characterization and Metrology ULSI Technology,
          New York,* AIP 550, Melville, 2001, p. 42.
[4.79]   M. Traving, G. Schindler, G. Steinlesberger, W. Steinhögl, M. Engelhardt, *Proc. AMC
          2002*, p. 753
[4.80]   J. Chen, S. Parikh, T. Vo, S. Rengarajan, T. Mandrekar, P. Ding, L. Chen, R. Mosely, in
          Proc. Int.. Interconnect Technology Conf., San Francisco, IEEE, 2002, p. 185.
[4.81]   E. Zschech, W. Blum, I. Zienert, P.R. Besser, *Z. Metallkd.*, **92**, 803 (2001).
[4.82]   Z. Bian, E. O. Schaffner, R.E. Geer, in *Proc. Int.. Interconnect Technology Conf., San
          Francisco,* IEEE, 2002, p. 204.
[4.83]   R. Huebner, M. Hecker, N. Mattern, V. Hoffmann, K. Wetzig, C. Wenger, H. J.
          Engelmann, C. Wenzel, E. Zschech, J.W. Bartha, *Thin Solid Films* **437**, 248 (2003).
[4.84]   M. Stavrev, D. Fischer, C. Wenzel, K. Drescher, N. Mattern, *Thin Solid Films,* **307**, 79
          (1997).
[4.85]   N.N. Greenwood, A. Earnshaw, *Chemie der Elemente,* VCH, Weinheim, 1988, p. 534.
[4.86]   D. Edelstein, C. Uzoh, C. Cabral, Jr., P. De Haven, P. Buchwalter, A. Simon, E. Cooney,
          S. Malhotra, D. Klaus, H. Rathore, B. Agarwala, D. Nguyen, in *Proc. ULSI XVII Conf.,
          Pittsburgh, Pennsylvania,* AIP, 2002, p. 541.
[4.87]   R. Huebner, K. Wetzig, *Acta Crystallogr.,Sect. A,,* **58**, C347 (2002).
[4.88]   AMC 2002, p. 829
[4.89]   H.J. Engelmann, E. Zschech, in *Proceedings Int. Conf. Characterization and Metrology
          ULSI Technology, New York,* AIP 550, Melville, 2001, p. 491.
[4.90]   E. Zschech, H.J. Engelmann, H. Stegmann, H. Saage, Q. deRobillard, *Future Fab Int.,* **14**,
          127 (2003).
[4.91]   H. Stegmann, H.J. Engelmann, E. Zschech, *Microelectron. Eng.,* **65**, 171 (2003).
[4.92]   C. Prindle, B. Brennan, D. Denning, I. Shahvandi, S, Guggilla, L. Cheng, C. Marcadal, D.
          Deyo, U. Bhandary, in *Proc. Intn. Interconnect Technology Conf., San Francisco,* IEEE,
          2002, p. 182.
[4.93]   M. Eisenberg, *J. Vac. Sci. Technol. A* **,13**, 590 (1995).
[4.94]   T. Harada, *Proc. ULSI XIV Conf., Pittsburgh, Pennsylvania,* AIP, 1999, p. 329.
[4.95]   K.W. Kwon, C. Ryu, R. Sinclair and S. S. Wong, *Appl. Phys. Lett.,* **71**, 3069 (1997).
[4.96]   O. Sneh, R. B. Clark-Phelps, A. R. Londergan, J. Winkler, T. E. Seidel, *Thin Solid Films,*
          402, 248 (2002).
[4.97]   M. Ritala, M. Leskela, in *Handbook of Thin Film Materials,* ed. H. S. Nalwa, San Diego,
          2002, Vol. 1, p. 103.
[4.98]   S.M. George, A.W. Ott, J.W. Klaus, *J. Phys. Chem.,* **100**, 13121 (1996).
[4.99]   S.M. Rossnagel, A. Sherman, F. Turner, *J. Vac. Sci. Technol. B,* **18**, 2016 (2000).
[4.100]   T. Abell, D. Shamiryan, J. Schuhmacher, W. Besling, V. Sutcliffe, K. Maex, *Proc. AMC
          2002*, p. 717
[4.101]   O. van der Straten, Y. Zhu, E. Eisenbraun, A. Kaloyeros, in *Proc. Int. Interconnedt
          Technology Conf., San Francisco,* IEEE, 2002, p. 181.
[4.102]   W. Zeng, E. Eisenbraun, A. Kaloyeros, *Proc. AMC 2002,* p. 853
[4.103]   K.I. Choi, B.H. Kim, S.B. Kang, G.H. Choi, U.I. Chung, J.T. Moon, Proc. AMC 2002,
          p. 775
[4.104]   S. Lehtonen, V. Plessky, M. T. Honkanen, V. Ovrinnikov, J. Turunen, M.M. Salomaa, in
          Proc. 1999 IEEE Ultrasonics Symp., IEEE, Piscataway, 1999, p. 395.
[4.105]   C.C.W. Ruppel, N. Geng, A. Waldherr, R. Dill, *in Proc. Intern. Symp. On Acoustic Wave
          Devices for Future Mobile Communication Systems,* Chiba University, Chiba/JP, 2001,
          p. 9.

[4.106] H. Yatsuda, T. Horishima, T. Eimura, and T. Ooiwa, in *Proc. 1994 IEEE Ultrasonics Symp., Cannes,* IEEE, Piscataway, 1994, p. 159.

[4.107] C. Kaneshiro, T. Nakajima, Y. Aoki, K. Koh, K. Hohkawa, in *Proc. 2001 IEEE Ultrasonics Symp., Atlanta,* IEEE, Piscataway, 2001, p. 221.

[4.108] P. Wu, N.W. Emanetoglu, X. Tong, Y. Lu, in *Proc. 2001 IEEE Ultrasonics Symp., Atlanta,* IEEE, Piscataway, 2001, p. 211.

[4.109] R. Tucoulou, M. Brunel, D.V. Roshchupkin, I.A. Schelokov, J. Colin, J. Grilhe, *IEEE Trans. Utrason. Ferroelectric Frequency Control,* **46**, 211 (1999).

[4.110] C.K. Campbell, *Surface Acoustic Wave Devices for Mobile and Wireless Communications,* Academic Press, San Diego, 1998.

[4.111] H. Mughrabi, *Fatigue Fract. Eng. Mater. Struct.,* **22**, 633 (1999).

[4.112] O. Kraft, P. Wellner, M. Hommel, R. Schwaiger, E. Arzt, *Z. Metallk.,* **93**, 392 (2002).

[4.113] J.K. Howard, *J. Appl. Phys.,* **49**, 4083 (1978).

[4.114] S. Mayumi, T. Umemoto, M. Shishino, H. Nanatsue, S. Ueda, M. Inoue, in *Proc. 25th IEEE Int. Rel. Phys. Symp.,* 1987, p. 15.

[4.115] A. Yuhara, A. Watanabe, J. Yamada, *Jpn. J. Appl. Phys.,* **26-1**, 135 (1987).

[4.116] N. Hosaka, A. Yuhara, H. Watanabe, J. Yamada, M. Kajiyama, *Jpn. J. Appl. Phys.,* **27-1**, 175 (1988).

[4.117] G. Raml, W. Ruile, A. Springer, R. Weigel, in *Proc. 2001 IEEE Ultrasonics Symp., Atlanta,* IEEE, Piscataway, 2001, p. 153.

[4.118] S. Menzel, H. Schmidt, M. Weihnacht, K. Wetzig, in *Proc. 6th Int. Workshop on Stress Induced Phenomena in Metallizations, Ithaca,* AIP 612, Melville, 2002, p. 133.

[4.119] J.A. Greer, T.E. Parker, G.K. Montres, in *Proc. 1990 IEEE Ultrasonics Symp., Honolulu,* IEEE, New York, 1990, p. 483.

[4.120] A. Yuhara, H. Watanabe, N. Hosaka, J. Yamada, A. Iwama, *Jpn. J. Appl. Phys.,* **27-1**, 172 (1988).

[4.121] A. Yuhara, N. Hosaka, H. Watanabe, J. Yamada, M. Kajiyama, R. Fukaya, T. Kobayashi, in *Proc. 1990 IEEE Ultrasonics Symp., Honolulu,* IEEE, New York, 1990, p. 493.

[4.122] S. Menzel, H. Schmidt, K. Wetzig, M. Weihnacht, in *Proc. 12th Europ. Congr. on Electron Microscopy, Brno,* Czech. Soc. El. Micr., Brno, 2000, vol. 2, p. 541.

[4.123] M. Hofmann, Th. Gemming, S. Menzel, K. Wetzig, *Z. Metallk.,* **94**, 317 (2003).

[4.124] Y. Sato, T. Nishihara, O. Ikata, in *Proc. 1998 IEEE Ultrasonics Symp., Sendai,* IEEE, Piscataway, 1998, p. 17.

[4.125] T. Nishihara, *German Patent,* # 196 51 582 A1 (1996).

[4.126] R. Takayama, H. Nakanishi, T. Sakuragawa, T. Kawasaki, K. Nomura, in *Proc. 2000 IEEE Ultrasonics Symp., San Juan/Puerto Rico,* IEEE, Piscataway, 2000, p. 9.

[4.127] L. Berger, J.W. Mrosk, C. Ettl, H. J. Fecht, U. Wolff, in *Proc. 3rd Int. Conf. MicroMat 2000, Berlin,* ddp Goldenhagen, Dresden 2000.

[4.128] N. Matsukura, A. Kamijo, E. Oosuka, Y. Takahashi, N. Sakairi, Y. Yamamoto, *Jpn. J. Appl. Phys.,* **35**, 2983 (1996).

[4.129] Y. Ebata, M. Koshino, O. Furukawa, S. Ichikawa, in *Proc. 2000 IEEE Ultrasonics Symp., San Juan/Puerto Rico,* IEEE, Piscataway, 2000, p. 5.

[4.130] X. Federspiel, F. Voiron, M. Ignat, T. Marieb, H. Fujimoto, *Mater. Res. Soc. Symp. Proc.,* **514**, 547 (1998).

[4.131] H. Kimura, K. Sasamori, A. Inoue, *Mater. Trans. JIM,* **39**, 773 (1998).

[4.132] M. Hofmann, *private communication.*

[4.133] D.J. Kim, Y.T. Kim, J.W. Park, *J. Appl. Phys.,* **82**, 4847 (1997).

[4.134] C. Cabral, K. L. Saenger, *J. Mater. Res.,* **15**, 194 (2000).

[4.135] S. Baunack, S. Menzel, M. Pekarcikova, H. Schmidt, M. Albert, K. Wetzig, *Anal. Bioanal. Chem.,* **375**, 891 (2003).

[4.136] S. Strehle, Diploma thesis, Westsächsische Hochschule Zwickau, 2002.

[4.137] N. Kimura et al. in *Proc. 1998 IEEE Ultrasonics Symp., Sendai,* IEEE, Piscataway, 1998, p. 315.

[4.138]   J.-P. Laine, V. P. Plessky, M.M. Salomaa, in *Proc. 1996 IEEE Ultrasonics Symp. San Antonio/Texas*, IEEE Piscataway, 1996, p. 15.
[4.139]   H. Schmidt, S. Menzel, M. Weihnacht, R. Kunze, in *Proc. 2001 IEEE Ultrasonics Symp., Atlanta*, IEEE, Piscataway, 2001, p. 97.
[4.140]   H.-P. Feuerbaum, U. Knauer, H.-P. Grassl, R. Veith, *Electronics*, **56**, 132 (1983).
[4.141]   S.B. Menzel , M. Albert, D. Reitz, H. Wendrock, H. Schmidt, M. Weihnacht, K. Wetzig, J.W. Bartha, in *Proc. Mater. Res. Soc. Symp. Vol. 833(2005) G3.13.1, http://www.mrs.org*
[4.142]   M. Pekarčíková, Ph.D thesis, TU Dresden (2005).
[4.143]   H. Schmidt, R. Kunze, M. Weihnacht, S. Menzel, in *Proc. 2002 IEEE Ultrasonics Symp., München*, IEEE, Piscataway, 2002, p. 415.
[4.144]   U. Burges, Ph.D. thesis, TH Aachen, 1995.
[4.145]   V. Weihnacht, Ph.D. thesis, TU Bergakademie Freiberg, 2001.
[4.146]   H. Schmidt, Ph.D. thesis, TU Dresden, 2005.
[4.147]   M. Pekarčíková, M. Hofmann, S. Menzel, T. Gemming, H. Schmidt, K. Wetzig, *IEEE Trans. Ultrason. Ferroelectrics Frequency Control*, Vol. 52, No. 5 (2005), p. 911.
[4.148]   D. Reitz, *private communication.*
[4.149]   F. Kubat, Ph.D. thesis, University of Freiburg (2004).
[4.150]   C. Eberl, Ph.D. thesis, University of Stuttgart (2004).
[4.151]   F. Petroff, A. Barthelemy, D.H. Mosca, D.K. Lottis, A. Fert, P.A. Schroeder, W.P. Pratt, Jr., R. Loloee, S. Lequien, *Phys. Rev. B*, **44**, 5355 (1991).
[4.152]   S.S.P. Parkin, Z.G. Li, D.J. Smith, *Appl. Phys. Lett.*, **58**, 2710 (1991).
[4.153]   T. Lucinski, F.Stobiecki., D. Elefant, D. Eckert, G. Reiss, B. Szymanski, J. Dubowik, M. Schmidt, H. Rohrmann, K. Roell, *J. Magn. Magn. Mater.*, **174** , 192 (1997).
[4.154]   L. van Loyen, D. Elefant, D. Tietjen, C.M. Schneider, M. Hecker, J. Thomas, *J. Appl. Phys.*, **87**, 4852 (2000).
[4.155]   S.S.P. Parkin, *Appl. Phys. Lett.*, **60**, 512 (1992).
[4.156]   A. Hütten, S. Mrozek, S. Heitmann, T. Hempel, H. Brückl, G. Reiss, *Acta Mater.*, **47**, 424(1999).
[4.157]   D.J. Chakrabarti, D.E. Langhlin, S.W. Chen, Y.A. Chang, in *Binary Alloy Phase Diagrams*, ed. T.B. Massalski et al., ASM International, 1992, 2nd edn., Vol. 2, p. 1442.
[4.158]   M. Hecker, L. van Loyen, D. Tietjen, N. Schell, C.M. Schneider, *Mater.Sci. Forum*, **378–381**, 370 (2001).
[4.159]   M. Hecker, W. Pitschke, D. Tietjen, C.M. Schneider, *Thin Solid Films*, **411**, 234 (2002).
[4.160]   M. Bobeth, M. Hecker, W. Pompe, C.M. Schneider, J. Thomas, A. Ullrich, K. Wetzig, *Z. Metallkd.*, **92**, 7 (2001).
[4.161]   P. Bruno, *J. Phys.: Condens. Matter*, **11**, 9403 (1999).
[4.162]   D.E. Bürgler, S. Demokritov, P. Grünberg, M. Johnson, in *Handbook of Magnetic Materials*, ed. K.H.J. Buschow, Elsevier, Amsterdam, 2001, Vol. 13, p. 1.
[4.163]   K. Rätzke, M.J. Hall, D.B. Jardine, W.C. Shih, R.E. Somekh, A.L. Greer, *J. Magn. Magn. Mater.*, **204**, 61 (1999).
[4.164]   D.J. Larson, P.H. Clifton, N. Tabat, A. Cerezo, A.K. Petford-Long, R.L. Martens, T.F. Kelly, *Appl. Phys. Lett.*, **77**, 726 (2000).
[4.165]   H. Kikuchi, J.F. Bobo, L. Robert, *IEEE Trans. Magn.*, **33**, 3583 (1997).
[4.166]   M. Hecker, D. Tietjen, H. Wendrock, C.M. Schneider, N. Cramer, L. Malkinski, R.E. Camley, Z. Celinski, *J. Magn. Magn. Mater.*, **247**, 62 (2002).
[4.167]   W. Brückner, S. Baunack, M. Hecker, J.-I. Mönch, L. van Loyen, C.M. Schneider, *Appl. Phys. Lett.*, **77**, 358 (2000).
[4.168]   B. Heinrich, M. From, J.F. Cochran, M. Kowalewski, D. Atlan, Z. Celinski, K. Myrtle, *J. Magn. Magn. Mater.*, **140**, 545 (1995).
[4.169]   S.S. Parkin, R. Bhadra, K.P. Roche, *Phys. Rev. Lett.*, **66**, 2152 (1991).
[4.170]   D.H. Mosca, F. Petroff, A. Fert, P.A. Schroeder, W.P. Pratt Jr., R. Laloee, *J. Mag. Magn. Mater.*, **94**, L1 (1991).

[4.171]   D. Elefant, D. Tietjen, R. Schaefer, D. Eckert, R. Kaltofen, M. Mertig, C. M. Schneider,
          J. Appl. Phys., **91**, 8590 (2002).
[4.172]   D. Elefant, D. Tietjen, L. van Loyen, I. Moench, C. M. Schneider, *J. Appl. Phys.*, **89**, 7118
          (2001).
[4.173]   M.T. Johnson, S.T. Purcell, N.W.E. McGee, R. Coehoorn, J. aan de Stegge, W. Hoving,
          Phys. Rev. Lett., **68**, 2688 (1992).
[4.174]   M.T. Johnson, R. Coehoorn, J.J. de Vries, N.W.E. McGee , J. aan de Stegge, P. J. H.
          Bloemen, *Phys. Rev. Lett.*, **69**, 969 (1992).
[4.175]   P.J.H. Bloemen, M.T.Johnson, M.T. H. van de Vorst, R. Coehoorn, J.J. de Vries, R.
          Jungblut, J. aan de Stegge, A. Reinders, W.J.M. de Jonge, *Phys. Rev. Lett.*, **72**, 764 (1994).
[4.176]   P.J. H. Bloemen, M.T. Johnson, M.T.H. van de Vorst, R. Coehoorn, A. Reinders, J. aan de
          Stegge, R. Jungblut, W.J.M. de Jonge, *J. Magn. Magn. Mater.*, **148**, 193 (1995).
[4.177]   E.H. Sondheimer, *Adv. Phys.*, **1**, 1 (1952).
[4.178]   J.R. Sambles, T.W. Preist, *J. Phys. F: Met. Phys.*, **12**, 1971 (1971).
[4.179]   H.A.M. van den Berg, in *Magnetic Multilayers and Giant Magnetoresistance*, Springer
          Series in Surface Sciences, ed. U. Hartmann, Springer, Berlin, 2000, Vol. 37, p. 241.
[4.180]   M. Mao, S. Funada, C.-Y. Hung, T. Schneider, M. Miller, H.-C. Tong, C. Qian, L.
          Miloslavsky, *IEEE Trans. Magn.*, **35**, 3913 (1999).
[4.181]   H.N. Fuke, K. Saito, M. Yoshikawa, H. Iwasaki, M. Sahashi, *Appl. Phys. Lett.*, **75**, 3680
          (1999).
[4.182]   J.P. Nozieres, S. Jaren, Y.B. Zhang, A. Zeltser, K. Pentek, V.S. Speriosu, *J. Appl. Phys.*,
          87, 3920 (2000).
[4.183]   S. Groudeva-Zotova, D. Elefant, R. Kaltofen, D. Tietjen, J. Thomas, V. Hoffmann, C.M.
          Schneider, *J. Magn. Mater*, **263**, 57 (2003)
[4.184]   J.S. Moodera, G. Mathon, *J. Magn. Magn. Mater.*, **200**, 248 (1999).
[4.185]   S.S.P. Parkin, K.P. Roche, M.G. Samant, P.M. Rice, R.B. Beyers, R.E. Scheuerlein,
          E.J.O'Sullivan, S.L. Brown, J. Bucchigano, D.W. Abraham, Y. Lu, M. Rooks, R.L.
          Trouilloud, R.A. Wanner, W.J. Gallagher, *J. Appl. Phys.*, **85**, 5828 (1999).
[4.186]   S. Tehrani, B. Engel, J.M. Slaughter, E. Chen, M. DeHerrera, M. Durlam, P. Naji, R. Whig,
          J.Janesky, J. Calder, *IEEE Trans. Magn.*, **36**, 2752 (2000).
[4.187]   L. Néel, *C. R. Acad. Sci.*, **255**, 1545 (1962).
[4.188]   L. Néel, *C. R. Acad. Sci.*, **255**, 1676 (1962).
[4.189]   J.C.S. Kools, W. Kula, D. Mauri, T. Lin, *J. Appl. Phys.*, **85**, 4466 (1999).
[4.190]   S. Tegen, I. Mönch, J. Schumann, H. Vinzelberg, C.M. Schneider, *J. Appl. Phys.*, **89**, 8169
          (2001).
[4.191]   E.R. Nowak, P. Spradling, M.B. Weissman, S.S.P. Parkin, *Thin Solid Films*, **377-378**, 699
          (2000).
[4.192]   M. Sato, K. Kobayashi, *IEEE Trans. Magn.*, **33**, 3553 (1997).
[4.193]   J.R. Childress, M.M. Schwickert, R.E. Fontana, M.K. Ho, P.M. Rice, B.A. Gumey,
          J. Appl.Phys., **89,** 7353 (2001).
[4.194]   M.J. Carey, N. Smith, B.A. Gurney, J.R. Childress, T. Lin, *J. Appl. Phys.*, **89**, 6579 (2001).
[4.195]   S. Cardoso, P.P. Freitas, Z.G. Zhang, P. Wei, N. Barradas, J.C. Soares, *J. Appl. Phys.* **89**,
          6650 (2001).
[4.196]   N. Matsukawa, A. Odagawa, Y. Sugita, Y. Kawashima, Y. Morinaga, M. Satomi, M.
          Hiramoto, J. Kuwata, A*ppl. Phys. Lett.*, **81**, 4784 (2002).
[4.197]   E. Spiller, *Soft X-ray Optics,* SPIE Optical Engineering Press, 1994.
[4.198]   D. Attwood, *Soft X-rays and Extreme Ultraviolet Radiation: Principles and Applications,*
          Cambridge University Press, 1999.
[4.199]   T.W. Barbee Jr., S. Mrowka, M. C. Hettrick, *Appl. Opt.*, **24**, 883 (1985).
[4.200]   G.S. Lodha, K. Yamashita, T. Suzuki, K. Tamura, T. Ishigami, S. Takahama, Y. Namba,
          Appl. Opt., **33**, 5869 (1994).
[4.201]   D.L. Windt, *Appl. Phys. Lett.*, **74**, 2890 (1999).

[4.202]   P.H. Mao, F.A. Harrison, Y.Y. Platonov, D. Broadway, B. Degroot, F.E. Christensen,
          W.W. Craig, C.J. Hailey, *Proc. SPIE*, **3114**, 526 (1997).
[4.203]   D.L. Windt, *Proc. SPIE*, **3448**, 280 (1998).
[4.204]   D.L. Windt, F. E. Christensen, W. W. Craig, C. Hailey, F. A. Harrison, M. Jimenez-Garate,
          R. Kalyanaraman, P. H. Mao, *Proc. SPIE*, **4012**, 442 (2000).
[4.205]   J.C. Peffen, E. Ziegler, *Soc. Francaise du Vide*, **54**, 467 (1999).
[4.206]   R. Dietsch, St. Braun, Th. Holz, H. Mai, R. Scholz, L. Brügemann, *Proc. SPIE*, **4144**, 137
          (2000).
[4.207]   H. Mai, R. Dietsch, Th. Holz, S. Voellmar, S. Hopfe, R. Scholz, P. Weissbrodt,
          R. Krawietz, B. Wehner, H. Eichler, H. Wendrock, *Proc. SPIE*, **2253**, 268 (1994).
[4.208]   R. Dietsch, Th. Holz, D. Weißbach, R. Scholz, *Appl. Surf. Sci.*, **7993**, 1 (2002).
[4.209]   J. Wood, Y.Y. Platonov, L. Gomez, D. Broadway, *Proc. SPIE*, **4782**, 152 (2002).
[4.210]   T. Salditt, D. Lott, T.H. Metzger, J. Peisl, G. Vignaud, P. Hoghoj, O. Scharpf, P. Hinze,
          R. Lauer, *Phys. Rev. B*, **54**, 5860 (1996).
[4.211]   D.L. Windt, E.M. Gullikson, C.C. Walton, *Opt. Lett.*, **27**, 2212 (2002).
[4.212]   J. Birch, F. Eriksson, G.A. Johansson, H.M. Hertz, *Vacuum*, **68**, 275 (2002).
[4.213]   T. Kuhlmann, S. Yulin, T. Feigl, N. Kaiser, T. Gorelik, U. Kaiser, W. Richter, *Appl. Opt.*,
          41, 2048 (2002).
[4.214]   C. Michaelsen, J. Wiesmann, R. Bormann, C. Nowak, C. Dieker, S. Hollensteiner,
          W. Jaeger, *Proc. SPIE*, **4501**, 135 (2001).
[4.215]   J.M. Slaughter, B.S. Medower, R.N. Watts, C. Tarrio, T.B. Lucatorto, C.M. Falco, *Opt.
          Lett.*, **19**, 1786 (1994).
[4.216]   C. Montcalm, B.T. Sullivan, S. Duguay, M. Ranger, W. Steffens, H. Pepin, M. Chaker,
          Opt. Lett., **20**, 1450 (1995).
[4.217]   C. Montcalm, S. Bajt, P. Mirkarimi, E. Spiller, F. Weber, J. Folta, *Proc. SPIE*, **3331**, 42
          (1998).
[4.218]   K.M. Skulina, C.S. Alford, R.M. Bionta, D.M. Makowiecki, E.M. Gullikson, R. Soufli
          J.B. Kortright, J.H. Underwood, *Appl. Opt.*, **34**, 3727 (1995).
[4.219]   S. Braun, H. Mai, M. Moss, R. Scholz, A. Leson, *Jpn. J. Appl. Phys.*. **41**, 4074 (2002).
[4.220]   S. Bajt, J. Alameda, T. W. Barbee Jr., W. M. Clift, J. A. Folta, B. Kauffman, E. Spiller,
          Proc. SPIE, **4506**, 65 (2001).
[4.221]   E. Louis, A.E. Yakshin, P.C. Gorts, S. Oestreich, R. Stuik, E.L. Maas, M.J. Kessels,
          F. Bijkerk, M. Haidl, S. Müllender, M. Mertin, D. Schmitz, F. Scholze, G. Ulm, *Proc.
          SPIE*, **3997**, 406 (2000).
[4.222]   T. Feigl, H. Lauth, S. Yulin, N. Kaiser, *Microelectron. Eng.*, **57-58**, 3 (2001).
[4.223]   M. Grigonis, E. Knystautas, *Appl. Opt.*, **36**, 2839 (1997).
[4.224]   Y.A. Uspenskii, V.E. Levashov, A.V. Vinogradov, A.I. Fedorenko, V.V. Kondratenko,
          Y.P. Pershin, E.N. Zubarev, V.Y. Fedotov, *Opt. Lett.*, **23**, 771 (1998).
[4.225]   U. Kleineberg, T. Westerwalbesloh, O. Wehmeyer, M. Sundermann, A. Brechling,
          U. Heinzmann, M. Haidl, S. Müllender, *Proc. SPIE*, **4506**, 113 (2001).
[4.226]   E. Spiller, *Opt. Eng.*, **29**, 609 (1990).
[4.227]   A. Kloidt, Ph.D. thesis, Universität Bielefeld, 1993.
[4.228]   St. Braun, H. Mai, M. Moss, R. Scholz, A. Leson, *Proc. SPIE*, **4782**, 185 (2002).
[4.229]   R. Schlatmann, A. Keppel, Y. Xue, J. Verhoeven, C. H. M. Mareé, F. H. P. M. Habraken,
          J. Appl. Phys., **80**, 2121 (1996).
[4.230]   A.K. Freund, in *Complementarity between Neutron and Synchrotron X-ray Scattering*,
          ed. A. Furrer, World Scientific, Singapore 1998, p. 329.
[4.231]   C. Morawe, J.-C. Pfeffen, E. Ziegler, A.K. Freund, *Proc. SPIE*, **4145**, 61 (2001).
[4.232]   J.H. Underwood, T.W. Barbee, *Appl. Opt.*, **20**, 3027 (1981).
[4.233]   R. Dietsch, Th. Holz, H. Mai, S. Hopfe, R. Scholz, B. Wehner, H. Wendrock, *Mater. Res.
          Soc. Symp. Proc.*, **384**, 345 (1995).
[4.234]   A. Baranov, R. Dietsch, Th. Holz, M. Menzel, D. Weißbach, R. Scholz, V. Melov,
          J. Schreiber, *Proc. SPIE*, **4782**, 160 (2002).

[4.235]   A. Wootton, J. Arthur, T. Barbee, R. Bionta, A. Jankowski, R. London, D. Ryutov,
          R. Shepherd, V. Shlyaptsev, R. Tatchyn, A. Toor, *Nucl. Instrum. Methods Phys. Res. Sect.
          A*, **483**, 345 (2002).
[4.236]   M. Schuster, H. Göbel, L. Brügemann, D. Bahr, F. Burgaezy, C. Michaelsen, M. Störmer,
          P. Ricardo, R. Dietsch, T. Holz, H. Mai, *Proc. SPIE*, **3767**, 183 (1999).
[4.237]   R. Soufli, E. Spiller, M.A. Schmidt, J.C. Davidson, K.F. Grabner, E.M. Gullikson,
          B.B. Kaufmann, S. Mrowka, S.L. Baker, H.N. Chapman, R.M. Hudyma, J.S. Taylor,
          C.C. Walton, C. Montcalm, J.A. Folta, *Proc. SPIE*, **4343**, 51 (2001).
[4.238]   R. Ulrich, in *IEEE Proceedings of the 52nd Electronic Components and Technology
          Conference, San Diego,* IEEE, Piscataway, 2002, p. 772.
[4.239]   W. Brückner, S. Baunack, D. Elefant, G. Reiss, *J. Appl. Phys.*, **79**, 8516 (1996).
[4.240]   J.J. Vandenbroek, J.J. Donkers, R.A.F. Vanderrijt, J.T.M. Janssen, *Philips J.Res.*, **51**, 429
          (1998).
[4.241]   H.M. Clearfield, S. Wijeyesekera, E. Logan, A. Lu, D. Gieser, C. Lin, J. Jing, in *IEEE
          Proceedings of the 1998 International Conference on Multichip Modules and High Density
          Packaging, Denver,* IEEE, Piscataway, 1998, p. 478.
[4.242]   H. Dintner, R. Riesenberg, H. Bartuch, in *Proceedings of the 1991 ELMAT Conference
          "Materials in Microelectronics", Stuttgart,* VDI-Verlag, Berlin, 1991, p. 153.
[4.243]   T. Lenihan, L. Schaper, Y. Shi, G. Morcan, J. Parkerson, in *IEEE Proceedings of the 46th
          Electronic Components and Technology Conference, Orlando,* IEEE, Piscataway,1996,
          p.119.
[4.244]   A. Heinrich, J. Schumann, H. Vinzelberg, U. Brüstel, C. Gladun, *Thin Solid Films*, **223**,
          311 (1993).
[4.245]   A. Heinrich, H. Vinzelberg, W. Brückner, G. Sobe, J. Schumann, in *Proceedings of the
          1991 ELMAT Conference "Materials in Microelectronics", Stuttgart,* VDI-Verlag, Berlin,
          1991, p. 173.
[4.246]   H. Vinzelberg, personal communication, IFW Dresden (Germany), 2002.
[4.247]   J. Rector, J. Dougherty, V. Brown, J. Galvagni, J. Prymak, in *IEEE Proceedings of the 47th
          Electronic Components and Technology Conference, San Jose,* IEEE, Piscataway,1997,
          p. 713.
[4.248]   S.K. Bhattacharya, R.R. Tummala, *J. Mater. Sci.-Mater. Electron.*, **11**, 253 (2000).
[4.249]   J.P. Seidel, H. Friedli, H. Auer, *Chip- Business and Technical News from Unaxis
          Semiconductors*, **6**, 51 (2002).
[4.250]   U. Brüstel, A. Heinrich, J. Schumann, H. Vinzelberg, *BMBF-Project Report "Hochstabile
          elektrische Funktionsschichten für miniaturisierte Chipbauelemente"-Förderkennzeichen
          03N 1020E4,* NMT-FZ Jülich, Germany, 1999.
[4.251]   G. Sobe, H.D. Bauer, J. Henke, A. Heinrich, H. Schreiber, R. Grötzschel, *J. Less Common
          Met.*, **169**,331 (1991).
[4.252]   K. Wetzig, *Metalloberfläche*, **47**, 11 (1993).
[4.253]   D.J.H. Cockayne, D.R. McKenzie, *Acta Crystallogr. Sect. A*, **44**, 870 (1988).
[4.254]   W. Scholz, Ph.D.thesis, ZFW Dresden , 1987.
[4.255]   Z. Horita, T. Sano, M. Nemeto, *J. Electron. Microsc.*, **35**, 324 (1986).
[4.256]   E. van Capellen, J.C. Doukan, *Ultramicroscopy*, **53**, 343 (1994).
[4.257]   H.-D. Bauer, J. Thomas, K. Wetzig, *Phys. Status Solidi A*, **150**, 141 (1995).
[4.258]   J. Schumann, W. Brückner, A. Heinrich, *Thin Solid Films*, **228**, 44 (1993).
[4.259]   W. Bergmann, in *Werkstofftechnik, Volume 1: Grundlagen,* Carl Hanser Verlag, München,
          1989, p.246.
[4.260]   J.P. Seidel, P. Muralt, W. Rietzler, R. Hall, *Hybrid Circuit Technol.*, **7**, 17 (1990).
[4.261]   W. Brückner, S. Baunack, D. Elefant, G. Reiss, *J. Appl. Phys.*, **79**, 8516 (1996).
[4.262]   W. Brückner, J. Schumann, A. Heinrich, J. Thomas, W. Hinüber, *DVS-Berichte*, **141**,
          20(1992).

[4.263]   H. Griessmann, A. Heinrich, J. Schumann, D. Elefant, W. Pitschke, J. Thomas, in *Proceedings of the 18th Conference on Thermoelectrics, Baltimore,* IEEE, Piscataway, 1999, p. 662.

[4.264]   C.E. Allevato, C.B. Vinning, in *Proceedings of the 28th Intersociety Energy Conversion Engineering Conference, Washington,* American Chemical Society, Washington, 1993, vol. 1, p. 239.

[4.265]   D.A. Lange, G.A. Gibson, C.M. Falco, *Proc. SPIE,* **2021**, 67 (1993).

[4.266]   T. Rodriquez, A. Almendra, M.F. da Silva, J.C. Soares, H. Wolters, A. Rodriquez, J. Sanz-Maudes, *Nucl. Instrum. Methods Phys. Res. B,* **113**, 279 (1996).

[4.267]   R. Kurt, Ph.D.thesis, TU Dresden, 1998.

[4.268]   R. Kurt, W. Pitschke, A. Heinrich, J. Schumann, J. Thomas, K. Wetzig, A. Burkov, *Thin Solid Films,* **310**, 8 (1997).

[4.269]   W. Pitschke, R. Kurt, A. Heinrich, J. Schumann, H. Grießmann, H. Vinzelberg, *J. Mater. Res.,* **15**, 772 (2000).

[4.270]   H. Lange, *Phys. Status Solidi B,* **201**, 3 (1997).

[4.271]   D. Hofman, C. Kleint, J. Thomas, K. Wetzig, *Ultramicroscopy,* **81**, 271 (2000).

[4.272]   J. Thomas, D. Hofman, C. Kleint, J. Schumann, K. Wetzig, *Anal. Bioanal. Chem.,* **374**, 695 (2002).

[4.273]   J. Thomas, J. Schumann, W. Pitschke, *Fresenius' J. Anal. Chem.,* **358**, 325 (1997).

[4.274]   W. Pitschke, D. Hofman, J. Schumann, C. Kleint, A. Heinrich, A. Burkov, *J. Appl. Phys.,* **89**, 3229 (2001).

5 Devices

5.1 Device Related Aspects for Si-Based Electronics

5.1.1 Interconnect Technology and Materials Trends for Memory and Logic Products

The interconnect technology for dynamic random access memory (DRAM) chips reflects currently the most aggressive (tight) metal pitch and highest aspect ratio contacts, because of array wiring requirements. The introduction of low-k dielectric materials and of copper is required to meet the performance of high-speed memory products, and therefore, a broader usage in DRAMs is expected. However, the pricing sensitivity of these products could delay the introduction of these new materials. As a consequence, the capabilities for aluminum processing must be continuously improved and extended.

The technical product driving for the smallest feature size and the tighter first-wiring-level contacted pitch remains the DRAM. An emerging number of applications including system-on-chip (SoC), however, will challenge microprocessors [5.1]. Table 5.1 summarizes key technology parameters for state-of-the-art logic and DRAM interconnect technologies [5.2]. Inlaid copper process flows already dominate the high performance microprocessor (HP MPU) fabrication. While current inlaid copper processes utilize physical vapor deposition (PVD), Ta-based barriers, and Cu nucleation layers, continued scaling of feature sizes requires the development of other materials and nucleation layer deposition techniques. The 2003 ITRS [5.1] considers the difficulty in integrating new low-k dielectric materials by a less aggressive scaling of the dielectrics than put forward in earlier roadmaps.

Table 5.1: Key 130 nm technology node parameters for logic and DRAM [5.1, 5.2].

Parameter	Logic	DRAM
No. metal levels	8	3
Local wiring pitch (nm)	350	260
Wiring	Cu	Al
Eff. resistivity ($\mu\Omega$ cm)	2.2	3.3
Interconnect architecture	damascene	metal RIE
Eff. dielectric constant	3.0–3.6	4.1

Managing the rapid rate of materials introduction (see Sections 2.1 and 4.1) and the increasing system complexity represents the short-term challenge for the semiconductor industry. In the long run, material innovations with traditional scaling will no longer satisfy the overall performance requirements. In particular, interconnect innovations have first to deliver solutions for global interconnects [5.1].

Metal Based Thin Films for Electronics, Second Edition. Klaus Wetzig and Claus M. Schneider (Eds.)
Copyright © 2006 WILEY-VCH Verlag GmbH & Co. KGaA, Weinheim
ISBN: 3-527-40650-6

5.1.2   Copper Inlaid Process:
Process Integration and Materials Related Topics

Figure 2.2 in Section 2.1 shows the comparison between the conventional aluminum technology and the copper inlaid (or damascene) technology. The raw aluminum process flow starts with the deposition of aluminum on a dielectric layer. The aluminum lines are fabricated from the blanket layer using lithographic patterning and a subsequent reactive ion etching (RIE) process. Finally, the aluminum interconnects are embedded in another dielectric layer. The copper single inlaid process starts with a lithographic patterning and etching of vias and/or trenches into a dielectric layer. Subsequently, a barrier and a nucleation layer are deposited into the etched structure, followed by the filling of the structure with copper. Finally, the structure is planarized by chemical-mechanical polishing (CMP), i.e., excess copper is removed to isolate neighboring interconnect lines, followed by additional dielectric deposition processes. The advantages of the metal inlaid technology include easier metal patterning (less sensitive to metal composition), easier lithographic alignment, improved planarity, better tool clustering logistics, and fewer process steps (particularly in the case of the dual inlaid technology).

Using copper/low-k process schemes, not only the permittivity of the dielectric material (dielectric constant k) has to be low, but also one of the primary integration challenges is the adhesion of barrier and capping materials to the low-k dielectric during planarization. Other requirements are resistance to processing chemicals, low moisture absorption and thermomechanical stability. The effective permittivity k_{eff} encountered by the signal in the interconnect structure is the most important parameter and it depends not only on the permittivity of the bulk interlayer dielectric, but essentially on the integration scheme as well. Each material in the interlayer dielectric (ILD) stack contributes to k_{eff}. Simulations demonstrate the significant influence of etch stop layers, copper capping layers and hard masks on the k_{eff} value [5.3]. These simulations show also that the implementation of low-k and ULK materials seems to be required for higher metallization levels rather than for local interconnects. However, new integration schemes could be required, in addition to these ultra-low-k (ULK) materials, to realize a k_{eff} value of 2.0 or less.

Manufacturability and defect reduction are important tasks besides the technology and materials issues that have to be solved. They involve plasma damage, contamination, cleaning of high-aspect ratio features and defect-tolerant processes. Increased integration challenges associated with etch, strip and clean steps will require new approaches. For example, the etch processes for a dual inlaid structure with and without an embedded trench etch stop are different. Requirements for pre-etch and post-etch clean, which utilize new process concepts such as supercritical CO_2, have been expanded. Alternative hydrogen-reducing gas chemistries may be needed for stripping photoresist from porous silicon oxide or similar ULK materials. Planarization continues to be a critical step for the interconnect technology. The change-over to the copper inlaid technology requires copper CMP processes for at least three different material combinations depending on the technology nodes: Cu/SiO_2, Cu/low-k dielectrics, Cu/ULK dielectrics. Each of these back-end-of-line material sets brings new metal planarization challenges. One of the primary integration challenges with low-k materials is the adhesion failure of barrier or capping materials on the dielectric layer during planarization. Porous ULK materials are even more problematic and are therefore one of the key focus areas for planarization development efforts.

Particularly for thick metal films used for either global wiring or inductors, solutions that include novel process tools combining copper electrochemical deposition (ECD) and planarization are under discussion.

5.1.3   Wiring Hierarchy for Copper/Low-*k* on-Chip Interconnects

In the 2003 ITRS [5.1], the technology requirements for interconnects in two specific classes of products are addressed: high performance microprocessors (HP MPU) and dynamic random access memories (DRAM). The roadmap continues to reflect the hierarchical wiring design trend featuring reduced aspect ratios. Local, intermediate and global wiring levels (wiring pitches/aspect ratios) are differentiated to highlight a hierarchical scaling methodology for potential interconnect solutions. This hierarchical wiring approach includes steadily increasing pitches and thicknesses at each metal level to alleviate the impact of interconnect delay on performance (see Fig. 5.1) [5.1].

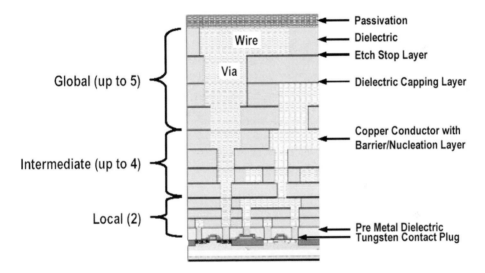

Figure 5.1: Cross-section of hierarchical scaling [5.1]

This hierarchical grouping of interconnects into local, intermediate and global ones is more important for HP MPU chips, since they usually use a higher number of metallization levels than memory chips of the same lithography generation: The number of metal levels for HP MPUs will increase to 10 and more, compared to 3 or 4 metal levels for DRAMs. As an example, the current generation of AMD's microprocessors manufactured in the 90 nm technology node has 9 levels of copper interconnects.

One major difference between interconnects of different hierarchical levels is that local interconnects scale with the scaling factor of the devices for a certain technology node, but intermediate and global interconnect lengths do not. This fact requires a separate assessment of the different groups of interconnects. Local interconnects connect transistor elements within an execution unit or within a functional block in the first level of the multi-level metallization system. The overall performance requirements seem to be achievable by technology and material innovations, particularly by the implementation of copper and low-k materials, as long as so-called size effects dominate the metal resistivity (see Section 2.1). Intermediate and global interconnects provide the connections within a functional block with typical lengths up to 3–4 mm and between the functional blocks with total wire lengths up to the chip dimensions, respectively. Whereas copper/low-k dielectric back-end-of-line structures could possibly be applied for the intermediate wiring level to reduce the RC delay, global interconnects require a more general innovation [5.4].

The global interconnect performance needed for future generations of ICs according to the 2003 ITRS [5.1], particularly the reduced RC delay, cannot be achieved by materials changes alone, even when assuming the most optimistic values of metal resistivity and dielectric constant. The signal delay time for global wires will continue to increase with the scaling of the interconnect dimensions, primarily due to the increasing resistance of the wires and their increasing length. The power distribution at constant voltage through equipotential wires to all V_{dd} bias points requires an increasingly low resistance of the global wires to avoid the voltage drop problem. The increasing power supply current, related to the decreasing V_{dd}, causes an increased voltage drop between power supply and the bias point for the fixed global wire resistance. Consequently, this requirement demands increasingly low resistance paths from the power supply to the V_{dd} bias points [5.1].

Figure 5.2 shows the delay of the local and global wiring in future generations [5.1]. The increase in the relative delay in global interconnects can be substantially reduced by repeaters, but these devices consume power and need chip area for additional active units and their vertical wiring. As a consequence, interconnects have to move to additional higher levels with larger dimensions, if they are too long, i.e., if they approach the delay issue.

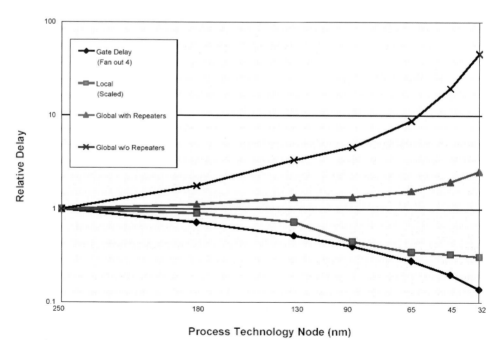

Figure 5.2: Delay of local and global wiring versus feature size [5.1]

At a time when technology and materials for the traditional metal/dielectric system will have been pushed to their limits, optimized integrated system level solutions will be necessary to solve the on-chip global interconnect problem. This means that design, process technology, and materials engineering, as well as more innovative packaging and board approaches, will need to come together.

5.1.4 New Global Interconnect Concepts

In the long term, innovations such as wafer-scale packaging and package intermediated intra- and inter-chip interconnects will be needed to minimize the problems associated with global interconnects. Many new design and technology options are currently under investigation to overcome the performance limitation of traditional interconnects, such as cooled superconductors, package intermediated interconnects, RF/microwave interconnects, 3D interconnects [5.5], optical interconnects [5.6]. The considerations include even more radical solutions, such as carbon nanotubes with their outstanding electrical and mechanical properties [5.7], spin coupling, and molecular interconnects. All these approaches, however, are bringing even more material and process integration challenges. According to the 2003 ITRS [5.1], it is expected that one or more alternative interconnect approaches will begin to be used within the next five years. For on-chip interconnects, nanotubes are expected, by several research and industry groups, to be the most probable solution after the copper age. Optical interconnects in which photons rather than electrons will transfer signals will be beneficial for longer-distance applications like board-to-board and chip-to-chip connections, but experts yet disagree on whether optical interconnects will be employed to connect subsystems within a chip.

5.2 SAW High Frequency Filters, Resonators and Delay Lines

5.2.1 Introduction

As already pointed out in the introduction of Section 2.2, SAW devices play a unique role in modern telecommunication systems. They are also used in consumer electronics and could possibly be applied in remote control and sensing. There are a variety of main types of SAW devices the construction and basic function of which are considered in this section. They are based on the functional elements such as interdigital transducers (IDTs), reflective gratings, waveguides described in Section 2.2. Not included here are devices using the interplay of SAW structures with electronic circuits such as oscillators [5.8]. As special cases of high frequency applications new types of acoustic devices with bulk wave character such as film bulk acoustic resonators (FBARs) and solidly mounted resonators (SMRs) have appeared recently [5.9]. These also will not be considered here. As already pointed out in Section 2.2 we can give only an introduction to the different types of SAW devices. For exhaustive descriptions the reader is referred to the literature cited in Section 2.2.

5.2.2 Transversal Filters

A transversal filter divides an (electronic) input signal into individual signals with simultaneous shifting of their phases, and a subsequent summation process for the output signal. This very behavior is exhibited by a SAW device comprising two IDTs (see Fig. 5.3). This follows from the principle of excitation of single SAWs by each finger pair of a first IDT and their interference in a second IDT. Regarding Fig. 5.3 and having in mind the delta function model (Section 2.2.3) it is easily seen that the electric signal impinging on the input IDT is subdivided into single SAWs which are phase shifted according to the individual propagation distance when finally summed in the output IDT.

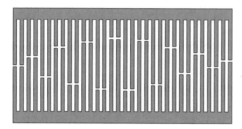 

Figure 5.3 Schematic view of a SAW transversal filter

Basics

The bandpass behavior of a transversal filter becomes evident when considering the relationship between the spatial distribution of the individual SAW sources of the input IDT and the overall transmission behavior given by the Fourier transform (see Fig. 2.27). The performance of bandpass filters is described by parameters such as insertion loss, bandwidth, side lobe suppression, shape factor, etc.

Insertion loss is, by definition, the energy loss caused by inserting a two-port network (in our case a SAW filter) into a transmission line and is measured in logarithmic units (decibels, dB). Due to the bidirectionality of simple IDTs, the minimum insertion loss of an input–output configuration is 6 dB. Means to overcome this situation are described below in the section on recent developments in low-loss filters.

As follows from the Fourier transform, the bandwidth varies roughly with the reciprocal IDT length. One also has to consider, however, the piezoelectric strength of the material used as a SAW substrate. The higher the SAW coupling factor (see Section 2.2.3) the larger the fractional bandwidth of a transversal filter that will be achieved without simultaneous increase in the insertion loss. This correlation between fractional bandwidth and coupling factor can be understood in terms of an equivalent circuit model for IDTs and the requirement of impedance matching in the case of a minimum insertion loss.

Side lobe suppression and shape factor are terms to describe features of the filter transmission curve in addition to the bandwidth. Side lobe suppression means the reduction of the signal amplitude outside the passband. According to Fig. 2.27 the transmission curve for a rectangular-shaped IDT follows a sinc-function with its typical side-lobes. Further suppression is only achieved by IDT weighting (see next subsection). Independently, different kinds of spurious signals can contribute to the degradation of the out-of-passband selectivity. It is important to note that all processes which reduce the homogeneity of the material or the geometrical form of the electrodes will contribute to such unwanted signals. The shape factor of the transmission curves characterizes the steepness of the skirts.

Similarly to the side lobe suppression it also depends on both the IDT construction and the influence of spurious signals.

Overlap weighting. As demonstrated in Fig. 2.27 the use of weighting functions for the IDTs is needed to fulfil special filter requirements. Different weighting techniques for tuning the excitation strength of finger pairs have been developed. Overlap (or apodizing) weighted IDTs (Fig. 5.4 (a)) result in the appearance of sources for SAW generation with large differences in length. Due to the existence of many sources with small overlap, SAW diffraction and finger end effects will become important for the filter behavior. Apparently, all these geometrical particularities have to be carefully incorporated in the device simulation and, vice versa, any deviations, with respect to geometry or material behavior, from the designed structure will affect the filter performance.

A homogeneous metallization ratio for all parts of the overlap weighted IDT is introduced by adding dummy fingers. Thus the same SAW propagation conditions for inactive regions as for the overlap regions are ensured.

A direct combination of two weighted IDTs with an overall frequency response given by the product of single IDT frequency responses is impossible. To circumvent this problem a multistrip coupler (MSC) can be used (see Fig. 2.33).

A further problem, relevant for all weighting techniques, arises from the wish to achieve the desired transmission behavior with a minimum IDT length. The introduction of appropriate window functions to be multiplied by the weighting function will bring about an improved solution. As a result, a small finger overlap will be further decreased.

IDT structures primarily serve as electrodes for the SAW excitation. In addition, the IDT fingers also reflect SAWs more or less according to the reflection coefficient (see Section 2.2.2). In many cases of transversal filters these reflections are unwanted and are suppressed by replacing the $\lambda/4$-fingers with split-fingers ($\lambda/8$ strips). Both IDTs in Fig. 5.3 contain split-fingers. The main reason to use split-fingers is the suppression of the triple-transit-echo (TTE) of SAWs which appears time-delayed by running twice the distance between input and output IDT and causes unwanted ripples in the transmission curve. Furthermore, internal reflections can distort the shape of the transmission curve. The benefits of $\lambda/8$ strips for the filter performance have to be balanced against the drawbacks in the fabrication process and reliability.

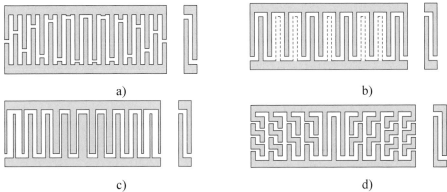

a)   b)

c)   d)

Figure 5.4: Weighting methods for SAW transversal filters. a) Overlap weighting, b) withdrawal weighting, c) width weighting, d) series-weighting

Withdrawal weighting. Several disadvantages of overlap-weighted IDTs can be avoided when using a weighting method with constant overlap length. A widely applied principle called withdrawal weighting [5.10] makes use of the variation of the source density in the SAW propagation direction. A small weighting factor means a thinned distribution of sources and is achieveded by selective removal of fingers (Fig. 5.4(b)). The large weighting steps in comparison with the possible fine tuning of finger length do not restrict the practicability of this method. Thinned regions of IDTs can be filled with dummy fingers to ensure homogeneity of the metallization ratio.

Special weighting. The variation of the finger width results in a variation of the excitation strength, because of the dependence of the SAW coupling factor on the metallization ratio (see Section 2.2.3). This method, designated width-weighting, (Fig. 5.4(c)) needs a highly reproducible manufacturing process for the metallization structures.

Another kind of weighting is based on scaling-down of the exciting electric field by subdividing the finger pairs into dogleg-like [5.11] sections (Fig. 5.4(d)). Contrary to simple overlap weighting the SAW excitation is distributed across the whole IDT aperture.

Low-Loss Filters

It is a common feature of low-loss filters that the 6 dB insertion loss of two birectional IDTs can be appreciably reduced or even totally avoided. Diverse means described in the following are used for that purpose.

Multi IDT. The simplest way to circumvent the birectionality loss of at least one IDT is to introduce additional IDTs to pick up unused SAW energy streams. In a 3-IDT configuration [5.12] the SAWs launched from the center IDT to both sides will impinge on the output IDTs. A more complicated structure is shown in Fig. 5.5, where input and output transducers are interlaced, suggesting the name interdigitated interdigital transducer (IIDT) [5.13]. In order to conserve the launched acoustic energy completely, the IIDT structure is embedded between reflector gratings.

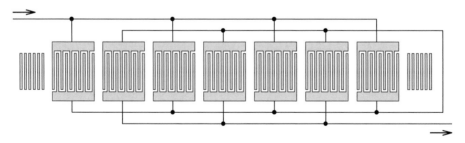

Figure 5.5: Schematic view of interdigitated interdigital transducers (IIDT)

SPUDT filters. The unidirectivity of single phase unidirectional transducers (SPUDTs, Section 2.2.3) avoids bidirectional SAW excitation by concept. However, the SPUDT optimization [5.14] with respect to weighting functions for excitation and reflection coefficients of the fingers includes a minimization of the insertion loss and of the TTE at the same time. A TTE suppression for an IDT is achieved for the case where the SAW reflection is compensated by the SAW regeneration.

Low-loss behavior and optimal approximation to a given transmission curve can be combined favorably by another elaborate design strategy [5.15]. In so-called RSPUDTs (resonant single phase unidirectional transducer) reflections between and within the IDTs are not suppressed, but used for the prolongation of the overall pulse response. One can say that as a result of these many SAW reflections, the IDT length is virtually enlarged, enabling the frequency characteristics to be fitted well.

Z path filters. The use of weighted reflectors with oblique incidence [5.16] (Fig. 5.6) is advantageous with respect to saving chip size due to the geometrical compactness. Moreover, similarly to RSPUDTs, the weighting of reflectors is more effective because of the longer SAW propagation paths through these structures.

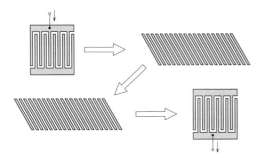

Figure 5.6: Geometric structure of a Z path filter [5.16]

5.2.3 Resonators

In view of acoustomigration effects (Section 4.3) which describe the degradation of metallization structures caused by the SAW dynamic stress fields, resonators appear as qualitatively different elements compared with transversal filters. Whereas in transversal filters the stress fields propagate across the surface, in resonators standing waves dominate. They exhibit nodes and antinodes of particle motion and mechanical stress, and thus cause a material degradation depending on their location with respect to the metal structures. High power durability can be achieved when using types of waves which have a reduced density of acoustic energy due to a large penetretion depth, as in the case of STW (see Section 2.2.2).

Basic Constructions

One-port resonator. SAW resonators (Fig. 5.7) can be compared to bulk resonators widely used for a long time, such as quartz tuning forks, microbalances or piezoceramic resonators. As known from bulk resonators, the depicted equivalent circuit is a consequence of the interplay of inertia, elasticity, piezoelectricity, and capacitive behavior of the electrodes used to excite the vibrations. SAW resonators behave quite similarly. Instead of the crystal surfaces, here the gratings act as acoustic wave reflectors. The reflection behavior is understandable in terms of Bragg reflection (Section 2.2.2), provided that the reflectors have a sufficient number of fingers for a given reflection coefficient per finger. Both reflectors act as mirrors and form a cavity. The resonance condition follows from the picture that the standing SAW must fit the distance between the mirrors.

According to the analysis of the equivalent circuit, the real part of the admittance as a function of frequency has a resonance peak and, just above this peak, a minimum called antiresonance. The peak location is given by the SAW propagation velocity, the width increases with energy losses, and the frequency difference between resonance and antiresonance increases with the SAW coupling factor. Thus, admittance curves of SAW resonators can give helpful information about changes in the material properties as well as geometric changes.

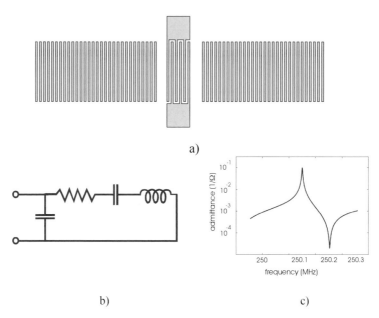

a)

b)   c)

Figure 5.7: One-port SAW resonator: a) construction, b) electrical equivalent circuit, c) admittance curve

Two-port resonator. Two-port resonators (Fig. 5.8) have reflector gratings on both sides of a two-IDT configuration. Instead of the admittance curve the transmission characteristic informs about the SAW resonance behavior. The equivalent circuit is similar to the case of a one-port-resonator, augmented by elements representing the SAW transmission between the IDTs. Because of the electrical separation of input and output, two-port resonators are advantageously applied as SAW oscillators. Two-port resonators are basically easier to use for circuitries due to just this circumstance. Furthermore, multimode resonators, described next, can be successfully constructed only in a two-port resonator configuration.

Multimode Resonators

Longitudinally coupled resonator filters (LCRF). For the two-port resonator of Fig. 5.8 the appearance of a symmetric and an antisymmetric mode is shown [5.9]. As is visible from the SAW amplitude distribution these modes correspond to the two variants of strongest energy transmission through both IDTs.

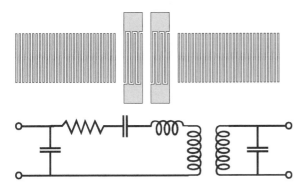

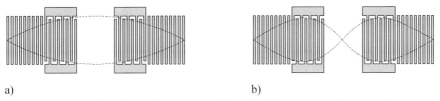

Figure 5.8: Two-port SAW resonator. Construction and equivalent circuit

Certainly the conditions of Bragg reflection and cavity resonance have also to be fulfilled. The frequencies for the symmetric and antisymmetric mode are slightly different, because they belong to similar, but clearly distinct states of the resonator. Such longitudinally coupled resonator filters (LCRF) can have more complicated structures, for example, by inserting additional structures in the center, resulting in the occurrence of higher order modes.

a)   b)

Figure 5.9: Longitudinally coupled resonator filter (LCRF); a) symmetric connection, b) anti-symmetric connection

Transversally coupled resonator filters (TCRF). By analogy to LCRF, transversally coupled resonator filters (TCRF) [5.18] can also be constructed. In Fig. 2.32 we have demonstrated the formation of 4 transversal modes in a waveguide structure. In a next step, the combination of both principles can be envisaged, as depicted in Fig. 5.10. The non-symmetric geometry of IDTs in the SAW propagation direction allows excitation and detection of both the lowest symmetric and antisymmetric longitudinal modes. The same is possible for the transversal modes, because of the separation of input (upper trace) and output IDT (lower trace). From the transmission curve the 4-fold splitting of the resonance peak is recognizable. In practice, this curve will be smoothed by the use of matching networks. Thus, the intended widening of the passband is achieved eventually.

Impedance Element Filters

One-port resonators are suitable for connection in filter networks on one chip. Due to their simplicity and the short SAW propagation distances, they ensure a low insertion loss of the order of 1 dB. In addition, high power durability is achieved, because the energy is distributed over many elements. Depending on the power partitioning between different

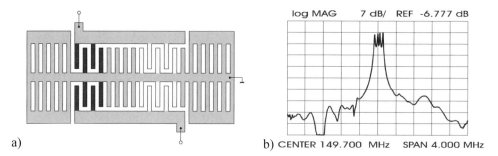

a)   b) CENTER 149.700 MHz   SPAN 4.000 MHz

Figure 5.10: Combination of TCRF and LCRF [5.18]; a) schematic geometry, b) transmission curve. Appearance of 4 resonances corresponding to the combination of 1st and 2nd transversal and 1st and 2nd longitudinal modes

resonators, the durability can be improved more easily by appropriate measures. There are different variants of the element configuration. The most investigated and applied types are ladder filters [5.19] (see Fig. 5.11).

Figure 5.11: Impedance filters: Variant of ladder-type filters

5.2.4   Filters with Spread Spectrum

This section describes SAW components which are distinguished from transversal filters and resonators in such a way that the transducer or reflector fingers do not have a constant period. Hence, the signals generated by these structures have a special time dependence.

Chirp Filters

The pulse response of the structures depicted in Fig. 5.12 is a frequency modulated pulse having increasing (up-chirp) or decreasing (down-chirp) frequency with time. Instead of

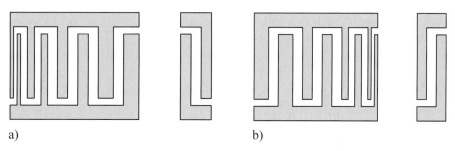

a)   b)

Figure 5.12: Basic structures of a chirp filter: a) up-chirp and b) down-chirp

IDTs arrays of grooves can be used for the formation of chirp signals, e.g. in a similar arrangement as the Z-path filter of Fig. 5.6. The importance of chirp signals arises from the pulse compression behavior when the signal is received by another chirp filter. A simple understanding of this property can be given when imaging that the received signal matches the filter structure. Chirp filters are used in radar techniques for measuring distances to reflecting targets by electromagnetic pulses which are chirp modulated for the enhancement of detectability by the subsequent pulse-compression process. RACs (reflective array compressors) are a favorable technical solution for this purpose.

Phase Shift Keying
PSK filters can be operated in a similar way to chirp filters. The generated waveforms correspond to the IDT geometry which exhibits taps of different polarity which can be programmable as shown in Fig. 5.13. The signals are therefore "symbols" with 180°-phase shifted "chips", as used for the generation of data streams in spread spectrum communication.

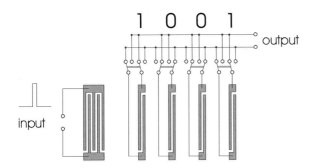

Figure 5.13: Programmable PSK filter

5.2.5 Delay Lines

We can consider a SAW delay line as a 2-IDT configuration such as shown in Fig. 5.3, but with some gap between input and output IDT. In SAW sensors [5.20] the free space can act as a sensing area for physical or chemical processes. Delay lines are used as part of oscillators containing an amplifier in the feedback connection from output to input. Depending on the particular use, the delay line may be equipped with unidirectional IDTs, taps, auxiliary layers for sensing etc., as demonstrated in Fig. 5.14. Delay lines can also be

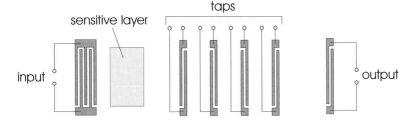

Figure 5.14: Scheme of a delay line with a sensitive layer and taps

employed as identification devices (ID tags). In this case, taps or reflecting elements representing a special phase code or frequency are placed on the substrate surface.

5.3   Sensor Devices

5.3.1   Introduction

In a wide variety of modern systems, from household appliances to fully automated industrial production lines, sensors are needed to control the operating function or to initiate a wanted action. In this section, examples will come from the automotive industry. Future automotive electronic systems will further improve driver safety, as well as comfort, engine efficiency and performance. The electronic stability program (ESP) prevents the car from spinning, new precrash sensors help to initiate the airbag as early as necessary, a force meter at the passenger seat detects whether it is occupied by a child or an adult in order to fire the airbag with the appropriate power, air quality sensors control the ventilation of the car, several acceleration sensors enable an active suspension system, the air-intake mass flow meter ensures combustion of the fuel with low emissions of pollutants and magnetic sensors detect the correct position of the crankshaft or camshaft for the stratified direct fuel injection of modern gasoline motors with low fuel consumption.

The rapidly increasing application of these systems will create a strong demand for reliable, high performance and low cost sensors, leading to the development of new technologies. According to market surveys automotive sensors will evolve into a multi-billion dollar business by 2005 [5.21].

5.3.2   Requirements for Thin Films to be Used as Transducers

Transducers transfer the quantity to be measured into a certain electrical value, which will then be the input for electronic control units. Different metal-based thin films may be prepared in such a way that their electrical resistance depends predominantly on the physical signal to be monitored. Special material compositions and deposition techniques make it possible that the cross sensitivities of these thin films are below the wanted accuracy of the transducer. The measured value must not depend on the history of the previous signals and therefore hysteresis or memory effects have to be minimized.

Besides the tailor-made design of the materials there are some other advantages of thin films for sensors. They may have a very reliable contact with the part to be measured, e.g., for sputtered strain gauges on mechanical parts. The integration of transducer elements and the electronic evaluation circuit yields better signal-to-noise ratios, improves performance and reliability and very often leads to lower production costs. Miniaturization of the sensor elements, as is done in microsystem technology, allows for systems with more functions, such as self-tests, reference elements, monitoring of several quantities and sensor arrays to see the spatial distribution. Therefore, several thin film techniques belong to the basic technologies for microsystem fabrication. Batch production of large numbers of small elements on a silicon wafer gives low cost sensors with light weight and high performance.

Due to preparation techniques like magnetron sputtering, thin films are very often in a meta-stable state. In order to avoid aging or degradation of the sensitivity of the sensor a postgrowth annealing step has to be applied. Thermal treatments enhance diffusion. Especially in the case of nanostructured films or multilayers with nanometer thickness scales, a lot of efforts are necessary to ensure long term stability of the material and the sensor's function.

For automotive applications, versatile sensors are required which have good accuracy, high functionality and safe operation under harsh environmental conditions: temperatures from –40 °C to +150 °C, temperature shock, moisture and salt fog or even use in motor oil, mechanical vibration and acceleration values of up to 200 times the gravitational acceleration. The car market demands component reliability of less than 10 ppm failures (one failure per 100,000 parts) during lifetimes of more than 15 years or 200,000 kilometers.

5.3.3   Thin Film Strain Gauges for Pressure Sensors and Force Meters

Metal-based thin film resistors may be expanded together with the mechanical part on which they are deposited. The length of the resistor increases and the area of cross section decreases. Only due to this variation in shape, the resistance of a rectangular pure metal or metal alloy thin film changes with a strain gauge factor of two. Since most materials have an elastic behavior without permanent deformation only for an elongation of less than 2 per mil small changes of the resistor values, less than 4 per mil, have to be measured by a so-called *Wheatstone* bridge.

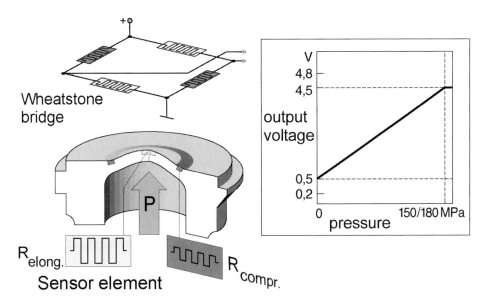

Figure 5.15: Measurement principle of the high pressure sensor with a *Wheatstone* bridge of thin film resistors

For semiconductors, certain cermets or conducting oxides, the electrical resistivity depends on the applied stress itself and the strain gauge factor will be one, two or three orders of magnitude higher than for pure metals. On the other hand, the resistivity also depends on the temperature and alters comparatively strongly with oxidation, film structure, diffusion and other aging effects. Taking into account all these cross sensitivities, the pure metal or metal alloy is a good choice for an accurate and reliable strain gauge sensor.

Knowing the elastic modulus of the mechanical part, the applied force may be calculated from the measured deformation. Different placements of strain gauges on specially designed elastic bodies allow one to measure even torque, stress distributions and pressure.

The rail pressure sensor for diesel motors of cars is explained as an example. It measures the pressure of up to 180 MPa in the rail of the common rail injection system. The high-pressure sensor consists of a steel diaphragm with ultra-thin strain gauges using thin-film techniques, as can be seen in Figs. 5.15 and 5.16.

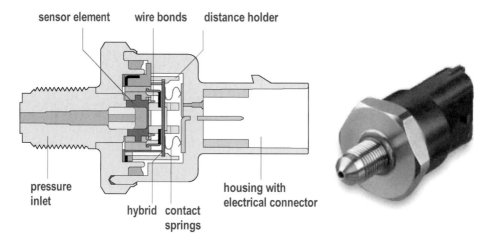

Figure 5.16: Sketch of the cross section of the high pressure sensor and photograph of the sensor

The piezo-resistive nickel–chrome–silicon resistor bridge on the steel diaphragm delivers a primary signal of 0–10 mV and is conducted to the evaluation circuit by bond wires and electric conductors in the sensor-internal spacer. The evaluation ASIC is situated on the hybrid board. The output signal is directly conducted to the connector pins. Protection against electro-magnetic interferences is realized on the hybrid.

5.3.4 Thin Film Thermometer in a Micromachined Air-Mass Flow Meter for Automotive Applications

For temperature measurements, thin films are used for thermopiles or as temperature dependent resistors. Thermoelements based on the Seebeck effect consist of two materials in contact at the point where the temperature is measured ("hot zone") and a similar contact pair at a "cold" reference site. Several pairs connected in a linear chain give a thermopile with a higher signal. In this way, it is possible to measure even small temperature differences, as necessary for bolometers measuring the infrared radiation of objects. For highly sensitive and rapid thermal sensors, the thermal mass of the transducer element should be minimized. This can be achieved e.g. by small thin film elements on thermal insulating membranes made by silicon micromachining techniques. Nowadays, miniaturized thermopile sensors are even used for contact-free measurement of the human body temperature. Thermoelectric multiple element infrared sensors for spectroscopic applications may have 256 or more cells (pixels) and use semimetallic (Bi, Sb, $Bi_{1-x}Sb_x$) and semiconducting materials (mixed crystal system Bi–Sb–Te–Se) with high thermoelectric figures of merit.

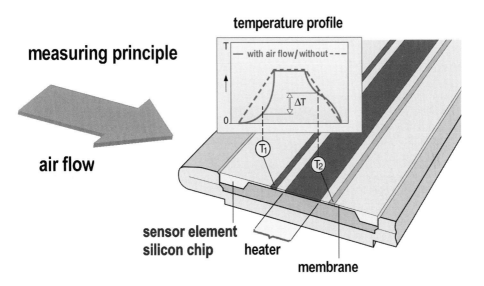

Figure 5.17: Working principle of micromachined air-mass flow sensor

Thermal sensors based on the temperature dependence of the electrical resistance of metals are easier to manufacture than thermopiles since they consist of one metal only - which is mostly platinum or nickel. In a variety of new sensors, the physical signal to be measured generates or modifies a temperature profile, which is then evaluated by means of several small temperature dependent resistors. Gas pressures, mass or volume fluxes or the true mean power of radio frequency signals may be measured.

One of the examples for automotive applications is a micromachined air-mass flow meter. If the quantity of air which a motor takes in is known, the right amount of fuel may be injected into the cylinder to ensure a total combustion with less pollutants. Previous generations of air-mass flow sensors used heated wires spanned across the air intake duct or heaters on a ceramic substrate and measured the electric power which was necessary to maintain the temperature of the hot zone. In 1995, micromachined versions have been onto the market. Figure 5.17 explains the functional principle of a micromachined air flow sensor.

A dielectric membrane, about 1 μm thin siliconnitride/siliconoxide, is used for low heat capacity and good thermal isolation. A heating current increases the temperature in the center of the membrane to a few hundred degrees centigrade. At zero air flow, the thermal profile is symmetric on both sides of the heating element. Two temperature sensors are located to the left and to the right of the heating element. An air flow from the left side decreases the temperature on this side of the membrane and the temperature sensor T_1 detects a lower temperature. The measured temperature difference between the left and the right temperature sensors T_1 and T_2, respectively, is a direct indicator of the air flow over the chip. Due to the low heat capacity of the membrane, the sensor exhibits very small response times of less than 15 ms. Even pulsation of the air inside the manifold can be detected, which was not possible with solutions in conventional technology. The heater and the temperature sensors are platinum thin films.

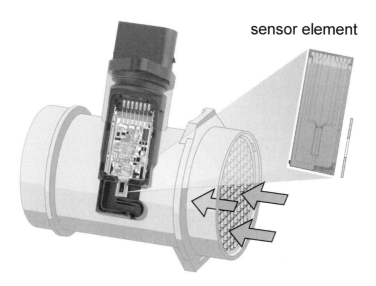

sensor element

Figure 5.18: Air-mass flow sensor HFM5 for engine control systems

The sensor can be equipped with an additional temperature sensor to measure the temperature of the intake air, if desired. These mass flow sensors, shown in Fig. 5.18, have found widespread use for gasoline engine control systems in Europe and the US. For diesel applications, they are used to govern the exhaust gas recirculation.

5.3.5 Magnetic Thin Films for Measuring Position, Angle, Rotational Speed and Torque

Due to their versatility and reliability, magnetic-field sensors are widely used in automotive applications to measure mechanical quantities like position, angle, or speed. They allow contact-less and thus wear-free measurement of these quantities. In comparison to optical and capacitive sensing principles, magnetic-field sensors are nearly insensitive to exposure to dirt or moisture. Manufactured by using thin-film and microsystem technologies, they also meet the low cost expectations in the automobile industry. In total, about one third of today's automotive sensors are based on the magnetic principle [5.21]. Figure 5.19 sketches the development of the magnetic sensor market for some automotive applications.

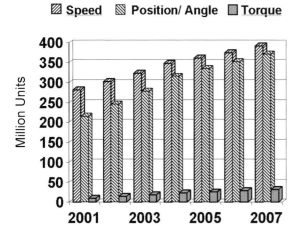

Figure 5.19: Market development for selected magnetic sensor applications

Different Magnetic Thin Film Systems for Different Applications

AMR sensors are based on the anisotropic magneto-resistive effect (AMR effect), which occurs in ferromagnetic transition metals, for example, permalloy NiFe. In these materials, the electrical resistance is a function of the angle between the electrical current and the magnetization direction. An external magnetic field can change the direction of the magnetization and thus the electrical resistance, allowing one to use the AMR as a transducer in magnetic-field sensors.

Although already discovered in 1857 by *Thomson* [5.22], only the progress in thin-film deposition techniques in the late 20th century allowed systematic studies of the AMR effect, resulting in first applications in read heads of hard-disk drives [5.23, 5.24] and magnetic sensors for industrial purposes in the 1980s and 1990s [5.25–5.27]. The breakthrough of AMR sensors for automotive applications occurred in the mid 1990s with their use as rotational speed sensors in anti-lock braking systems.

In the late 1980s, scientists in France and Germany observed large magnetoresistance in multilayers of alternating ferromagnetic and non-ferromagnetic materials [5.28, 5.29]. Since the changes in resistance reached up to 70 %, this effect was named giant magnetoresistance or GMR. Since then, intensive research and development has led to further improvements of the materials and already in 1995 the first devices were available on the market [5.30]. At the end of the 1990s, GMR-based sensors were installed as read heads in hard-disk drives [5.24]. In this field of application, GMR has nowadays replaced the formerly used AMR technology due to the better sensitivity, allowing much higher storing densities. Recent prototypes reach densities of 100 Gbit in^{-2} [5.31]. GMR angle sensors were first reported and introduced into the market by Siemens/Infineon in 1998 [5.32, 5.33]. GMR-based sensors also found their way into the automotive business: Several suppliers of automotive sensors, e. g. MELCO [5.34], Philips [5.35], Bosch [5.36] and Infineon [5.37] developed GMR sensors for various applications, e. g. steering wheel angle sensor, sensors for cam and crank shaft position and rotating speed sensors for anti-lock braking systems.

Magneto-resistive thin film sensors have to compete with monolithic Hall sensors, which are based on the Hall effect discovered by *Hall* in 1879. Typical sensor materials are n-type silicon, GaAs, InSb, and InAs. Most of the *Hall* sensors are based on silicon, because it is cheaper and the sensor can be easily integrated with an evaluation circuit. Since all sensors need some kind of electronics, the integration of silicon *Hall* elements can be done at nearly zero cost, and this is the key advantage in many cost-driven automotive applications. There are two other advantages of monolithic integration: first, the better immunity against electromagnetic interference and second, the very compact size, allowing standard and single mold packages coming from the semiconductor industry.

The sensitivity of silicon *Hall* elements ranges from some mV/T up to 100 mV/T and is about a factor two to four smaller than that of AMR sensors and a factor ten smaller than that of GMR sensors. Severe disadvantages of silicon *Hall* sensors are the huge offsets, the massive variations in sensitivity (± 20%), and the large temperature drifts. Therefore, they are not often used for precise measurements of magnetic fields. Since only the normal field component is detected, *Hall* sensors are typically used to measure the varying strength of an external magnetic field. In order to measure also the field direction, supporting measures, e.g., a special magnetic circuit, have to be added. However, most *Hall* applications in cars are restricted to simple switching or linear field measurements. Hall sensors for automotive applications are manufactured by Infineon as well as Honeywell, Allegro, Melexis and Micronas.

The *Hall* and AMR technologies, and recently also the GMR technologies are solid-state magnetic sensor technologies covering most of today's automotive applications together with inductively working sensors.

Rotating Speed Sensors

One high-volume application is the wheel speed detection for anti-lock braking (ABS) systems. The wheel speed information is also needed in modern vehicle dynamics control (VDC) and navigation systems. A classic field of application is the "powertrain" of cars, in which magnetic sensors deliver information about the cam and crank shaft position as well as the transmission speed.

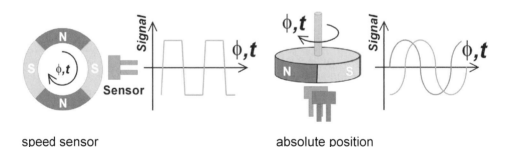

speed sensor   absolute position
incremental position sensor   angle sensor

Figure 5.20: Working principles of magnetic sensors

The sensor principle shown in Fig. 5.20 delivers a defined number of counts per turn of a rotating axis. Typically, a magnetic multipole wheel is used in order to create a periodically varying magnetic field. Alternatively, a gear wheel made of ferromagnetic material and an external magnet are used. Each of the different working principles of field sensors, ranging from inductive coil and Hall sensors to AMR and GMR, has its specific advantage, from low cost to high performance.

Figure 5.21: Silicon wafer with chips for GMR rotational speed sensors, packed elements, right: internal gradiometer layout

Figure 5.21 shows the layout of a GMR-type rotational speed sensor, which is designed as a gradiometer. The Wheatstone bridge is split into two half-bridges which are separated from each other by a distance of 2.5 mm. The gradiometer is only sensitive to field differences or gradients. Homogeneous interfering stray fields shift the operating point of the sensor, but do not generate any signal. Both GMR multilayers and spin valves can be used as sensing material.

In the case of GMR multilayers, as shown by the transfer curve in Fig. 5.22, the operating point should be shifted into the steep linear region with high sensitivity. This can be achieved by placing a supporting magnet behind the sensor, similar to AMR sensors, or by integrating this magnet into the GMR stack as an additional layer. It is crucial that the integrated layer possesses a very good and stable coercivity, otherwise the speed sensor will react differently depending on its distance from the target wheel. The GMR speed sensor shown here has an excellent sensitivity of more than 10 mV/VmT and a very good tolerance against interfering fields. It can be operated at temperatures of up to + 190 °C. The offset stability which is important for the signal quality is similar to AMR sensors (< 2 µV/VK).

When using spin valves as sensing material, it has to be ensured that the transfer curve is symmetric around zero with minimized hysteresis and that it has a sufficient linear working range. Otherwise even small interfering fields would saturate the sensor. This tailoring of the transfer curve can be achieved by optimizing the spin valve stack and the sensor design. The tilting of the transfer curve to expand the working range, for example, can be obtained by utilizing the form anisotropy in narrow GMR stripes.

Due to the higher signal and the better sensitivity compared to the established Hall and AMR technologies, the GMR speed sensor allows larger air gaps to the target wheel. At Robert Bosch GmbH working distances of about 7 mm were demonstrated, which have to be compared to 2–3 mm for a standard AMR sensor with the same commonly used multipole wheel. That is even more impressive when considering that the strength of the magnetic field of the target wheel decays nearly exponentially with increasing distance. On the other hand, keeping the air gap constant, the higher sensitivity of GMR makes it possible to reduce the diameter of the target wheels, to increase the number of poles on these wheels, to enhance the accuracy in timing applications or to use weaker, and consequently cheaper, magnets.

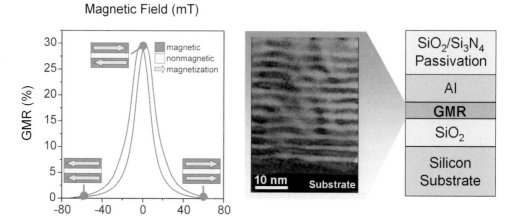

Figure 5.22: GMR curve and TEM picture of coupled multilayers

Angle Sensors

Modern vehicle dynamics control (VDC) and navigation systems require not only the wheel speed, but also the steering angle as input values, which are often provided by magnetic sensors. Other examples for angle sensor applications are pedal positions for e-gas, throttle valve angle, camshaft angle, seat position, fuel level indicators, or the angle position of e-motors and others. Recently developed electrical steering aids need to know the torque provided by the electrical motor. It can be measured by the angle difference of both ends of a built-in torsion element.

AMR angle sensors are operated in the saturation mode at magnetic fields of the order of 100 mT. For large fields the AMR effect turns into a pure angle dependence. Variations of the applied field do not affect the output signal. This measuring principle is fairly independent from assembling and magnet tolerances, as well as from aging- and temperature-induced changes in the magnetic field strength.

Figure 5.23 shows the principle setup and layout of a conventional AMR-180° sensor, consisting of eight AMR resistors combined into two Wheatstone bridges. One of these bridges is turned by 45°. Rotating an in-plane sensing field, for example, by simply rotating a permanent magnet above the sensor, the first bridge delivers a sine as an output signal, the other bridge a cosine signal. Using the arctangent-function, the angle of the external magnetic field can be calculated. In an evaluation circuit, this calculation is done using the so-called CORDIC algorithm. The amplitudes of both bridges are almost the same, since the bridges are placed into one another and thus have passed the same manufacturing processes and will be operated at the same temperature. Therefore, in this operation the temperature-dependent signal amplitudes are canceled, delivering the pure angle information largely without any temperature drifts. Of course, small asymmetries in the bridges always remain and lead to offset voltages with their own temperature dependences.

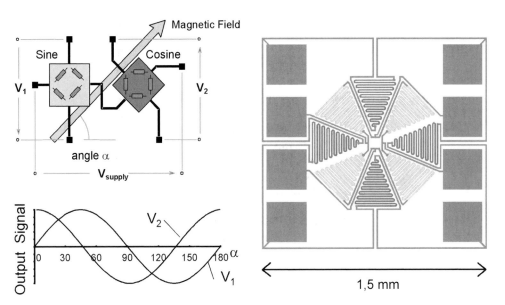

Figure 5.23: Principle setup and layout of an AMR-180° sensor

The AMR output signal depends on twice the external magnetic field angle, restricting the absolute measuring range to 180°. Since many applications, e.g. in steering systems, need the absolute angle information over a full rotation, Bosch developed an AMR sensor with an extended measuring range of 360° [5.38]. This AMR 360° element consists of an additional planar coil, which is placed above the AMR resistors (Fig. 5.24). An electrical current of the order of a few mA generates an auxiliary field, resulting in small changes in the output signal of both AMR bridges, which allow one to perform a range discrimination. The required intensity of the auxiliary field is determined on the one hand by the maximum external field, and on the other hand by thermal effects. In addition, power consumption should be as low as possible. A signal change of the order of only 1% of the signal amplitude, corresponding to a current of only a few mA, is already sufficient to ensure a faultless range discrimination.

Today's angle sensors based on the AMR effect allow one to obtain an accuracy of better than 1° over the full measuring range and for temperatures up to +150°C. The main cause for the errors are offset drifts (< 2 μV/VK per bridge). Inaccuracies due to form and uniaxial or crystal anisotropy are only in the range of 0.1°–0.2° and 0.02°, respectively. Their contribution only becomes important when using weaker magnetic fields. In the application, of course, the engineer has to pay attention in order to correctly design the magnetic circuit. Non-centric alignment of the external magnet, whose angle is to be measured, can easily produce rather high inaccuracies.

In GMR angle sensors, only spin valves can be used as the sensing material since the GMR effect in coupled multilayers is isotropic. Analogous to AMR-based sensors, the GMR resistors are arranged in two Wheatstone bridges. The main advantage of GMR is that it is a uniaxial effect, i.e. the resistance is proportional to the cosine of the angle Θ of the external field with respect to the pinning or reference direction. Therefore, GMR angle sensors have a 'natural' 360° measuring range in contrast to the 180° covered by AMR sensors. The angle signal calculated from the arctangent does not repeat after 180° as is the case in AMR sensors. Thus, no additional sensor or planar coils are necessary to sense a complete rotation of an external field.

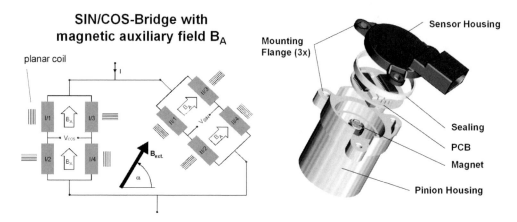

Figure 5.24: Sensing principle of 360°-AMR-element with switched auxiliary fields on same chip, right: steering wheel sensor LWS4 (Robert Bosch GmbH)

To create a sine and a cosine GMR bridge the single resistors need at least two pinning directions with 90° phase difference and an additional two directions to double the output signal (see Fig. 5.25). After deposition of the spin valve thin film layers, the pinning by an anti-ferromagnet usually has only one preferred direction. To change it, the material has to be heated above the Néel temperature of the antiferromagnet while applying an external field in the desired pinning direction. After cooling the pinning direction is frozen. The challenge is to do this imprinting of different pinning directions on the μm scale of the devices. Two different methods have been developed. Either, a current flowing through the GMR resistors or through an additional circuit provides the local heating or short laser pulses are focussed on the selected sensor region. [5.39, 5.40].

In Fig. 5.25 an example layout of a GMR angle sensor is shown. The full bridges are split into 16 single elements which are arranged on a circle. The rotational symmetry further minimizes errors. An accuracy better than 1.5° over the complete temperature range from – 40 to +150 °C can be achieved. Since the spin valve material is extremely robust, another feature of the sensor is its tolerance against interfering fields, exceeding 100 mT at 200 °C.

Compared to AMR, the GMR technology allows the absolute angle measurement over one complete rotation without any additional measure. The inherent 360° measuring range is also advantageous for small-angle applications, due to the larger linear response range. In addition, analogous to the speed sensor, weaker and thus more cost-efficient magnets can be used.

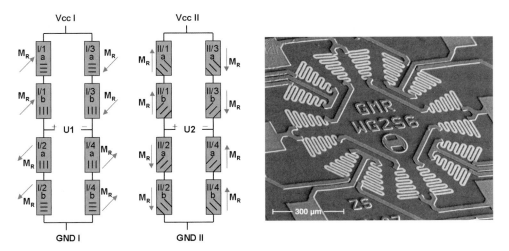

Figure 5.25: GMR 360° angle sensor, principle and chip layout

Conclusion and Outlook

Metal-based thin films enable the production of a wide variety of sensors in microsystems technology. Based on thin films, silicon micromachining and the processes for semiconductor integrated circuits, such as different kinds of film deposition, patterning by photolithography and etching, the microsystems technology has evolved from a niche technology to an important mainstream technology for many applications. High functionality, the possibility of integration with electronics, cost efficient batch processing and small size are key advantages. With the help of microsystems many new automotive electronic systems improve driver safety, as well as comfort, engine efficiency and performance.

5.4 X-Ray Optical Systems

5.4.1 Basic Properties of the Combination of X-Ray Optical Elements

Already in 1923 the glancing incidence total reflection of X-rays had been discovered by *Compton* [5.41] but, in particular, attempts to construct powerful systems for producing sharp X-ray images failed for a long period. Severe image defects resulted in poor image quality. The first useful solutions involving the aberration compensation of ray paths appeared in the middle of the last century. *Kirkpatrick and Baez* [5.42] proposed an optical system that corrects for the astigmatism of single spherical mirrors to a substantial extent by combining two of such elements perpendicularly to each other. They used this principle to realize the first successful X-ray reflection microscope, but its images still showed spherical aberration and coma. The correction of these image defects would require a change of the mirror cross sections to parabolic or elliptical shapes and the integration of additional mirror components for coma compensation, respectively.

At nearly the same time *Wolter* [5.43] proposed X-ray optical systems of toroidal shape which make use of elliptical or parabolic primary elements for compensation of the spherical aberration and subsequent confocally arranged hyperbolic secondary elements providing for coma correction.

These systems have also been used to a certain extent for microscopy applications, in particular for plasma diagnostics in laser fusion experiments [5.44-5.46], but a much broader use was found for these "*Wolter* optics" in X-ray astronomy for observations of the sun and other celestial objects emitting X-rays (X-ray and neutron stars, remnants of supernovae etc.) from orbiting spacecrafts.

"The renaissance of X-ray optics" [5.47] happened with the introduction of nm-scale multilayers as X-ray reflectors at the beginning of the 1980s. Since then a much wider domain of applications has grown. It is distinctly dominated by the development of high resolution projection systems for the new generation lithography (NGL) in the EUV region.

The combination of two or more X-ray multilayer mirrors in systems being able to collimate or focus X-ray beams in various analytical apparatuses or even to realize the stigmatic imaging of a particular object is a demanding task. It has to take into consideration, as a basic prerequisite, that the mirrors must be designed for exactly the same working wavelength, i.e.:

- the materials, from which the layer stacks are made, should be identical,
- the stack configuration has to be prepared with the same number of periods,
- the elements should have similar reflectivity and spectral resolution,
- figure tolerances of the mirror surfaces should be adequate to the characteristics of the system design.

Thus, systems of high quality, using either glancing incidence or normal incidence reflection mirrors, can be combined and will provide satisfactory optical throughput with minimum influence of imaging errors. Hitherto for a wide spectrum of applications various optical devices have been realized.

Particular solutions have been described in the literature for:

- materials science and plasma diagnostics by X-ray spectrometers, X-ray microscopes, X-ray diffractometers / reflectometers,
- space research and astronomy by X-ray telescopes,
- EUV lithography by X-ray projection systems and
- beam line applications by monochromators and polarizers.

A few selected examples of application will represent the working fields of prime importance and will be used to explain some typical results for:

- astronomy
- microscopy
- lithography and
- diffractometry / reflectometry.

5.4.2   X-Ray Astronomy

Undisturbed investigation of celestial X-ray sources needs observation from space, since the earth's atmosphere absorbs X-rays. Sounding rockets or artificial satellites are therefore the carriers of highly sophisticated telescope systems. These have to be constructed with high stability to withstand the load of the launching phase and the long-term influence of a hostile space environment. For their particular mission the telescopes are designed either to survey the entire sky (e.g., primary task of ROSAT) or to make observations of individual objects like selected stars or the chromosphere and corona of the sun.

Space Telescopes with Glancing-Incidence Optics

Glancing-incidence systems are usually composed of two single surface mirrors of cylindrical shape and curved cross section. The majority of the X-ray telescopes sent into space were thus of the *Wolter*-type (Fig. 5.26).

To increase their collecting area and wavelength coverage, nested and co-aligned telescopes of identical shape, but graded diameter, each consisting of a combination of paraboloid and hyperboloid are installed. The Roentgen-Satellite ROSAT [5.48], as a typical example, was launched in 1990. It uses four nested shells for an all-sky survey in the X-ray band from 0.1 to 2.4 keV. The reflector quality is characterized by a surface roughness of 0.28 nm and an angular resolution of the imaging system of 3.6 arcsec. It is distinguished by its excellent sensitivity to detect even very faint X-ray sources. Other advanced systems of this type (CHANDRA, XMM, see e.g. [5.49]) followed in 1999. For high reflectivity and spatial resolution (e.g. CHANDRA < 1 arcsec) the surface quality of the imaging elements is improved by steadily diminishing the roughness of the substrate surface by optical polishing and a high precision coating with noble metals (XMM with Au, CHANDRA with Ir).

These glancing-incidence telescopes delivered an important, but also limited success in space astronomy, leading to a demand for imaging with larger collection angles, better spatial resolution and reduced spectral bandpass. A solution of this problem was offered as soon as effective preparation techniques for interference multilayers working in the soft X-ray region emerged.

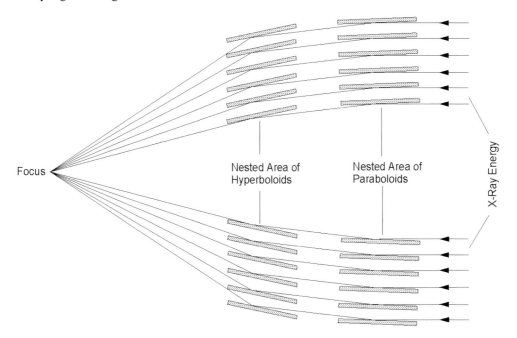

Figure 5.26: Schematic design of a *Wolter* type I telescope for glancing-incidence imaging

Space Telescopes with Multilayer Optics

A survey of possible applications of interference multilayers in astronomy was already given in 1979 [5.50]. The first use of soft X-ray multilayer-coated optics in astronomy [5.51] was published in 1987. Compared to GI-techniques the success was rather limited. The images obtained from the solar corona were of poor quality, as a very provisional optics was applied. Substantial improvements were soon achieved with dual-reflection telescopes. A pseudo-Cassegrain-type system provided, for the first time high, quality XUV coronal images with a spatial resolution of typically 1.5 arcsec [5.52]. The very convincing results of these early missions demonstrated the power of interference multilayer optics, in particular for the investigation of soft X-ray and EUV emission of the sun atmosphere. Very sophisticated and complex systems have been developed subsequently. The multi-spectral solar telescope array (MSSTA), flown first as a rocket-borne solar observatory in 1991 [5.53], is equipped with thirteen soft X-ray telescopes (Ritchey-Chretien, Cassegrain, see Fig. 5.27, and Herschel-types) all comprising multilayer-coated mirrors.

The various systems cover selected intervals in the spectral region from typically 4 nm to 35 nm. The material combinations selected for the reflector stacks are Rh/C (at $\lambda \approx 4 - 5$ nm), Mo/Si (at $\lambda \approx 13 - 21.5$ nm) and Mo/Mg$_2$Si (at $\lambda \approx 30 - 35$ nm). At the working wavelengths reflectances R from 2 to 30 % have been measured for line widths from 0.4 to 2.23 nm FWHM. In practice, the best spatial resolution of 0.7 arcsec is deduced e.g. from an image taken at $\lambda = 19.3$ nm.

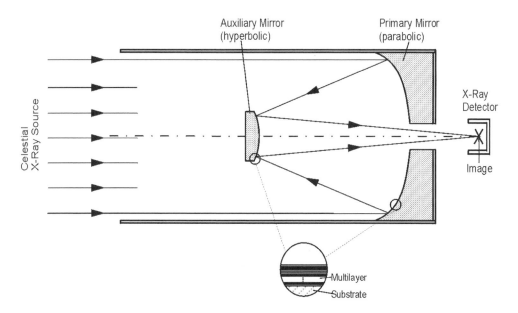

Figure 5.27: Schematic diagram of a Cassegrain-type telescope comprising multilayer-coated optics for normal-incidence imaging

5.4.3 X-Ray Microscopy

The well-known *Schwarzschild* objective [5.54] originally designed for astronomical observations in the visible region is a powerful focusing system for microscopic applications, too. In materials science and nano-fabrication techniques soft X-ray wavelengths are a prerequisite for surface analysis or for surface patterning purposes to achieve a spatial resolution in the sub-µm range.

a)

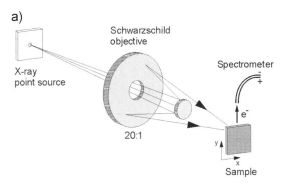

Figure 5.28: Photoemission microscope (after [5.55])
a) principle of X-ray scanning system involving Schwarzschild optics
b) basic design of a *Schwarzschild* objective

b)

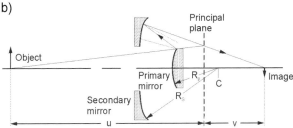

The schematic representation of an X-ray photoemission microscope is given in Fig. 5.28 a). Photoemission surface microscopy provides very detailed distributions of elemental concentrations and chemical bonding states in micro-volumina of the solid specimen.

Its principal part consists of the demagnifying Schwarzschild optics which images the monochromatic X-rays – projecting the beam defining pinhole – into a sub-µm spot size on the specimen surface. A typical demagnification is of the order of 10 to 20 times. Surface analysis is carried out by energy analysis of the photoelectrons emitted from the illuminated micro-volumina of the sample. Lateral scanning delivers spatial element distributions. The *Schwarzschild* objective (Fig. 5.28 b)) is comprised of two spherical elements of opposite curvature. High figure accuracy (typically $\lambda/50$) guarantees an excellent imaging quality. Multilayer coatings of Mo/Si, Ru/B_4C and NiCr/C have been employed to operate at selected wavelengths of $\lambda_X = 13.5$ nm, 6.8 nm and 4.48 nm, respectively [5.56–5.58]. For narrow Bragg peaks of the lower wavelength a graded *d*-spacing across mirrors of strong curvature is used.

The advantages of the *Schwarzschild* optics are:

- high lateral resolution due to a sharply focused beam, as opposite element curvature results in aberrations of opposite sign compensating each other,
- spheres, being simple optical surfaces, can be prepared to the highest quality and
- large numerical aperture gives a very small diffraction limit.

5.4.4   Extreme Ultraviolet Lithography (EUVL)

Presently, the semiconductor industry uses optical projection lithography with laser light of $\lambda = 248$ nm (KrF laser) to transfer patterns of integrated circuits from the mask to the chip. The working principle of projection lithography is similar to the well-known slide projection, just with the difference that the structures on the mask are demagnified. In addition, a significantly larger number of lenses has to be applied to demagnify the pattern without imaging errors. Typically, a pattern on the mask is imaged with a reduction of 4:1 by highly accurate projection optics onto a silicon wafer coated with photoresist. With this technology feature sizes of down to 130 nm can be realized. The challenge of the industry is to increase the density of chips on the wafer according to Moore's law which predicts the number of chips to double every 18–24 months [5.59]. However, the reduction of the pattern size is limited by the wavelength of the light used for the illumination. Therefore, there are worldwide ongoing efforts to decrease the wavelength of the illuminating light. The next candidates for shorter wavelengths for optical lithography are $\lambda = 193$ nm (ArF laser) and $\lambda = 157$ nm (F$_2$ laser). However, below 157 nm a physical limit is reached, because light of shorter wavelengths is absorbed by any material and the use of transmitting lenses is impossible. Consequently, a slightly different approach must be used for wavelengths shorter than $\lambda = 157$ nm: the transmitting lenses of the illumination system have to be replaced by reflecting mirrors [5.55, 5.60] (Fig. 5.29).

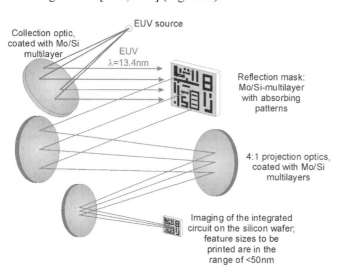

Figure 5.29: Schematic view of the principle of extreme ultraviolet lithography. Transmissive lenses used for photolithography have to be replaced by reflection optics based on Mo/Si multilayer coatings

Mo/Si Multilayers as Reflectors in EUVL Projection Optics

The search for short-wavelength light for which high-reflection normal incidence mirrors are available resulted in the choice of $\lambda = 13.4$ nm.

For these photons in the extreme ultraviolet (EUV) spectral range, multilayers consisting of periodical Mo/Si stacks can act as reflectors with reflectances of typically 70 % [5.61, 5.62]. Since at least six mirrors are necessary to demagnify the mask pattern and to correct imaging errors, already a reflectance increase of ≈ 1 % of every mirror results in an improvement of the overall throughput of ≈ 10 %. Therefore, continued investigations are being made to increase the reflectance. Recently, a further increase in the reflectance was

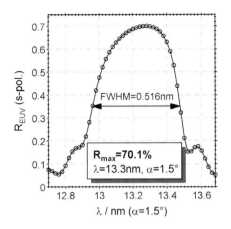

Figure 5.30: EUV reflectance of typical Mo/Si multilayers with tiny C barrier layers on the interfaces (multilayer deposition: IWS Dresden, EUV reflectance measurement: PTB/BESSY2 Berlin)

achieved by introducing tiny C or B_4C barrier layers onto the interfaces between the Mo and Si layers (Fig. 5.30, [5.61, 5.62]).

Besides the reflectance of the mirrors, the accuracy of the multilayer period over the entire mirror surface with diameters of up to 400 mm plays a decisive role for the image quality on the chip. Despite the stringent requirements that are needed for the mirrors, it has been shown that it is possible to fulfill the specifications using magnetron sputter deposition (MSD) [5.63].

Mo/Si Multilayers for EUVL Mask Blanks

From today's point of view, one of the most challenging problems of EUVL is the development of defect-free masks. The masks consist of patterned EUV absorbers placed on top of the so-called mask blanks consisting of Mo/Si multilayer reflectors with a size of 150×150 mm^2. Since the EUV beam is focused on the mask surface, the most critical issue in the production of such masks is their defect density. Any defect would be directly printed onto the wafer and may render the chip useless. Therefore, a technique must be developed that enables the deposition of Mo/Si multilayers with a low level of defects. It has been found that multilayers prepared by MSD are suitable for projection optics, but the defect density is several orders too high for use as mask blanks. Hence, mask blanks are usually deposited by ion beam deposition, where considerably lower defect densities can be reached [5.64]. Best values of the defect density on the mask blanks are in the range of 0.01 defects per cm^2 [5.65]. It is believed that it is not possible to reliably suppress all defects. Therefore, techniques have been investigated which can repair the different kinds of defects [5.66].

Differing from the requirements for illumination optics, where the overall throughput is proportional to the reflectance of one mirror to the power of N (with N = number of optical elements), the reflectance of the mask blanks has to fulfill less stringent requirements. Since the mask is only a single optical element in the optical path, the overall throughput is only proportional to the actual reflectance to the power of unity. Typical reflectances being reached with multilayers deposited by ion beam sputter deposition are in the range of 65 % [5.67, 5.68].

5.4.5   X-Ray Reflectometry and Diffractometry

For a long time X-ray methods have been used to investigate the structure of crystals. This can be done by employing laboratory or synchrotron X-ray sources. Synchrotron beamlines are characterized by well-defined X-ray beams with high brilliance and intensity. However, limited availability and high costs of synchrotron sources result in laboratory sources also being widely used despite their lower intensities and beam qualities. Since laboratory sources emit X-ray radiation (characteristic lines and bremsstrahlung) in all directions, optical elements are needed to extract monochromatic X-rays and to shape the beam. For this purpose diffraction of X-rays on the lattice planes of single crystals, total reflection on metallic layers, or reflection on multilayer structures can be used.

According to the Bragg equation for single crystals (see also Section 2.5) only one wavelength is diffracted at a given angle of incidence. Hence, single crystals act as monochromators for X-rays, but due to the high resolving power of crystals (see also Section 4.6.3) the angle of acceptance for the primary beam is low. Furthermore, if parallel beams or small X-ray spots are needed, the size of the beam has to be limited by a proper arrangement of slits within the optical path. A drawback of this approach is that most of the radiation coming from the source cannot be used. Hence, collection optics with improved angles of acceptance, similar to parabolic mirrors for visible light, would largely enhance the useable intensity of laboratory sources. With the development of nm-scale multilayer mirrors as reflectors for X-rays, the possibility arose to design such optical elements also for X-rays.

Parallel-Beam Optics

X-rays emitted by an isotropic source can be converted into parallel or focused beams using parabolically or elliptically curved multilayer mirrors [5.68]. The replacement of single crystal monochromators by multilayer mirrors can result in measurement advances for powder diffraction, grazing incidence diffraction, reflectometry, high-resolution diffraction and protein crystallography.

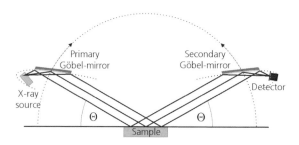

Figure 5.31: Twin *Goebel* mirror arrangement: two parabolic X-ray mirrors coated with Ni/C multilayers are combined

Reflectometry is the standard method of characterizing nm-thick multi- and single layers. From measurement of the specular reflection as a function of the grazing angle, information about reflectivity, stack regularity, multilayer period, layer thickness ratio, and interface quality of the layers can be deduced. In order to decrease the measurement time and increase the measurement accuracy, in addition to the approach of a primary *Goebel* mirror directly behind the X-ray source, a new concept for the X-ray optical path has been developed [5.69]. With the so-called twin *Goebel* mirror (TGM) arrangement two parabolic shaped X-ray mirrors coated with Ni/C gradient multilayers are combined. The divergent beam of a commercial X-ray source is collimated by the primary mirror, the resulting parallel beam is then used to scan the sample and the reflected parallel beam is focused to the detector entrance slit (Fig. 5.31).

Using the TGM arrangement, the following parameters become typical values for Cu-Kα reflectrometry:

- Detection of relative thickness deviations < 0.02 % $(1 - \sigma)$,
- Primary beam intensity $> 1.0 \cdot 10^{+9}$ cps (Cu-Kα_1 and -Kα_2),
- Suppression of Cu-Kβ radiation at a ratio of 1000000:1,
- Beam divergence $< 0.02°$.

5.4.6 X-ray Fluorescence Analysis

For many years X-ray fluorescence analysis (XRFA) has been a powerful tool for the qualitative and quantitative determination of chemical elements within unknown samples [5.70].

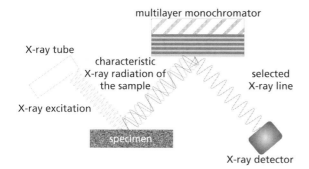

Figure 5.32: Schematic representation of the wavelength dispersive X-ray fluorescence analysis. Multilayer monochromators are used to analyze the characteristic radiation of an unknown specimen.

Besides different basic research applications, XRFA is also used in a wide field of materials research and development as well as in production processes (e.g. metallurgy, chemical, cement and pharmaceutical industries), in quality assurance (e.g. oil industry) and environmental studies (e.g. monitoring of wastewater contamination). All of these application fields require the steady improvement of XRFA tools in order to extend the range of detectable chemical elements, to improve their detection limits, and to reduce the measurement time.

With wavelength dispersive X-ray fluorescence analysis (WD-XRFA), in particular for light elements a remarkable performance enhancement can be obtained by increasing the reflectance and resolving power of multilayer monochromators that are used to analyze the characteristic radiation of unknown specimens (Fig. 5.32). Similar monochromators just with spherically curved surfaces are also used in electron probe microanalysis (EPMA), a further technique for the chemical analysis of materials (see also Section 2.5.3).

WD-XRFA using LSM as monochromators is predominantly applied for the detection of light elements, namely from Beryllium to Silicon (Section 2.5.3). The emission lines of these elements cover a photon energy range from 108.5 eV to 1740 eV, which corresponds to a wavelength range from 11.5 nm to 0.7 nm. The practical problem arises from the fact that it is not possible to find an all-purpose multilayer system, that is equally useful for all the emission lines. Commonly used multilayer combinations for the different emission lines are summarized in Table 5.2.

Table 5.2: Multilayer systems used for the WD-XRF analysis.

Chemical element	K_α emission line λ_x / nm	Materials
Be	11.43	Mo/Be, Mo/B$_4$C, La/B$_4$C
B	6.76	La/B$_4$C, Mo/B$_4$C
C	4.48	Cr/C, Ni/C
N	3.16	Cr/Sc
O	2.36	W/Si, Ni/BN
F, Na, Mg, Al, Si	1.83, 1.19, 0.989, 0.834, 0.713	W/Si, W/B$_4$C, Mo/Si, Mo/B$_4$C

In addition to the material selection problem, a reasonable tradeoff has to be found between the demand for high reflectance and for high resolving power of the monochromators. By changing the parameters of the multilayer (materials, period thickness, ratio between absorber and spacer layer thickness, number of periods) almost every time a solution exists that fits the needs of the XRFA spectrometer. In Fig. 5.33 an overview of reflectance and

a)

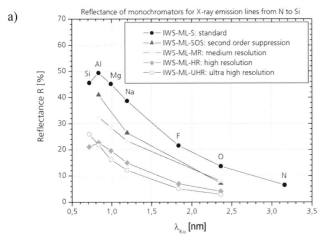

b)

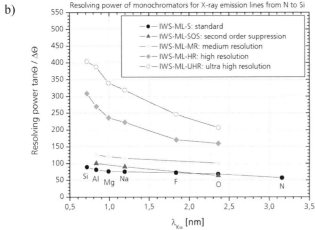

Figure 5.33: a) Reflectance of various multilayer systems for WD-XRFA, b) resolving power of the same multilayer monochromators as shown in a)

resolution of state-of-the-art multilayer monochromators for multiple lines from nitrogen to silicon is shown. For the detection of the elements from Be to O highly spezialized multilayers are used that work only for one single emission line (e.g. IWS-B for boron, IWS-C for carbon or IWS-N for nitrogen determination).

5.5 Thermoelectric Sensors and Transducers

5.5.1 Introduction

Thermoelectric sensors and transducers are based on the principle of thermoelectric energy conversion. This represents the transformation of thermal energy into electrical energy or the inverse process. In principle, there are two basic thermoelectric effects, the *Seebeck* and the *Peltier* effects, available for this purpose. These effects are successfully employed in thermoelectric signal detection and generation devices, as well as in thermoelectric refrigeration and heating systems. Depending on the energy level to which a practical system is exposed, we distinguish sensor devices in the low power range and generator or cooler units in the case of a high power level. Generally, thermoelectric devices may cover a very large energy range, i.e. from μW up to MW. Thermoelectric thin film systems, however, are predestinated for sensor and low-power generation applications.

5.5.2 Thermoelectric Energy Conversion – Some Basic Considerations

The practical realization of thermoelectric energy conversion requires particular devices, so-called thermocouple arrays. Their basic units consist of two conductors ("legs") made from thermoelectric materials with different types of conductivity which are connected by metallic bridges and electrodes, as shown schematically in Fig. 5.34. As seen in this figure the technological exploitation of the *Seebeck* effect requires a junction between n-type and p-type materials, i.e. between materials exhibiting dominating electron and hole conductivity, respectively. In passing we note that this carrier diffusion-based effect is basically an intrinsic bulk phenomenon which may be experimentally observed also in a homogeneous material as the so-called differential thermoelectric power.

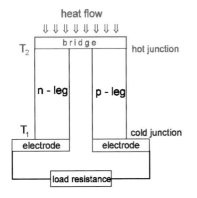

Figure 5.34: Diagram of a thermocouple unit for the demonstration of thermoelectric generation. The properties of the n- and p-leg are described by two sets of parameters A_1, S_1, ρ_1, λ_1, and A_2, S_2, ρ_2, λ_2, respectively, i.e. by the leg cross section, the Seebeck coefficient, electrical resistivity and thermal conductivity.

The diagram also shows that with respect to the *heat flow* arising from the existing temperature gradient the two branches with different conduction type form a parallel circuit, whereas with respect to the *current flow* the branches are connected in series. From this short description of the operational principle of a thermoelectric conversion unit it follows immediately that optimization strategies should not only encompass the thermoelectric material parameters, but also device design considerations. In detail, the construction of an efficient thermoelectric device involves the following tasks:

- choice of the optimum thermoelectric materials,
- optimization of the electrical contacts,
- optimization of the thermal management, i.e. sustaining the temperature gradient and heat transfer between thermocouples and environment.

5.5.3   Thermoelectric Sensors

For the purpose of temperature detection different working principles have proved to be applicable. They make use of the following specific features in the electrical transport properties of thin films:

- The ***temperature dependence of resistivity***. On this basis, thin film resistance bolometers have been developed reaching a relatively high specific sensitivity of up to 10^9 cm Hz$^{1/2}$ W^{-1} (see [5.71]).
- The ***abrupt resistivity change in superconducting materials*** at the transition temperature into the normal state. This pronounced switching effect has found application in the construction of superconducting bolometers. This type of device is characterized by a very low noise level, however, the operation of these bolometers needs cryogenic temperatures (see [5.72]).
- The ***non-linear current–voltage characteristic*** of semiconducting thin films embedded in ohmic contacts. This used in thermistor bolometers (see, e.g., [5.73]).
- The ***surface charge variation*** in planar capacitor devices with polar dielectric materials. This used for the detection of temperature changes. This effect is also known as pyroelectricity [5.74]
- The generation of a ***thermo-voltage*** on the basis of the ***Seebeck effect***. In this case, the difference between the object temperature (to be determined) and a reference point temperature (well-known) yields a proportional voltage signal (the integral thermoelectric power).

All the thermal detection principles mentioned above have found specific application fields. While resistance bolometers and superconducting bolometers exhibit the highest known sensitivity values (10^9 cm Hz$^{1/2}$ W^{-1}), they need a dedicated experimental environment, i.e. cryogenic temperatures to sustain the superconducting state. Thermoelectric sensors can be manufactured by thin film deposition procedures compatible with integrated circuit (IC) technology. This is a significant technological advantage. The wide range of the signal amplitudes, the good sensitivity and stability levels as well as the straightforward integration into microelectronic circuitry are the reasons for a widespread application of thermoelectric sensors. These sensors will be described in the following in more detail.

Main Parameters of Thermoelectric Sensors

For the evaluation of the performance of a thermal or radiation detector the following parameters generally have to be analyzed [5.75]:

- signal voltage U_S

$$U_S = (S_1 - S_2)(T_2 - T_1) \tag{5.2}$$

with $S_{1,2}$ denoting the *Seebeck* coefficients of the n- and p-leg, and $T_{1,2}$ the temperatures of the cold and hot sides.

- responsivity E

$$E = U_S/N = (S_2 - S_1)R_T \tag{5.3}$$

with the incident power N, and R_T denoting the thermal resistance of the absorber-thermocouple device : $R_T N = T_2 - T_1$.

- noise equivalent power NEP

$$NEP = U_N/E = \sqrt{4kT_1 R_{tot} \Delta f} / (S_2 - S_1)R_T \tag{5.4}$$

with the noise voltage in the case of *Nyquist* noise U_N, the Boltzmann constant k, the noise bandwidth Δf, the total resistance $R_{tot} = l\rho_1/A_1 + l\rho_2/A_2$ determined by the specific resistivities $\rho_{1,2}$ and leg cross sections $A_{1,2}$, and the leg length l.

- The detectivity represents the inverse noise equivalent power ($D = 1/NEP$). However, for comparison of sensors it is more convenient to analyze the specific detectivity D^* related to the absorber area A and a 1 Hz bandwidth:

$$D^* = D\sqrt{A\Delta f} = (S_1 - S_2)R_T A^{1/2} / \sqrt{4kT_1 R_{leg}} \tag{5.5}$$

- The response time τ

$$\tau = R_T C_T \tag{5.6}$$

with C_T the heat capacity of the thermocouple.

In order to obtain the maximum specific detectivity the geometrical design must fulfill the condition $A_1/A_2 = \sqrt{\lambda_2 \rho_1/\lambda_1 \rho_2}$, the heating power in the thermocouple should be equal to the power generated by the absorber, and the leg heat fluxes should be identical. In this case the detectivity is given by the following expression [5.75]:

$$D^*_{max} = \sqrt{Z_{1/2} T_1} / \sqrt[8]{\sigma_s k T_1^5} \tag{5.7}$$

where $Z_{1/2}$ is the total efficiency of the thermocouple according to the expression

$$Z_{1/2} = (S_1 - S_2)^2 / \left(\sqrt{\lambda_1 \rho_1} + \sqrt{\lambda_2 \rho_2}\right)^2 \tag{5.8}$$

and σ_s is the Stefan–Boltzmann constant.

From this discussion of the performance parameters it can be concluded for the quality of thermoelectric sensors that besides size optimization the materials properties *Seebeck* coefficient, thermal and electrical conductivity play the most important roles. This is represented by the thermoelectric figure of merit $Z = S^2/\rho\lambda$.

In the following some examples for the applications of thermoelectric sensors will be briefly addressed.

Planar Multi-Junction Thermal Converter (PMJTC)

Thermal converters are employed as highly sensitive and accurate standards for the measurement of alternating electrical quantities in the frequency range between 10 Hz and 1 MHz. Such converters are necessary, because the reference standards based on the Josephson and Quantum Hall effects are applicable for direct current situations only. Thermal converters operate on the principle of equivalence of the heating power (Joule's heat) of alternating (ac) and direct currents (dc) in a resistor. In a PMJTC the Joule's heat of the electrical current causes a temperature increase of the resistor (heater) measured with high precision and resolution by serially connected thermocouples. In the PMJTC device the temperature produced by Joule's heat from a known dc current through a resistive NiCrSi layer is compared to the heating effect of an unknown ac current. In principle, the temperature should be equal for the two signals at the same root-mean-square (rms) value. Thus, an unknown rms ac current can be converted into an equivalent dc current value. However, in a real converter a transfer difference δ is observed, being defined as

$$\delta = \left(U_{in}^{ac} - U_{in}^{dc} \right) / U_{in}^{dc} \tag{5.9}$$

for the case where $U_{out}^{ac} = U_{out}^{dc}$ [5.76]. The converter depicted in Figs. 5.35 and 5.36 provides a calibration process with a measuring uncertainty of $2 \cdot 10^{-6}$ in the frequency range $40\,\text{Hz} \leq f \leq 50\,\text{kHz}$ due to the very low transfer difference δ in the 10^{-6} range [5.77].

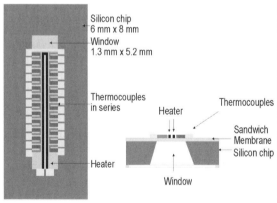
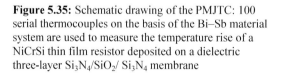

Figure 5.35: Schematic drawing of the PMJTC: 100 serial thermocouples on the basis of the Bi–Sb material system are used to measure the temperature rise of a NiCrSi thin film resistor deposited on a dielectric three-layer $Si_3N_4/SiO_2/ Si_3N_4$ membrane

Figure 5.36: Photograph of the PMJTC chip on its Al_2O_3 carrier. Aluminum bond wires connect the Al-pads on the chip to the Au-pads on the carrier

Deviations from these high-precision properties are found in the low-frequency range caused by nonlinearities in the input–output relation of the PMJTC (see Fig. 5.37). A compensation decrease of the effect of these nonlinearities can be achieved either by adding a thermal mass underneath the heater to increase the thermal time-constant or by adjusting the temperature dependence of the input power to output voltage ratio. For this purpose, the properties of the thermoelectric films and the thin-film resistor must be carefully controlled during preparation in such a way that the thermocouple materials and the resistor are directly engineered with a well-defined temperature dependence of the *Seebeck* coefficient and the temperature coefficient of resistance of the heater.

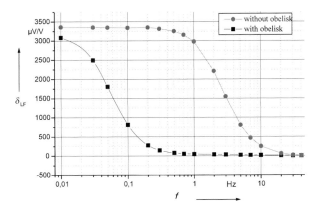

Figure 5.37: Measured ac–dc transfer difference δ of two types of PMJTC: (■) with silicon obelisk underneath the heater, and (●) without obelisk

Thermal Infrared (IR) Sensor

Recently, intensive efforts have been devoted to the development of CMOS compatible high efficiency infrared sensors for the wavelength range between 7 μm and 14 μm. In addition, the materials used as sensor materials should also ensure a stable operation at temperatures above room temperature. The classical thermoelectric materials on a Bi–Sb base which show the highest known efficiency values at room temperature are not compatible with the Si-based microelectronic technology. Si–Ge is established as a high-temperature thermoelectric material, whereas it exhibits only medium efficiency values around room temperature. In addition, however, this material is characterized by a very weak temperature variation of the efficiency in the room temperature regime. This is an attractive feature, if stable device parameters around room temperature are the objective. By contrast, the chalcogenide compounds exhibit a strong efficiency decay above room temperature which is quite disadvantageous for the application envisaged.

The compound semiconductor Si–Ge can be easily implemented into the CMOS production process. The growth of preparation of polycrystalline Si–Ge films can be achieved by means of plasma enhanced chemical vapor deposition (PECVD). This method has already been established in the course of a standard IC manufacturing process for several years. The deposition process parameters (temperature and reactant fluxes) and the post-deposition annealing conditions have been proved to be of significant importance for the optimization of thermoelectric performance.

A design proposal of a thermal non-contact IR sensor to be prepared within standard IC technology is shown in Fig. 5.38 [5.78]. The high thermal conductivity of the silicon wafer (ca. 150 W mK^{-1}) used as a support material for the thermoelectric sensor requires a dedicated design for the thermocouple array. The hot sides of the thermoelements arranged in close contact with an absorbing layer are placed on an artificial membrane acting as a thermal barrier to the silicon substrate. The latter serves as a heat sink and carries the cold side of the thermocouples. The membrane preparation is the most crucial technology step in the sensor fabrication. Due to the different expansion coefficients of silicon, silicon dioxide and silicon nitride, a stress-free membrane with reliable adhesion to the silicon wafer can be obtained only by a multi-layered structure stabilized by an internal stress compensation. Furthermore, the membrane layer should act as a stop in the back-side etching process and ensure stable adhesion of the thermoelectric, contact and interconnection layers on the front side. Finally, the sensor arrays must withstand the large strain applied to the membranes during the micro-machining processes, e.g., wafer cutting during the chip separation.

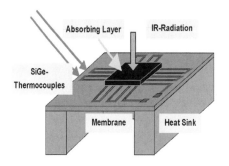

Figure 5.38: Schematic drawing of a thermoelectric IR sensor in IC technology

The performance of a thermal IR sensor is determined by the thermal management within the sensor and by the thermoelectric properties of the thermosensitive materials used. Therefore, finite element (FE) and network analyses are absolutely essential tools for thermal optimization in the design and simulation processes of the sensor arrays. The precise control of the deposition process is of great importance for the definition of the thermoelectric properties, too. As may be learned from Table 5.3 which compiles data from state of the art SiGe layers, the efficiency level achieved in SiGe bulk applications [5.79] is not yet within reach for thin films.

Table 5.3: Exemplary thermoelectric data of PECVD $Si_{0.7}Ge_{0.3}$ layers crystallized by a post-deposition annealing at 650 °C [5.78]

Doping	Seebeck coefficient S, $\mu V\ K^{-1}$	Thermal conductivity λ, $W\ m^{-1}\ K^{-1}$	Resistivity ρ, $m\Omega\ cm$	Figure of merit Z, $10^{-3}\ K^{-1}$
p-type	274	1.8	38	0.110
n-type	-150	5.0	3	0.150

Thermoelectric Flow Sensor

The functional principle of flow sensors is based on the transfer of heat to moving gases and fluids. As in all thermal sensors the device operation involves three subsequent steps [5.80]: (i) transformation of the non-thermal quantity, the gas flow, into a thermal heat signal, (ii) conversion of the heat power into a temperature difference signal, and (iii) transduction of the temperature difference into an electrical voltage. The first step in the operational process is the most complicated one, because the result strongly depends on the type of flow and the housing of the sensor. The heat transfer as a function of the flow velocity differs for laminar flow and turbulent flow. Parasitic heat fluxes within the sensor can easily change the heat transfer characteristics. Therefore, a careful encapsulation is required for a stable measuring effect and reliable operation of the device.

Figure 5.39 shows the schematic layout of a thin film thermocouple gas flow sensor. The sensor device consists of a thermocouple array, an integrated thin film heater and an optional thin film resistance thermometer. As carrier for this thin film device glass and ceramic substrates can be used.

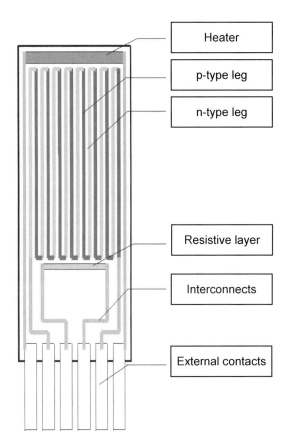

Figure 5.39: Schematic drawing of a thin film gas flow sensor

By means of a constant heating power a defined temperature gradient is established over the device length. In the presence of a gas flow this gradient is changed and can be calibrated as a function of the flow speed. Metal and silicide layers constitute the materials for the thermoelement legs. The sensor described can be produced in a low-cost version with limited accuracy. In order to obtain a higher precision of the device the sensor can be completed with an optional Pt thermometer film. It probes the temperature of the cold side and improves the accuracy of the gradient maesurement. Typical technical data are given in Table 5.4.

5.5.4 Thermoelectric Transducers

Thin film devices based on the exploitation of the *Seebeck* and *Peltier* effects as well as employing thermionic emission have found applications in low-power generators and microcoolers. These are transducer elements which convert heat fluxes in an electrical current flow or which pump heat from one point to the other in order to obtain a local cooling.

Table 5.4: Technical parameters of a thin film gas flow sensor

Length mm	Width mm	Heater resistance Ω	Nominal output voltage (ΔT=100 K) mV	Output voltage range mV	Internal resistance kΩ	Number of junctions	Spacing on the terminal block mm
25	5	200±2	37	35–39	1.6–2.0	14	1.27

Miniaturized devices of this type are of great interest for microelectronics technology, because they offer many advantages. The power generation and cooling can take place immediately on the chip surface and may be seamlessly integrated into the chip architecture. In this way, most of the manufacturing techniques employed in the IC technology can be used for the device fabrication.

Low-Power Generators

Discrete generators. In context with the increasing application of electronic microsystems in all areas of modern society, a huge number of low energy consumption components have been developed. Furthermore, the development of new microdevices is accompanied by a permanent reduction in the energy consumption down into the µW-range. Additionally, an increasing demand for self-sufficient systems can be noted. This means that the traditional way of supplying power by means of batteries or solar cells is supplemented by alternative approaches. Strong efforts are made to improve the energy balance of systems by local trapping of waste heat or regeneration of dissipated loss power immediately at the place of emergence. For such a purpose, conventional thermoelectric generators on the base of bulk materials are inappropriate due to their large size and performance parameters [5.82].

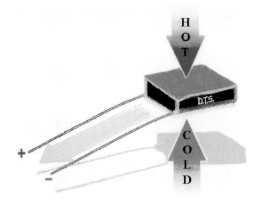

Figure 5.40: Functional principle and 1st prototype of a low power thermoelectric generator (LPTG)

Table 5.5: Geometrical parameters of the recent LPTG

Parameter	Value/tolerance
Length, mm	9.0 ± 0.05
Width, mm	5.2 ± 0.05
Height, mm	1.8 ± 0.1
Connection wire, mm	$2 \times (80{\pm}5)$, $\varnothing$ 0.2
Mass, mg	250 ± 10

The low power thermoelectric generator (LPTG, 1st generation) shown in Fig. 5.40 provides a power output of about 23.5 µW at a voltage of about 4.2 V under load, if the temperature difference between the hot and cold sides is set to $\Delta T \approx 20$ K (for the recent performance level see the data given in Table 5.6). This small and compact thermoelectric device is able to directly convert the thermal energy in a temperature gradient or heat flux into electrical energy. A good thermal contact of the generator to the heat source and sink is an essential condition for the efficient exploitation of the temperature gradient. The operational temperature of the above device should be in the regime $20\ °C \le T \le 100\ °C$.

Table 5.6: Typical electrical data of the recent low power thermoelectric generator ($R_{source} = R_{load}$), see also [5.83]

Temperature difference ΔT, K	Output voltage U_0, V	Output current I_0, µA	Output power P_0, µW
5	2.5	7.8	20.0
10	5.1	15.6	80.0
20	10.2	31.4	320.0
40	22.0	68.0	1 500
60	31.5	95.2	3 000

The fabrication process of the generator described above is based on the use of highly efficient thermoelectric thin films (predominantly the compound semiconductor class Bi–Te) deposited by physical vapor deposition onto polyimide substrates. The manufacturing process involves a multi-step route employing a variety of methods established in thin film technology: film deposition, photolithography, pattern generation, wafer (foil) cutting, device assembling. The heart of the LPTG comprises a complex layer stack consisting of a large number of thin foil segments arranged between two thermal coupling plates. The latter are usually made of ceramics to keep a good thermal contact to the heat sink and source. The film pattern size on the segments (leg width) amounts to 50 µm, the output voltage of an individual segment takes values of about 60 mV K^{-1}. The final step in the manufacturing process consists in an appropriate housing to protect the sensor against mechanical and environmental influences.

The LPTGs have a broad user potential as self-sufficient energy sources, e.g., in electronic heat cost allocators, electronic wristwatches, active transponders, self-sufficiently powered temperature warning and registration systems, or displays.

Integrated (on-chip) generators. The conversion of waste heat dissipated by integrated circuits or printed wiring boards into electrical power has proved to be an attractive measure for the improvement of the energy consumption balance in electronic systems. Therefore, recently, novel micro-scale thermoelectric generators (TEG) have been developed for different applications. In principle, CMOS technology offers the means to manufacture small-size and low-power thermoelectric systems at low cost. However, the well-known specific features of silicon-based integrated circuits consisting of a small vertical extension of their active components in the µm-range and a "top–bottom" external temperature gradient on the wafer require a specific thermo-management. The external cross-plane temperature gradient should be transformed into a local in-plane temperature gradient. In the latter case the thermocouples could be arranged across the entire chip area. This problem can be solved by a proper architecture of the solid state generator, combining the BiCMOS technology with surface micro-machining as shown in Fig. 5.41 [5.84]. In contrast to conventional large scale TEGs the noticeable feature of micro-scale TEGs consists of a small internal thermal resistance as compared to the large thermal resistance of the enclosing components. As a consequence, an integrated thermoelectric generator as described here works in the constant heat flow regime, contrary to the usual constant temperature difference operation mode of conventional TEGs.

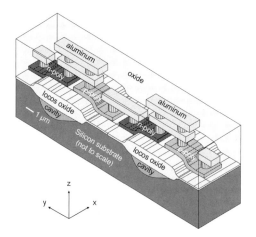

Figure 5.41: Schematic view of two thermoelectric couples of the BiCMOS realization. Additionally, the cavities which can be etched into the Si-substrate are shown

The generator shown in Fig. 5.41 has been fabricated in a BiCMOS production facility and consists of 59.400 thermocouples arranged on a chip area of about 6 mm^2. Poly-Si and poly-SiGe have been used as n- and p-leg materials doped by phosphorus and boron implantation, respectively. In the case of the SiGe-legs prepared by chemical vapor deposition the growth of large grains needs to be suppressed by means of a thin polysilicon seed layer. The ohmic contact bridges for the leg junctions are realized by aluminum layers. Additionally, every other junction is joined by a further Al bridge in order to improve the thermal coupling to the surface. Besides the material properties of the semiconducting leg the thermal insulation between the cold and the hot junctions has a significant influence on the performance of the TEG. Therefore, in the present case, the LOCOS oxide layers with a thickness of 1.6 µm, acting as thermal barrier between the hot and cold ends, are completed by defined cavities prepared by micro-machining processes using wet and dry etching techniques. Tests of the generator structures have shown that the cavity approach for the heat flux optimization resulted in a significant improvement of the output voltage level reaching a range from 100 μV K^{-1} to 200 μV K^{-1}. Because of the large number of thermocouples the total resistance of the array can rise up to 10 MΩ due to the huge number of contact interfaces (4 contacts per couple). By optimization of the material parameters and fabrication technology the manufacture of generators with an output voltage of 1V at load of 1 μW for a chip area of less than 25 mm^2 at a temperature difference of 20 K seems feasible.

Microcoolers

Peltier-effect based solid state heat pumps and refrigerator units have been established as compact and reliable cooling equipment for a variety of applications. Design flexibility, maintenance-free operation and arbitrary mounting position are some of the advantages of thermoelectric cooling devices. In general, the commercially available solid state coolers are designed on the base of bulk thermoelectric materials characterized by a figure of merit

$$Z \times T = S^2 T / \rho \lambda_t \leq 1 \tag{5.10}$$

limiting the COP of thermoelectric devices to the range of 5 % to 10 %. On the one hand, despite tremendous efforts in the materials science of thermoelectric compounds the progress in the *ZT* enhancement of room temperature bulk materials during the last years is only marginal. On the other hand, several approaches for the enhancement of the efficiency based on thin film technologies were proposed recently. The three main research directions are the following:

- Multi-quantum well structures with quantum confinement effects resulting in an enhancement of the density of states in the vicinity of the Fermi energy [5.85].
- Thin-film superlattice structures characterized by an acoustic mismatch between superlattice layers. This reduces the lattice thermal conductivity on the basis of phonon-blocking within these layered stacks [5.86].
- Heterostructures with thermionic emission allowing an improvement of the cooling efficiency by the selective ejection of hot carriers across the barrier layers [5.87].

Solid state thermoelectric devices realized by thin film preparation offer the remarkable advantage of localized and rapid cooling due to their capacity for miniaturization. Thus, the cooling power can be fed immediately into active micro- and opto-electronic devices with a response time about 10^4 times faster than comparable bulk devices. *Fan et al.* [5.88] could show that by means of Si–Ge superlattices, as a material compatible with the IC technology, it is possible to fabricate micro-cooler layer stacks of 2 μm thickness consisting of 200 layer pairs of Si$_{0.89}$Ge$_{0.10}$C$_{0.01}$/Si and a Si cap layer on an area of 40 × 40 μm^2. At room temperature (298 K) a cooling effect of 2.8 K was measured, whereas at 373 K the cooling effect was 6.9 K, demonstrating the known efficiency enhancement of SiGe at increasing temperature. Even though the performance of such a cooling device is still far from the required cooling power for the application in integrated circuits, the way is opened for further technological developments. Material optimization, thermal network simulation, and device design are the fields of research activity promising significant progress in the efficiency level of integrated micro-coolers [5.87].

More recently, *Ghoshal et al.* [5.89] proposed a new innovative cooling device, the cold point thermoelectric cooler, comprising a micromechanical set-up manufactured on the basis of thin film technology. In order to confine the thermoelectric transport processes at the surfaces, the authors have designed the cold ends as a double-sided array of metallic point contacts placed between two thermoelectric materials with different types of conductivity.

The point contacts form a network of PVD-prepared Cu–Ni micro-pyramids capped by a Au layer with a tip radius of 0.6 μm and arranged with a pitch width of 20 μm into an 80 × 80 array. This assembly mounted together with a thermocouple onto an appropriate carrier serves as cold end contact. The hot sides of the devices were made from polycrystalline $Bi_{0.5}Sb_{1.5}Te_3$ and $Bi_2Te_{2.9}Se_{0.1}$ as p-type and n-type legs, respectively. The vacuum operation test of a prototype has shown that for device currents between 100 mA and 500 mA an increasing temperature difference with respect to the hot side (room) temperature of 20 K and 70 K is reached, respectively. The estimation of the related figure of merit yields the range $ZT = 1.4$–1.7, exceeding significantly the values known from the bulk chalkogenides. The reason for this improvement is obviously seen in the reduction of the thermal conductivity at the n/p-interface and the weak electron–phonon coupling at the junction.

5.5.5   Outlook

At present the thermoelectricity technology has very successfully provided products for power supply, cooling, and sensor systems, which are mainly utilized in space exploration and military applications. The consumer market represents as yet only a niche segment in these initiatives. For a long time, a strong stimulus for innovative projects and developments was missing. The main reason for this situation was the low efficiency level of thermoelectric materials $(ZT < 1)$. In addition, the expensive fabrication technology forbade mass production. During the last 10 years, however, an essential change in the situation can be noted in thermoelectricity science. This may be substantiated by the following three issues. Firstly, the feasibility to significantly improve ZT has been confirmed. The novel compound Zn_4Sb_3 and the skutterudite-like materials $CeFe_4Sb_{12}$ and $CoSb_3$ have shown ZT values in the range 1.0 to 1.4 and are available for prototype manufacture. A maximum thermal/electrical conversion efficiency of about 15% for the temperature gradient 300 °C/975 °C is predicted [5.90]. Secondly, the mass production of thermoelectric components using microsystem technologies is under development [5.91]. Thirdly, thermoelectric solid-state coolers for local thermal management and temperature stabilization of microelectronic and optoelectronic devices can be prepared by means of the integrated circuit fabrication technology on the basis of SiGeC/Si superlattice structures [5.88].

These latest developments open the pathway for introducing the thermoelectricity-based technology and devices into the microelectronics and automotive industries as branches promising wide fields of application. The recent progress in materials science and technology of thermoelectric devices will promote the attractiveness and applicability of this solid state energy conversion technique in the near future.

5.6   References

[5.1]   The International Technology Roadmap for Semiconductors (ITRS), 2003 edition.
[5.2]   K. Buchanan, in *Proceedings of the 2002 GaAs MANTECH Conference*, GaAs MANTECH Inc., St. Louis, 2002, p. 7.
[5.3]   M. Engelhardt, G. Schindler, and C. Werner, *Proc. of the 1st International Conference on Semiconductor Technology*, 2001, vol. 2, p. 206.
[5.4]   M. Engelhardt, G. Schindler, G. Steinlesberger, W. Steinhögl, *Microelectron. Eng.*, **64**, 11 (2002).
[5.5]   S. Thomas, *AMC 2002, p. 3*

[5.6]   D. A. B. Miller, IEEE *J. Selected Topics in Quantum Electronics* 6, 1312 (2002)

[5.7]   F. Kreupl, A.P. Graham, G.S. Duesberg, W. Steinhögl, M. Liebau, E. Unger, W. Hönlein, *Microelectron. Eng.*, **64**, 399 (2002).

[5.8]   G.K.Montress, T.E.Parker, D. Andres, in *Proceedings of the 1994 IEEE Ultrasonics Symp., Cannes,* IEEE, Piscataway, 1994, p. 43.

[5.9]   P.Muralt in *Piezoelectric Materials and Devices*, ed. N.Setter, N.Setter Ceramics Laboratory, Lausanne, 2002, p. 303.

[5.10] C.S.Hartmann, in *Proceedings of the 1982 IEEE Ultrasonics Symp., San Diego,* IEEE, Piscataway, 1982, p. 423.

[5.11] M.Feldmann, J.Henaff, in *Proceedings of the 1978 IEEE Ultrasonics Symp, Cherry Hill,* IEEE, Piscataway, 1978, p. 720.

[5.12] M.F.Lewis, *Electron. Lett.*, **8**, 553 (1972).

[5.13] M.F.Lewis, in *Proceedings of the 1982 IEEE Ultrasonics Symp., San Diego,* IEEE, Piscataway, 1982, p. 12.

[5.14] C.S.Hartmann, B.P.Abbott, in *Proceedings of the 1989 IEEE Ultrasonics Symp., Montreal,* IEEE, Piscataway, 1989, p. 79.

[5.15] P.Ventura, M.Solal, P.Dufilie, J.Desbois, M.Doisy, J.M.Hodé, in *Proceedings of the 1992 IEEE Ultrasonics Symp., Tucson,* IEEE, Piscataway, 1992, p. 71.

[5.16] J.Machui, W.Ruile, in *Proceedings of the 1992 IEEE Ultrasonics Symp., Tucson,* IEEE, Piscataway, 1992, p. 147.

[5.17] T.Morita, Y.Watanabe, M.Tanaka, Y.Nakazawa, in *Proceedings of the 1992 IEEE Ultrasonics Symp., Tucson,* IEEE, Piscataway, 1992, p. 95.

[5.18] G.Martin, in *Proceedings of the 1999 IEEE Ultrasonics Symp., Caesars Tahoe,* IEEE, Piscataway, 1999, p. 15.

[5.19] O.Ikata, T.Miyashita, T.Matsuda, T.Nishikawa, Y.Satoh, in *Proceedings of the 1992 IEEE Ultrasonics Symp., Tucson,* IEEE, Piscataway, 1992, p. 112.

[5.20] D.S. Ballantine, R.M. White, S.J. Martin, A.J. Ricco, E.T Zellers, G.C. Frye, H. Wohltjen, *Acoustic Wave Sensors* in Applications of Modern Acoustic, ed. R.Stern and M.Levy, Academic Press, Lausanne, 1997, Ch. 3.2.

[5.21] Strategy Analytics, Market Study *Active Sensor Market Trends, Part I* , March 2001.

[5.22] W. Thomson, *Proc. R. Soc. London, Ser. A*, **8**, 546 (1857).

[5.23] M.H. Kryder, *An Introduction to Magnetic Recording Heads*, Proceedings of the NATO Advanced Study Institute, Kluwer, Dordrecht, 2001, p. 449.

[5.24] C.H. Tsang, R. E. Fontana Jr., T. Lin, D.E. Heim, B.A. Gurney, M.L. Williams, *IBM J. Res. Develop.*, **42**, 103 (1998).

[5.25] U. Dibbern, *Magnetoresistive Sensors*, in Sensors, Vol. 5, Magnetic Sensors, VCH, Weinheim, 1989.

[5.26] K. Dietmayer, *Technisches Messen*, **68**, 269 (2001).

[5.27] S. Tumanski, *Thin Film Magnetoresistive Sensors*, Institute of Physics Publishing, Bristol, 2001.

[5.28] M. Baibich, J. Brot, A. Fert, F. Nguyen Van Dau, F. Petroff, P. Etienne, G. Creuzet, A. Friederich, J. Chazelas, *Phys. Rev. Lett.*, **61**, 2472 (1988).

[5.29] J. Barnas, A. Fuss, R. Camley, P. Grünberg, W. Zinn, *Phys. Rev. B*, **42**, 8110 (1990).

[5.30] J. Daughton, A.V. Pohm, R.T. Fayfield, C.H. Smith, *J. Phys. D*, **32**, R169 (1999),

[5.31] H. Kanai, K. Noma, and J. Hong, *Fujitsu Sci. Tech. J.*, **37**, 174 (2001).

[5.32] H. A. M. van den Berg, W. Clemens, G. Gieres, G. Rupp, M. Vieth, J. Wecker, S. Zoll, *J. Magn. Magn. Mater.*, **165**, 524 (1997).

[5.33] Infineon Technologies Semiconductor Data Book 04-99, 175 (1999).

[5.34] *SAE Automotive Engineering 2002*, September Issue (2002).

[5.35] K.-M.H. Lensson, D.J. Adelerhof, H.J. Gassen, A.E.T. Kuiper, G.H.J. Somers, J.B.A.D. van Zon, *Sens. Actuators*, **85**, 1 (2000).

[5.36] C.P.O. Treutler, H. Siegle, *Technisches Messen*, **68**, 280 (2001).

[5.37]   D. Hammerschmidt, E. Katzmaier, D. Tatschl, W. Granig, J. Zimmer, B. Vogelgesang,
         R. Rettig, *Giant Magneto Resistors – Sensor Technology and Automotive Applications,*
         SAE 2005 World Congress & Exhibition, April 2005, Detroit, MI, USA.
[5.38]   U. Caduff, H. Schweren, H. Kittel, *Low Cost Angle Sensors for Multi-Purpose
         Applications,* SAE-Conference, Detroit, USA, 2000, Pat. DE 19839446 A1, US 6.433.535
         B1.
[5.39]   J.K. Spong, V.S. Speriosu, R.E. Fontana, Jr., M.M. Dovek., T.L. Hylton, *IEEE Trans.
         Magn.,* **32**, 366, (1996).
[5.40]   A. Johnson, G. Mörsch, H. Gunther, Intern. Patent WO 02082111 (2001).
[5.41]   A. Compton, *Phil. Mag.,* **45**, 1121 (1923).
[5.42]   P. Kirkpatrick, A. V. Baez, *J. Opt. Soc. Am.,* **38**, 766 (1948).
[5.43]   H. Wolter, *Ann. Phys.,* **10**, 94 (1952).
[5.44]   D.T. Attwood, *J. Quant. Electron.,* **14**, 909 (1978).
[5.45]   R.H. Price, in *Low-Energy X-ray Diagnost. 1981, AIP Conf. Proc.,* **75**, 189 (1981).
[5.46]   J.K. Silk, *Proc. SPIE,* **184**, 40 (1979).
[5.47]   J.H. Underwood, D.T. Attwood, *Physics Today,*, April 1984, 44 (1984).
[5.48]   B. Aschenbach, in *UV and X-ray Spectroscopy of Laboratory and Astrophysical
         Plasmas,*Cambridge University Press, 1993, p. 434.
[5.49]   http://imagine.gsfc.nasa.gov/index.html.
[5.50]   J.H. Underwood, T.W. Barbee, D.C. Keith, *Proc. SPIE,* **184**, 123 (1979).
[5.51]   J.H. Underwood, M.E. Bruner, B.M. Haisch, W.A. Brown, L.W. Acton, *Science,* **238**, 61
         (1987).
[5.52]   A.B.C. Walker, T.W. Barbee, R.B. Hoover, J.F. Lindblom, *Science,* **241**, 1781 (1988).
[5.53]   A.B.C. Walker, R.B. Hoover, T.W. Barbee, in *Physics of Solar and Stellar Coronae,*
         Kluwer, Dordrecht, The Netherlands, 1993, p. 83.
[5.54]   K. Schwarzschild, *Astr. Mitteil. Königl. Sternwarte Göttingen,* **10**, 9 (1905).
[5.55]   D. Attwood, *Soft X-rays and Extreme Ultraviolet Radiation,* Cambridge University Press,
         Cambridge, 1999.
[5.56]   F. Cerrina, G. Margaritondo, J.H. Underwood, M. Hettrick, M.A. Green, L.J. Brillson,
         A. Franciosi, H. Hochst, P.M. Deluca, M.N. Gould, *Nucl. Instrum. Methods Phys. Res.
         Sect.A,* **266**, 303 (1988).
[5.57]   J.B. Kortright, E.M. Gullikson, P.E. Denham, *Appl. Opt.,* **32**, 6961 (1993).
[5.58]   K. Murakami, T. Oshino, H. Nakamura, M. Ohtani, H. Nagata, *Appl. Opt.,* **32**, 7057 (1993).
[5.59]   G.E. Moore, *Electronics,* **38**, 114 (1965).
[5.60]   G.D. Kubiak, D.R. Kania, in *Extreme Ultraviolet Lithography in OSA, Trends in Optics
         and Photonics,* Optical Society of America, Washington, DC, 1996, Vol. 4.
[5.61]   St. Braun, H. Mai, M. Moss, R. Scholz, A. Leson, *Jpn. J. Appl. Phys.,* **41**, 4074 (2002).
[5.62]   S. Bajt, J. Alameda, T.W. Barbee Jr., W.M. Clift, J.A. Folta, B. Kauffman, E. Spiller, *Proc.
         SPIE,* **4506**, 65 (2001).
[5.63]   R. Soufli, E. Spiller, M.A. Schmidt, J.C. Davidson, K.F. Grabner, E.M. Gullikson,
         B.B. Kaufmann, S. Mrowka, S.L. Baker, H.N. Chapman, R.M. Hudyma, J. S. Taylor,
         C.C. Walton, C. Montcalm, J.A. Folta, *Proc. SPIE,* **4343**, 51 (2001).
[5.64]   M. Bujak, S. Burkhart, C. Cerjan, P. Kearney, C. Moore, S. Prisbrey, D. Sweeney,
         W. Tong, S. Vernon, C. Walton, A. Warrick, F. Weber, M. Wedowski, K. Wilhelmsen,
         J. Bokor, S. Jeong, G. Cardinale, A. Ray-Chaudhuri, A. Stivers, E. Tenjil, P. Yan,
         S. Hector, K. Nguyen, *Proc. SPIE,* **3665**, 30 (1999).
[5.65]   J.A. Folta, J. Davidson, C.C. Larson, C C. Walton, P.A. Kearney, *Proc. SPIE,* **4688**, 173
         (2002).
[5.66]   A. Barty, P.B. Mirkarimi, D.G. Stearns, D.W. Sweeney, H.N. Chapman, M. Clift, S.D.
         Hector, Y. Moonsuk, *Proc. SPIE,* **4688**, 385 (2002).
[5.67]   B. Rauschenbach, *Vacuum,* **69**, 3 (2002).
[5.68]   M. Schuster, H. Göbel, L. Brügemann, D. Bahr, F. Burgaezy, C. Michaelsen, M. Störmer,
         P. Ricardo, R. Dietsch, T. Holz, H. Mai, *Proc. SPIE,* **3767**, 183 (1999).

[5.69]   Th. Holz, R. Dietsch, H. Mai, L. Brügemann, *Trans. Tech. Publications*, **321-324**, 179 (2000).
[5.70]   R. Jenkins, *X-Ray Fluorescence Spectrometry*, John Wiley & Sons Inc, New York, 1999.
[5.71]   N. Butler, R. Blackwell, R. Murphy, R. Silva, C. Marshall, *Proc. SPIE*, **2552**, 583 (1995).
[5.72]   L. M´echin, J-C. Vill´egier, *J. Appl. Phys.*, **81**, 7039 (1997).
[5.73]   T. Ichihara, K. Aizawa, *J. Non-Crystalline Solids*, **B230**, 1345 (1998).
[5.74]   P. Muralt, *Rep. Prog. Phys.*, **64**, 1339 (2001).
[5.75]   H. Baltes, D. Moser, F. Voelklein, in S*ensors: A Comprehensive Survey*, Mechanical Sensors, ed. H.H. Bau, VCH , Weinheim, 1994, Vol. 7, p. 13.
[5.76]   H. Laiz, M. Klonz, *Microelectron. J.*, **30**, 1155 (1999).
[5.77]   M. Klonz, *PTBnews*, **98.2**, 1 (1998).
[5.78]   W. Brode, K. Bruehne, M. Rakhlin, M. Schubert, T. Pannek, H. Beyer, D. Eberhard, C. Kuenzel, A. Lambrecht, I. Stark, M. Stordeur, G. Paul, V. Sauchuk, J. Schumann, in *Proceedings of the 6th European Workshop on Thermoelectrics, Freiburg i.B.,* Fraunhofer-Institute of Physical Measurement Techniques IPM, Freiburg, 2001, p. 110.
[5.79]   C.B. Vining, in *CRC Handbook of Thermoelectrics*, ed. D..M. Rowe, CRC Press, Boca Rota, 1995, p. 329.
[5.80]   S. van Herwarden, *Sens. Mater.*, **8**, 373 (1996).
[5.81]   W. Brode, Firmenmitteilung aus der TFT Siegert GmbH Hermsdorf (Germany) 2002.
[5.82]   M. Stordeur, and I. Stark, in *Proceedings of the 9th International Trade Fair and Conference for Sensors, Tansducers & Systems, Nürnberg,* AMA Fachverband, Nürnberg, 1999, Vol. II, p. 193.
[5.83]   F. Schmidt, in *Proceedings of the 2nd GMM-workshop 2002, Dresden,* VDE/VDI-Gesllschaft Mikroelektronik, Mikro-und Feinwerktechnik, Frankfurt a.M.,2002, p. 71.
[5.84]   M. Strasser, R. Aigner, and G. Wachutka, in *Proceedings of the 6th European Workshop on Thermoelectrics, Freiburg i.B.,* Fraunhofer-Institute of Physical Measurement Techniques IPM, Freiburg, 2001, p. 89.
[5.85]   L.D. Hicks, and M.D. Dresselhaus, *Phys. Rev. B*, **47**, 12727 (1993).
[5.86]   R. Venkatasubramanian, in *Proceedings of the 1st National Thermogenic Cooler Workshop, Fort Belvoir,* Center for Night Vision and Electro-Optics, Fort Belvoir, 1992, p. 196.
[5.87]   G.D. Mahan, L.M. Woods, *Phys. Rev. Lett.*, **80**, 4016 (1998).
[5.88]   X. Fan, g. Zeng, C. Labounty, J.E. Bowers, E. Croke, C.C. Ahn, S. Huxtable, A. Majumdar, A. Shahouri, *Appl. Phys. Lett.*, **78**, 1580 (2001).
[5.89]   U. Ghoshal, S. Ghoshal, C. McDowell, L. Shi, S. Cordes, M. Farinelli, *Appl. Phys. Lett.*, **80**, 3006 (2002).
[5.90]   T. Caillat, J.-P. Fleurial, J. Snyder, J. Sakamoto, in *Proceedings of the 7th European Workshop on Thermoelectrics, Pamplona, Spain,* Universidad Publica de Navarra, Pamplona, 2002, p. 17.
[5.91]   H. Böttner, A. Schubert, K.H. Schlereth, D. Eberhard, A. Gavrikov, M. Jägle, G. Kühner, C. Künzel, J. Nurnus, G. Plescher, in *Proceedings of the 6th European Workshop on Thermoelectrics,* Freiburg i.B., Fraunhofer-Institute of Physical Measurement Techniques IPM, Freiburg, 2001, p. 7.

6   Outlook

The material compiled in this book gives a clear impression of the wide variety of application fields in which metal films are employed. The main driving force for the future development of metal-based thin films and thin film systems, however, is certainly the continuing advancement in microelectronics and sensor technology. In this context, at least four issues of major importance may be identified:

- Incorporation of new functionalities
- Development of novel materials
- Reduction of feature sizes
- Development of improved analysis approaches

All of these four issues pose numerous substantial challenges to research and development which are met by ongoing intense research efforts. It must be pointed out, in particular, that the above aspects are closely interconnected. For example, the development of new materials and functionalities usually requires adapted analysis techniques in order to be able to reliably characterize materials properties and to identify the causes of device failures. On the other hand, these new functionalities may lead to novel sensor devices which in turn can then be used for an improved analysis approach in other fields.

6.1   New Functionalities

The primary impetus for the advent of novel devices with new functionalities comes from the exploitation of quantum physics. This means that the fundamental properties of individual charge carriers – charge and spin – and their characteristic behavior on nanometer length scales may be utilized in the device design and operation. The future will therefore see a continuous transition from traditional microelectronics to the more advanced nanoelectronics with its multiple facets [6.1]. One of the long-standing issues is the so-called single electron transistor (SET). Its operational principle is based on the phenomenon of Coulomb blockade, i.e., the current through a quantum dot may be sensitively controlled by its charging state. Since the quantum dot comprises discrete electronic levels, increasing or decreasing the number of electrons in the quantum dot yields a strong effect on the current. For a reliable operation of these devices at room temperature, the level spacing must be of the order of the thermal energy, i.e. $E_{th} \sim 25$ meV at 300 K. This can be achieved only by extremely small dots with lateral dimensions in the 10 nm range. The main problem for the technical realization of SETs is thus the reliable and reproducible fabrication of nm-sized quantum dots. This is another driving force for the ongoing downsizing efforts (see also below). Alternative approaches on the borderline to molecular electronics, for example, on the basis of carbon nanotubes are also pursued [6.2–6.5].

Metal Based Thin Films for Electronics, Second Edition. Klaus Wetzig and Claus M. Schneider (Eds.)
Copyright © 2006 WILEY-VCH Verlag GmbH & Co. KGaA, Weinheim
ISBN: 3-527-40650-6

In Section 2.4 we have introduced the field of magnetoelectronics or *spintronics*. The exploitation of spin-dependent transport phenomena constitutes a novel facet in microelectronics which opens the avenue to numerous new devices and applications. Passive elements, such as magnetic sensors, are already in use as read heads in hard disk drives. The magnetic random access memory (MRAM) represents a more complex spintronics device. It promises a nonvolatile DRAM compatible memory concept and is expected to hit the market soon. A commercial success of the MRAM will significantly enhance the interest in spintronics and will increase the desire to exploit its huge potential. In addition, strong research activities are currently focusing on spin injection phenomena. Projecting the progress having been made in this particular field into the future, first demonstrators of active spintronic devices may be expected within the next 5 – 10 years.

Another topic which is relevant for microelectronics and is directly exploiting quantum phenomena is the so-called *Quantum Computing* (QC). Although still in its very infancy QC is thought to revolutionize data processing and logical operations. In contrast to the current binary logic which uses two distinct states ("0" and "1") to encode information, quantum computing allows a "continuous" encoding. This is achieved by using the interaction between at least two well-defined quantum objects (quantum bits, qubits). Currently, very different solid state based approaches are being investigated in order to realize these qubits. They range from semiconductor quantum dots over superconducting devices (Josephson junctions, magnetic flux quanta) to magnetic spin systems. Showing the feasibility of quantum computing will be one of the major challenges in the near future.

6.2 Materials-Related Aspects

The continuous striving for an improvement in device performance – for example, with respect to the speed of microprocessors or capacity of memory units – also drives the search for new materials with optimized properties. However, the trade-off with new materials is often that their use may also create new problems which need to be solved. In Si-based technology these problems are mostly caused by the high reactivity of Si with many other elements. Therefore, the introduction of a new interconnect technology – as in the Cu case – or of new low- and high-k dielectrics is most often accompanied by the development of new barrier layers and refined strategies to avoid interdiffusion or intermixing of the various materials.

This aspect also becomes very important when combining Si-based components with other metal-based thin film systems, for example, magnetic systems. One of the main tasks for the general introduction of spintronics will therefore be to ensure the stability and functionality of the magnetic film systems within a silicon environment. This can be achieved by appropriate passivation and barrier layers. A particular aspect which must be considered in this context is also the thermal stability of the magnetic components during a silicon back-end processing. Another aspect in this field concerns the interconnects in spinelectronic devices. The magnetic switching in MRAM architectures is achieved by passing a sequence of high current pulses through dedicated "write" lines which creates the magnetic field at the position of the MRAM cell to be switched. Increasing the integration density in MRAMs will also increase the current density in the write lines. The realization of high-density MRAMs may therefore lead to new challenges for the interconnect technology. Alternative interconnects are already discussed in the context of silicon devices, e.g. carbon nanotubes. They have a very high current-carrying capacity and may therefore also be of interest for spinelectronic devices.

A final materials-related aspect concerns the improvement of the systems with respect to electro- and acoustomigration effects. These phenomena are responsible for a steady degradation of the device performance and become progressively more relevant with the increase of the power density in the device. Considering the fact that the function of many modern microelectronic devices – not only those discussed in this book – is based on nanometer-scaled films, electromigration may take a new role. The same is true for acoustomigration if new materials for surface acoustic wave devices and the demand for high-power high-frequency filters are taken into account.

A wide-spread use of thermoelectric devices is as yet hampered by the limited efficiency of the currently available materials. Therefore, a strong improvement can be expected from materials science, which explains the intense research effort in this field. Various concepts such as, for example, multi-quantum well structures, are currently investigated and promise a significant progress in the improvement of the thermoelectric efficiencies in the near future.

Another fascinating field – which because of reasons of space could not be treated in this book – is the combination of micro- and nanoelectronic components with neuronal systems (neuroelectronics) which is expected to have a tremendous potential in medical applications. In this case, the biocompatibility of the materials used is of vital importance. This approach raises completely new demands on the properties of passivation layers and the chemical stability of devices in a wet environment.

In view of these considerations, the developments of new materials and the solution of materials-related problems will be one of the large working fields in microelectronics, in particular, with respect to metal-based thin films.

6.3 Microelectronics – Quo Vadis?

A reliable guide for developments and trends in microelectronics is the International Technology Roadmap for Semiconductors (ITRS) [4.1]. It is the result of a worldwide consensus building process and predicts the main trends in semiconductor industry. A persistent and extremely important trend is in integration level, which is usually expressed as Moore's law, stating that the number of components per chip doubles every 18 months. The growing systems integration is characterized by smart objects, by rising wafer packaging and system-in-package. Polymer electronics and flexible wafers are future developments. The most significant trend for human society is the decreasing cost per interconnect function which is reflected by proliferation of computers, electronic communication and consumer electronics. Another scaling trend is the enhancement of both speed and power, characterized by microprocessor clock rates in the GHz range and by laptop or cell phone battery life prolongation. Growing compactness is accompanied by a continuing miniaturization in microelectronics, as characterized by technology nodes. The production of the 100 nm node has just been reached, with a projection to 10 nm within the next decade. This acts as the driving force for nanotechnology developments in microelectronics [6.6]. Related to front-end processes this includes topics like MOSFET architecture and future gate dieelectrics, up to the application of quantum devices. Nanometer structures become also significant in back end of line processes. Barrier layers with a thickness below 10 nm are required by the transition from Al to Cu interconnects. Novel interline dielectrics are under development, with nanopores to decrease the dielectric constant in order to enable high speed operation. Trends in functionality are given by demands for single electron detectors and spintronic devices, as presented in the preceding paragraph. Long-term demands are nanotubes electronics and quantum computing, but also nonvolatile memory devices.

Special developments and trends in metallizations are characterized by new thin film deposition techniques such as atomic layer deposition (ALD) of Cu layers and of barriers. Further topics are the optimization of ultrathin diffusion barrier layers and the introduction of Cu alloy metallization with optimized electromigration reliability. This requires a deeper understanding of damage effects such as electromigration and stress migration. Cu damascene nano-interconnects and optical interconnects could be the next and following interconnect technologies.

6.4   What You See is What You Get

The development of novel material systems and innovative device concepts relies heavily on appropriate analysis techniques which permit the determination of the critical physical and material parameters. In view of the continuing downscaling of microelectronic devices, highly resolving microscopy approaches receive particular interest. In addition to a mere image with high lateral resolution, chemical information on a small scale is also of significant importance. The latter is necessary to detect intermixing and segregation at interfaces. This can be achieved by sophisticated transmission electron microscopy techniques with energy filtering of the transmitted electrons (EF-TEM) or from the energy loss near-edge structure (ELNES). A further improvement of the lateral resolution can be achieved by aberration-corrected electron optical columns.

With the advent of spintronics also comes the need to characterize the magnetic properties of a film system or a device, in particular, on a small lateral scale. This can be achieved, for example, by magnetic force microscopy (MFM) or scanning electron microscopy with spin polarization analysis (SEMPA). The chemical complexity of magnetic systems and the different types of magnetic order (ferromagnetism, antiferromagnetism) add further complications to the problem. In this case, techniques based on polarized synchrotron radiation in the soft X-ray regime promise a solution. For example, X-ray photoelectron microscopy (XPEEM) has been shown to be a versatile approach to imaging micromagnetic structures with a lateral resolution down to about 50 nm [6.7]. Both ferromagnetic and antiferromagnetic layered systems have been successfully investigated with chemical selectivity. The magnetic contrast mechanisms are provided by magnetic circular (MXCD) and linear X-ray dichroism (MLXD) in the case of ferromagnets and antiferromagnets, respectively. A complementary technique is provided with X-ray transmission microscopy (XTM) [6.8]. The magnetic contrast mechanisms are the same as those used in XPEEM, but instead of the emitted electrons the transmitted X-rays are used to form an image by means of imaging zone plates. The samples used in XTM must fulfill the same requirements as TEM samples. By variation of the X-ray photon energy, local spectroscopic information about the chemical element or state may also be obtained. The current research in this field aims at a considerable improvement of the spatial resolution down to the 10 nm regime.

Another crucial aspect in magnetic systems is the dynamics of magnetization reversal or the magnetic switching speed. In order to be able to characterize the magnetic switching, the process of magnetization reversal must be analyzed by time-resolving approaches. The time scale for magnetization reversal phenomena may range down to the 10 ps region. An already established technique in this area is the time-resolving magneto-optical Kerr (MOKE) microscopy [6.9]. Much shorter time-scales can be accessed by employing femtosecond laser systems.

The investigation of layered magnetic systems can also be based on X-ray scattering or reflectivity techniques. If these studies are carried out with polarized light, one may obtain simultaneously structural and magnetic information, supplied by the MXCD and MXLD effects mentioned above. In addition, by analyzing the diffuse scattering, the interfacial roughness may be quantified.

All of the synchrotron radiation based techniques mentioned above are subject to ongoing improvement. The unique combination of magnetic sensitivity and chemical selectivity renders these techniques extremely useful.

When moving to smaller and smaller structures, as is the case with nanoelectronics and molecular electronics, manipulation techniques on these length scales must also be available. Therefore scanning probe techniques, for example, scanning tunneling microscopy (STM) and particularly atomic force microscopy (AFM) have become increasingly important. They are not only used for the investigation of the morphology or electronic situation at a sample surface. Nowadays individual atoms and molecules may be moved around on a surface and may be grouped into larger ensembles by means of an STM or AFM tip. Of course, this approach is only usable for single devices and will certainly not lead to a viable fabrication process of industrial relevance. It shows clearly, however, the challenges that will have to be met in the progress of electronics.

6.5 References

[6.1]   *Nanoelectronics and Information Technology*, ed. R. Waser, Wiley-VCH, Weinheim, 2003.
[6.2]   S. Tans, M. Devoret, H. Dai, A. Thess, R. Smalley, L. Geerligs, C. Dekker, *Nature (London)*, **386**, 474 (1997).
[6.3]   M. Bockrath, D. Cobden, P. McEuen, N. Chopra, A. Zettl, A. Thess, R. Smalley, *Science*, **275**, 1922 (1997).
[6.4]   S. Heinze, J. Tersoff, R. Martel, V. Derycke, J. Appenzeller, P. Avouris, *Phys. Rev. Lett.*, **89**, 106801 (2002).
[6.5]   A. Bachtold, P. Hadley, T. Nakanishi, C. Dekker, *AIP Conf. Proc.*, **633**, 502 (2002).
[6.6]   J. W. Bartha, in *Proceedings of Nanofair 2002*, Strasbourg, France, 2002.
[6.7]   C. M. Schneider, G. Schönhense, *Rep. Prog. Phys.*, **65**, R1785 (2002).
[6.8]   P. Fischer, T. Eimüller, G. Schütz, G. Schmahl, P. Guttmann, D. Raasch, in *X-Ray Microscopy and Spectromicroscopy*, ed. J. Thieme, G. Schmahl, D. Rudolph, E. Umbach, Springer-Verlag, Berlin, 1998.
[6.9]   W. K. Hiebert, A. Stankiewicz, M. R. Freeman, *Phys. Rev. Lett.*, **79**, 1134 (1997).

Index

Metal Based Thin Films for Electronics, Second Edition. Klaus Wetzig and Claus M. Schneider (Eds.)
Copyright © 2006 WILEY-VCH Verlag GmbH & Co. KGaA, Weinheim
ISBN: 3-527-40650-6